MOLECULAR VIBRATIONS

The Theory of Infrared and Raman Vibrational Spectra

MOLECULAR VIBRATIONS

The Theory of Infrared and Raman Vibrational Spectra

E. Bright Wilson, Jr.
Theodore William Richards Professor of Chemistry, Emeritus, Harvard University

J.C. Decius
Professor of Chemistry, Oregon State University

Paul C. Cross
Late President, Mellon Institute

DOVER PUBLICATIONS, INC.
NEW YORK

Published in Canada by General Publishing Company, Ltd., 30 Lesmill Road, Don Mills, Toronto, Ontario.
Published in the United Kingdom by Constable and Company, Ltd., 10 Orange Street, London WC2H 7EG.

This Dover edition, first published in 1980, is an unabridged and corrected republication of the work originally published in 1955 by McGraw-Hill Book Company, Inc.

International Standard Book Number: 0-486-63941-X
Library of Congress Catalog Card Number: 79-055912

Manufactured in the United States of America
Dover Publications, Inc.
180 Varick Street
New York, N.Y. 10014

In memoriam

EMILY BUCKINGHAM WILSON

PREFACE

In recent years a very considerable body of theory has been developed which makes it possible to understand and utilize the infrared and Raman spectra of a large number of polyatomic molecules. The purpose of this book is to develop essential elements of this theory, starting from its simplest form and advancing to fairly elaborate and powerful theorems useful in more complicated applications. In order to hold the volume to a reasonable size, only vibrational spectra are treated in detail. However, vibrational (as opposed to rotational) spectra constitute the great bulk of published infrared and Raman data. The theory herein presented applies to gases and is a useful approximation for liquids but requires some extension for crystals.

The treatment of this book is intended to give a consistent system, but is not meant to be an exhaustive survey of the literature, which has now grown to a considerable size. Individual molecules are introduced only as examples, the whole aim being to lead the reader through an increasingly powerful series of mathematical techniques with a completeness sufficient to enable him not only to use these tools in analyzing experimental data but also to understand their derivation and to be prepared to extend and adapt them to new problems.

The mathematical demands on the reader are quite light in the early chapters and, as more powerful tools are required for the more efficient treatment of larger molecules, these tools, such as matrix algebra and group theory, are introduced and explained. The arrangement is such that the reader whose interest is in the qualitative assignment of spectral lines assisted by symmetry considerations and selection rules can read those chapters dealing with this problem without requiring the chapters which lead into the isotope effects, or further into the calculation of force constants.

Unfortunately, the manuscript was prepared too early to make possible strict adherence to the recommendations of the Joint Commission on Spectroscopy of the International Astronomical Union and the International Union of Pure and Applied Physics. So far as possible, however, we have followed the generally accepted conventions employed by G. Herzberg in his widely used book on "Infrared and Raman Spectra of Polyatomic Molecules."

It has not been an easy task to complete this text, but the process has been greatly aided by the large number of our students and associates who have read and criticized parts of earlier versions. In the preparation of the final manuscript we had the benefit of assistance and criticism from Dr. H. C. Allen, Dr. E. J. Bair, G. A. Crosby, Dr. D. F. Eggers, J. T. Lund, and Dr. E. L. Wagner.

E. Bright Wilson, Jr.
J. C. Decius
Paul C. Cross

CONTENTS

MOLECULAR VIBRATIONS

The Theory of Infrared and Raman Vibrational Spectra

CHAPTER 1

INTRODUCTION

An immense amount of experimental data has been accumulated from investigations of the infrared absorption spectra and of the Raman effect in polyatomic molecules. Only an extremely small fraction of this material has been subjected to analysis, although the theoretical tools for such an analysis are quite well developed and the results which could be obtained are of considerable interest. One reason for this situation is the amount of labor required to unravel the spectrum of a complex molecule, but an additional deterrent has been the unfamiliar mathematics, such as group theory, in terms of which the most powerful forms of the theory of molecular dynamics have been couched. When only the necessary parts of these mathematical techniques are considered, the difficulty of understanding the theory of the vibrational and rotational spectra of polyatomic molecules is greatly reduced.

In this first chapter, a short general survey of the background of the subject will be given, to serve as an introduction to the more mathematical treatment which follows.

1-1. Infrared Spectra

The absorption or emission spectrum arising from the rotational and vibrational motions of a molecule which is not electronically excited is mostly in the infrared region. A small molecule having an electric moment emits and absorbs light of frequency below about 250 wave numbers[1] because of its rotational motion. Molecules which are absorbing or emitting 1 quantum of vibrational energy show bands in the region from about 200 to 3,500 cm^{-1}, while bands due to several vibrational quantum jumps are detected all the way from a few hundred to many

[1] The symbol ω will be used throughout to represent the reciprocal of the wavelength λ. The unit for ω is the reciprocal centimeter (cm^{-1}) or the wave number, since ω represents the number of waves per centimeter. Inasmuch as the frequency ν is connected with ω by the relationship $\nu = \omega c$, where c is the velocity of light, it will frequently be said that the frequency is ω wave numbers. Experimentalists often express their results in terms of the wavelength in microns (μ), 1 micron being 10^{-4} cm. If λ_μ is the wavelength in microns, $\omega = 10^4/\lambda_\mu$. Recently the symbol K, for Kayser, has been proposed to replace cm^{-1}.

thousand wave numbers, sometimes being observable in the visible portion of the spectrum (13,000 to 26,000 cm^{-1}). See Fig. 1-1.

Infrared spectra may be observed either in emission or in absorption, although the latter method is by far the more common. In absorption experiments light from a suitable source is passed through a tube containing the gas to be studied and thence into the spectrograph.[1] If the spectrograph is of low resolving power, a series of wide bands is observed which correspond to the vibrational transitions, but if a spectrograph of

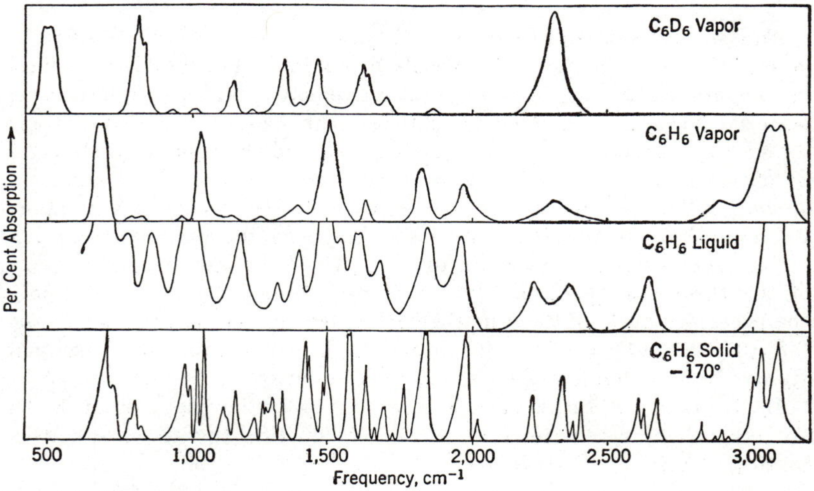

FIG. 1-1. Vibrational absorption spectra in the infrared as illustrated by benzene.

higher resolution is used, these bands may break up into lines which can be correlated with the energy levels of rotation. In practice, only a few light molecules (H_2O, NH_3, CH_4, CO_2, etc.) have been observed in the

[1] Information concerning experimental methods can be obtained from the following sources:

W. W. Coblentz, "Glazebrook's Dictionary of Applied Physics," Vol. 4, p. 136, Macmillan, London, 1923.

G. R. Harrison, R. C. Lord, and J. R. Loofbourow, "Practical Spectroscopy," Prentice-Hall, New York, 1948.

G. Laski, "Handbuch der Physik," Vol. 19, p. 802, Springer, Berlin, 1928.

J. Lecomte, "Le Spectre infrarouge," Recueil des conférences-rapports de documentation sur la physique, Vol. 14, Les Presses Universitaires de France, Paris, 1928.

F. I. G. Rawlins and A. M. Taylor, "Infrared Analysis of Molecular Structure," Cambridge, New York, and London, 1929.

C. Schaefer and F. Matossi, "Das ultrarote Spektrum," Springer, Berlin, 1930.

G. B. B. M. Sutherland, "Infrared and Raman Spectra," Methuen, London, 1935.

V. Z. Williams, *Rev. Sci. Instr.*, **19**: 135 (1948).

infrared region with high enough resolving power to resolve the rotational structure. Figure 1-1 shows some observed spectra.

Liquids and solids are also studied, and they yield interesting results, but except in so far as they give vibrational spectra in close agreement with those found for the corresponding gases they will not be discussed in this book, in which interactions between separate molecules will be neglected.

1-2. Raman Spectra[1]

If the substance being studied (as a gas, liquid, or solid) is strongly illuminated by monochromatic light in the visible or ultraviolet region and the scattered light observed in a spectrograph,[2] a spectrum is obtained (see Fig. 1-2) which consists of a strong line (the *exciting line*) of the same frequency as the incident illumination together with weaker lines on either side shifted from the strong line by frequencies ranging from a few to about 3,500 wave numbers. The pattern of lines is symmetrical about the exciting line except with regard to intensities, the lines on the high-frequency side being considerably weaker than the others. In fact, they are frequently too weak to be observed. The lines of frequency less than the exciting lines are called Stokes lines, the others anti-Stokes lines.

These lines differing in frequency from the exciting line, the Raman lines, have their origin in an interchange of energy between the light quanta and the molecules of the substance scattering the light. The lines which appear very near the exciting line are correlated with changes in the

[1] The original papers dealing with this effect are:

A. Smekal, *Naturwissenschaften*, **11**: 873 (1923).

H. A. Kramers, *Nature*, **113**: 673 (1924).

H. A. Kramers and W. Heisenberg, *Nature*, **114**: 310 (1924).

C. V. Raman, *Indian J. Phys.*, **2**: 1 (1928).

G. Landsberg and L. Mandestamm, *Naturwissenschaften*, **28**: 557 (1928).

[2] Information concerning the experimental methods for the study of the Raman effect can be obtained from the following sources:

S. Bhagavantam, "Scattering of Light and the Raman Effect," Chemical Publishing, New York, 1942.

P. Daure, "Introduction à l'étude de l'effêt Raman," Éditions de la Revue d'optique, théorique et instrumentale, 1933.

G. Glockler, *Revs. Mod. Phys.*, **15**: 112 (1943).

J. H. Hibben, "The Raman Effect and Its Chemical Applications," Reinhold, New York, 1939.

G. Joos, "Handbuch der Experimentalphysik," Vol. 22, p. 413, Akademische Verlagsgesellschaft, Leipzig, 1929.

K. W. F. Kohlrausch, "Der Smekal-Raman-Effekt," Springer, Berlin, 1931. See also "Erganzungsband 1931–37," Springer, Berlin, 1938.

K. W. F. Kohlrausch, "Ramanspektren," Vol. 9, Sec. VI of Eucken-Wolf, "Hand- und Jahrbuch der Chemischen Physik," Akademische Verlagsgesellschaft, Leipzig, 1943. Reprinted by Edwards, Ann Arbor, Mich., 1945.

Symposium, *Trans. Faraday Soc.*, **25**: 781 (1929).

rotational energy states of the molecules without changes in the vibrational energy states and form the *pure rotation* Raman spectrum. The lines farther from the exciting line are really bands of unresolved lines and are associated with simultaneous changes in the vibrational and rotational energy states.

The frequency shifts, that is, the differences between the frequencies of the Raman lines and the exciting line, are independent of the frequency of the exciting line. A mercury arc is usually used for illumination, and there are a number of the strong mercury lines which are used to excite

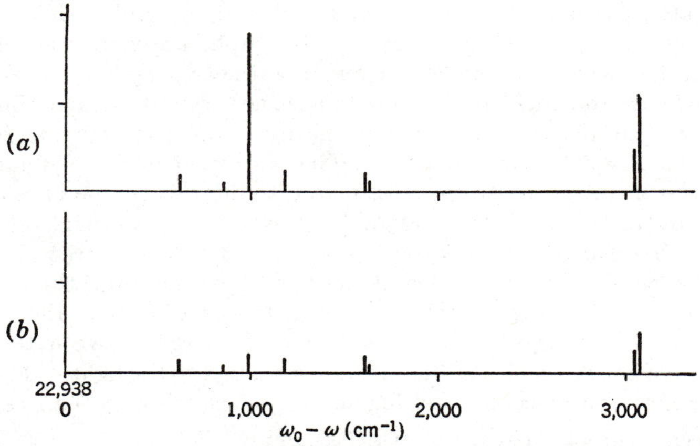

FIG. 1-2. Plot of the Raman spectrum of benzene. The heights of the lines indicate relative intensity, (a) being the parallel spectrum and (b) the perpendicular spectrum ($b/a = \rho_n$; see Sec. 3-6). The scale is in cm^{-1} and the Raman shift is measured from the mercury exciting line at 22,938.04 cm^{-1} (4,358 A). (*These data, used in the example discussed in Chap.* 10, *are from Angus, Ingold, and Leckie, J. Chem. Soc.*, 1936, *p.* 925.)

Raman spectra. The frequencies for a given molecule found in infrared absorption frequently agree with the frequency *shifts* found in the Raman effect, but this is not always true and depends on the symmetry of the molecule in a way which is now well understood.

By polarizing the incident light or in other ways, it is possible to find the *degree of depolarization* of each frequency shift in the Raman spectrum, a quantity which will be important in the interpretation of the experimental results. This quantity is the ratio, for the scattered light, of the intensities of the components polarized perpendicular and parallel, respectively, to the direction of polarization of the incident illumination.

1-3. The Molecular Model

In attempting to account for the observed infrared and Raman spectra of real molecules, a certain simplified model for such molecules is adopted, and then the spectra which this model would exhibit are calculated. The specification of the model involves certain parameters such as size, stiffness of valence bonds, etc., which can be varied within limits set by other types of experimental evidence until the best agreement with experiment is obtained. An attempt is usually made to select a number of such parameters which is much smaller than the number of experimental quantities so that the success of the theory can be tested by the agreement which it provides with experiment.

The model which will be used in this book consists of particles held together by certain forces. The particles, which are to be endowed with mass and certain electrical properties, represent the atoms and are to be treated as if all the mass were concentrated at a point. It is assumed that the atoms may be electrically polarized by an external electrical field, such as that of a beam of light, and that they may or may not be permanently polarized by their mutual interactions in such a manner that the whole molecule has a resultant electric moment.[1] Both the polarizability and the electric moment of the model may vary as the particles (hereafter called atoms) change their relative positions. Finally, the atoms may possess an internal degree of freedom or nuclear spin which introduces certain symmetry restrictions.

The forces between the particles may be crudely thought of as weightless springs which only approximately obey Hooke's law and which hold the atoms in the neighborhood of certain configurations relative to one another. This picture of the forces as springs is useful for visualization, but is not sufficiently general for all cases. For example, it does not cover cases of restricted rotation about single bonds such as may occur in ethane. The nature of these interatomic forces is one of the chief problems still being studied and will be discussed in Chap. 8. The search for a potential function which involves a small number of parameters and which at the same time permits good agreement with experiment is by no means ended.

The statement that the model obeys the laws of quantum mechanics is an essential part of its specification. However, since atoms are fairly heavy particles (compared to electrons), it will sometimes be true that classical mechanics when properly used gives results which are good approximations to those of quantum mechanics.

[1] The *electric moment* $\boldsymbol{\mu}$ of an electrically neutral molecule is a vector quantity whose direction is that of a line joining the center of charge of the negative charges with the center of charge of the positive charges and whose magnitude is the length of that line multiplied by the total negative or the total positive charge, these being equal. An atom or molecule is said to be *polarized* by an electric field when the displacements of charges caused by the electric field produce or alter the electric moment of the atom or molecule. If the electric field strength is $\boldsymbol{\mathcal{E}}$ and the induced moment is $\boldsymbol{\mu}$, then $\boldsymbol{\mu} = \alpha\boldsymbol{\mathcal{E}}$ defines the *polarizability* α for an isotropic molecule.

Since the atoms of this model have been regarded as point masses with certain electrical properties, there is an apparent disagreement with the fact that many experiments require that atoms be made of electrons and nuclei. It is possible to reconcile these two points of view. If the wave equation for a molecule made up of electrons and nuclei is set up, a procedure[1] exists whereby this equation may be separated into two equations, one of which governs the electronic motions and yields the forces between the atoms, whereas the other is the equation for the rotation and vibration of the atoms and is identical with the equation for the model adopted here. In principle, therefore, the forces between the atoms can be calculated *a priori* from the electronic wave equation, but in practice this is not mathematically feasible (except for H_2), and it is necessary to postulate the forces in such a manner as to obtain agreement with experiment. Therefore, although it is theoretically possible to start with a model consisting of electrons and nuclei interacting coulombically and obeying the laws of quantum mechanics, in practice it is necessary to assume the nature of the equilibrium configuration and of the forces between the atoms, so that it seems more desirable to start with the model in which the atoms are the units.

This separation of the electronic motion and the nuclear motions is only an approximation which may break down in certain cases, especially for high electronic states. If there were no interaction between the two types of motions, there would be no Raman effect of any importance. However, the coupling is small for the lowest electronic state.

1-4. Classical Theory of Vibrational and Rotational Spectra

Classical electromagnetic theory[2] requires that an accelerated charged particle emit radiant energy. On this basis a rotating molecule with an electric moment should emit light of the same frequency as the frequency of rotation. Because of the Maxwellian distribution of rotational velocities, a collection of gas molecules should emit a band of frequencies possessing an intensity maximum which is related to the most probable frequency of rotation. In practice, this prediction of classical theory

[1] M. Born and J. R. Oppenheimer, *Ann. Physik,* **84:** 457 (1927).

R. L. Kronig, "Band Spectra and Molecular Structure," Cambridge, New York and London, 1930.

[2] For presentations of classical radiation theory, see the following:

L. Page, "Introduction to Theoretical Physics," Chap. 12, Van Nostrand, New York, 1934.

F. K. Richtmyer and E. H. Kennard, "Introduction to Modern Physics," Chap. 2, McGraw-Hill, New York, 1947.

A. E. Ruark and H. C. Urey, "Atoms, Molecules, and Quanta," Chap. 6, McGraw-Hill, New York, 1930.

F. Zerner, "Handbuch der Physik," Vol. 12, p. 1, 1927, Springer, Berlin.

is quite closely verified experimentally for heavy molecules and low resolution.

The molecular model of the previous section can move as a whole, rotate about its center of mass, and vibrate. The translational motion does not ordinarily give rise to radiation. Classically, this follows because acceleration of charges is required for radiation. The rotational motion causes practically observable radiation if, and only if, the molecule has an electric (dipole) moment. The vibrational motions of the atoms within the molecule may also be associated with radiation if these motions alter the electric moment. A diatomic molecule has only one fundamental frequency of vibration so that if it has an electric moment its infrared emission spectrum will consist of a series of bands, the lowest of which in frequency corresponds to the distribution of rotational frequencies for nonvibrating molecules. The other bands arise from combined rotation and vibration; their centers correspond to the fundamental vibration frequency and its overtones. A polyatomic molecule has more than one fundamental frequency of vibration so that its spectrum is correspondingly richer.

The Raman effect can also be explained classically. The electric vector of the incident illumination induces in the molecule an oscillating electric moment which emits radiation. If the molecule is at rest, the induced moment, and therefore the scattered light, has the same frequency as the incident light. If, however, the molecule is rotating or vibrating, this is not necessarily the case, because the amplitude of the induced electric moment may depend on the orientation of the molecule and the relative positions of its atoms. Since the configuration changes periodically because of rotation and vibration, the scattered radiation is "modulated" by the rotational and vibrational frequencies so that it consists of light of frequencies equal to the sum and to the difference of the incident frequency and the frequencies of the molecular motions, in addition to the incident frequency.

Thus, the classical theory of radiation and classical mechanics provides an explanation of the general features of both infrared and Raman spectra. It cannot, however, account for the details and is to be regarded as only a rough approximate method of treatment.

1-5. The Quantum Viewpoint

When the molecular spectra of a few light molecules are observed with spectrographs of high resolving power, the bands previously discussed are resolved into a series of closely spaced lines. Classical theory is unable to explain this phenomenon. The explanation, of course, depends on the use of quantum theory, in which the molecule is restricted to definite, discrete energy levels of rotation and vibration. Radiation occurs only when a molecule undergoes a transition from one stationary state to another of different energy. The Bohr frequency rule gives the frequency ν of the light radiated or absorbed on transitions between states of energies $W_{n''}$ and $W_{n'}$. It is

$$\nu_{n''n'} = \nu_{n'n''} = \frac{W_{n'} - W_{n''}}{h} \tag{1}$$

where h is Planck's constant.[1]

Not every transition can occur with the emission or absorption of radiation. The rules which tell which transitions may occur are called *selection rules.*

Although the classical theory is not correct in predicting that the observed radiation will consist of frequencies occurring in the motion of the system, there is an asymptotic relationship between the frequencies predicted by the classical and by the quantum theory, known as Bohr's correspondence theorem for frequencies.[2] According to this theorem the frequencies emitted and absorbed by a quantum system approach asymptotically the classical frequencies of the system as the quantum numbers of the initial and final states are increased. The intensities of the quantum transitions will likewise asymptotically approach the intensities calculated classically, as the quantum numbers increase.

From the quantum viewpoint, the band of lowest frequency in the infrared spectrum (it may extend into the microwave region) of a molecule with an electric moment consists of discrete lines, each of which corresponds to a transition between two different rotational energy levels of the nonvibrating molecule (or rather, of the molecule in its lowest vibrational energy level). The other bands with higher frequencies correspond to transitions involving simultaneous changes of rotational and vibrational energies. The spacing between adjacent vibrational levels is considerably greater than that between adjacent rotational levels so that, although the various vibrational bands are usually fairly

[1] Throughout this book, in dealing with transition phenomena, the general array of quantum numbers specifying the *upper* state will be indicated by n' and that specifying the lower state by n''. Whenever a double subscript is used to specify a transition, the symbols will be written in the order initial, final. Equation (1) thus states that the frequency, $\nu_{n''n'}$, of the absorption process $n'' \rightarrow n'$ is equal to the frequency, $\nu_{n'n''}$, of the emission process $n' \rightarrow n''$, and that this frequency is given by $(W_{n'} - W_{n''})/h$.

[2] See the following sources:

N. Bohr, *Z. Physik,* **2:** 423 (1920); **13:** 117 (1923).

W. Pauli, "Handbuch der Physik," Vol. 23, p. 1, Springer, Berlin, 1926.

A. Sommerfeld, "Atombau und Spektrallinien," 5th ed., Vol. 1, pp. 671*ff*., Vieweg, Brunswick, 1931.

J. H. Van Vleck, "Quantum Principles and Line Spectra," Bulletin of the National Research Council, No. 54, Chap. 3, Washington, D.C., 1926.

widely spaced, it requires a spectrograph of very high resolving power to separate the rotational lines.

The fundamental frequency of the classical explanation corresponds to a quantum transition from one vibrational state to the next, while the overtone frequencies correspond to transitions to other than adjacent levels. Since the vibrational levels are nearly but not quite evenly spaced, the vibrational bands will fall into series with frequencies which are almost but not quite multiples of the fundamental frequencies.

The quantum picture of the Raman effect is that a photon of energy $h\nu_0$ (ν_0 being the frequency of the incident light) comes up to a molecule in a given stationary state, causing a transition to another higher (or lower) energy level different in energy by an amount $h\nu_{n''n'}$. This amount of energy is subtracted from (or added to) the incident photon so that the emitted or "scattered" photon then has the energy $h\nu_0 \mp h\nu_{n''n'}$ and therefore has the frequency $\nu_0 \mp \nu_{n''n'}$. Since in general more molecules are in the lower than in the higher energy states, there will be more cases in which the photon gives up some of its energy than vice versa, so that the Stokes lines will be stronger than the anti-Stokes lines.

In calculating the energy levels and selection rules, the principles of quantum mechanics must be used. This is usually done through the medium of the Schrödinger equation and wave mechanics, but the equivalent mathematical techniques of matrix mechanics and the operator calculus are frequently useful.[1]

1-6. Applications

There are three main applications of the interpreted results of infrared and Raman studies. These are the study of the nature of the forces acting between the atoms of a molecule, the determination of molecular structure, and the calculation of thermodynamic quantities.

The fundamental frequencies of vibration obtained from infrared and Raman spectra have provided considerable information about the interatomic forces in various molecules. It is found that different types of valence bonds exhibit different degrees of resistance to stretching and bending which are roughly independent of the molecule in which the bond occurs. Further, empirical relations between the length of a bond and

[1] In this book an elementary knowledge of quantum mechanics will be assumed. Any material, however, which is not covered in L. Pauling and E. B. Wilson, Jr., "Introduction to Quantum Mechanics," McGraw-Hill, New York, 1935, will be developed in the text or appendixes. For a more advanced treatment of the subject, see E. C. Kemble, "Fundamental Principles of Quantum Mechanics," McGraw-Hill, New York, 1937; also L. I. Schiff, "Quantum Mechanics," 2d ed., McGraw-Hill, New York, 1955; or K. S. Pitzer, "Quantum Chemistry," Prentice-Hall, New York, 1953.

its resistance to stretching have been found which promise to have useful applications.[1]

There are several ways in which information about molecular structure can be obtained from infrared and Raman spectra. Probably the most important is the determination of moments of inertia from the spacing of the rotational lines. This remains one of the most reliable methods known for the determination of molecular sizes of simple molecules although with present experimental techniques it cannot be used for any but very light molecules. In recent years this method has been enormously extended by the development of techniques for the use of the millimeter and centimeter wavelength regions, *i.e.*, the regions of microwave spectroscopy. The vibrational spectrum can also be used to provide clues as to the structure of a molecule, especially with regard to its symmetry.

In many ways the most valuable application of the data of infrared and Raman studies is to the calculation of the heat capacity, entropy, and free energy of gaseous molecules. For such calculations a knowledge of the moments of inertia and vibration frequencies of the molecule is necessary. Calculations of this sort have been carried out for a large number of simple molecules with results which usually surpass in accuracy those of any other method. If the value for the heat of reaction is known at any temperature, spectroscopic data can be used to find the heat of the reaction at any other temperature, the free energy and entropy changes, and the equilibrium constant at any temperature.

All these applications require a careful consideration of the principles underlying the interpretation of the spectral data. Because of the failure to recognize the importance of some of these principles, many false conclusions have been drawn in the past from spectroscopic experiments.

[1] R. M. Badger, *J. Chem. Phys.*, **2**: 128 (1934); **3**: 710 (1935).
W. Gordy, *J. Chem. Phys.*, **14**: 305 (1946).

CHAPTER 2

THE VIBRATION OF MOLECULES

The study of molecular vibrations will be introduced by a consideration of the elementary dynamical principles applying to the treatment of small vibrations. In order that attention may be focused on the dynamical principles rather than on the technique of their application, this chapter will employ only relatively familiar and straightforward mathematical methods, and the illustrations will be simple. This will serve adequately as an introduction to the applications of quantum mechanics and group theory to the problem of molecular vibrations. Since, however, these straightforward methods become cumbersome and impractical, even for simple molecules, equivalent but more powerful techniques using matrix and vector notations will be discussed in Chap. 4.

2-1. Separation of Rotation and Vibration[1]

The logical way to begin the mathematical treatment of the vibration and rotation of a molecule is to set up the classical expressions for the kinetic and potential energies of the molecule in terms of the coordinates of the atoms, and then to use these expressions to obtain the wave equation for vibration, rotation, and translation. Following this, it should be proved that when the proper coordinate system is used, the complete wave equation can be approximately separated into three equations, one for translation, one for rotation, and one for vibration. Unfortunately, this procedure is not a very simple one and utilizes more quantum-mechanical technique than is required for the discussion of the vibrational equation itself. Consequently, the actual carrying out of the separation will be deferred until Chap. 11, and only a summary of the results thus obtained will be presented at this point. The reader who prefers to follow the more logical order may turn to Chap. 11 before continuing with the present sections.

It is found that the proper coordinates to use are the following: the three cartesian coordinates of the center of mass of the molecule, the three Eulerian angles[2] of a rotating system of cartesian coordinates, the axes of which coincide with the principal axes of inertia of the undistorted molecule, and finally the cartesian coordinates of the atoms with respect

[1] References to the original papers dealing with this topic will be found in Chap. 11.

[2] Eulerian angles are described in Appendix I.

to the rotating coordinate system. Since there are only $3N$ degrees of freedom for a molecule of N atoms, there are six too many coordinates in the above list, so that all of them cannot be independent; six conditions connecting them must exist. However, just six conditions are required (for nonlinear molecules) to define the rotating coordinate system. Three of these locate the origin of the rotating system at the center of mass of the molecule, thus assuring that the rotating system moves with the molecule. The other three conditions tie the coordinate system to the molecule so that they rotate together.

The effect of these conditions is to enable the vibrations to be treated in terms of the coordinates of the moving system of axes just as if the molecule were not rotating or undergoing translation. The $3N$ cartesian coordinates of the moving system are used, together with the six conditions above which prevent translation or rotation with respect to the moving axes.

Let x_α, y_α, z_α be the coordinates of the αth atom in terms of the moving system, and a_α, b_α, c_α be the values of the coordinates of the equilibrium position of the αth atom; *i.e.*, the values assumed by x_α, y_α, z_α when the molecule is at rest in its equilibrium position. Displacements from equilibrium will be measured by $\Delta x_\alpha = x_\alpha - a_\alpha$, $\Delta y_\alpha = y_\alpha - b_\alpha$, and

$$\Delta z_\alpha = z_\alpha - c_\alpha$$

The condition that the origin be at the center of mass yields the equations

$$\sum_{\alpha=1}^{N} m_\alpha x_\alpha = 0$$
$$\sum_{\alpha=1}^{N} m_\alpha y_\alpha = 0 \qquad (1)$$
$$\sum_{\alpha=1}^{N} m_\alpha z_\alpha = 0$$

in which m_α is the mass of the αth atom. Similar expressions must hold for the equilibrium configuration, in which $x_\alpha = a_\alpha$, $y_\alpha = b_\alpha$, $z_\alpha = c_\alpha$. Consequently, the following relations will be valid:

$$\sum_{\alpha=1}^{N} m_\alpha \, \Delta x_\alpha = 0$$
$$\sum_{\alpha=1}^{N} m_\alpha \, \Delta y_\alpha = 0 \qquad (2)$$
$$\sum_{\alpha=1}^{N} m_\alpha \, \Delta z_\alpha = 0$$

The other three conditions on the moving system are not as simple and obvious as the three just given. They are chosen so that the axes will rotate with the molecule, but it is not easy to define what is meant by "rotating with the molecule" when all the atoms in the molecule are moving relative to one another in their vibrational motions. It might, for example, be specified that there should be no angular momentum with respect to the translating-rotating coordinate system. This is not a convenient definition of the rotating system, but the definition which is adopted, for reasons given in Chap. 11, is closely related. Thus, the components $\mathfrak{m}_x$, $\mathfrak{m}_y$, and $\mathfrak{m}_z$ of the angular momentum in the moving system are

$$\begin{aligned} \mathfrak{m}_x &= \sum_{\alpha=1}^{N} m_\alpha(y_\alpha \dot{z}_\alpha - z_\alpha \dot{y}_\alpha) \\ \mathfrak{m}_y &= \sum_{\alpha=1}^{N} m_\alpha(z_\alpha \dot{x}_\alpha - x_\alpha \dot{z}_\alpha) \\ \mathfrak{m}_z &= \sum_{\alpha=1}^{N} m_\alpha(x_\alpha \dot{y}_\alpha - y_\alpha \dot{x}_\alpha) \end{aligned} \tag{3}$$

A dot over a symbol means the time derivative, that is, $\dot{x}_\alpha = dx_\alpha/dt$, etc. For small displacements, Δx_α, Δy_α, etc., are small, so that x_α, y_α, and z_α can be replaced by a_α, b_α, and c_α, respectively, these being the coordinates of the equilibrium position of the atom α. Under these circumstances,

$$\begin{aligned} \mathfrak{m}_x &\cong \sum_{\alpha=1}^{N} m_\alpha(b_\alpha \dot{z}_\alpha - c_\alpha \dot{y}_\alpha) \\ \mathfrak{m}_y &\cong \sum_{\alpha=1}^{N} m_\alpha(c_\alpha \dot{x}_\alpha - a_\alpha \dot{z}_\alpha) \\ \mathfrak{m}_z &\cong \sum_{\alpha=1}^{N} m_\alpha(a_\alpha \dot{y}_\alpha - b_\alpha \dot{x}_\alpha) \end{aligned} \tag{4}$$

The conditions actually employed in defining the rotating system of axes are (see also Sec. 2-5)

$$\begin{aligned} \sum_{\alpha=1}^{N} m_\alpha(b_\alpha\,\Delta z_\alpha - c_\alpha\,\Delta y_\alpha) &= 0 \\ \sum_{\alpha=1}^{N} m_\alpha(c_\alpha\,\Delta x_\alpha - a_\alpha\,\Delta z_\alpha) &= 0 \\ \sum_{\alpha=1}^{N} m_\alpha(a_\alpha\,\Delta y_\alpha - b_\alpha\,\Delta x_\alpha) &= 0 \end{aligned} \tag{5}$$

If these are differentiated with respect to the time, it is seen that they become equivalent to the equations obtained when the approximate expressions for $\mathfrak{m}_x$, $\mathfrak{m}_y$, and $\mathfrak{m}_z$ in (4) are equated to zero, since

$$\left(\frac{d\,\Delta x_\alpha}{dt}\right) = \dot{x}_\alpha$$

2-2. Small Vibrations in Classical Mechanics[1]

As a consequence of the conclusions set forth in the previous section, the problem of the vibration of a molecule may be treated independently of its rotation[2] by using a system of coordinates moving with the molecule and satisfying the six conditions of Eqs. (2) and (5), Sec. 2-1. Since classical mechanics yields a solution of the problem of small vibrations which is easier to visualize than the quantum mechanical solution, it will be employed first.

The kinetic energy is given by

$$2T = \sum_{\alpha=1}^{N} m_\alpha \left[\left(\frac{d\,\Delta x_\alpha}{dt}\right)^2 + \left(\frac{d\,\Delta y_\alpha}{dt}\right)^2 + \left(\frac{d\,\Delta z_\alpha}{dt}\right)^2\right] \tag{1}$$

It is very convenient to replace the coordinates $\Delta x_1, \ldots, \Delta z_N$ by a new set of coordinates $q_1, \ldots, q_{3N}$ defined as follows

$$q_1 = \sqrt{m_1}\,\Delta x_1 \quad q_2 = \sqrt{m_1}\,\Delta y_1 \quad q_3 = \sqrt{m_1}\,\Delta z_1 \quad q_4 = \sqrt{m_2}\,\Delta x_2, \text{etc.} \tag{2}$$

and known as mass-weighted cartesian displacement coordinates. In terms of the time derivatives of these coordinates the kinetic energy[3] is

$$2T = \sum_{i=1}^{3N} \dot{q}_i^2 \tag{3}$$

The potential energy will be some function of the displacements and therefore of the q's. For small values of the displacements, the poten-

[1] General treatments will be found in the following:

J. H. Jeans, "Theoretical Mechanics," p. 348, Ginn, Boston, 1907.

E. T. Whittaker, "Analytical Dynamics," 3d ed., Chap. 7, Cambridge, New York and London, 1927.

For the application to molecules, see the following:

N. Bjerrum, *Verhandl. deut. physik Ges.*, **16**: 737 (1914).

D. M. Dennison, *Revs. Mod. Phys.*, **3**: 280 (1931).

[2] For a further discussion of the justification of the method used in this section, see Appendix II.

[3] The subscripts α and β (running from 1 to N) will be used to enumerate *atoms*, while italic subscripts $i, j; k, l; m, n; q, r; s, t;$ and u, v will be used to enumerate coordinates and will run from 1 to $3N$ or from 1 to $3N - 6$ or $3N - 5$.

tial energy V may be expressed as a power series in the displacement q_i:

$$2V = 2V_0 + 2\sum_{i=1}^{3N}\left(\frac{\partial V}{\partial q_i}\right)_0 q_i + \sum_{i,j=1}^{3N}\left(\frac{\partial^2 V}{\partial q_i\,\partial q_j}\right)_0 q_iq_j + \text{higher terms}$$

$$= 2V_0 + 2\sum_{i=1}^{3N} f_iq_i + \sum_{i,j=1}^{3N} f_{ij}q_iq_j + \text{higher terms} \tag{4}$$

By choosing the zero of energy so that the energy of the equilibrium configuration is zero, V_0 may be eliminated. Furthermore, when all the q's are zero, the atoms are all in their equilibrium positions so that the energy must be a minimum for $q_i = 0$, $i = 1, 2, 3, \ldots$. Therefore[1]

$$\left(\frac{\partial V}{\partial q_i}\right)_0 = f_i = 0 \qquad i = 1, 2, \ldots, 3N$$

For sufficiently small amplitudes of vibration, the higher terms (cubic, quartic, etc., in the q's) can be neglected, so that

$$2V = \sum_{i,j=1}^{3N} f_{ij}q_iq_j \tag{5}$$

in which the f_{ij}'s are constants given by

$$f_{ij} = \left(\frac{\partial^2 V}{\partial q_i\,\partial q_j}\right)_0 \tag{6}$$

with $f_{ij} = f_{ji}$.

Newton's equations of motion can be written in the form

$$\frac{d}{dt}\frac{\partial T}{\partial \dot{q}_j} + \frac{\partial V}{\partial q_j} = 0 \qquad j = 1, 2, \ldots, 3N \tag{7}$$

since T is a function of the velocities only (in this coordinate system) and V is a function of the coordinates only. Substitution of the expressions for T and V given above yields the equations

$$\ddot{q}_j + \sum_{i=1}^{3N} f_{ij}q_i = 0, j = 1, 2, \ldots, 3N \tag{8}$$

This is a set of $3N$ simultaneous second-order linear differential equations. One possible solution is

$$q_i = A_i \cos(\lambda^{\frac{1}{2}}t + \epsilon) \tag{9}$$

[1] Here again the treatment disregards the fact that the coordinates q_i are not all independent. For justification of this method, see Appendix II.

where A_i, λ, and ϵ are properly chosen constants. If this expression is substituted in the differential equations, a set of algebraic equations results:

$$\sum_{i=1}^{3N} (f_{ij} - \delta_{ij}\lambda)A_i = 0 \qquad j = 1, 2, \ldots, 3N \tag{10}$$

in which δ_{ij}, the Kronecker delta symbol, equals unity if $i = j$ and is zero otherwise. Equation (10) is a set of simultaneous homogeneous linear algebraic equations in the $3N$ unknown amplitudes A_i.

Only for special values of λ does (10) have nonvanishing solutions; for all other values of λ the solution is the trivial one $A_i = 0$, $i = 1, 2, \ldots, 3N$, corresponding to no vibration.

The special values of λ are those which satisfy the determinantal or *secular equation*[1]

$$\begin{vmatrix} f_{11} - \lambda & f_{12} & f_{13} & \cdots & f_{1,3N} \\ f_{21} & f_{22} - \lambda & f_{23} & \cdots & f_{2,3N} \\ \cdots & \cdots & \cdots & \cdots & \cdots \\ f_{3N,1} & f_{3N,2} & f_{3N,3} & \cdots & f_{3N,3N} - \lambda \end{vmatrix} = 0 \tag{11}$$

The elements of this determinant are the coefficients of the unknown amplitudes A_i in the set of equations (10). When a fixed value of λ, say λ_k, is chosen so as to cause the determinant to vanish, the coefficients of the unknown A_i in (10) become fixed, and it is then possible to obtain a solution, A_{ik}, for which the additional subscript k will be used to indicate the correspondence with the particular value of λ_k. Such a system of equations does not determine the A_{ik} uniquely, but gives only their ratios: an arbitrary set A'_{ik} may be obtained by putting $A_{1k} = 1$. A convenient and unique mathematical solution may be designated by the quantities l_{ik} which are defined in terms of an arbitrary solution, A'_{ik}, by the formula[2]

$$l_{ik} = \frac{A'_{ik}}{\left[\sum_i (A'_{ik})^2\right]^{\frac{1}{2}}} \tag{12}$$

[1] For proofs of the mathematical theorems used in this chapter without proof, see the following:

G. Birkhoff and S. MacLane, "A Survey of Modern Algebra," Macmillan, New York, 1944.

M. Bocher, "Higher Algebra," Macmillan, New York, 1929.

R. A. Frazer, W. J. Duncan, and A. R. Collar, "Elementary Matrices," Cambridge, New York and London, 1938.

[2] The l_{ik} do not depend upon the value assumed for A_{1k}. Suppose $A''_{1k} = C$; then $A''_{ik} = CA'_{ik}$ and

$$\frac{A''_{ik}}{\left[\sum_i (A''_{ik})^2\right]^{\frac{1}{2}}} = \frac{CA'_{ik}}{C\left[\sum_i (A'_{ik})^2\right]^{\frac{1}{2}}} = l_{ik}$$

Note that these amplitudes are normalized in the sense that

$$\sum_i l_{ik}^2 = 1 \tag{13}$$

The solution of the actual physical problem then can be obtained by putting

$$A_{ik} = K_k l_{ik} \tag{14}$$

where the K_k are constants determined by the initial values of the coordinates q_i and velocities $\dot{q}_i$ (Sec. 2-4).

The secular equation (11) is of such fundamental importance in the study of vibration that it merits further attention. It consists of $3N$ rows and columns since there are $3N$ unknowns A_i. Consequently, when expanded, it yields an algebraic equation apparently of the $3N$th degree, inasmuch as the first term is $\pm\lambda^{3N}$. It will be shown later, however, that six of the roots are zero so that the equation reduces to one of the $3N - 6$th degree. There are thus $3N - 6$‡ nonzero roots of the secular equation. Each root, λ_k, corresponds to a set of amplitudes A_{ik} and consequently to a solution (9) of the original equations of motion.

2-3. Normal Modes of Vibration

Properties. It is of considerable importance to examine the nature of the solutions obtained above. It is evident from Eq. (9), Sec. 2-2, that each atom is oscillating about its equilibrium position with a simple harmonic motion of amplitude $A_{ik} = K_k l_{ik}$, frequency $\lambda_k^{\frac{1}{2}}/2\pi$, and phase ϵ_k. Furthermore, corresponding to a given solution λ_k of the secular equation, the frequency and phase of the motion of *each coordinate* is the same, but the amplitudes may be, and usually are, different for each coordinate. On account of the equality of phase and frequency, each atom reaches its position of maximum displacement at the same time, and each atom passes through its equilibrium position at the same time. A mode of vibration having all these characteristics is called a *normal mode of vibration,* and its frequency is known as a *normal,* or *fundamental, frequency* of the molecule.

Figure 2-1 shows the three normal modes of vibration of the water molecule. The arrows represent the relative displacements of the atoms, in a mass-weighted coordinate system, when the molecule is vibrating in the particular mode of vibration. From the nature of a normal mode the displacements of the different atoms remain in the same ratio to one another throughout the motion. The atoms are also constrained to move back and forth along straight lines.

‡ $3N - 5$ for linear molecules (see Sec. 2-8).

Degeneracy. From the discussion of the secular equation, it is seen that there are $3N - 6$ λ's which are not zero and therefore $3N - 6$ modes of vibration and frequencies (for nonlinear molecules). However, the frequencies are not necessarily all distinct; some of the roots of the secular equation may occur more than once. Such frequencies are said to be *degenerate*. When a degenerate value of λ is substituted in Eq. (10), Sec. 2-2, the resulting equations do not suffice to determine uniquely the quantities l_{ik}; instead there will be an infinite number of sets of values for

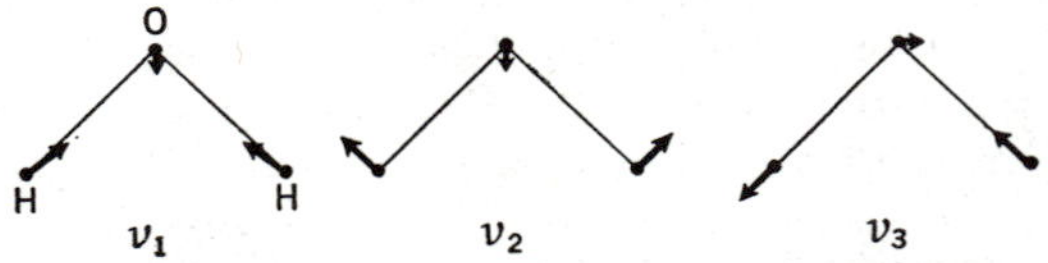

FIG. 2-1. Normal modes of vibration of the water molecule in mass-weighted coordinates. To represent actual relative motions in space, the arrows representing displacements of the oxygen atom should be only one-fourth as long as here shown.

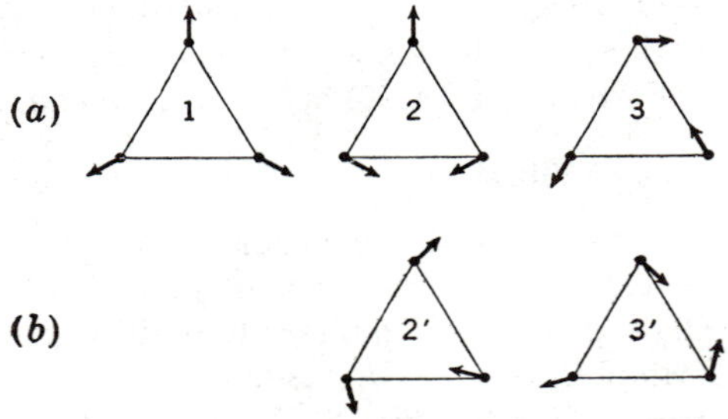

FIG. 2-2. Normal modes of vibration for equilateral triangular type molecules. (*a*) One choice of the normal modes of vibration. (*b*) Alternative choice of the normal modes of motion for the degenerate frequencies. $2' = 3 + 2$, $3' = 3 - 2$.

the l_{ik}'s which will satisfy the equations. These sets are, however, related. If λ_k is doubly *degenerate, i.e.*, occurs twice, then there will be only two independent sets of coefficients l_{ik} and therefore two independent normal modes of vibration with this frequency, but there are an infinite number of ways in which the two independent modes may be chosen. A number of modes of vibration are said to be *independent* if the motion represented by no one of them can be reproduced by superimposing the others with any choice of amplitudes and phases.

An example of a system with degeneracy is a molecule with three identical atoms at the corners of an equilateral triangle (the carbon atoms in cyclopropane, for instance). Figure 2-2*a* shows one choice of normal modes for this system, the last two modes having the same frequency. Figure 2-2*b* shows another choice for the degenerate vibrations, the first being obtained from 2 and 3 of Fig. 2-2*a* by adding 2 to 3, the second by subtracting 2 from 3. The motion similar to 2 but rotated 120° is also a normal mode, but is not independent of 2 and 3. It is to be emphasized that the sum or difference of two normal modes of vibration is itself a normal mode only when the two original motions have the same frequency; *i.e.*, are degenerate.

If modes having the same frequency are superimposed with different phases, the atoms no longer move in straight lines, but in ellipses around the equilibrium positions. In special cases the ellipses become small circles. Figure 2-3, which may be verified by the reader, shows the motion resulting from the superposition of modes 2 and 3 of Fig. 2-2*a* with a phase difference of $\pi/2$. In this special case, the radii of the circles are all equal and depend on the amplitudes of vibration.

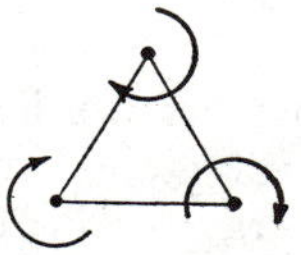

FIG. 2-3. Motion resulting from the combination, with a phase difference of $\pi/2$, of the normal modes of vibration corresponding to the degenerate frequencies of an equilateral triangular type of molecule.

General Solution. If two modes of vibration having different frequencies are superimposed, the resulting motion is more complicated and is not a normal mode of vibration. Since the equations of motion are linear differential equations, the sum of two or more solutions of the type given in Eq. (9), Sec. 2-2, with arbitrary constant coefficients is also a solution. The most general solution is therefore given by

$$q_i = \sum_{k=1}^{3N} l_{ik} K_k \cos (\lambda_k^{\frac{1}{2}} t + \epsilon_k) \qquad (1)$$

which has $6N$ arbitrary constants, the amplitudes K_k, and the phases ϵ_k. In writing the range of k as 1 to $3N$, it is implied that there are $3N$ sets l_{ik}, whereas there are only $3N - 6$ normal modes of vibration. The other six sets l_{ik} are obtained when the value 0 is substituted for λ_k in Eq. (10), Sec. 2-2. Since this root of the secular equation occurs six times, there will be six independent sets of l_{ik}'s, or six independent modes of motion with zero frequency. It will be shown that these correspond to the three translations and three rotations of the molecule. Equation (1) is the mathematical way of expressing the result obtained when the normal modes of motion are superimposed with arbitrary amplitudes K_k and arbitrary phases ϵ_k.

The values of the K's and ϵ's depend on the initial conditions; *i.e.*, the way in which the motion was imparted to the molecule. Since there is the proper number of independent arbitrary constants, (1) is the most general solution of the problem.

2-4. Normal Coordinates

Definition. In order to carry out the quantum mechanical treatment of molecular vibrations, it is necessary to introduce a new set of coordinates Q_k, $k = 1, 2, \ldots, 3N$, called *normal coordinates*. It will be shown that there is one normal coordinate associated with each normal mode of motion, and vice versa. The normal coordinates are defined in terms of the mass-weighted cartesian displacement coordinates q_i by the linear equations

$$Q_k = \sum_{i=1}^{3N} l''_{ki} q_i \qquad k = 1, 2, \ldots, 3N \tag{1}$$

in which the coefficients l''_{ki} have been chosen so that in terms of the new coordinates the kinetic and potential energies have the forms

$$2T = \sum_{k=1}^{3N} \dot{Q}_k^2 \qquad 2V = \sum_{k=1}^{3N} \lambda'_k Q_k^2 \tag{2}$$

In other words, the potential energy in terms of the normal coordinates involves no cross products but only squares of Q's, while the kinetic energy retains its original form.

It will now be shown how the l''_{ki} are related to the l_{ik} of Eq. (12), Sec. 2-2, and how the constants λ'_k are related to the normal frequencies λ. Note that the indices i and j refer to the q coordinates, while k and l refer to the Q coordinates.

Linear Transformations. A set of linear algebraic equations connecting two sets of quantities such as the q's and Q's is called a *linear transformation*. If the numerical values of the original coordinates are given, those of the new coordinates (the Q's) can be obtained from (1). The set of equations

$$q_i = \sum_{k=1}^{3N} l'_{ik} Q_k \qquad i = 1, 2, \ldots, 3N \tag{3}$$

which give the values of the q's when those of the Q's are known, is called the *inverse* of the transformation (1). By substitution of (3) in (1), and vice versa, it is seen that

$$\sum_{i=1}^{3N} l''_{ki} l'_{il} = \delta_{kl} \qquad \sum_{k=1}^{3N} l'_{ik} l''_{kj} = \delta_{ij} \tag{4}$$

in which δ_{kl} is the Kronecker delta symbol.

The notation $(l^{-1})_{ik}$ is often used for the coefficients of the transformation which is the inverse of the transformation with the coefficients l_{ki}.

Equations of Motion. If normal coordinates are used, the equations of motion become

$$\frac{d}{dt}\frac{\partial T}{\partial \dot{Q}_k} + \frac{\partial V}{\partial Q_k} = \ddot{Q}_k + \lambda'_k Q_k = 0 \qquad k = 1, 2, \ldots, 3N \tag{5}$$

the solutions of which are

$$Q_k = K'_k \cos(\lambda'^{\frac{1}{2}}_k t + \epsilon'_k) \qquad k = 1, 2, \ldots, 3N \tag{6}$$

where K'_k and ϵ'_k are arbitrary constants. The solution in terms of the q's can be obtained by using (3), the result being

$$q_i = \sum_{k=1}^{3N} l'_{ik} K'_k \cos(\lambda'^{\frac{1}{2}}_k t + \epsilon'_k) \tag{7}$$

Comparison of this form of the general solution with that given in Eq. (1), Sec. 2-3, shows that

$$l'_{ik} = l_{ik} \qquad \text{and} \qquad \lambda'_k = \lambda_k \tag{8}$$

Consequently, the coefficients l_{ik} which specify the normal modes of motion are identical with the coefficients l'_{ik} of the transformation from the normal coordinates Q_k to the original coordinates q_i, while the roots λ_k of the secular equation are the coefficients of Q_k^2 in the expression for $2V$.

Orthogonality. The transformation to the normal coordinates has a further very important property. A transformation which transforms the sum of the squares of one set of coordinates into the sum of the squares of the other set is called an *orthogonal* transformation. Since $\Sigma \dot{q}_i^2 \rightarrow \Sigma \dot{Q}_k^2$, the transformation giving the q's in terms of the Q's is an orthogonal transformation. From (3) there is obtained

$$\sum_{i=1}^{3N} \dot{q}_i^2 = \sum_{i,k,l} l_{ik} l_{il} \dot{Q}_k \dot{Q}_l = \sum_k \dot{Q}_k^2 \tag{9}$$

the equality of the first and last terms being part of the definition of the normal coordinates. For (9) to be true, the following relation must hold:

$$\sum_i l_{ik} l_{il} = \delta_{kl} \tag{10}$$

Furthermore, by similar arguments involving the transformation giving the Q_k in terms of the q_i [coefficients $l''_{ki} = (l^{-1})_{ki}$] it is found that

$$\sum_k (l^{-1})_{ki} (l^{-1})_{kj} = \delta_{ij} \tag{11}$$

Comparison with (4) and (8) shows that

$$(l^{-1})_{ki} = l_{ik} \tag{12}$$

This property of orthogonal transformations is extremely convenient. Thus, a table of transformation coefficients such as Table 2-1 serves to

TABLE 2-1. TRANSFORMATION COEFFICIENTS FOR AN ORTHOGONAL TRANSFORMATION

	Q_1	Q_2	Q_3	. . .	Q_{3N}
q_1	l_{11}	l_{12}	l_{13}	. . .	$l_{1,3N}$
q_2	l_{21}	l_{22}	l_{23}	. . .	$l_{2,3N}$
q_3	l_{31}	l_{32}	l_{33}	. . .	$l_{3,3N}$
. . .	. . .	. . .	. . .	. . .	. . .
q_{3N}	$l_{3N,1}$	$l_{3N,2}$	$l_{3N,3}$	. . .	$l_{3N,3N}$

give both the transformation from the q's to the Q's (read down) and the inverse transformation from the Q's to the q's (read across).

The diagrams which represent normal modes of motion can also be used to represent normal coordinates if the arrows are drawn so as to represent displacements, not in ordinary units, but in the mass-adjusted scale of coordinates q_i. Then the component of an arrow along the direction of the coordinate q_i is proportional to l_{ik}, and since $(l^{-1})_{ki} = l_{ik}$, the diagram represents not only the relative amplitudes of the atoms during the normal mode of motion k but also the coefficients in the transformation

$$Q_k = \sum_i l_{ik} q_i \tag{13}$$

defining the normal coordinate Q_k (see Fig. 2-1, for example).[1]

2-5. Modes of Motion with Zero Frequency

A Special Set of Coordinates. It has been previously mentioned that six of the $3N$ roots of the secular equation (11), Sec. 2-2, have the value zero. This will now be proved. The basis of the proof is that there are six modes of motion of zero frequency, namely, the three translations and three rotations. The roots of the secular equation are properties of the molecule and not of the particular coordinate system used to set up the equation. Consequently, a special set of coordinates $\mathcal{R}_1, \mathcal{R}_2, \ldots, \mathcal{R}_{3N}$ may be used, having the following properties: (*a*) it is defined in terms of the q's by an orthogonal transformation; (*b*) six of the $\mathcal{R}$'s are

$$\begin{aligned}
\mathcal{R}_1 &= \mathfrak{N}_1 \sum_{\alpha=1}^{N} m_\alpha^{\frac{1}{2}} q_{x\alpha} \qquad \mathcal{R}_2 = \mathfrak{N}_2 \sum_{\alpha=1}^{N} m_\alpha^{\frac{1}{2}} q_{y\alpha} \\
\mathcal{R}_3 &= \mathfrak{N}_3 \sum_{\alpha=1}^{N} m_\alpha^{\frac{1}{2}} q_{z\alpha} \\
\mathcal{R}_4 &= \mathfrak{N}_4 \sum_{\alpha=1}^{N} m_\alpha^{\frac{1}{2}} (b_\alpha q_{z\alpha} - c_\alpha q_{y\alpha}) \\
\mathcal{R}_5 &= \mathfrak{N}_5 \sum_{\alpha=1}^{N} m_\alpha^{\frac{1}{2}} (c_\alpha q_{x\alpha} - a_\alpha q_{z\alpha}) \\
\mathcal{R}_6 &= \mathfrak{N}_6 \sum_{\alpha=1}^{N} m_\alpha^{\frac{1}{2}} (a_\alpha q_{y\alpha} - b_\alpha q_{x\alpha})
\end{aligned} \tag{1}$$

[1] If the quantities involved are complex, the analogous transformation with the property

$$(l^{-1})_{ki} = l^*_{ik}$$

is called a *unitary* transformation. It transforms $\Sigma \dot{q}_i \dot{q}_i^*$ into $\Sigma \dot{Q}_k \dot{Q}_k^*$.

in which $q_{x\alpha} = m_\alpha^{\frac{1}{2}} \Delta x_\alpha$, etc., and the index α refers to atoms, not coordinates. This notation is merely another way of numbering the regular coordinates $q_1, q_2, \ldots, q_{3N}$ so as to distinguish those associated with the three coordinate directions. a_α, b_α, c_α are the coordinates of the equilibrium position of the αth atom. The $\mathfrak{N}$'s are normalizing constants chosen to make the transformation orthogonal. The above conditions do not completely specify the rest of the $\mathfrak{R}$'s but this is not necessary. In terms of the $\mathfrak{R}$'s, the kinetic and potential energies will have the forms

$$2T = \sum_{u=1}^{3N} \dot{\mathfrak{R}}_u^2 \qquad 2V = \sum_{uv} F_{uv}\mathfrak{R}_u\mathfrak{R}_v \tag{2}$$

and the same procedure as in Sec. 2-2 leads to the secular equation

$$\begin{vmatrix} F_{11} - \lambda & F_{12} & F_{13} & \cdots & F_{1,3N} \\ F_{21} & F_{22} - \lambda & F_{23} & \cdots & F_{2,3N} \\ F_{31} & F_{32} & F_{33} - \lambda & \cdots & F_{3,3N} \\ \cdots & \cdots & \cdots & \cdots & \cdots \\ F_{3N,1} & F_{3N,2} & F_{3N,3} & \cdots & F_{3N,3N} - \lambda \end{vmatrix} = 0 \tag{3}$$

Effect of Translation. It will next be shown that an infinitesimal translation τ of the whole molecule in the x direction changes $\mathfrak{R}_1$ into $\mathfrak{R}_1 + \mathfrak{N}_1\tau \sum_{\alpha=1}^{N} m_\alpha$, but does not affect the other $\mathfrak{R}$'s. The translation τ adds τ to every x_α, and therefore, since $q_{x\alpha} = m_\alpha^{\frac{1}{2}} \Delta x_\alpha$, it adds $m_\alpha^{\frac{1}{2}}\tau$ to every $q_{x\alpha}$. Then, from the definition of $\mathfrak{R}_1$,

$$\mathfrak{R}_1 = \mathfrak{N}_1 \sum_\alpha m_\alpha^{\frac{1}{2}} q_{x\alpha} \rightarrow \mathfrak{N}_1 \sum_\alpha m_\alpha^{\frac{1}{2}}(q_{x\alpha} + m_\alpha^{\frac{1}{2}}\tau) = \mathfrak{R}_1 + \mathfrak{N}_1\tau \sum_\alpha m_\alpha \tag{4}$$

The transformation defining the $\mathfrak{R}$'s may be written as

$$\mathfrak{R}_u = \sum_\alpha (l_{u\alpha}q_{x\alpha} + m_{u\alpha}q_{y\alpha} + n_{u\alpha}q_{z\alpha}) \tag{5}$$

in which the quantities $l_{u\alpha}$, $m_{u\alpha}$, and $n_{u\alpha}$ are the transformation coefficients. Since the transformation is orthogonal,

$$\sum_{\alpha=1}^{N} (l_{u\alpha}l_{v\alpha} + m_{u\alpha}m_{v\alpha} + n_{u\alpha}n_{v\alpha}) = 0 \qquad u \neq v \tag{6}$$

Insertion of the values of $l_{1\alpha}$, $m_{1\alpha}$, and $n_{1\alpha}$ taken from (1) gives

$$\mathfrak{N}_1 \sum_{\alpha=1}^{N} m_\alpha^{\frac{1}{2}} l_{v\alpha} = 0 \qquad v \neq 1 \tag{7}$$

Consequently, the translation τ in the x direction has the following effect on $\mathcal{R}_v$:

$$\mathcal{R}_v \to \sum_\alpha [l_{v\alpha}(q_{x\alpha} + m_\alpha^{\frac{1}{2}}\tau) + m_{v\alpha}q_{y\alpha} + n_{v\alpha}q_{z\alpha}]$$
$$= \mathcal{R}_v + \tau \sum_\alpha m_\alpha^{\frac{1}{2}} l_{v\alpha} = \mathcal{R}_v \qquad v \neq 1 \quad (8)$$

Therefore, $\mathcal{R}_1$ is the coordinate which represents translation in the x direction, while the other coordinates are independent of such a translation.

Exactly similar arguments show that $\mathcal{R}_2$ and $\mathcal{R}_3$ represent translations in the y and z directions, respectively. By considering the effect of infinitesimal rotations about the x, y, and z axes, it is found by similar methods that these motions are measured by $\mathcal{R}_4$, $\mathcal{R}_5$, $\mathcal{R}_6$, respectively, and that they do not affect the other $\mathcal{R}$'s.

Form of Potential Energy. Since the potential energy depends only on the internal configuration of the molecule, it is unchanged by a translation or rotation. The translation τ in the x direction has the effect

$$\begin{aligned} 2V = \sum_{u,v} F_{uv}\mathcal{R}_u\mathcal{R}_v &\to \sum_{v=2}^{3N} F_{1v}(\mathcal{R}_1 + \Upsilon)\mathcal{R}_v \\ &+ \sum_{u=2}^{3N} F_{u1}\mathcal{R}_u(\mathcal{R}_1 + \Upsilon) + F_{11}(\mathcal{R}_1 + \Upsilon)^2 \\ &+ \sum_{u\neq 1, v\neq 1} F_{uv}\mathcal{R}_u\mathcal{R}_v \\ &= \sum_{u,v} F_{uv}\mathcal{R}_u\mathcal{R}_v + 2\Upsilon \sum_v F_{1v}\mathcal{R}_v + \Upsilon^2 F_{11} \end{aligned} \qquad (9)$$

where $\Upsilon = \mathfrak{N}_1\tau \sum m_\alpha$. Since $2V$ must be unchanged by such a translation, *i.e.*, must be independent of Υ, no matter what the values of the $\mathcal{R}$'s, it must be true that

$$F_{1v} = 0 \qquad v = 1, 2, \ldots, 3N \qquad (10)$$

By identical arguments involving translations in the y and z directions and infinitesimal rotations about the x, y, and z axes, it follows that

$$F_{1v} = F_{2v} = F_{3v} = F_{4v} = F_{5v} = F_{6v} = 0 \qquad v = 1, 2, \ldots, 3N \quad (11)$$

so that in terms of this special coordinate system the secular equation has the form

$$\begin{vmatrix} -\lambda & 0 & 0 & 0 & 0 & 0 & 0 & 0 & \cdots \\ 0 & -\lambda & 0 & 0 & 0 & 0 & 0 & 0 & \cdots \\ 0 & 0 & -\lambda & 0 & 0 & 0 & 0 & 0 & \cdots \\ 0 & 0 & 0 & -\lambda & 0 & 0 & 0 & 0 & \cdots \\ 0 & 0 & 0 & 0 & -\lambda & 0 & 0 & 0 & \cdots \\ 0 & 0 & 0 & 0 & 0 & -\lambda & 0 & 0 & \cdots \\ 0 & 0 & 0 & 0 & 0 & 0 & F_{77}-\lambda & F_{78} & \cdots \\ 0 & 0 & 0 & 0 & 0 & 0 & F_{87} & F_{88}-\lambda & \cdots \\ \cdots & \cdots & \cdots & \cdots & \cdots & \cdots & \cdots & \cdots & \cdots \end{vmatrix} = 0 \quad (12)$$

or

$$\lambda^6 \begin{vmatrix} F_{77}-\lambda & F_{78} & \cdots & F_{7,3N} \\ F_{87} & F_{88}-\lambda & \cdots & F_{8,3N} \\ \cdots & \cdots & \cdots & \cdots \\ F_{3N,7} & F_{3N,8} & \cdots & F_{3N,3N}-\lambda \end{vmatrix} = 0 \quad (13)$$

Consequently, there are six roots with value zero as originally stated, and the six corresponding normal modes of motion are the translations and rotations. Furthermore, $\mathcal{R}_1, \ldots, \mathcal{R}_6$ are a set of corresponding normal coordinates.

Comparison of (1) with Eqs. (2) and (5), Sec. 2-1, shows that the conditions required to define the translating-rotating system are

$$\mathcal{R}_u = 0 \qquad u = 1, 2, 3, 4, 5, 6 \quad (14)$$

2-6. Other Types of Coordinates

The coordinates q_i previously introduced are much the most useful for theoretical purposes, because of the simple form they give to the kinetic energy, and the related fact that the transformation to normal coordinates is orthogonal. However, in practical applications, other types of coordinates are frequently more convenient. Thus, it is usually easier to use ordinary x, y, z coordinates in place of q's with mass-adjusted scales.

Cartesians. For convenience in writing, the symbols $\xi_1, \xi_2, \ldots, \xi_{3N}$ will be used for $\Delta x_1, \Delta y_1, \Delta z_1, \Delta x_2, \ldots, \Delta z_N$; and m_i will represent the mass of the atom to which ξ_i refers. Then

$$2T = \sum_{i=1}^{3N} m_i \dot{\xi}_i^2 \quad (1)$$

and

$$2V = \sum_{i,j=1}^{3N} f'_{ij} \xi_i \xi_j \quad (2)$$

The secular equation can be obtained in the same way as before and has the form

$$\begin{vmatrix} f'_{11} - m_1\lambda & f'_{12} & f'_{13} & \cdots & f'_{1,3N} \\ f'_{21} & f'_{22} - m_2\lambda & f'_{23} & \cdots & f'_{2,3N} \\ f'_{31} & f'_{32} & f'_{33} - m_3\lambda & \cdots & f'_{3,3N} \\ \cdots & \cdots & \cdots & \cdots & \cdots \\ f'_{3N,1} & f'_{3N,2} & f'_{3N,3} & \cdots & f'_{3N,3N} - m_{3N}\lambda \end{vmatrix} = 0 \quad (3)$$

The constants f'_{ij}, called the force constants, do not involve the masses of the atoms, in contrast to the constants f_{ij} in terms of the mass-adjusted coordinates q_i. The relationship between the two sets of constants is

$$f_{ij} = \frac{f'_{ij}}{(m_i m_j)^{\frac{1}{2}}} \quad (4)$$

General Case. It is not necessary to use cartesian coordinates. Any coordinate system in terms of which the kinetic and potential energies are homogeneous quadratic forms in the velocities and coordinates, respectively, can be used. In the general case the coefficients t_{ij} in the kinetic energy

$$2T = \sum_{i,j} t_{ij}\dot{\eta}_i\dot{\eta}_j \quad (5)$$

may be functions of the coordinates η_i. The proper procedure in that event is to expand them as power series in the coordinates:

$$t_{ij} = t_{ij}^0 + \sum_{k=1}^{3N} t_{ij}^{(k)}\eta_k \cdots \quad (6)$$

Since the present theory is designed to apply only to infinitesimal vibrations, all but the first term can be neglected. If the potential energy has the form

$$2V = \sum_{ij} f''_{ij}\eta_i\eta_j \quad (7)$$

in which higher terms have been neglected, then the usual considerations involving the equations of motion yield a secular equation

$$\begin{vmatrix} f''_{11} - t^0_{11}\lambda & f''_{12} - t^0_{12}\lambda & \cdots & f''_{1,3N} - t^0_{1,3N}\lambda \\ f''_{21} - t^0_{21}\lambda & f''_{22} - t^0_{22}\lambda & \cdots & f''_{2,3N} - t^0_{2,3N}\lambda \\ \cdots & \cdots & \cdots & \cdots \\ f''_{3N,1} - t^0_{3N,1}\lambda & f''_{3N,2} - t^0_{3N,2}\lambda & \cdots & f''_{3N,3N} - t^0_{3N,3N}\lambda \end{vmatrix} = 0 \quad (8)$$

This is the most general form of the secular equation.

Internal Coordinates. In all the coordinate systems so far introduced, the six conditions of no rotation and no translation were applied after the solution of the secular equation, by placing the six normal coordinates corresponding to translation and rotation equal to zero (see Sec. 2-5). In many cases, however, it is more convenient to apply these six conditions at the very beginning, using them to eliminate six of the original coordinates η_i. (The symbols η_i will be used for the most general set of $3N$ displacement coordinates and will not imply that the coordinates are necessarily noncartesian.) This is possible because the six conditions of Eqs. (2) and (5), Sec. 2-1, are six independent relations among the $3N$ coordinates η_i.

The reduction in the number of original coordinates may be carried out in two ways. One method is to use the six conditions to express six of the η_i's in terms of the remaining $3N - 6$ coordinates. The other method is to introduce a new set of $3N - 6$ coordinates $S_1, S_2, \ldots, S_{3N-6}$ which are defined by means of the six conditions and $3N - 6$ relations connecting the S's with the η's. Such coordinates are known as *internal* coordinates because they describe the internal configuration of the molecule without regard for its position as a whole in space.

In either case it is necessary to obtain the kinetic and potential energies as quadratic forms in the velocities $\dot{S}_i$ and the coordinates S_i, respectively, using only the constant part of the coefficients as before. The secular equation will then have the same form as previously, (8), except that it will consist of only $3N - 6$ rows and columns ($3N - 5$ for linear molecules). It is this reduction in the size of the secular equation which makes the use of internal coordinates useful, inasmuch as in most applications of the method of normal coordinates one of the most troublesome steps is the solution of the secular equation, a difficulty which increases rapidly with the degree of the equation.

It should be emphasized that when coordinates are used which do not have mass-adjusted scales, so that the transformation to normal coordinates is not an orthogonal transformation, the very useful relationship in Eq. (12), Sec. 2-4, cannot be employed.

2-7. An Illustration

Description of System. With the elementary methods described in this chapter, even as simple a molecule as water is rather too cumbersome to be used as an illustration. In later chapters much more powerful methods (which are, however, based on those in this chapter) will be developed. Until then, an artificial example may prove helpful in illustrating the idea of normal vibrations and normal coordinates. Such an example is provided by a linear system of two point masses and three weightless springs as shown in Fig. 2-4. The springs 1 and 2 are fastened to fixed points so that no question of rotation or translation enters. Furthermore, only linear motions will be considered. Therefore, only two coordinates need be used, which may be called Δx_1 and Δx_2, the displacements of particles 1 and 2, respectively, from their equilibrium positions.

Solution with Cartesian Coordinates. The potential energy of the system is given by

$$2V = f_1(\Delta r_1)^2 + f_{12}(\Delta r_{12})^2 + f_2(\Delta r_2)^2 \tag{1}$$

in which f_1, f_{12}, f_2 are the *force constants* (Hooke's law constants) for the springs and Δr_1, Δr_{12}, Δr_2 are the *extensions* of the springs from their equilibrium lengths.

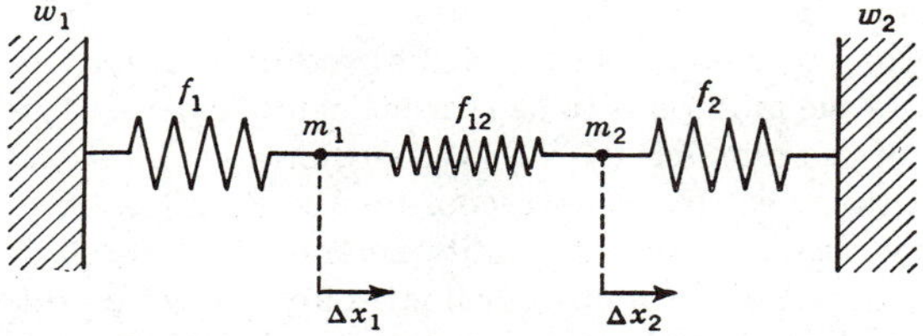

FIG. 2-4. Diagram of the linear two-body vibrator discussed in Sec. 2-7. w_1 and w_2 are fixed supports. Masses m_1 and m_2 are connected to w_1 and w_2 through springs having force constants f_1 and f_2, and are connected together by a spring having the force constant f_{12}. The motion is considered to be restricted to a direction along the x axis, and Δx_1 and Δx_2 are displacements, in the direction indicated, of m_1 and m_2 from their equilibrium positions.

In this case it is seen that

$$\Delta r_1 = \Delta x_1 \qquad \Delta r_2 = -\Delta x_2 \qquad \Delta r_{12} = \Delta x_2 - \Delta x_1 \tag{2}$$

so that

$$2V = (f_1 + f_{12})(\Delta x_1)^2 - 2f_{12}\,\Delta x_1\,\Delta x_2 + (f_2 + f_{12})(\Delta x_2)^2 \tag{3}$$

The kinetic energy is simply

$$2T = m_1(\Delta \dot{x}_1)^2 + m_2(\Delta \dot{x}_2)^2 \tag{4}$$

The equations of motion, Eqs. (7), Sec. 2-2, are then

$$\begin{aligned} m_1\,\Delta\ddot{x}_1 + (f_1 + f_{12})\,\Delta x_1 - f_{12}\,\Delta x_2 &= 0 \\ m_2\,\Delta\ddot{x}_2 - f_{12}\,\Delta x_1 + (f_2 + f_{12})\,\Delta x_2 &= 0 \end{aligned} \tag{5}$$

The substitution

$$\begin{aligned} \Delta x_1 &= A_1 \cos(\lambda^{\frac{1}{2}}t + \epsilon) \\ \Delta x_2 &= A_2 \cos(\lambda^{\frac{1}{2}}t + \epsilon) \end{aligned} \tag{6}$$

leads to the pair of algebraic equations

$$\begin{aligned} (f_1 + f_{12} - m_1\lambda)A_1 - f_{12}A_2 &= 0 \\ -f_{12}A_1 + (f_2 + f_{12} - m_2\lambda)A_2 &= 0 \end{aligned} \tag{7}$$

for the amplitudes A_1 and A_2. The equations have nonzero solutions only if the determinant of the coefficients vanishes, *i.e.*, if

$$\begin{vmatrix} f_1 + f_{12} - m_1\lambda & -f_{12} \\ -f_{12} & f_2 + f_{12} - m_2\lambda \end{vmatrix} = 0 \tag{8}$$

This is the secular equation for the problem. Its roots (there are two of them) give the fundamental frequencies of vibration of the system (ν) since $\lambda = 4\pi^2\nu^2$. Because the determinant here has only two rows, it is easily expanded and yields

$$m_1m_2\lambda^2 - [m_1(f_2 + f_{12}) + m_2(f_1 + f_{12})]\lambda + f_1f_2 + (f_1 + f_2)f_{12} = 0 \quad (9)$$

a quadratic equation in λ. This has the two roots

$$\lambda = \frac{1}{2m_1m_2}\Big([m_1(f_2 + f_{12}) + m_2(f_1 + f_{12})] \pm \{[m_1(f_2 + f_{12}) - m_2(f_1 + f_{12})]^2 + 4m_1m_2f_{12}^2\}^{\frac{1}{2}}\Big) \quad (10)$$

The normal modes of vibration may be found by substituting first one, then the other, of the above values of λ into (7) and solving for the ratios of the amplitudes A_i in the two cases.

FIG. 2-5. Normal modes of motion of the linear two-body vibrator of Sec. 2-7 for the special case, $m_1 = m_2$ and $f_1 = f_2$.

A Special Case. It is enlightening to consider some special cases. First, let $m_1 = m_2 = m$ and $f_1 = f_2 = f$. Then (10) becomes

$$\lambda = \frac{1}{m}(f + f_{12} \pm f_{12}) \quad (11)$$

and the equations for A_i yield

$$A_1 = -A_2 \text{ for } \lambda = \frac{1}{m}(f + 2f_{12}) \quad (12)$$

and

$$A_1 = A_2 \text{ for } \lambda = \frac{1}{m}f \quad (13)$$

The modes of motion in this case are very simple. For the first (the highest frequency, since here f_{12} is necessarily positive as it is the force constant of a spring), the two particles are oscillating with the same frequency, equal amplitudes, but with opposite phases (see Fig. 2-5*a*). In the low-frequency motion, the particles have the same phase, as shown in Fig. 2-5*b*. Here the center spring is not stretched during the motion so that f_{12} does not enter the expression for λ.

If f_{12} is reduced to zero, the two modes have the same frequency, which is just that of an individual mass attached to the fixed point by its own spring. If on the other hand f is reduced to zero, one frequency vanishes and the other is given by $\lambda = (2/m)f_{12}$, or in the general case

$$\lambda = \left[\frac{m_1 + m_2}{m_1 m_2}\right] f_{12}$$

This is the result for a free diatomic molecule.

It should be noted that it is not necessary to write down the equations corresponding to (5), (6), and (7) for each problem, since it is seen that the elements of the secular equation can be taken directly from the expressions for $2T$ and $2V$ [see Eqs. (6) to (8), Sec. 2-6]. Note, however, that the element in the first row and second column is $-f_{12}$, which is $\frac{1}{2}$ the coefficient of $\Delta x_1 \, \Delta x_2$ in $2V$. This is correct because the term $-2f_{12} \, \Delta x_1 \, \Delta x_2$ of (3) is really $-f_{12} \, \Delta x_1 \, \Delta x_2 - f_{12} \, \Delta x_2 \, \Delta x_1$.

Use of Mass-weighted Coordinates. In this treatment ordinary cartesian displacement coordinates have been used, whereas in the earlier part of the chapter mass-weighted cartesian coordinates were employed. If $q_1 = (m_1)^{\frac{1}{2}} \, \Delta x_1$ and $q_2 = (m_2)^{\frac{1}{2}} \, \Delta x_2$, the kinetic and potential energies become

$$\begin{aligned} 2V &= \frac{f_1 + f_{12}}{m_1} q_1^2 - 2 \frac{f_{12}}{(m_1 m_2)^{\frac{1}{2}}} q_1 q_2 + \frac{f_2 + f_{12}}{m_2} q_2^2 \\ 2T &= \dot{q}_1^2 + \dot{q}_2^2 \end{aligned} \tag{14}$$

so that the secular equation has the form

$$\begin{vmatrix} \dfrac{f_1 + f_{12}}{m_1} - \lambda & \dfrac{-f_{12}}{(m_1 m_2)^{\frac{1}{2}}} \\ \dfrac{-f_{12}}{(m_1 m_2)^{\frac{1}{2}}} & \dfrac{f_2 + f_{12}}{m_2} - \lambda \end{vmatrix} = 0 \tag{15}$$

This differs from the previous form, (8), only in that each row and each column has been divided by $(m_i)^{\frac{1}{2}}$, i being the number of the row or column. The roots λ are therefore unchanged.

Normal Coordinates. Mass-weighted coordinates are particularly suitable when it is desired to find the normal coordinates, inasmuch as the transformation connecting these two sets is orthogonal. From Eqs. (12) and (10), Sec. 2-2, it is seen that the transformation coefficients l_{ik} are determined by the equations

$$\begin{aligned} \left(\frac{f_1 + f_{12}}{m_1} - \lambda_k\right) l_{1k} - \frac{f_{12}}{(m_1 m_2)^{\frac{1}{2}}} l_{2k} &= 0 \\ \frac{-f_{12}}{(m_1 m_2)^{\frac{1}{2}}} l_{1k} + \left(\frac{f_2 + f_{12}}{m_2} - \lambda_k\right) l_{2k} &= 0 \end{aligned} \tag{16}$$

in which the appropriate root λ_k has been substituted. These equations determine only the ratios of the l's; their absolute values are fixed by the normalization conditions [Eq. (13), Sec. 2-2]

$$l_{11}^2 + l_{21}^2 = 1 \qquad l_{12}^2 + l_{22}^2 = 1 \tag{17}$$

In the special case previously treated where $f_1 = f_2$ and $m_1 = m_2$, use of (11), (16), and (17) leads to the result

$$l_{11} = l_{12} = l_{22} = -l_{21} = 2^{-\frac{1}{2}} \tag{18}$$

so that the normal coordinates are

$$Q_1 = 2^{-\frac{1}{2}}(q_1 - q_2) \qquad Q_2 = 2^{-\frac{1}{2}}(q_1 + q_2) \tag{19}$$

In terms of these coordinates the potential and kinetic energies have the forms

$$2V = \lambda_1 Q_1^2 + \lambda_2 Q_2^2 \qquad 2T = \dot{Q}_1^2 + \dot{Q}_2^2 \tag{20}$$

where λ_1 and λ_2 are the two roots of the secular equation. The reader should verify this result for the special case considered above.

2-8. Linear Molecules

Throughout the previous sections there have been references to the fact that linear molecules (such as CO_2, C_2H_2, etc.) are exceptional in that they have only five modes of motion of zero frequency and consequently $3N - 5$ normal modes of vibration. In a linear molecule, the three conditions on the moving coordinate system given in Eq. (5), Sec. 2-1, reduce to the two equations

$$\begin{aligned} \sum_{\alpha=1}^{N} m_\alpha c_\alpha \, \Delta y_\alpha &= 0 \\ \sum_{\alpha=1}^{N} m_\alpha c_\alpha \, \Delta x_\alpha &= 0 \end{aligned} \tag{1}$$

if the axis of the molecule is taken as the z direction. This is true because the equilibrium positions of all the atoms lie on the z axis so that of the coordinates a_α, b_α, c_α of these equilibrium positions, only c_α is different from zero. There are thus only five conditions altogether and therefore only five motions of zero frequency. The condition which drops out is the prohibition of rotation about the z axis, since a linear molecule, considered as made up of point masses, cannot rotate about its axis, unless it is distorted.

However, when the molecule is bent out of line during the course of a vibration, it then has meaning to discuss the possibility of rotation about its axis. Such a rotation is not forbidden by any of the conditions above and may give rise to an angular momentum about the axis. For the present, however, the displacements from equilibrium are being considered as infinitesimal so that this question does not arise.

The quadratic terms in the potential energy of a linear molecule do not involve any cross products of displacements along the axis with displacements perpendicular to the axis; *i.e.*, terms of the type $\Delta x_\alpha \, \Delta z_\beta$ or $\Delta y_\alpha \, \Delta z_\beta$ (where α may equal β). This results from the fact that the molecule and its potential energy are both symmetrical about the axis. Therefore the coefficient of a term such as $\Delta x_\alpha \, \Delta z_\beta$ must vanish because such a term causes the potential energy to be different when Δx_α is positive from its value when Δx_α is negative. By a similar argument, it is seen that there can be no cross products between Δx_α and Δy_β. The secular equation for a linear molecule will thus have many zero elements, and if the numbering of the coordinates is chosen so that all the x's come first, then the y's and finally the z's, the secular equation will have the general form shown in Fig. 2-6, in which the shaded areas indicate parts of the equation containing nonzero elements, while the unshaded areas indicate parts in which all the elements are zero. A secular equation which has this form is said to be *factored,* the shaded blocks being called the *factors,* since when the whole determinant is expanded it can be written as the product of several factors, each of which is the expanded form of one of the shaded blocks considered as a small, separate determinant. A secular equation which factors in this manner is consequently equivalent to a number of separate, smaller determinantal equations corresponding to the shaded blocks.

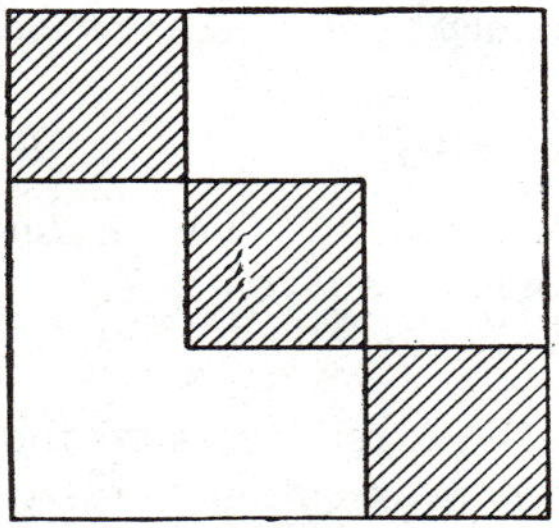

FIG. 2-6. Form of the secular equation for linear molecules, using cartesian displacement coordinates. This illustrates the factoring made possible by the symmetry of the potential function. All nonvanishing elements fall in the shaded areas, and the three blocks correspond to the Δx's, Δy's, and the Δz's, respectively.

Since the secular equation for a linear molecule breaks up into three factors, one involving the coordinates Δx_α, another the Δy_α, and the third the Δz_α, it follows from the connection between the secular equation and the normal modes of vibration (see Sec. 2-2) that there will be three directions x, y, and z. Normal vibrations will thus involve displacements either along the axis of the molecule or at right angles to it, but never both in the same normal mode of vibration. (This, of course, does not prevent the molecule from vibrating in such a mixed manner, but such a motion is not a normal mode of vibration but a superposition of such modes.)

Another consequence of the symmetry of linear molecules about their axes is that normal vibrations involving displacements at right angles to the axis occur in pairs with a common degenerate frequency. This is true because of the physical indistinguishability of the x and y directions in such a molecule.

Molecules all of whose atoms lie in a single plane have the property that the quadratic part of the potential energy contains no cross terms between displacements in the plane and those perpendicular to it, the reason for this being the same as for the similar result for linear molecules. Consequently, the secular equation for a planar molecule has at least two factors, and the normal modes of vibration are motions either in the plane or perpendicular to it. A planar molecule does not necessarily display any degeneracy, however.

In Chap. 6 it will be shown that factoring of the secular equation is connected with the symmetry of the molecule and may occur in molecules which are neither linear nor planar.

CHAPTER 3

WAVE MECHANICS AND THE VIBRATION OF MOLECULES

3-1. The Wave Equation for the Vibration of the Harmonic Oscillator Model

In the previous chapter it was indicated that when classical mechanics is employed the rotation and vibration of a molecule can be treated separately. The proof of this will be given in Chap. 11, where it will likewise be shown that rotation and vibration are also approximately separable when wave mechanics is used. To this degree of approximation the total wave function ψ for the motions of the atoms can be written as a product of a vibrational wave function ψ_V and a rotational wave function ψ_R; that is,

$$\psi \cong \psi_V \psi_R \tag{1}$$

The function ψ_R, which is a function of the Eulerian angles θ, ϕ, and χ (Appendix I) describing the orientation in space of the rotating coordinate system, is obtained by solution of the rotational wave equation. Some results from the rotational problem are summarized in Appendix XVI.

The vibrational wave function ψ_V is a function of the internal coordinates (the normal coordinates are usually used) and is a solution of the vibrational wave equation. In Sec. 2-4 it was shown that the kinetic and potential energies of vibration are given by the expressions

$$T = \tfrac{1}{2} \sum_{k=1}^{3N-6} \dot{Q}_k^2 \qquad V = \tfrac{1}{2} \sum_{k=1}^{3N-6} \lambda_k Q_k^2 \tag{2}$$

in terms of the normal coordinates Q_k. The vibrational wave equation will then have the form[1]

$$\frac{-h^2}{8\pi^2} \sum_{k=1}^{3N-6} \frac{\partial^2 \psi_V}{\partial Q_k^2} + \tfrac{1}{2} \sum_{k=1}^{3N-6} \lambda_k Q_k^2 \psi_V = W_V \psi_V \tag{3}$$

where W_V is the vibrational energy and h is Planck's constant.

[1] If the possible complications due to rotation are ignored, the wave equation (3) follows directly from the expression (2) for the kinetic and potential energies, either by the standard method (see Sec. 11-4) or more naïvely by treating the problem just as if the Q_k's were ordinary cartesian coordinates. The more exact methods given in Chap. 11 show that (3) is correct (to the desired degree of approximation) even when rotation is considered.

The advantage of using normal coordinates will now be evident, since the wave equation (3) in this form is separable into $3N - 6$ equations, one for each normal coordinate. Let

$$W_V = W(1) + W(2) + \cdots + W(3N - 6) \tag{4}$$

and

$$\psi_V = \psi(Q_1)\psi(Q_2) \cdots \psi(Q_{3N-6}) \tag{5}$$

Then it will be seen that the wave equation (3) is satisfied if the functions $\psi(Q_k)$ and the energies W_k satisfy equations of the type

$$\frac{-h^2}{8\pi^2}\frac{d^2\psi(Q_k)}{dQ_k^2} + \tfrac{1}{2}\lambda_k Q_k^2\psi(Q_k) = W(k)\psi(Q_k) \tag{6}$$

Each of these equations is a total differential equation in one variable, Q_k. In fact (6) is the well-known wave equation for the linear harmonic oscillator, expressed in terms of the normal coordinate Q_k instead of the usual linear coordinate x. The solution ψ_V of the vibrational problem is therefore expressible as a product of harmonic oscillator functions $\psi(Q_k)$, one for each normal coordinate, while the total vibrational energy W_V is the sum of the energies of $3N - 6$ harmonic oscillators.

3-2. Description of the Energy Levels of the Harmonic Oscillator

Quantum Numbers and Normal Frequencies. The nature of the energy levels will be discussed before that of the wave functions themselves, since the former are more important. As is well known[1] the energy levels of a linear harmonic oscillator are given by the expression

$$W = (v + \tfrac{1}{2})h\nu \qquad v = 0, 1, 2, \ldots \tag{1}$$

where v is the quantum number which can take on any positive integral value including zero, while ν is the classical frequency of the system; h is Planck's constant. Consequently, the vibrational energy of a molecule with several classical frequencies ν_k is, from Eq. (4), Sec. 3-1, of the form

$$W = (v_1 + \tfrac{1}{2})h\nu_1 + (v_2 + \tfrac{1}{2})h\nu_2 + \cdots + (v_{3N-6} + \tfrac{1}{2})h\nu_{3N-6} \tag{2}$$

that is, every normal coordinate Q_k has associated with it a quantum number v_k and a *normal frequency* ν_k, these latter frequencies being the classical normal frequencies of vibration.

Nomenclature. As an illustration, Fig. 3-1 shows the lower energy

[1] The harmonic oscillator is treated in practically all elementary books on wave mechanics. See, for example, L. Pauling and E. B. Wilson, Jr., "Introduction to Quantum Mechanics," Sec. 11, McGraw-Hill, New York, 1935. The original wave mechanical treatment was given by E. Schrödinger, *Naturwissenschaften,* **14:** 664 (1926).

levels of vibration of the water molecule. It will be noted that the lowest energy level, called the *ground level,* for which all the quantum numbers are zero, is 4,500 cm^{-1} above the zero of energy. This quantity is the zero-point energy of the molecule and is equal to $\frac{1}{2}h \sum_{k=1}^{3N-6} \nu_k$. In polyatomic molecules the zero-point energy may be of considerable magnitude and is an important quantity. The energy levels for which all quantum numbers are zero except one, which has the value unity, are called *fundamental levels.* When only one normal vibration is excited—*i.e.*, when only one v_k is different from zero, but that quantum number is greater than one—the corresponding energy levels are called *overtone levels.* When two or more quantum numbers have nonzero values, the resulting levels are known as *combination levels.* It is clear from the example of H_2O that the energy-level diagram of a polyatomic molecule becomes increasingly and rapidly complicated as the energy is increased. Figure 3-1 is calculated from (2) and therefore does not show the convergence of the energy levels observed in actual molecules where the higher powers of the displacements in the expansion of the potential energy, as in Eq. (4), Sec. 2-2, are not negligible. The effect of these so-called *anharmonic* terms and other deviations from the simple picture will be discussed in Chap. 8.

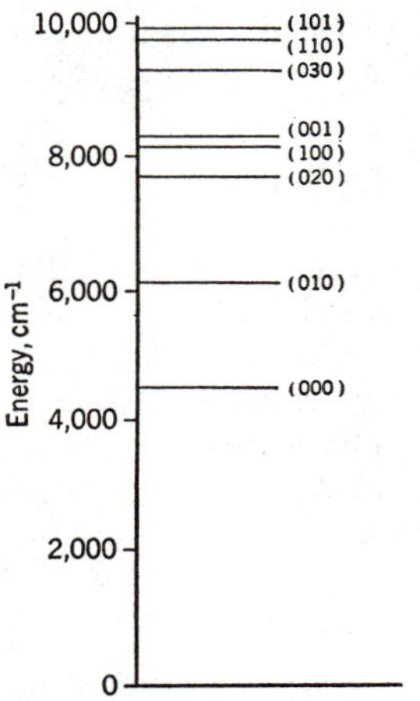

FIG. 3-1. Lower energy levels of vibration of water molecule, assuming harmonic oscillator model with $\omega_1 = (\nu_1/c)$ 3,650 cm^{-1}, $\omega_2 =$ 1,595 cm^{-1}, and $\omega_3 =$ 3,756 cm^{-1}. Levels are indexed by the quantum numbers (n_1,n_2,n_3).

By the Bohr frequency rule [Eq. (1), Sec. 1-5], the frequency of the light absorbed or emitted by a molecule is given by

$$\nu_{n''n'} = \nu_{n'n''} = \frac{W_{n'} - W_{n''}}{h} \tag{3}$$

where $W_{n'}$ is the energy of the *upper* state and $W_{n''}$ is the energy of the *lower* state. By combination with (2) it follows that a transition from the ground state to the state in which $v_k = 1$, all other v's zero, will have the frequency ν_k, which is the classical frequency of the kth normal mode of vibration. Such frequencies are known as *fundamental* frequencies. Similarly, transitions from the ground level to overtone levels are called *overtones,* and those from the ground level to combination levels are *combination frequencies* or *combinations.* In absorption spectra, transitions arising from the ground state are the most important because that state has, in general, the greatest population, but some transitions will start at excited states and end higher. These are known as *difference frequencies* or *difference combinations.*

Vibrational Degeneracy. In some molecules there will be energy levels to which belong more than one wave function, in other words, *degenerate* energy levels. For example, if the molecule has a doubly degenerate normal frequency $\nu_k = \nu_l$, to which therefore two normal coordinates Q_k and Q_l correspond, it is evident that the quantum mechanical energy levels corresponding to states in which these normal modes are excited (that is, v_k or $v_l \neq 0$) will be degenerate. The terms in the energy formula which depend on v_k and v_l can be combined to the single term

$$(v_k + v_l + 1)h\nu_k \tag{4}$$

and the energy will depend not on the individual values of v_k and v_l (which determine ψ_V) but on their sum. The states $v_k = 0$, $v_l = 1$, and $v_k = 1$, $v_l = 0$, have the same energy but different wave functions. The energy level for which $v_k + v_l = 2$ is triply degenerate, since the sets of quantum numbers (2,0), (0,2), and (1,1), correspond to it. In general, the level with $v_k + v_l = v$ will be $(v + 1)$-fold degenerate. In a similar manner a triply degenerate classical frequency will contribute the term

$$(v_k + v_l + v_m + \tfrac{3}{2})h\nu_k = (v + \tfrac{3}{2})h\nu_k \tag{5}$$

to the energy. The level with $v_k + v_l + v_m = v$ will have a degeneracy $\frac{1}{2}(v + 1)(v + 2)$. In addition to this type of degeneracy, arising from strict degeneracy in the classical frequencies, there is another type known as accidental degeneracy. This type occurs when the numerical values of the classical frequencies are such as to cause two energy levels to coincide more or less closely. The consequences of such an accidental coincidence will be discussed in Sec. 8-7.

3-3. Nature of the Wave Functions

The solutions of the harmonic oscillator equation, Eq. (6), Sec. 3-1, have been described in many places.[1] They are called the Hermite orthogonal functions and are of the form

$$\psi_{v_k}(Q_k) = \mathfrak{N}_{v_k} e^{-\frac{1}{2}\gamma_k Q_k^2} H_{v_k}(\gamma_k^{\frac{1}{2}} Q_k) \tag{1}$$

in which $\mathfrak{N}_{v_k}$ is the normalizing factor,

$$\mathfrak{N}_{v_k} = \left[\left(\frac{\gamma}{\pi}\right)^{\frac{1}{2}} \frac{1}{2^{v_k}(v_k!)}\right]^{\frac{1}{2}} \tag{2}$$

[1] See footnote on p. 35.

γ_k is the quantity $4\pi^2\nu_k/h$, and $H_{v_k}(\gamma_k^{\frac{1}{2}}Q_k)$ is a polynomial of degree v_k in Q_k. The first three polynomials are

$$H_0(z) = 1 \qquad H_1(z) = 2z \qquad H_2 = 4z^2 - 2 \tag{3}$$

where $z = \gamma_k^{\frac{1}{2}}Q_k$.
The higher polynomials may be obtained from the recursion formula

$$H_{v+1}(z) - 2zH_v(z) + 2vH_{v-1}(z) = 0 \tag{4}$$

Certain integrals involving these functions will be used later. These can be obtained by the methods described in detail in Sec. 11 of Pauling and Wilson, "Introduction to Quantum Mechanics." See also Secs. 7-3 and 8-6 of the present volume. The normalization integral is

$$\int_{-\infty}^{\infty} \psi_v^*\psi_{v'}\,dQ = \delta_{vv'} \tag{5}$$

In Appendix III a number of these important integrals have been tabulated.

3-4. Selection Rules in Wave Mechanics

The intensity of a spectral line is determined by the probability of the transition which gives rise to the line. It has been found[1] that the probability $A_{n'n''}$ of a spontaneous transition in 1 sec from a higher state n' to a lower state n'' with the emission of light of frequency

$$\nu_{n'n''} = \frac{W_{n'} - W_{n''}}{h}$$

is

$$A_{n'n''} = \frac{64\pi^4\nu_{n'n''}^3}{3hc^3}\,[|(\mu_X)_{n'n''}|^2 + |(\mu_Y)_{n'n''}|^2 + |(\mu_Z)_{n'n''}|^2] \tag{1}$$

The coefficient of absorption equals the coefficient of induced emission $B_{n'n''}$ and both are given by the equation

$$B_{n''n'} = B_{n'n''} = \frac{8\pi^3}{3h^2}\,[|(\mu_X)_{n'n''}|^2 + |(\mu_Y)_{n'n''}|^2 + |(\mu_Z)_{n'n''}|^2] \tag{2}$$

In these equations, h is Planck's constant, c is the velocity of light, and $(\mu_X)_{n'n''}$ is the integral

$$(\mu_X)_{n'n''} = \int \psi_{n'}^*\mu_X\psi_{n''}\,d\tau \tag{3}$$

[1] See Pauling and Wilson, "Introduction to Quantum Mechanics," Sec. 40, for a derivation of these results. The quantities $A_{n'n''}$ and $B_{n''n'}$ are called Einstein coefficients.

in which $\psi_{n'}^*$ is the complex conjugate of the complete wave function for the state n', $d\tau$ is the volume element of configuration space, and μ_X is the expression for the X component of the electric moment[1] of the system. The integration is over the entire configuration space of the system.

If $(\mu_X)_{n'n''}$, $(\mu_Y)_{n'n''}$, and $(\mu_Z)_{n'n''}$ are all zero for a given transition $n' \rightarrow n''$, that transition will not give rise to a spectral line in either absorption or emission.[2] Rules which specify which transitions can occur with the emission or absorption of radiation are called *selection rules.*

3-5. Infrared Selection Rules and Intensities for the Harmonic Oscillator Model[3]

Factoring of the Wave Function. In applying the methods of the previous section to the problem of determining the vibrational selection rules, it has to be remembered that in Eq. (3), Sec. 3-4, ψ_n is the complete wave function, while the axes X, Y, and Z are space-fixed axes. The complete wave function for a molecule is approximately of the form

$$\psi = \psi_E \psi_V \psi_R \psi_T \tag{1}$$

where ψ_E, ψ_V, ψ_R, ψ_T, are, respectively, the electronic, vibrational, rotational, and translational wave functions. It is possible to show that the electronic wave function does not enter in the problems treated here; moreover, the model which is the basis of the discussions in this book (see Sec. 1-4) is one in which the question of the structure of the atoms themselves is avoided. Consequently, ψ_E will be ignored.

[1] The electric moment $\boldsymbol{\mu}$ of a system is a vector, defined in the footnote, p. 5. Analytically, $\boldsymbol{\mu}$ has the components μ_X, μ_Y, and μ_Z given by the expressions

$$\mu_X = \sum_\alpha e_\alpha X_\alpha$$

$$\mu_Y = \sum_\alpha e_\alpha Y_\alpha$$

$$\mu_Z = \sum_\alpha e_\alpha Z_\alpha$$

where e_α is the charge and X_α, Y_α, and Z_α are the cartesian coordinates (space-fixed axes) of the αth particle, the sum being over all the particles.

[2] This statement is not strictly true in that there can be radiation due to magnetic dipole or electric quadrupole effects, but these are usually weaker by several orders of magnitude and hence are at present rarely observed in infrared or Raman spectra. See, for example, the following:

G. Herzberg, *Z. physik. Chem.*, (*B*), **4**: 223 (1929).

G. Herzberg, *Can. J. Research*, (*A*), **28**: 144 (1940).

H. M. James and A. S. Coolidge, *Astrophys. J.*, **87**: 438 (1938).

E. C. Kemble, "Fundamental Principles of Quantum Mechanics," Sec. **54***g*, McGraw-Hill, New York, 1937.

[3] D. M. Dennison, *Revs. Mod. Phys.*, **3**: 280 (1931).

Derivation of Translational Selection Rule $\Delta n_T = 0$. The wave function for translation of the molecule as a whole ψ_T does not need to be discussed in detail for the following reasons. The X component of the electric moment μ_{X} may be written as

$$\mu_{\text{X}} = \sum_{\alpha=1}^{N} e_\alpha \text{X}_\alpha = \sum_{\alpha=1}^{N} e_\alpha(\text{X}_0 + X_\alpha)$$
$$= \text{X}_0 \sum_{\alpha=1}^{N} e_\alpha + \sum_{\alpha=1}^{N} e_\alpha X_\alpha = \sum_{\alpha=1}^{N} e_\alpha X_\alpha = \mu_X \quad (2)$$

with similar expressions for μ_Y and μ_Z. Here X_α is the X coordinate of atom α in the space-fixed system, X_α is the coordinate of α in a system of axes which moves with the center of gravity of the molecule but remains parallel to the X, Y, Z set, X_0 is the X coordinate of the center of gravity, and e_α is the effective charge on the αth atom. Since the molecule is assumed to be neutral, $\Sigma e_\alpha = 0$. The function ψ_T is a function of X_0, Y_0, and Z_0, the coordinates of the center of gravity, so that the integral for μ_{X} may be written as

$$(\mu_{\text{X}})_{n'n''} = \int \psi_{T'}^* \psi_{T''} \, d\tau_T \int \psi_{V'}^* \psi_{R'}^* \mu_X \psi_{V''} \psi_{R''} \, d\tau_{VR} \quad (3)$$

since $\mu_{\text{X}} = \mu_X$ is independent of X_0, Y_0, Z_0. But the functions ψ_T are mutually orthogonal so that $(\mu_{\text{X}})_{n'n''}$ vanishes unless $T' = T''$, that is, unless the initial and final states of the molecule have the same translational quantum numbers.

Separation of Rotational and Vibrational Factors. There remains the problem of separating the effects due to ψ_R and ψ_V. To do this, it is necessary to introduce the rotating coordinate system discussed earlier, Sec. 2-1. This system moves with the molecule in that its origin is fixed to the center of gravity of the molecule and also in that it rotates with the molecule as previously discussed. It, therefore, has the same origin as the moving system X, Y, Z but rotates with respect to it. Consequently, there will exist a relationship between the coordinates of a particle in the two systems of the form

$$\begin{aligned} X_\alpha &= \Phi_{Xx} x_\alpha + \Phi_{Xy} y_\alpha + \Phi_{Xz} z_\alpha \\ Y_\alpha &= \Phi_{Yx} x_\alpha + \Phi_{Yy} y_\alpha + \Phi_{Yz} z_\alpha \\ Z_\alpha &= \Phi_{Zx} x_\alpha + \Phi_{Zy} y_\alpha + \Phi_{Zz} z_\alpha \end{aligned} \quad (4)$$

in which x_α, y_α, z_α are the coordinates of the particle in the rotating system, X_α, Y_α, Z_α are the coordinates in the system which is moving but not rotating, while the quantities Φ_{Xx}, etc., are the direction cosines connecting the various pairs of axes. These quantities Φ_{Xx}, etc., are functions of the angles which define the position of the rotating system of axes, the Eulerian angles, but it is not necessary to inquire further into their nature at this point, since they determine the rotational but not the vibrational selection rules. The rotational selection rules are summarized in Appendix XVI.

From the nature of μ_X, etc., it is clear that the relation connecting μ_X and μ_x (the latter being the electric moment component in terms of the rotating axis system) is the same as that connecting X and x; that is,

$$\mu_X = \Phi_{Xx}\mu_x + \Phi_{Xy}\mu_y + \Phi_{Xz}\mu_z \tag{5}$$

with similar expressions for μ_Y and μ_Z. The integral in (3) thus becomes

$$\begin{aligned}\int \psi_{V'}^*\psi_{R'}^*\mu_X\psi_{V''}\psi_{R''}\,d\tau_{VR} &= \int \psi_{R'}^*\Phi_{Xx}\psi_{R''}\,d\tau_R \int \psi_{V'}^*\mu_x\psi_{V''}\,d\tau_V \\ &+ \int \psi_{R'}^*\Phi_{Xy}\psi_{R''}\,d\tau_R \int \psi_{V'}^*\mu_y\psi_{V''}\,d\tau_V + \int \psi_{R'}^*\Phi_{Xz}\psi_{R''}\,d\tau_R \int \psi_{V'}^*\mu_z\psi_{V''}\,d\tau_V\end{aligned} \tag{6}$$

with similar equations for the integrals involving μ_Y and μ_Z. The conclusion is therefore reached that a transition with the emission or absorption of radiation can occur between the vibrational states V' and V'' if one or more of the integrals

$$\int \psi_{V'}^*\mu_x\psi_{V''}\,d\tau_V \qquad \int \psi_{V'}^*\mu_y\psi_{V''}\,d\tau_V \qquad \int \psi_{V'}^*\mu_z\psi_{V''}\,d\tau_V$$

is different from zero. It has been assumed that no change in electronic state takes place, and it has been shown that the translational quantum number does not change. There may, however, be simultaneous changes in the rotational quantum numbers; these are discussed in Appendix XVI.

Expansion of the Electric Moment. The definition of the electric moment previously given, $\mu_x = \sum_\alpha e_\alpha x_\alpha$, shows directly the dependence of μ_x on the positions of the particles if these particles are electrons and nuclei, inasmuch as the charges e_α are then constants. If, however, the particles are the atoms themselves, the charges e_α must be considered as "effective charges" which may vary as the atoms move, inasmuch as when the positions of the atoms are changed the electron distribution may also change. This does not alter the previous results regarding the transformation properties of μ_x, but it means that the electric moment is not necessarily a linear function of the coordinates of the atoms. In general, the electric moment can be expanded as a power series in the coordinates of the atoms, the normal coordinates are most convenient for this purpose, with the result that

$$\mu_x = \mu_x^0 + \sum_{k=1}^{3N-6} \mu_x^{(k)}Q_k + \text{higher terms} \tag{7}$$

Similar equations describe μ_y and μ_z. μ_x^0, the x component of the electric moment possessed by the molecule in its equilibrium position, is practically equal to the x component of the permanent electric moment of the molecule. $\mu_x^{(k)} = (\partial\mu_x/\partial Q_k)_0$ is the coefficient of the normal coordinate Q_k in the expansion. Very little is known of the relative magnitudes of $\mu_x^{(k)}$ and the higher terms, although it is usually assumed that the latter are smaller.[1]

Vibrational Selection Rules for the Harmonic Oscillator. If the higher terms in (7) are neglected, and if the vibrational wave function ψ_V is assumed to be strictly of the form described earlier in this chapter, that is, a product of harmonic oscillator functions, then the selection rules for vibrational transitions are very restrictive indeed. The integral for μ_x becomes

$$\int \psi_{V'}^* \mu_x \psi_{V''}\, d\tau_V = \mu_x^0 \int \psi_{V'}^* \psi_{V''}\, d\tau_V + \sum_{k=1}^{3N-6} \mu_x^{(k)} \int \psi_{V'}^* Q_k \psi_{V''}\, d\tau_V \tag{8}$$

The first term vanishes unless $V' = V''$ because of the orthogonality of the functions ψ_V. Therefore, the permanent electric moment μ^0 has no influence on the intensity of vibrational transitions; it does, however, determine the intensity of the pure rotation spectrum. The integral in the second term can be split up into factors as shown below:

$$\int \psi_{V'}^* Q_k \psi_{V''}\, d\tau_V = \int \psi_{v_1'}^*(Q_1)\psi_{v_1''}(Q_1)\, dQ_1 \int \psi_{v_2'}^*(Q_2)\psi_{v_2''}(Q_2)\, dQ_2 \cdots \int \psi_{v_k'}^*(Q_k) Q_k \psi_{v_k''}(Q_k)\, dQ_k \cdots \tag{9}$$

in which the product-type wave functions of Eq. (5), Sec. 3-1, have been introduced for ψ_V. Because of the orthogonality of the functions $\psi_v(Q)$, the integral in (9) will vanish unless $v_1' = v_1''$, $v_2' = v_2''$, etc., with the exception of v_k' and v_k''. For these quantum numbers, it must be true that $v_k'' = v_k' + 1$ or $v_k' - 1$ if the factor

$$\int \psi_{v_k'}^*(Q_k) Q_k \psi_{v_k''}(Q_k)\, dQ_k$$

is to be different from zero (see Appendix III). The net result of all these considerations can be stated as follows: If it is assumed that only linear terms in the expansion of the electric moment are important, and if it is assumed that the vibrational wave functions are products of harmonic oscillator functions, then the only vibrational transitions which can occur with the emission or absorption of radiation are those in which only one quantum number changes and that quantum number changes by one unit only. Furthermore, the quantum number v_k which changes must be one for which $\mu_x^{(k)}$, $\mu_y^{(k)}$, or $\mu_z^{(k)}$ is different from zero.

Qualitative Intensities for Real Molecules. In other words, on the basis of the harmonic oscillator model, only fundamental frequencies should

[1] B. L. Crawford and H. L. Dinsmore, *J. Chem. Phys.*, **18**: 983, 1682 (1950). D. F. Eggers and B. L. Crawford, *J. Chem. Phys.*, **19**: 1554 (1951).

appear in the absorption or emission spectrum. Not all fundamentals will be allowed, but only those associated with normal modes of vibration which cause a change in the electric moment of the molecule ($\mu_x^{(k)}$, $\mu_y^{(k)}$, and $\mu_z^{(k)}$ measure this change). Experimentally, *i.e.*, for actual molecules, it is found that the fundamental frequencies are usually the most intense, but other transitions such as overtones and combinations also appear, sometimes quite strongly. It is evident that some of the assumptions of the above derivation must therefore be invalid. It is certain from other evidence (the convergence of the energy levels) that the harmonic oscillator approximation is not perfect, as indeed is to be expected since the cubic, quartic, etc., terms in the potential energy [Eq. (4), Sec. 2-2] are not zero, but in addition it is probably true that higher terms in the expansion of the electric moment (7) must also be considered. Regardless of these complications, it remains true that a fundamental frequency will not occur in the absorption or emission spectrum of a gas unless the corresponding normal vibration changes the electric moment.

3-6. Classical Intensities of Radiation from an Induced Dipole

The greater complexity of the mathematical treatment of scattered radiation as compared with ordinary emission or absorption makes it worthwhile to carry through in some detail the classical formulation prior to the quantum mechanical treatment which will be given in outline in Sec. 3-7. Just as a permanent oscillating dipole in a molecule can act as a radiator classically, so can an induced dipole, the mean rate of total radiation being given, in either case, by[1]

$$I = \frac{16\pi^4\nu^4}{3c^3}\mu_0^2 \tag{1}$$

in which ν is the frequency of oscillation and of the emitted light, c the velocity of light, and μ_0 the amplitude in the expression

$$\mu = \mu_0 \cos 2\pi\nu t \tag{2}$$

If one introduces the nonrotating axes depicted in Fig. I-1, App. I, the total radiation emitted per unit solid angle in the X direction is given by

$$I = \frac{2\pi^3\nu^4}{c^3}(\mu_{0Y}^2 + \mu_{0Z}^2) \tag{3}$$

Polarizability Tensor. If the molecule in which an electric dipole $\boldsymbol{\mu}$ is induced is isotropic, then the relation between $\boldsymbol{\mu}$ and the electric field vector $\boldsymbol{\mathcal{E}}$ of the incident radiation is simply given by

$$\boldsymbol{\mu} = \alpha\boldsymbol{\mathcal{E}} \tag{4}$$

[1] See, for example, J. C. Slater and N. H. Frank, "Introduction to Theoretical Physics," p. 293, McGraw-Hill, New York, 1933.

where α is a constant called the polarizability of the molecule. In such a case, it is clear that the scattered radiation whose intensity is given by (3) would be polarized with its electric vector lying in the direction of the projection of $\boldsymbol{\mathcal{E}}$ upon the YZ plane. In general, however, for an *anisotropic* molecule, (4) must be replaced by the more complex expression

$$\begin{aligned} \mu_X &= \alpha_{XX}\mathcal{E}_X + \alpha_{XY}\mathcal{E}_Y + \alpha_{XZ}\mathcal{E}_Z \\ \mu_Y &= \alpha_{YX}\mathcal{E}_X + \alpha_{YY}\mathcal{E}_Y + \alpha_{YZ}\mathcal{E}_Z \\ \mu_Z &= \alpha_{ZX}\mathcal{E}_X + \alpha_{ZY}\mathcal{E}_Y + \alpha_{ZZ}\mathcal{E}_Z \end{aligned} \tag{5}$$

in which the quantities $\alpha_{FF'}$ are independent of the components of the electric vector, but are dependent upon the orientation of the molecule relative to the nonrotating axes, X, Y, Z. The quantities $\alpha_{FF'}$ are called the components of a tensor[1] by virtue of their transformation under changes of the coordinate system. In matrix symbolism,[2] (5) may be written in the abbreviated form

$$\boldsymbol{\mu} = \boldsymbol{\alpha}\boldsymbol{\mathcal{E}} \tag{6}$$

where multiplication of the vector $\boldsymbol{\mathcal{E}}$ by $\boldsymbol{\alpha}$ is to be carried out in accordance with the rules of matrix algebra. It is evident from (5) that, in general, the direction of the induced dipole is not parallel with the direction of the incident electric field vector. It can be shown, however, that a set of axes in the molecule exists such that the relation between $\boldsymbol{\mu}$ and $\boldsymbol{\mathcal{E}}$, when referred to these axes, assumes the simple form

$$\begin{aligned} \mu_1 &= \alpha_1\mathcal{E}_1 \\ \mu_2 &= \alpha_2\mathcal{E}_2 \\ \mu_3 &= \alpha_3\mathcal{E}_3 \end{aligned} \tag{7}$$

Such axes are called *principal axes of polarizability* and the associated α_i, $i = 1, 2, 3$, the *principal values* of the polarizability.[3] For the isotropic case, it is apparent that $\alpha_1 = \alpha_2 = \alpha_3 = \alpha$. These axes are clearly the most convenient ones to use in the discussion of induced dipoles, and will be used extensively throughout the remainder of this section.

[1] A knowledge of tensors will not be required for the understanding of this book, but the interested reader will find an introduction to this subject in H. Margenau and G. M. Murphy, "The Mathematics of Physics and Chemistry," Van Nostrand, New York, 1943.

[2] An introduction to matrix algebra will be found in Appendix V.

[3] The determination of the principal values of the polarizability requires the solution of a third-order secular determinant, entirely analogous to the secular determinant described in Chap. 2.

Intensity of the Scattered Light in Terms of Polarizability. We desire to obtain expressions for the scattered light intensity by the use of (3) and substitution of appropriate expressions for μ_{0Y}^2, etc. The components of the electric vector will thus appear, but can later be related to the incident intensity. It will now be assumed that the direction of propagation of the incident light coincides with the Y axis. If the incident light is plane-polarized with the electric vector parallel to the direction of observation,

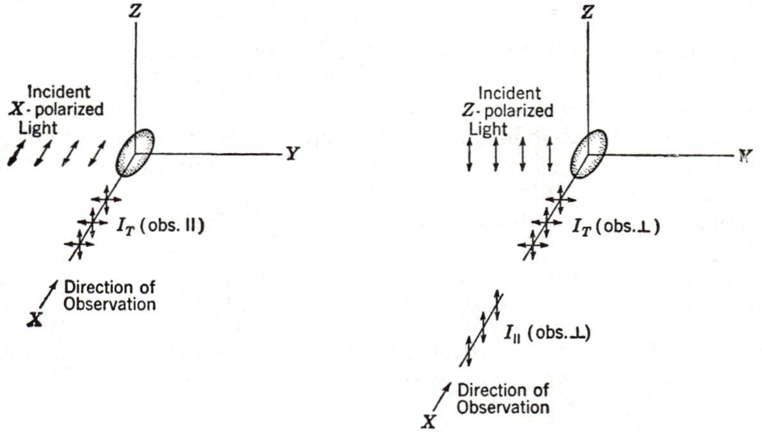

FIG. 3-2. Schematic representation of experimental conditions defining I_T (obs. ||), I_T (obs. ⊥), and $I_{||}$ (obs. ⊥).

i.e., along the X axis, the scattered intensity will be designated as I_T (obs. ||) whereas the intensity in case the electric vector lies along the Z axis will be designated by I_T (obs. ⊥) (Fig. 3-2). According to (3) and (5)

$$I_T \text{ (obs. } \|) = \frac{2\pi^3\nu^4}{c^3}(\alpha_{YX}^2 + \alpha_{ZX}^2)\mathcal{E}_0^2 \tag{8}$$

$$I_T \text{ (obs. } \perp) = \frac{2\pi^3\nu^4}{c^3}(\alpha_{YZ}^2 + \alpha_{ZZ}^2)\mathcal{E}_0^2 \tag{9}$$

That part of the light which is polarized parallel to $\mathcal{E}$ is, in the second case:

$$I_{\|} \text{ (obs. } \perp) = \frac{2\pi^3\nu^4}{c^3}\alpha_{ZZ}^2\mathcal{E}_0^2 \tag{10}$$

It is convenient to make use of the relation from electromagnetic theory

$$I_0 = \frac{c}{8\pi}\mathcal{E}_0^2 \tag{11}$$

where I_0 is the intensity of (incident) radiation whose electric vector has the amplitude $\mathcal{E}_0$. Substituting (11) in (8), (9), and (10),

$$I_T\ (\text{obs.}\ \|) = \frac{16\pi^4\nu^4}{c^4} I_0(\alpha_{YX}^2 + \alpha_{ZX}^2) \tag{12}$$

$$I_T\ (\text{obs.}\ \perp) = \frac{16\pi^4\nu^4}{c^4} I_0(\alpha_{YZ}^2 + \alpha_{ZZ}^2) \tag{13}$$

$$I_\|\ (\text{obs.}\ \perp) = \frac{16\pi^4\nu^4}{c^4} I_0\alpha_{ZZ}^2 \tag{14}$$

Averaging Over All Orientations. The last three equations are appropriate for a single radiator; to obtain the expression for a gas whose molecules are free to assume all orientations with respect to the observer's axes with equal probability, (12), (13), and (14) should be multiplied by the number of molecules and the terms $\alpha_{FF'}^2$ averaged over all orientations of the principal axes with respect to the fixed X, Y, Z axes. The general formulas of transformation for the $\alpha_{FF'}$ are of the form

$$\alpha_{FF'} = \sum_{gg'} \alpha_{gg'}\Phi_{Fg}\Phi_{F'g'} \tag{15}$$

where the Φ_{Fg} are the direction cosines between the F axes and the g axes. In case the g axes are *principal axes,* (15) assumes the form

$$\alpha_{FF'} = \sum_{i=1}^{3} \alpha_i\Phi_{Fi}\Phi_{F'i} \tag{16}$$

The quantities α_i are, of course, constants under the averaging process, so that the averages required in (12) to (14) are expressible as

$$\begin{aligned}\overline{\alpha_{FF'}^2} &= \overline{\left(\sum_i \alpha_i\Phi_{Fi}\Phi_{F'i}\right)^2} \\ &= \sum_i \alpha_i^2\overline{\Phi_{Fi}^2\Phi_{F'i}^2} + 2\sum_{i<j} \alpha_i\alpha_j\overline{\Phi_{Fi}\Phi_{F'i}\Phi_{Fj}\Phi_{F'j}}\end{aligned} \tag{17}$$

In Appendix IV the necessary averages of the direction cosines are found to be

$$\overline{\Phi_{Fi}^2\Phi_{F'i}^2} = \begin{cases} \frac{1}{5} & F = F' \\ \frac{1}{15} & F \neq F' \end{cases}$$

$$\overline{\Phi_{Fi}\Phi_{F'i}\Phi_{Fj}\Phi_{F'j}} = \begin{cases} \frac{1}{15} & F = F' \\ -\frac{1}{30} & F \neq F' \end{cases}$$

Using these results in (17), Eqs. (12) to (14) then yield for the intensities scattered by N molecules which are freely orientable in space:

$$I_T \text{ (obs. } ||) = \frac{1}{15} \frac{16\pi^4\nu^4}{c^4} NI_0 \left(2 \sum_i \alpha_i^2 - 2 \sum_{i<j} \alpha_i\alpha_j \right) \tag{18}$$

$$I_T \text{ (obs. } \perp) = \frac{1}{15} \frac{16\pi^4\nu^4}{c^4} NI_0 \left(4 \sum_i \alpha_i^2 + \sum_{i<j} \alpha_i\alpha_j \right) \tag{19}$$

$$I_{\|} \text{ (obs. } \perp) = \frac{1}{15} \frac{16\pi^4\nu^4}{c^4} NI_0 \left(3 \sum_i \alpha_i^2 + 2 \sum_{i<j} \alpha_i\alpha_j \right) \tag{20}$$

The two kinds of summations appearing in these last three equations may be conveniently replaced by functions of the *spherical part* of the polarizability, α, and the *anisotropy*, β, which are defined by

$$\alpha = \tfrac{1}{3}(\alpha_1 + \alpha_2 + \alpha_3) \tag{21}$$

$$\beta^2 = \tfrac{1}{2}[(\alpha_1 - \alpha_2)^2 + (\alpha_2 - \alpha_3)^2 + (\alpha_3 - \alpha_1)^2] \tag{22}$$

In terms of these quantities, Eqs. (18) to (20) become

$$I_T \text{ (obs. } ||) = \frac{16\pi^4\nu^4}{c^4} NI_0 \frac{2\beta^2}{15} \tag{23}$$

$$I_T \text{ (obs. } \perp) = \frac{16\pi^4\nu^4}{c^4} NI_0 \frac{45\alpha^2 + 7\beta^2}{45} \tag{24}$$

$$I_{\|} \text{ (obs. } \perp) = \frac{16\pi^4\nu^4}{c^4} NI_0 \frac{45\alpha^2 + 4\beta^2}{45} \tag{25}$$

Depolarization Ratios. The quantities most frequently observed experimentally are the *depolarization ratios*, ρ_l or ρ_n, defined as the ratio of the scattered intensity which is polarized perpendicular to $\mathbf{\varepsilon}$, that is, in the direction of propagation of the incident light, to the intensity parallel to $\mathbf{\varepsilon}$. When *linear* (plane) polarized incident light is used, and the observations are made in a direction perpendicular to $\mathbf{\varepsilon}$ (as well as to the propagation direction, which has been assumed throughout the previous derivations), the ratio of intensities is given by

$$\rho_l = \frac{I_T \text{ (obs. } \perp) - I_{\|} \text{ (obs. } \perp)}{I_{\|} \text{ (obs. } \perp)} = \frac{3\beta^2}{45\alpha^2 + 4\beta^2} \tag{26}$$

If the incident light is *natural* (unpolarized), the depolarization ratio may be computed by considering the scattered light to represent the sum of the intensities of observations made parallel and perpendicular to the incident electric vector of a polarized beam. That part of the light from the parallel observation, being unpolarized, contributes one-half its intensity to the scattered light polarized, respectively, parallel and perpendicular to $\mathbf{\varepsilon}$:

$$\rho_n = \frac{I_T \text{ (obs. } \perp) - I_{\|} \text{ (obs. } \perp) + \frac{1}{2}I_T \text{ (obs. } ||)}{I_{\|} \text{ (obs. } \perp) + \frac{1}{2}I_T \text{ (obs. } ||)} = \frac{6\beta^2}{45\alpha^2 + 7\beta^2} \tag{27}$$

Expansion of the Polarizability in the Normal Coordinates. Thus far, the components of the polarizability have been assumed to be independent of the time. For a molecule executing small vibrations, however, the polarizability may be expanded in terms of the normal coordinates. Thus the principal values could be written as

$$\alpha_i = \alpha_i^0 + \sum_k \left(\frac{\partial \alpha_i}{\partial Q_k}\right)_0 Q_k + \text{higher terms} \tag{28}$$

just as in the case of the permanent dipole moment [Eq. (7), Sec. 3-5]. The corresponding component of the oscillating electric moment would, therefore, be of the form:

$$\mu_i = \mathcal{E}_{0i} \left\{ \alpha_i^0 \cos 2\pi\nu_0 t + \sum_k \left(\frac{\partial \alpha_i}{\partial Q_k}\right)_0 Q_k^0 \left[\frac{1}{2}\cos 2\pi(\nu_0 + \nu_k)t + \frac{1}{2}\cos 2\pi(\nu_0 - \nu_k)t\right]\right\} \tag{29}$$

in which Q_k^0 is the amplitude of the kth normal coordinate; it is apparent that the oscillating dipole will contain the shifted frequencies $\nu_0 \pm \nu_k$ as well as the incident frequency ν_0. Equations (23) to (25) will apply separately for the Rayleigh frequency, $\nu = \nu_0$, and for the Raman frequencies, $\nu = \nu_0 \pm \nu_k$. It is to be noted that the α_i^0 determine the Rayleigh intensities, whereas $(\partial\alpha_i/\partial Q_k)_0$ determine the Raman intensities. According to this classical treatment, the intensities of the so-called Stokes lines $(\nu_0 - \nu_k)$ to the anti-Stokes lines $(\nu_0 + \nu_k)$ should be in the ratio

$$\frac{I_{\text{Stokes}}}{I_{\text{anti-Stokes}}} = \frac{(\nu_0 - \nu_k)^4}{(\nu_0 + \nu_k)^4} \tag{30}$$

which will certainly not exceed unity. This is in contradiction to the experimental findings, which are consistent with the prediction of quantum mechanical theory which will be given in the following section.

3-7. Quantum-mechanical Theory of Raman Intensities

The rigorous derivation of the intensity formulas for the Raman effect or, indeed, for the infrared would require a completely satisfactory quantum mechanical theory of radiation, which does not appear to have been developed at this time. There seems to be little doubt, however, that essentially correct treatments of both these phenomena have been obtained, using as a foundation either the Dirac theory of radiation or the correspondence principle. There is not space to give the detailed derivation of these formulas here,[1] but an attempt will be made to outline the principles on which it is based and to summarize the results.

In Sec. 3-6 the classical equations are given for the emission of radiation by an oscillating dipole.[2] In order to apply the quantum theory, the hypothesis is made (supported by the Dirac radiation theory and by many successful applications) that the classical formulas can be converted into quantum formulas by replacing μ_0^2 by $4|\mu_{n''n'}|^2$, in which

$$\boldsymbol{\mu}_{n''n'} = \int \psi_{n''}^* \boldsymbol{\mu} \psi_{n'} \, d\tau \tag{1}$$

In the Raman effect, the electric vector of the incident illumination, acting on the charges of the molecule, perturbs its wave function. If the integrals $\boldsymbol{\mu}_{n''n'}$ are calculated using the perturbed wave functions (including the time factors), it is found that for $n'' = n'$ there are terms of the same frequency (ν_0) as the incident light, but for $n'' \neq n'$ there are terms with the frequency $\nu_0 \pm \nu_{n''n'}$. The terms with unshifted frequency are responsible for the Rayleigh scattering, while those of frequency $\nu_0 \pm \nu_{n''n'}$ give rise to the Raman effect. The intensity of the radiation can be calculated by obtaining the terms in $\boldsymbol{\mu}_{n''n'}$, of the given frequency, multiplying by four and inserting in Eq. (3), Sec. 3-6.

The formula obtained in this way contains so many quantities which are ordinarily not known that it is not directly useful. However, it can be shown that under certain conditions, it can be reduced to a relatively simple formula, involving the polarizability of the molecule. These con-

[1] The formulas are fully derived by G. Placzek, "Marx Handbuch der Radiologie," 2d ed., Vol. VI, Part II, pp. 209–374, 1934. Earlier references on the theory include the following:

J. Cabannes, *Trans. Faraday Soc.*, **25**: 813 (1929).

J. Cabannes and Y. Rocard, *J. phys. radium*, **10**: 52 (1929).

H. A. Kramers and W. Heisenberg, *Z. Physik*, **31**: 681 (1925).

C. Manneback, *Z. Physik*, **62**: 224 (1930).

G. Placzek, *Z. Physik*, **70**: 84 (1931): *Leipziger Vorträge*, 71 (1931).

G. Placzek and E. Teller, *Z. Physik*, **81**: 209 (1933).

F. Rasetti, *Leipziger Vorträge*, 59 (1931). (These articles appear also in English in "The Structure of Molecules," P. Debye, editor, Blackie, Glasgow, 1932.)

A. Smekal, *Naturwissenschaften*, **11**: 873 (1923).

J. H. Van Vleck, *Proc. Nat. Acad. Sci.*, **15**: 754 (1929).

[2] For presentations of the quantum theory of radiation, see the following sources:

G. Breit, *Revs. Mod. Phys.*, **4**: 504 (1932).

P. A. M. Dirac, *Proc. Roy. Soc. (London)*, *(A)*, **112**: 661 (1926); *(A)*, **114**: 243 (1927).

E. Fermi, *Revs. Mod. Phys.*, **4**: 87 (1932).

W. Heitler, "The Quantum Theory of Radiation," Oxford, New York and London, 1954.

G. Placzek, "Marx Handbuch der Radiologie," 2d ed., Vol. VI, Part II, pp. 209–374, 1934.

ditions are first, *that the lowest electronic state of the molecule must be essentially nondegenerate;* and, second, *that the frequency shifts must be small compared with both the frequency of the incident light* ν_0, *and the frequency difference* $\nu_E - \nu_0$, when ν_E is the lowest electronic absorption frequency.

Using the polarizability matrix elements, $|(\alpha_{YX})_{n''n'}|^2$, $|(\alpha_{ZX})_{n''n'}|^2$, $|(\alpha_{YZ})_{n''n'}|^2$, and $|(\alpha_{ZZ})_{n''n'}|^2$, where

$$(\alpha_{FF'})_{n''n'} = \int \psi^*_{n''} \alpha_{FF'} \psi_{n'} \, d\tau \tag{2}$$

the intensities corresponding to Eqs. (12) to (14), Sec. 3-6, could be computed, at the same time multiplying by the factor of 4 introduced above. Two new considerations enter, however, when one proceeds to compute the intensities for a gas consisting of freely orientable molecules, whose individual scattering would be given by such formulas. In the first place, not all the molecules can undergo the transition $W_{n''} \rightarrow W_{n'}$, but only a number, $N_{n''}$, which are initially in the energy level described by the quantum number n''. In the second place, one must take account of the orientational degeneracy described by the magnetic quantum number‡ M. The total intensity will depend upon a summation over all final magnetic quantum numbers, M', and an averaging over all $g_{n''}$ degenerate initial states characterized by the magnetic quantum number, M''; in other words, the intensities will be proportional to expressions of the form

$$\frac{N_{n''}}{g_{n''}} \sum_{M''M'} |(\alpha_{FF'})_{M''M'}|^2 \tag{3}$$

For each pair of magnetic quantum numbers, the tensor $(\boldsymbol{\alpha})_{M''M'}$ will have a corresponding set of principal axes and principal values, oriented in some manner relative to a preferred direction in space. But the physical observable, in this case the intensity, must be independent of the direction chosen as a "preferred" one; hence it is legitimate to average over all orientations of this direction relative to the observer's axes. Since it is permissible to change the order of the averaging over all orientations and the summation over M'' and M', the averaging over all orientations can be carried out for each pair of quantum numbers, M'' and M', followed by summation. This is equivalent, however, to averaging over all directions of the principal axes in the classical case, since for each given M'' and M', the principal axes are defined only relative to the preferred direction of quantization. Thus the quantum-mechanical result, corresponding to the classical equations, (23) to (25) of Sec. 3-6, is

‡ The magnetic quantum number will be considered in Appendix XVI where the rotational energy levels are discussed.

$$I_T\ (\text{obs. } \|) = \frac{64\pi^4\nu^4}{c^4}\frac{N_{n''}}{g_{n''}} I_0\left(\frac{2}{15}\sum_{M''M'} \beta^2_{M''M'}\right) \tag{4}$$

$$I_T\ (\text{obs. } \perp) = \frac{64\pi^4\nu^4}{c^4}\frac{N_{n''}}{g_{n''}} I_0\left(\sum_{M''M'} (\alpha_{M''M'})^2 + \frac{7}{45}\sum_{M''M'} \beta^2_{M''M'}\right) \tag{5}$$

$$I_\|\ (\text{obs. } \perp) = \frac{64\pi^4\nu^4}{c^4}\frac{N_{n''}}{g_{n''}} I_0\left(\sum_{M''M'} (\alpha_{M''M'})^2 + \frac{4}{45}\sum_{M''M'} \beta^2_{M''M'}\right) \tag{6}$$

in which

$$\alpha_{M''M'} = \tfrac{1}{3}\sum_i (\alpha_i)_{M''M'} \tag{7}$$

$$\beta^2_{M''M'} = \tfrac{1}{2}\sum_{i<j} [(\alpha_i)_{M''M'} - (\alpha_j)_{M''M'}]^2 \tag{8}$$

and $(\alpha_i)_{M''M'}$ is a principal value of the tensor $(\boldsymbol{\alpha})_{M''M'}$ whose components are to be computed in terms of the nonrotating axes, *i.e.*,

$$(\alpha_{FF'})_{M''M'} = \int \psi^*_{n''M''}\alpha_{FF'}\psi_{n'M'}\, d\tau \tag{9}$$

Note that the ratio of the Stokes to the anti-Stokes intensities is now given by

$$\frac{I_{\text{Stokes}}}{I_{\text{anti-Stokes}}} = \left(\frac{\nu_0 - \nu_{n''n'}}{\nu_0 + \nu_{n''n'}}\right)^4 \frac{N_{n''}}{N_{n'}}\frac{g_{n'}}{g_{n''}} \tag{10}$$

so that it may now be expected that the Stokes lines will be more intense than the anti-Stokes lines, since the population ratio is given by the Boltzmann factor,

$$\frac{N_{n''}}{N_{n'}}\frac{g_{n'}}{g_{n''}} = e^{(W_{n'}-W_{n''})/kT} \tag{11}$$

with $W_{n'} > W_{n''}$.

The depolarization ratios, corresponding to Eqs. (26) and (27), Sec. 3-6, will be

$$\rho_l = \frac{3\sum_{M''M'} \beta^2_{M''M'}}{45\sum_{M''M'} (\alpha_{M''M'})^2 + 4\sum_{M''M'} \beta^2_{M''M'}} \tag{12}$$

$$\rho_n = \frac{6\sum_{M''M'} \beta^2_{M''M'}}{45\sum_{M''M'} (\alpha_{M''M'})^2 + 7\sum_{M''M'} \beta^2_{M''M'}} \tag{13}$$

Selection Rules. The conditions under which the quantities $\alpha_{M''M'}$ and $\beta_{M''M'}$ can be nonvanishing will now be considered in turn. Although the components of the tensor $(\boldsymbol{\alpha})_{M''M'}$ are to be referred to the nonrotating axes, it is easily shown, as in Appendix XV, that

$$\alpha_{M''M'} = \frac{1}{3} \sum_i (\alpha_i)_{M''M'} = \frac{1}{3} \sum_F (\alpha_{FF})_{M''M'} = \frac{1}{3} \sum_g (\alpha_{gg})_{M''M'} \tag{14}$$

in which the last equation shows that it is legitimate to compute $\alpha_{M''M'}$ in terms of molecule-fixed axes. To that approximation in which the wave function can be written as a product of a rotational function, ψ_R, with quantum numbers R (including M) and a vibrational wave function, ψ_V, the integral $(\alpha_{gg})_{M''M'}$ is

$$(\alpha_{gg})_{M''M'} = \int \psi_{R''}^* \psi_{R'} \, d\tau_R \int \psi_{V''}^* \alpha_{gg} \psi_{V'} \, d\tau_V \tag{15}$$

since α_{gg} (in molecule-fixed axes) does not involve the coordinates of rotation. The orthogonality of the rotational wave functions therefore leads to the result

$$\alpha_{M''M'} = \delta_{R''R'} \alpha_{V''V'} \tag{16}$$

Consequently, $\alpha_{M''M'}$ is zero unless the rotational states are the same, and in addition it will be shown in Sec. 7-8 that it is also zero for all fundamental vibrational transitions except those for which the normal coordinate involved has the complete symmetry of the molecule. The vanishing of $\alpha_{M''M'}$ in these cases has important consequences. The depolarization ratios of (12) and (13) become $\rho_l = \frac{3}{4}$ and $\rho_n = \frac{6}{7}$ if $\alpha_{M''M'}$ is zero, no matter what the value of $\beta^2_{M''M'}$, and may therefore be used as a criterion for the symmetry type of the normal coordinate. On the other hand, the depolarization ratios will approach zero only if $\beta^2_{M''M'}$ vanishes, which occurs if the molecule is isotropic.

In order to consider the selection rules for the contribution of $\beta^2_{M''M'}$ to the Raman intensity in more detail, however, it is necessary to consider the relations between the integrals of $\alpha_{FF'}$ in terms of nonrotating axes with those in terms of molecule-fixed axes. This was simple for $\alpha_{M''M'}$ because it was independent of the axes used, but this is not true of $\beta^2_{M''M'}$, so that a relation similar to that used in the infrared case [Eq. (5), Sec. 3-5] is necessary.[1] This is

$$\alpha_{FF'} = \sum_{gg'} \Phi_{Fg} \Phi_{F'g'} \alpha_{gg'} \tag{17}$$

in which Φ_{Fg} is the direction cosine of the angle between the nonrotating axis F and the molecule-fixed or rotating axis g (see Appendix I).

The quantities $\alpha_{gg'}$ are functions of the normal coordinates only while the direction cosines are functions of the Eulerian angles, so that the

[1] Equation (8) is appropriate only for the principal value of $(\boldsymbol{\alpha})_{M''M'}$ which must be computed, starting with components of $(\boldsymbol{\alpha})_{M''M'}$ relative to the nonrotating axes of the observer.

integral $(\alpha_{FF'})_{M''M'}$ is a sum of terms, each of which factors into a rotational integral involving a product $\Phi_{Fg}\Phi_{F'g'}$ and a vibrational integral involving an $\alpha_{gg'}$. Consequently, the rotational selection rules will be determined by integrals such as

$$\int \psi_{R''}^{*} \Phi_{Fg} \Phi_{F'g'} \psi_{R'} \, d\tau_R \tag{18}$$

while the vibrational selection rules will be determined by integrals such as

$$\int \psi_{V''}^{*} \alpha_{gg'} \psi_{V'} \, d\tau_V \tag{19}$$

If all such integrals vanish for a given vibrational transition $V''V'$, that is, if

$$(\alpha_{xx})_{V''V'} = (\alpha_{yy})_{V''V'} = (\alpha_{zz})_{V''V'} = (\alpha_{xy})_{V''V'} = (\alpha_{yz})_{V''V'} = (\alpha_{zx})_{V''V'} = 0$$

the transition will be inactive in the Raman effect. The symmetry conditions under which these quantities vanish are discussed in Sec. 7-8. The rotational selection rules are summarized in Appendix XVI.

CHAPTER 4

MORE ADVANCED METHODS OF STUDYING VIBRATIONS

The basic principles given in the previous chapters underlie all methods of calculating vibration frequencies, but the actual procedures described so far are too inefficient to be useful for molecules of even moderate size. A number of more powerful techniques will be described in the present chapter. So far, no one approach has definitely emerged as superior to all others.

One of the most powerful tools in simplifying the treatment of larger molecules is the use of molecular symmetry in factoring the secular equation. This will be developed at length in the next three chapters. The methods of the present chapter are presented in a form suitable for application to the molecule as a whole, but it will be seen later that they can also be applied to the separate factors of the secular equation which can be obtained when there is symmetry. It is the combination of the developments of the present chapter with the symmetry considerations to be introduced later which provides the most effective approach now available.[1]

4-1. Internal Coordinates and Their Relation to Atomic Displacements

Changes in interatomic distances or in the angles between chemical bonds, or both, can be used to provide a set of $3N - 6$ (or $3N - 5$ for linear molecules) *internal* coordinates (Sec. 2-6), *i.e.*, coordinates which are unaffected by translations or rotations of the molecule as a whole. These are particularly important because they provide the most physically significant set for use in describing the potential energy of the molecule. The kinetic energy, on the other hand, is more easily set up in terms of cartesian displacement coordinates of the atoms (Sec. 2-6). A relation between the two types is therefore needed.

Since the whole treatment has so far been restricted to the case of infinitesimal amplitudes of vibration, all displacements can be considered

[1] The reader interested in the qualitative enumeration and assignment of vibration frequencies rather than their numerical computation in terms of molecular force constants may prefer to skip the present chapter or postpone it until after Chap. 5 and parts of Chaps. 6 and 7.

infinitesimal and only first-order or linear terms need be evaluated, so that a considerable simplification results. If S_t represents one of the $3N - 6$ internal coordinates and ξ_i one of $3N$ cartesian displacement coordinates, the relations sought will be of the form

$$S_t = \sum_{i=1}^{3N} B_{ti}\xi_i \qquad t = 1, 2, \ldots, 3N - 6 \tag{1}$$

where the coefficients B_{ti} are constants determined by the geometry of the molecule. Instead of using three cartesian coordinates to describe the displacement of an atom, it is convenient to introduce a vector $\boldsymbol{\varrho}_\alpha$ for each atom α whose components along the three axis directions are the cartesian displacement coordinates ξ_i, $\xi_{i'}$, $\xi_{i''}$ for that atom. Likewise it is useful to group the coefficients B_{ti} for a given S_t into sets of three, each

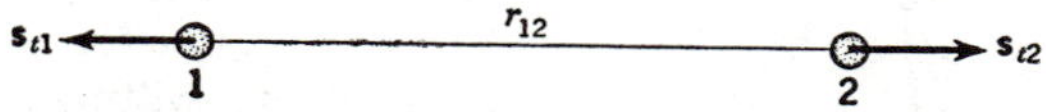

FIG. 4-1. Construction of vectors $\mathbf{s}_{t1}$ and $\mathbf{s}_{t2}$ for the increase in interatomic distance between atoms 1 and 2.

set B_{ti}, $B_{ti'}$, $B_{ti''}$ being associated with a given atom α. These quantities can be considered as the components of a vector $\mathbf{s}_{t\alpha}$ associated with the atom α *and* with the internal coordinate S_t. Then (1) above takes on the simpler form[1]

$$S_t = \sum_{\alpha=1}^{N} \mathbf{s}_{t\alpha} \cdot \boldsymbol{\varrho}_\alpha \tag{2}$$

where the dot represents the scalar product of the two vectors. This form has the advantage that it is now unnecessary to specify any axes for the displacement coordinates. Furthermore, simple rules can be worked out for writing down the vectors $\mathbf{s}_{t\alpha}$. The physical meaning of the vector $\mathbf{s}_{t\alpha}$ is as follows: Let all atoms except atom α be in their equilibrium positions. The direction of $\mathbf{s}_{t\alpha}$ is the direction in which a given displacement of atom α will produce the greatest increase of S_t. The magnitude, $|\mathbf{s}_{t\alpha}|$, of $\mathbf{s}_{t\alpha}$ is equal to the increase in S_t produced by unit displacement of the atom in this most effective direction. These statements follow from consideration of (2) above.

The simple types of coordinates in terms of which the potential functions for most molecules are usually expressed will now be worked out in detail.

Bond Stretching. Let S_t be the increase in the distance between atoms 1 and 2 (see Fig. 4-1). Clearly the most efficient direction to displace

[1] M. Eliashevich, *Compt. rend. acad. sci. U.R.S.S.*, **28**: 605 (1940).
E. B. Wilson, Jr., *J. Chem. Phys.*, **9**: 76 (1941).

the end atoms is along the line connecting them, but in directions away from each other. Furthermore, the vectors $\mathbf{s}_{t1}$ and $\mathbf{s}_{t2}$ should here be unit vectors. For the coordinate S_t, all other vectors $\mathbf{s}_{t\alpha}(\alpha \neq 1, 2)$ are zero since displacements of other atoms will not affect S_t.

It is often convenient to express the vectors $\mathbf{s}_{t\alpha}$ in terms of unit vectors along certain of the interatomic connecting lines, for example, along the chemical valence bonds. Let the unit vector directed from atom α toward atom β be denoted by $\mathbf{e}_{\alpha\beta}$. When S_t is the extension of the bond between atoms 1 and 2, one has

$$\mathbf{s}_{t1} = \mathbf{e}_{21} = -\mathbf{e}_{12} \qquad \mathbf{s}_{t2} = \mathbf{e}_{12} \tag{3}$$

Valence Angle Bending. Another example is the increase in the angle between two valence bonds attached to the same atom 3. Let S_t be this internal coordinate. Then (see Fig. 4-2) the $\mathbf{s}$ vector for one of the end atoms, say $\mathbf{s}_{t1}$, will be perpendicular to the side 3,1 of the bond angle, and pointed outward, since this is the direction in which displacements of atom 1 will be most effective in increasing the bond angle. Further, the length of $\mathbf{s}_{t1}$ is $1/r_{31}$, where r_{31} is the length of the side 3,1, since a unit (infinitesimal) displacement of 1 along $\mathbf{s}_{t1}$ will increase ϕ by an amount $1/r_{31}$. Similarly, $\mathbf{s}_{t2}$ is a vector of length $1/r_{32}$ in the plane of 1, 2, and 3 and perpendicular to 3,2, as shown. To find the $\mathbf{s}$ vector for the apex atom, use is made of the fact that a rigid displacement of the whole molecule does not alter the angles. Suppose that the apex atom is given a certain displacement. Then, by shifting the whole molecule rigidly by an equal but opposite amount, the apex atom can be brought back to its original position and the end atoms displaced by amounts equal and opposite to the original displacement of the apex atom. The effects of displacements of the end atoms have already been worked out so that this construction enables the effect of apex atom displacements to be determined, also. The result is that the $\mathbf{s}$ vector, $\mathbf{s}_{t3}$, for the apex atom is the sum of the $\mathbf{s}$ vectors for the end atoms, with reversed sign, or

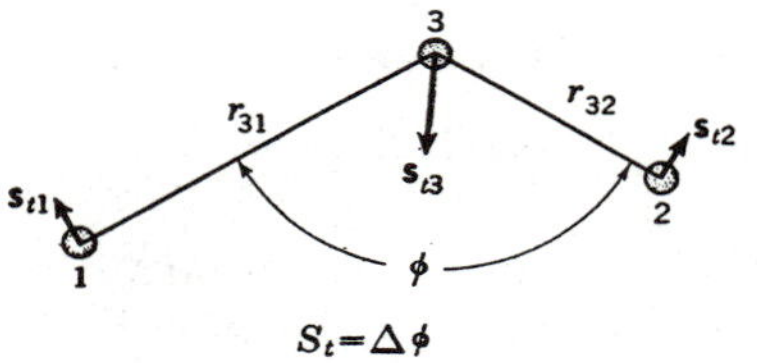

FIG. 4-2. Construction of vectors $\mathbf{s}_{t1}$, $\mathbf{s}_{t2}$, and $\mathbf{s}_{t3}$ for the increase in the angle ϕ formed by two bonds.

$$\mathbf{s}_{t3} = -\mathbf{s}_{t1} - \mathbf{s}_{t2} \tag{4}$$

Clearly the $\mathbf{s}$ vectors for all other atoms except those at the ends and apex of the angle in question will be zero.

The $\mathbf{s}$ vectors can also be written in terms of unit vectors along the bonds. Let $\mathbf{e}_{31}$ and $\mathbf{e}_{32}$ be unit vectors from the apex angle along the bonds, 3,1 and 3,2, respectively. Then, a little reflection shows that[1]

$$\mathbf{s}_{t1} = \frac{\cos\phi\mathbf{e}_{31} - \mathbf{e}_{32}}{r_{31}\sin\phi} \tag{5}$$

$$\mathbf{s}_{t2} = \frac{\cos\phi\mathbf{e}_{32} - \mathbf{e}_{31}}{r_{32}\sin\phi} \tag{6}$$

$$\mathbf{s}_{t3} = \frac{[(r_{31} - r_{32}\cos\phi)\mathbf{e}_{31} + (r_{32} - r_{31}\cos\phi)\mathbf{e}_{32}]}{r_{31}r_{32}\sin\phi} \tag{7}$$

In many cases it is more convenient to use $r_{31}\,\Delta\phi$, or $r_{32}\,\Delta\phi$, or $(r_{31}r_{32})^{\frac{1}{2}}\,\Delta\phi$ instead of $\Delta\phi$ itself as the internal coordinate since then the bending force constants will have the same dimensions as have the stretching force constants. The expressions for $\mathbf{s}_{t1}$, etc., given above would then need to be multiplied by r_{31}, r_{32}, or $(r_{31}r_{32})^{\frac{1}{2}}$.

An alternate method of deriving the vectors $\mathbf{s}_{t\alpha}$ is illustrated in the case of the bending coordinate as follows. The scalar product of unit vectors directed outwards from the central atom gives the cosine of the angle ϕ:

$$\cos\phi = \mathbf{e}_{31}\cdot\mathbf{e}_{32} \tag{8}$$

The expression for a small variation in ϕ may be obtained by differentiation of (8).

$$\Delta\cos\phi = -\sin\phi\,\Delta\phi = \mathbf{e}_{31}\cdot\Delta\mathbf{e}_{32} + \mathbf{e}_{32}\cdot\Delta\mathbf{e}_{31} \tag{9}$$

The small variation of the unit vectors appearing on the right side of (9) may be expressed in terms of the arbitrary displacement vectors of the atoms $\boldsymbol{\rho}_1$, $\boldsymbol{\rho}_2$, $\boldsymbol{\rho}_3$, and it is clear that we can now obtain the expressions for the vectors $\mathbf{s}_t$ in

$$S_t = \Delta\phi = \sum_{\alpha=1}^{N} \mathbf{s}_{t\alpha}\cdot\boldsymbol{\rho}_\alpha$$

by substituting appropriate expressions for $\Delta\mathbf{e}_{31}$ and $\Delta\mathbf{e}_{32}$ in (9).

With the use of the definition of a unit vector

$$\mathbf{e}_{3\alpha} = \frac{\mathbf{r}_{3\alpha}}{r_{3\alpha}} \qquad \alpha = 1, 2 \tag{10}$$

in which $\mathbf{r}_{3\alpha}$ is the vector from atom 3 to atom α and $r_{3\alpha}$ is the distance between the atoms 3 and α, differentiation yields

$$\Delta\mathbf{e}_{3\alpha} = \frac{r_{3\alpha}\,\Delta\mathbf{r}_{3\alpha} - \mathbf{r}_{3\alpha}\,\Delta r_{3\alpha}}{r_{3\alpha}^2} \tag{11}$$

[1] A useful alternative form for these equations is

$$\mathbf{s}_{t1} = \mathbf{e}_{31} \times \frac{\mathbf{e}_{31}\times\mathbf{e}_{32}}{r_{31}\sin\phi}, \text{ etc.}$$

in which $\mathbf{r}_{3\alpha}$ and $r_{3\alpha}$ may be given their values at equilibrium. Using a prime to distinguish the vector $\mathbf{r}_{3\alpha}$ in a displaced configuration (Fig. 4-3),

$$\mathbf{r}'_{3\alpha} = \mathbf{r}_{3\alpha} + \boldsymbol{\rho}_\alpha - \boldsymbol{\rho}_3 \tag{12}$$

From (12) it immediately follows that

$$\Delta\mathbf{r}_{3\alpha} = \boldsymbol{\rho}_\alpha - \boldsymbol{\rho}_3 \tag{13}$$

To obtain $\Delta r_{3\alpha}$ one computes the scalar product of each side of (12) by itself, neglecting second-order terms in the $\boldsymbol{\rho}$.

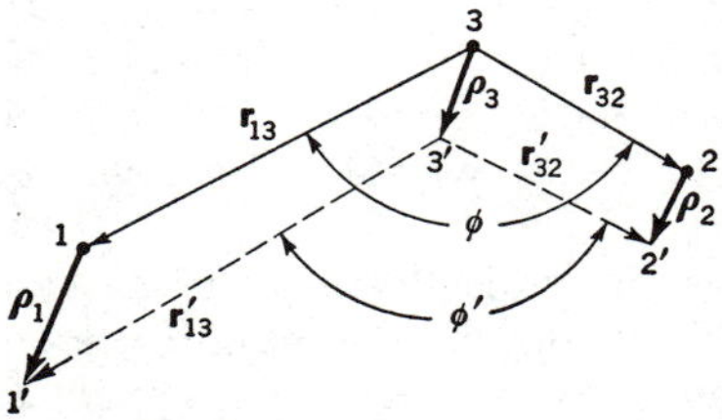

FIG. 4-3. Relation between displaced and equilibrium $\mathbf{r}_{\alpha\beta}$ vectors.

$$(r'_{3\alpha})^2 = (r_{3\alpha})^2 + 2\mathbf{r}_{3\alpha}\cdot(\boldsymbol{\rho}_\alpha - \boldsymbol{\rho}_3) \tag{14}$$

This equation shows that the small variation in the square of the bond length is $2\mathbf{r}_{3\alpha}\cdot(\boldsymbol{\rho}_\alpha - \boldsymbol{\rho}_3)$, or,

$$\Delta r_{3\alpha} = \frac{\Delta(r_{3\alpha})^2}{2r_{3\alpha}} = \mathbf{e}_{3\alpha}\cdot(\boldsymbol{\rho}_\alpha - \boldsymbol{\rho}_3) \tag{15}$$

Substitution of (15) and (13) into (11) gives the small changes in $\mathbf{e}_{31}$ and $\mathbf{e}_{32}$ in terms of the equilibrium values of the $\mathbf{e}_{3\alpha}$ and of the displacements $\boldsymbol{\rho}_1$, $\boldsymbol{\rho}_2$, $\boldsymbol{\rho}_3$, of the atoms, whereupon substitution of (11) into (9) and collection of the vectors which act as scalar product multipliers of the $\boldsymbol{\rho}$'s gives the result:

$$S_t = \Delta\phi = \left(\frac{\cos\phi\,\mathbf{e}_{31} - \mathbf{e}_{32}}{r_{31}\sin\phi}\right)\cdot\boldsymbol{\rho}_1 + \left(\frac{\cos\phi\,\mathbf{e}_{32} - \mathbf{e}_{31}}{r_{32}\sin\phi}\right)\cdot\boldsymbol{\rho}_2 + \left[\frac{(r_{31} - r_{32}\cos\phi)\mathbf{e}_{31} + (r_{32} - r_{31}\cos\phi)\mathbf{e}_{32}}{r_{31}r_{32}\sin\phi}\right]\cdot\boldsymbol{\rho}_3 \tag{16}$$

The vectors $\mathbf{s}_{t\alpha}$ may be immediately identified in (16) and are found to be identical with the expression previously obtained, (5) to (7).

Angle between a Bond and a Plane Defined by Two Bonds. Another[1] useful type of internal coordinate is the angle formed by a bond 4,1 and the plane of the three atoms 2, 3, and 4, all four atoms being in one plane in the equilibrium position (see Fig. 4-4). An example is the angle between the C═C bond and the plane of a CH_2 group in ethylene. From the general rules, the $\mathbf{s}$ vectors for all the atoms will be perpendicular to the equilibrium plane. The length of $\mathbf{s}_{t1}$ is also seen to be $1/r_{41}$. The effect of displacements of the other atoms 2, 3, and 4 can be found by applying rigid rotations and translations to the whole molecule so as to bring 2, 3, and 4 back to their original positions, with a consequent calculable additional displacement of atom 1, the effect of which is now

[1] The reader may prefer to skip the rest of this section and refer to it only when he has occasion to use the material in it.

known.[1] The result of this procedure turns out to be, for the absolute values $s_{t\alpha}$, of the vectors,

$$
\begin{aligned}
s_{t1} &= \frac{1}{r_{41}} && \text{“end atom”} \\
s_{t2} &= \frac{\sin\phi_2}{r_{42}\sin\phi_1} && \text{“anchor atom”} \\
s_{t3} &= \frac{\sin\phi_3}{r_{43}\sin\phi_1} && \text{“anchor atom”} \\
s_{t4} &= -\frac{1}{r_{41}} - \frac{\sin\phi_2}{r_{42}\sin\phi_1} - \frac{\sin\phi_3}{r_{43}\sin\phi_1} && \text{“apex atom”}
\end{aligned}
\tag{17}
$$

In case the four atoms are not coplanar, a generalization of the formulas for the **s** vectors may be made. The computation may be carried out in a

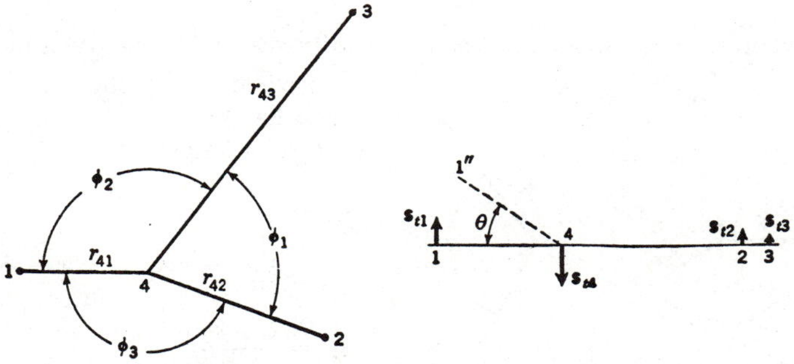

$S_t = \Delta\theta$

FIG. 4-4. Internal coordinate measuring bending of bond out of plane, top and side views. In the latter, the position marked 1″ is the location atom 1 would occupy if the distorted molecule is subjected to rigid rotations and translations which restore atoms 2, 3, and 4 to their original plane.

fashion similar to the second method applied to the valence angle bending coordinate. The basic definition of the angle involved may be given as

$$\sin\theta = \frac{\mathbf{e}_{42} \times \mathbf{e}_{43}}{\sin\phi_1} \cdot \mathbf{e}_{41} \tag{18}$$

The coordinate S_t is taken as $\Delta\theta$ and, by differentiation and use of the previous formulas (11) and (16) for $\Delta\mathbf{e}_{4\alpha}$ and $\Delta\phi_1$, respectively, the following rather complicated results are obtained:

[1] Formally this invariance to rigid translations and rotations is expressed by the conditions

$$\sum_\alpha \mathbf{s}_\alpha = 0 \qquad \sum_\alpha \mathbf{R}_\alpha \times \mathbf{s}_\alpha = 0$$

in which $\mathbf{R}_\alpha$ is the vector from an arbitrary origin to the equilibrium position of atom α. See R. J. Malhiot and Salvador M. Ferigle, *J. Chem. Phys.*, **22**: 717 (1954).

$$
\begin{aligned}
\mathbf{s}_{t1} &= \frac{1}{r_{41}}\left(\frac{\mathbf{e}_{42}\times\mathbf{e}_{43}}{\cos\theta\sin\phi_1} - \tan\theta\mathbf{e}_{41}\right)\\
\mathbf{s}_{t2} &= \frac{1}{r_{42}}\left[\frac{\mathbf{e}_{43}\times\mathbf{e}_{41}}{\cos\theta\sin\phi_1} - \frac{\tan\theta}{\sin^2\phi_1}(\mathbf{e}_{42} - \cos\phi_1\mathbf{e}_{43})\right]\\
\mathbf{s}_{t3} &= \frac{1}{r_{43}}\left[\frac{\mathbf{e}_{41}\times\mathbf{e}_{42}}{\cos\theta\sin\phi_1} - \frac{\tan\theta}{\sin^2\phi_1}(\mathbf{e}_{43} - \cos\phi_1\mathbf{e}_{42})\right]\\
\mathbf{s}_{t4} &= -\mathbf{s}_{t1} - \mathbf{s}_{t2} - \mathbf{s}_{t3}
\end{aligned}
\tag{19}
$$

If $\theta = 0$ (the planar case), the above expressions show that the $\mathbf{s}_{t\alpha}$ are all perpendicular to the plane, $\mathbf{s}_{t4}$ being directed in the sense opposite to that of the other three, and since expressions of the form

$$\frac{(\mathbf{e}_{42}\times\mathbf{e}_{43})}{\sin\phi_1}$$

etc., are unit vectors, the lengths are readily seen to be as given in (17).

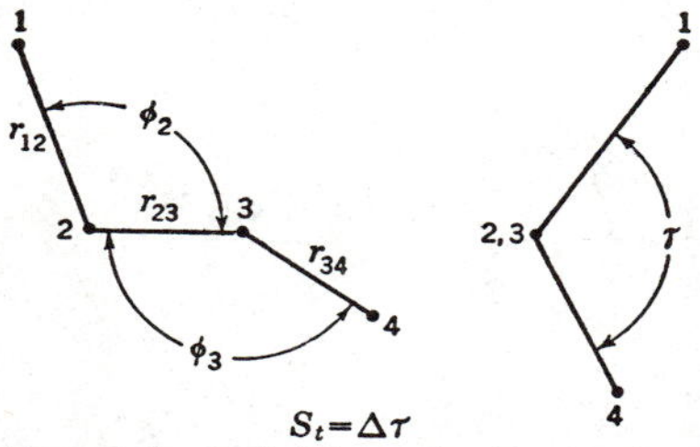

FIG. 4-5. Definition of the torsion angle, τ.

Even when $\theta \neq 0$, the expressions (19) can be shown to be equivalent to vectors of lengths:

$$
\begin{aligned}
s_{t1} &= \frac{1}{r_{41}}\\
s_{t2} &= \frac{\cos\phi_1\cos\phi_2 - \cos\phi_3}{r_{42}\cos\theta\sin^2\phi_1}\\
s_{t3} &= \frac{\cos\phi_1\cos\phi_3 - \cos\phi_2}{r_{43}\cos\theta\sin^2\phi_1}
\end{aligned}
$$

The direction of $\mathbf{s}_{t1}$ is perpendicular to the bond 4,1 and in a plane perpendicular to the plane 2, 4, 3 while the directions of $\mathbf{s}_{t2}$ and $\mathbf{s}_{t3}$ are each perpendicular to the plane 2, 4, 3. The positive sense of each vector must be chosen consistent with the convention defining the sense of the angle θ (Fig. 4-4).

Torsion. Another type of internal coordinate which may be useful is the change in the dihedral angle, τ, between the planes determined by atoms 1, 2, 3 and 2, 3, 4, respectively, when atoms 1 to 4 are bonded in sequence (Fig. 4-5). The magnitude of the dihedral angle may be determined in accordance with the following convention: let τ be restricted to the interval $-\pi < \tau \leq \pi$; then let τ be positive if, viewing the atoms along the bond 2,3 with 2 nearer the observer than 3, the angle from the projection of 2,1 to the projection of 3,4 is traced in the clockwise sense. The analytical definition of τ can be given as

$$\cos\tau = \frac{(\mathbf{e}_{12}\times\mathbf{e}_{23})\cdot(\mathbf{e}_{23}\times\mathbf{e}_{34})}{\sin\phi_2\sin\phi_3} \tag{20}$$

since $(\mathbf{e}_{12} \times \mathbf{e}_{23})/\sin\phi_2$ and $(\mathbf{e}_{23} \times \mathbf{e}_{34})/\sin\phi_3$ are unit vectors perpendicular, respectively, to the planes 1, 2, 3 and 2, 3, 4. Then by methods previously illustrated, one finds **s** vectors for the internal coordinate $S_t = \delta\tau$

$$\mathbf{s}_{t1} = -\frac{\mathbf{e}_{12} \times \mathbf{e}_{23}}{r_{12}\sin^2\phi_2} \tag{21}$$

$$\mathbf{s}_{t2} = \frac{r_{23} - r_{12}\cos\phi_2}{r_{23}r_{12}\sin\phi_2}\,\frac{\mathbf{e}_{12} \times \mathbf{e}_{23}}{\sin\phi_2} + \frac{\cos\phi_3}{r_{23}\sin\phi_3}\,\frac{\mathbf{e}_{43} \times \mathbf{e}_{32}}{\sin\phi_3} \tag{22}$$

$$\mathbf{s}_{t3} = [(14)(23)]\mathbf{s}_{t2} \tag{23}$$

$$\mathbf{s}_{t4} = [(14)(23)]\mathbf{s}_{t1} \tag{24}$$

where the expressions in brackets in (23) and (24) indicate that the latter vectors can be obtained by permutation of the atom subscripts 1 and 4, and 2 and 3 in the expressions for the first two vectors.

4-2. Construction and Properties of the G Matrix[1]

In what follows frequent use will be made of the set of quantities $G_{tt'}$ defined by the equations[2] (in the noncomplex case)

$$G_{tt'} = \sum_{i=1}^{3N} \frac{1}{m_i} B_{ti}B_{t'i} \qquad t, t' = 1, 2, \ldots, 3N - 6 \tag{1}$$

in which m_i is the mass of the atom to which the subscript i refers and the coefficients B_{ti} have previously [Eq. (1), Sec. 4-1] been introduced to relate the internal coordinates S_t with the cartesian displacement coordinates ξ_i, or

$$S_t \doteq \sum_{i=1}^{3N} B_{ti}\xi_i \qquad t = 1, 2, \ldots, 3N - 6 \tag{2}$$

Instead of the $3N$ coefficients B_{ti} (for each S_t), it is more convenient to use the N vectors $\mathbf{s}_{t\alpha}$, one for each atom, described in the last section. In terms of these

$$G_{tt'} = \sum_{\alpha=1}^{N} \mu_\alpha \mathbf{s}_{t\alpha} \cdot \mathbf{s}_{t'\alpha} \tag{3}$$

where the dot represents the scalar product of the two vectors and $\mu_\alpha = 1/m_\alpha$, the reciprocal of the mass of atom α.

[1] No knowledge of matrix algebra is required for the understanding of this section, but the simple principles given in Appendix V will be found very useful and eventually indispensable for the serious student of molecular dynamics.

[2] E. B. Wilson, Jr., *J. Chem. Phys.*, **7**: 1047 (1939).
M. Eliashevich, *Compt. rend. acad. sci. U.R.S.S.*, **28**: 605 (1940).
A. G. Meister and F. F. Cleveland, *Am. J. Phys.*, **14**: 13 (1946).

It is seen that there is one element $G_{tt'}$ for each pair of internal coordinates S_t and $S_{t'}$. No coordinate axes are needed to compute these quantities if the **s** vectors are available. Further, the results of the previous section give expressions for the **s** vectors in terms of bond vectors and other simple unit vectors so that the scalar products of **s** vectors required can be reduced to a sum of scalar products of certain unit vectors, particularly bond vectors, within the molecule. In practice, therefore, one would construct a table of scalar products of these unit vectors as part of the calculation procedure for a given molecule.

It is also possible to tabulate the elements $G_{tt'}$ which are of common occurrence. For example, if S_t represents the extension of a bond between atoms 1 and 2, Eq. (3), Sec. 4-1, and (3) show that G_{tt} is $\mu_1 + \mu_2$. Similarly if S_t is the extension of the bond which forms the side 3,1 of a simple valence angle ϕ and if $S_{t'}$ is the increase in ϕ, Eqs. (3), (5), and (7), Sec. 4-1, and (3) show that $G_{tt'}$ is given by $-(\mu_3 \sin \phi)/r_{32}$. Whenever such internal coordinate combinations occur, these results can be used. Appendix VI lists a considerable number of **G** matrix elements.

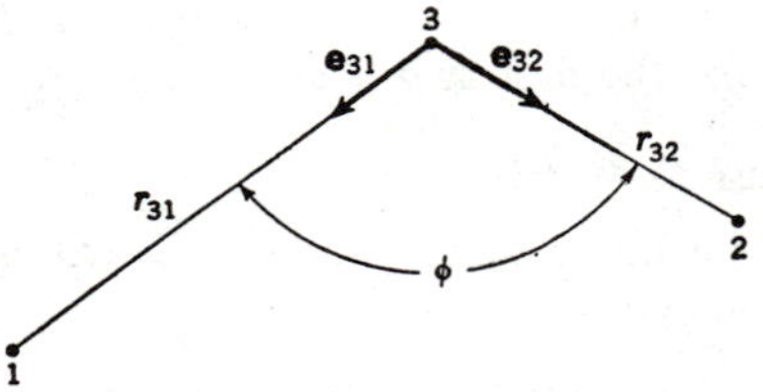

FIG. 4-6. Nonlinear triatomic molecule 1,3,2.

The quantities $G_{tt'}$ are closely related[1] to the expression for the kinetic energy of the molecule in terms of the internal coordinates S_t. They may therefore be used to construct certain forms of the secular equation (Secs. 4-3 to 4-5). Note that $G_{tt'} = G_{t't}$. If the coordinate S_t happens to be orthogonal to $S_{t'}$ (see Sec. 2-4), then $G_{tt'} = 0$. This will clearly be the case if the set of atoms for which $\mathbf{s}_{t\alpha} \neq 0$ and the set for which $\mathbf{s}_{t'\beta} \neq 0$ contain no atoms in common, and may be true under other circumstances as well.

As an example, the quantities $G_{tt'}$ for a nonlinear triatomic molecule 1, 3, 2 will be evaluated (see Fig. 4-6). The three internal coordinates to be used are S_1, the increase in the length of the bond 3,1; S_2, the increase in the length of the bond 3,2; and S_3, the increase in the angle ϕ. The two unit bond vectors $\mathbf{e}_{31}$ and $\mathbf{e}_{32}$ are shown in Fig. 4-6. The **s**

[1] Actually the numbers $G_{tt'}$ form a matrix **G** such that the vibrational kinetic energy T is given by

$$2T = \sum_{tt'} (G^{-1})_{tt'}\dot{S}_t\dot{S}_{t'}$$

$$= \sum_{tt'} G_{tt'}P_tP_{t'} \tag{4}$$

where $\mathbf{G}^{-1}$ is the matrix reciprocal to **G** and P_t is the momentum conjugate to S_t (see Appendix VII).

vectors can be expressed in terms of them, according to the rules derived in the previous section.

$$\begin{array}{lll} \mathbf{s}_{11} = \mathbf{e}_{31} & \mathbf{s}_{13} = -\mathbf{e}_{31} & \mathbf{s}_{12} = 0 \\ \mathbf{s}_{21} = 0 & \mathbf{s}_{23} = -\mathbf{e}_{32} & \mathbf{s}_{22} = \mathbf{e}_{32} \end{array}$$
$$\mathbf{s}_{31} = \frac{\cos\phi\mathbf{e}_{31} - \mathbf{e}_{32}}{r_{31}\sin\phi} \tag{5}$$
$$\mathbf{s}_{32} = \frac{\cos\phi\mathbf{e}_{32} - \mathbf{e}_{31}}{r_{32}\sin\phi}$$
$$\mathbf{s}_{33} = \frac{[(r_{31} - r_{32}\cos\phi)\mathbf{e}_{31} + (r_{32} - r_{31}\cos\phi)\mathbf{e}_{32}]}{r_{31}r_{32}\sin\phi}$$

Furthermore, $\mathbf{e}_{31}\cdot\mathbf{e}_{31} = 1$, $\mathbf{e}_{32}\cdot\mathbf{e}_{32} = 1$, $\mathbf{e}_{31}\cdot\mathbf{e}_{32} = \cos\phi$. Then

$$G_{11} = \mu_1 + \mu_3 \qquad G_{22} = \mu_2 + \mu_3$$
$$G_{33} = \frac{\mu_1}{r_{31}^2} + \frac{\mu_2}{r_{32}^2} + \mu_3\left(\frac{1}{r_{31}^2} + \frac{1}{r_{32}^2} - \frac{2\cos\phi}{r_{31}r_{32}}\right)$$
$$G_{12} = \mu_3\cos\phi \qquad G_{13} = \frac{-\mu_3\sin\phi}{r_{32}} \tag{6}$$
$$G_{23} = \frac{-\mu_3\sin\phi}{r_{31}}$$

4-3. The Secular Equation in Terms of Internal Coordinates

In Appendix VII, it is shown that the kinetic energy of vibration can be written in terms of internal coordinates in the form

$$2T = \sum_{tt'} (G^{-1})_{tt'}\dot{S}_t\dot{S}_{t'} \tag{1}$$

where the coefficients $(G^{-1})_{tt'}$ are related to the numbers $G_{tt'}$ introduced in the previous section by the equations[1]

$$\sum_{t'} (G^{-1})_{tt'}G_{t't''} = \delta_{tt''} \tag{2}$$

and

$$\sum_{t'} G_{tt'}(G^{-1})_{t't''} = \delta_{tt''} \tag{3}$$

where $\delta_{tt''}$ is unity if $t = t''$ and is zero otherwise. These equations can be solved (usually numerically[2]) for the coefficients $(G^{-1})_{tt'}$ if the values of the $G_{t't''}$ are given.

[1] In matrix language, the matrix $\mathbf{G}^{-1}$ is the inverse of the matrix $\mathbf{G}$. But note that the number $(G^{-1})_{tt'}$ is not usually the reciprocal of the individual element $G_{tt'}$.

[2] The numerical methods are fairly tedious for large values of N but are routine and often used. See, for example, R. A. Frazer, W. J. Duncan, and A. R. Collar, "Elementary Matrices," Macmillan, New York, 1946.

If the potential energy is expressed in the same internal coordinates so that

$$2V = \sum_{tt'} F_{tt'} S_t S_{t'} \tag{4}$$

the $F_{tt'}$ being the force constants, the vibrational problem leads to a secular equation

$$\begin{vmatrix} F_{11} - (G^{-1})_{11}\lambda & F_{12} - (G^{-1})_{12}\lambda & \cdots & F_{1n} - (G^{-1})_{1n}\lambda \\ F_{21} - (G^{-1})_{21}\lambda & F_{22} - (G^{-1})_{22}\lambda & \cdots & F_{2n} - (G^{-1})_{2n}\lambda \\ \cdots & \cdots & \cdots & \cdots \\ F_{n1} - (G^{-1})_{n1}\lambda & F_{n2} - (G^{-1})_{n2}\lambda & \cdots & F_{nn} - (G^{-1})_{nn}\lambda \end{vmatrix} = 0 \tag{5}$$

Here $\lambda = 4\pi^2\nu^2$ as usual and n is the number of internal coordinates, $3N - 6$ (or $3N - 5$ for the linear case). This equation follows directly from the considerations of Sec. 2-6. Symbolically it can also be written as

$$|\mathbf{F} - \mathbf{G}^{-1}\lambda| = 0 \quad (6$$

This form of the secular equation has the advantage of using force constants $F_{tt'}$ of direct physical significance (*i.e.*, in terms of bond distances and angles, or interatomic distances). These force constants also appear directly in the elements of the equation and thus do not need any preliminary algebraic treatment before use. On the other hand, it is often inconvenient to have the unknown λ appear off the principal diagonal (especially when numerical solution is employed) and the job of inverting the **G** matrix to get $\mathbf{G}^{-1}$ is somewhat tedious.

Another form of the secular equation can be obtained from the above by multiplying it through by the determinant of **G**; that is, by

$$|\mathbf{G}| = \begin{vmatrix} G_{11} & G_{12} & \cdots & G_{1n} \\ G_{21} & G_{22} & \cdots & G_{2n} \\ \cdots & \cdots & \cdots & \cdots \\ G_{n1} & G_{n2} & \cdots & G_{nn} \end{vmatrix}$$

using the known rules for the multiplication of determinants.[1] This yields the equation

$$\begin{vmatrix} \Sigma G_{1t}F_{t1} - \lambda & \Sigma G_{1t}F_{t2} & \cdots & \Sigma G_{1t}F_{tn} \\ \Sigma G_{2t}F_{t1} & \Sigma G_{2t}F_{t2} - \lambda & \cdots & \Sigma G_{2t}F_{tn} \\ \cdots & \cdots & \cdots & \cdots \\ \Sigma G_{nt}F_{t1} & \Sigma G_{nt}F_{t2} & \cdots & \Sigma G_{nt}F_{tn} - \lambda \end{vmatrix} = 0 \tag{7}$$

or symbolically

$$|\mathbf{GF} - \mathbf{E}\lambda| = 0 \tag{8}$$

[1] Which are the same as for the multiplication of matrices.

where **E** is the unit matrix. Because of the relation between the quantities $G_{tt'}$ and $(G^{-1})_{tt'}$ given in (2) and (3), the off-diagonal λ's have been eliminated and the diagonal λ's have unit coefficients.[1] Because both **F** and **G** are symmetrical (that is, $F_{tt'} = F_{t't}$), rows and columns in (8) can be interchanged to yield another form which can be written as

$$|\mathbf{FG} - \mathbf{E}\lambda| = 0 \tag{9}$$

These forms of the secular equation have the advantage of not requiring the construction of the inverse $\mathbf{G}^{-1}$. Further, λ occurs only on the diagonal. However, they are in general unsymmetrical, *i.e.*, usually,

$$\sum_{t''} F_{tt''}G_{t''t'} \neq \sum_{t''} F_{t't''}G_{t''t}$$

Furthermore, the force constants $F_{tt'}$ do not enter the terms as simply as in the form (5).

By multiplying through the secular equation in form (7) by the determinant whose elements $(F^{-1})_{tt'}$ satisfy the equation[2]

$$\sum_{t'} (F^{-1})_{tt'}F_{t't''} = \delta_{tt''} \tag{10}$$

another form symbolically represented by

$$|\mathbf{G} - \mathbf{F}^{-1}\lambda| = 0 \tag{11}$$

can be constructed.[3] This is usually simpler to obtain than (5) because the approximate potential functions often used may contain few off-diagonal elements (*i.e.*, few force constants $F_{tt'}$ with $t' \neq t$) and the problem of finding the numbers $(F^{-1})_{tt'}$ is in this case easier than that of getting the elements $(G^{-1})_{tt'}$ needed for (5).

4-4. A Form of the Secular Equation Convenient for Machine Solution

Another form of the secular equation[4] is

$$\begin{vmatrix} \sigma & 0 & \cdots & 0 & F_{11} & F_{12} & \cdots & F_{1n} \\ 0 & \sigma & \cdots & 0 & F_{21} & F_{22} & \cdots & F_{2n} \\ \cdots & \cdots & \cdots & \cdots & \cdots & \cdots & \cdots & \cdots \\ 0 & 0 & \cdots & \sigma & F_{n1} & F_{n2} & \cdots & F_{nn} \\ G_{11} & G_{12} & \cdots & G_{1n} & \sigma & 0 & \cdots & 0 \\ G_{21} & G_{22} & \cdots & G_{2n} & 0 & \sigma & \cdots & 0 \\ \cdots & \cdots & \cdots & \cdots & \cdots & \cdots & \cdots & \cdots \\ G_{n1} & G_{n2} & \cdots & G_{nn} & 0 & 0 & \cdots & \sigma \end{vmatrix} = 0 \tag{1}$$

[1] In matrix language (see Appendix V) $\mathbf{GG}^{-1} = \mathbf{E}$, whence

$$|\mathbf{G}||\mathbf{F} - \mathbf{G}^{-1}\lambda| = |\mathbf{GF} - \mathbf{E}\lambda| = 0$$

[2] That is, the $(F^{-1})_{tt'}$ form a matrix reciprocal to the matrix **F**.

[3] W. J. Taylor and K. S. Pitzer, *J. Research Natl. Bur. Standards,* **38:** 1 (1947).

[4] E. B. Wilson, Jr., *J. Chem. Phys.,* **15:** 736 (1947).

in which $\sigma = 2\pi\nu$, the $F_{tt'}$ are the force constants in terms of some set of coordinates and the $G_{tt'}$ are the coefficients described in Sec. 4-2, for the same coordinates. Usually, internal coordinates would be used. Symbolically, (1) could be written

$$\begin{vmatrix} \sigma\mathbf{E} & \mathbf{F} \\ \mathbf{G} & \sigma\mathbf{E} \end{vmatrix} = 0 \tag{2}$$

E being the unit matrix.

This form of the secular equation has twice as many rows and columns as the usual forms. This is a very serious drawback for numerical solution as is the fact that it is unsymmetrical, **F** appearing in one corner and **G** in the other. However, the fact that the force constants $F_{tt'}$ and the elements $G_{tt'}$ appear directly is an advantage. Furthermore, the unknown σ occurs only on the principal diagonal. In Sec. 9-10 an electrical machine will be described which can handle this form of the secular equation.

The equation as written has $2n$ instead of n roots but actually these occur in pairs $\pm\sigma$, that is, if $+\sigma$ is a root, so is $-\sigma$. This is proved by multiplying the first n rows and the last n columns of (1) by -1. The negative values are of course without physical meaning.

To show that (1) is indeed equivalent to the other forms of the secular equation, such as Eq. (8), Sec. 4-3, multiply (1) by the same equation in which $-\sigma$ replaces $+\sigma$. The result can be written as

$$\begin{vmatrix} \mathbf{FG} - \sigma^2\mathbf{E} & \mathbf{0} \\ \mathbf{0} & \mathbf{GF} - \sigma^2\mathbf{E} \end{vmatrix} = 0 \tag{3}$$

This factors into two equations

$$|\mathbf{FG} - \sigma^2\mathbf{E}| = 0 \tag{4}$$

and

$$|\mathbf{GF} - \sigma^2\mathbf{E}| = 0 \tag{5}$$

which are identical with Eqs. (9) and (8), Sec. 4-3, since $\sigma^2 = \lambda$.

Another way of deriving (1) is to start with the equations of motion in Hamiltonian form, using the result of the footnote to Sec. 4-2 that

$$2T = \sum_{tt'} G_{tt'}P_tP_{t'}$$

the P's being momenta conjugate to the S's. Then Hamilton's equations of motion

$$\dot{P}_t = -\frac{\partial H}{\partial S_t} \qquad \dot{S}_t = \frac{\partial H}{\partial P_t} \tag{6}$$

become

$$\dot{P}_t = -\sum_{t'} F_{tt'}S_{t'} \tag{7}$$

$$\dot{S}_t = \sum_{t'} G_{tt'}P_{t'} \tag{8}$$

since the Hamiltonian function is here

$$H = T + V \tag{9}$$

Substitution of

$$S_t = A_t \cos(\sigma t + \epsilon) \tag{10}$$

and

$$P_t = A'_t \sin(\sigma t + \epsilon) \tag{11}$$

leads to the $2n$ simultaneous equations

$$\sigma A'_t + \sum_{t'} F_{tt'}A_{t'} = 0 \qquad t = 1, 2, \ldots, n \tag{12}$$

$$\sum_{t'} G_{tt'}A'_{t'} + \sigma A_t = 0 \qquad t = 1, 2, \ldots, n \tag{13}$$

The equation of compatibility of these $2n$ equations in the $2n$ unknowns A'_t and A_t is the secular equation (1). Furthermore (12) and (13) together can be used to obtain the coefficients A'_t and A_t, once a suitable value of σ is inserted.[1] There are still other forms of the secular equation, some of which may be useful for special purposes. For example, for solution by the inductance-capacity electric network (Sec. 9-10), it is necessary (for practical reasons) to have the secular equation in a form which is symmetrical and has the unknown occurring only on the principal diagonal and with unit coefficient. This is the form which Eq. (5), Sec. 4-3, would take if the internal coordinates used were mutually orthogonal and properly normalized, because then $G_{tt'} = \delta_{tt'}$. Details of the orthogonalization procedure are given in Sec. 9-2.

Various other forms of the secular equation have been proposed and sometimes used. For them the reader is referred to the original papers.[2]

[1] Actually the coefficients A'_t and A_t are not independent but are connected by the relation

$$A'_t = -\sigma \sum_{t'} (G^{-1})_{tt'}A_{t'} \tag{14}$$

See Appendix VIII.

[2] For example, see the following:

O. Redlich, *Z. physik. Chem.*, (*B*), **28**: 371 (1935).

O. Redlich and H. Tompa, *J. Chem. Phys.*, **5**: 529 (1937).

E. B. Wilson, Jr., and B. L. Crawford, Jr., *J. Chem. Phys.*, **6**: 223 (1938).

4-5. Direct Expansion of the Secular Equation

The form of the secular equation [Eq. (1), Sec. 4-4] is convenient to use in showing how to get the secular equation in expanded form,[1] that is, in the form of an ordinary algebraic equation in the unknown quantity λ. A determinant is expanded by forming the sum of all products, with proper signs, constructed by selecting one and only one element from each row and column. One such product is obviously σ^{2n}, made up of the elements on the principal diagonal. All other terms in the expansion are obtained by substituting for two or more of these σ's one or more $F_{tt'}$'s and one or more $G_{tt'}$'s. Consider the terms in σ^{2n-2}. One of them is obtained by replacing the first σ by F_{11}, say. Then the $(n+1)$st σ must be replaced by G_{11} in order to satisfy the rule that one and only one element from every row and column must appear. In this way, one builds the whole coefficient of σ^{2n-2}, namely,

$$-\sum_{tt'} G_{tt'}F_{tt'} \tag{1}$$

In other words the coefficient of λ^{n-1} (or σ^{2n-2}) is obtained by summing (with minus sign) over all force constants $F_{tt'}$, each multiplied by the corresponding $G_{tt'}$.

One term in λ^{n-2} (or σ^{2n-4}) is obtained by replacing the first and second σ's by F_{11} and F_{22}; then one must replace the $(n+1)$st and $(n+2)$nd by G_{11} and G_{22} or by G_{12} and G_{21}. Likewise F_{12} and F_{21} could have been used instead of F_{11} and F_{22}. Generalization of this, including consideration of the rule for signs, shows that the whole term is

$$\lambda^{n-2}\Sigma G^{(2)}F^{(2)} \tag{2}$$

where $G^{(2)}$ is any two-rowed minor of $|\mathbf{G}|$ and $F^{(2)}$ is the corresponding two-rowed minor of $|\mathbf{F}|$. The sum is over all such two-rowed minors.

This can be carried further to give the rule for any term, namely,

$$(-1)^s\lambda^{n-s}\Sigma G^{(s)}F^{(s)} \tag{3}$$

where $G^{(s)}$ denotes any s-rowed minor of the determinant of $\mathbf{G}$, $F^{(s)}$ is the corresponding s-rowed minor of $|\mathbf{F}|$, and the sum is over all possible such s-rowed minors. In general, there will be $[n!/(n-s)!s!]^2$ s-rowed minors of $\mathbf{F}$ (or $\mathbf{G}$) and therefore terms in the sum above. However, many will be duplicates because of the symmetry of $\mathbf{F}$ and $\mathbf{G}$ (they must be all counted, however) and often many will be zero.

It should be pointed out that by the above rule the constant term in the expansion, *i.e.*, the coefficient of λ^0, is

$$|\mathbf{G}||\mathbf{F}|$$

i.e., the product of the determinant of $\mathbf{G}$ and the determinant of $\mathbf{F}$.

[1] See E. B. Wilson, Jr., *J. Chem. Phys.*, **7**: 1047 (1939), where a different proof is given.

This method of expansion provides a very convenient way of handling the secular equation when n is small, say not over four or five. The number of minors which have to be evaluated rises very rapidly with n, however, so that the labor and danger of error soon become impractically great.

A Modification. It will be seen that every term in the expanded secular equation contains a factor of degree s in the reciprocal masses as well as λ^{n-s} and a factor depending on the molecular geometry. It is convenient to extend the analysis further so as to have a system for calculating the geometric factors separately. If one defines the quantities

$$H_{tt'}^{(\alpha)} = \mathbf{s}_{t\alpha} \cdot \mathbf{s}_{t'\alpha} \tag{4}$$

then Eq. (3), Sec. 4-2, yields the expression

$$G_{tt'} = \sum_{\alpha=1}^{N} \mu_\alpha H_{tt'}^{(\alpha)} \tag{5}$$

or, in matrix form

$$\mathbf{G} = \sum_{\alpha=1}^{N} \mu_\alpha \mathbf{H}^{(\alpha)} \tag{6}$$

Furthermore, any s-rowed minor of $|\mathbf{G}|$ can be written as

$$\begin{vmatrix} G_{tt} & G_{tt'} & \cdots \\ G_{t't} & G_{t't'} & \cdots \\ \cdots & \cdots & \cdots \end{vmatrix} = \sum_{\alpha,\alpha'\cdots} \mu_\alpha \mu_{\alpha'} \cdots \begin{vmatrix} H_{tt}^{(\alpha)} & H_{tt'}^{(\alpha')} & \cdots \\ H_{t't}^{(\alpha)} & H_{t't'}^{(\alpha')} & \cdots \\ \cdots & \cdots & \cdots \end{vmatrix} \tag{7}$$

from the properties of determinants. Here the sum is over all sets of α's including permutations thereof.

4-6. An Example: The Nonlinear Triatomic Molecule

The **G** elements have been worked out in Sec. 4-2 for the case of the nonlinear molecule 1, 3, 2 shown in Fig. 4-6. These can therefore be used to give the expanded form of the secular equation for this molecule. The **G** matrix elements of Eq. (6), Sec. 4-2, give the result below

$$\mathbf{G} = \left\| \begin{matrix} \mu_1 + \mu_3 & \mu_3 \cos\phi & -\dfrac{\mu_3 \sin\phi}{r_{32}} \\ \mu_3 \cos\phi & \mu_2 + \mu_3 & -\dfrac{\mu_3 \sin\phi}{r_{31}} \\ -\dfrac{\mu_3 \sin\phi}{r_{32}} & -\dfrac{\mu_3 \sin\phi}{r_{31}} & \dfrac{\mu_1}{r_{31}^2} + \dfrac{\mu_2}{r_{32}^2} + \mu_3\left(\dfrac{1}{r_{31}^2} + \dfrac{1}{r_{32}^2} - \dfrac{2\cos\phi}{r_{31}r_{32}}\right) \end{matrix} \right\| \tag{1}$$

From this one writes for the matrices $\mathbf{H}^{(\alpha)}$, etc., the following

$$\mathbf{H}^{(1)} = \begin{Vmatrix} 1 & 0 & 0 \\ 0 & 0 & 0 \\ 0 & 0 & \dfrac{1}{r_{31}^2} \end{Vmatrix} \qquad \mathbf{H}^{(2)} = \begin{Vmatrix} 0 & 0 & 0 \\ 0 & 1 & 0 \\ 0 & 0 & \dfrac{1}{r_{32}^2} \end{Vmatrix}$$

$$\mathbf{H}^{(3)} = \begin{Vmatrix} 1 & C & -\dfrac{S}{r_{32}} \\ C & 1 & -\dfrac{S}{r_{31}} \\ -\dfrac{S}{r_{32}} & -\dfrac{S}{r_{31}} & \dfrac{r_{12}^2}{r_{31}^2 r_{32}^2} \end{Vmatrix} \tag{2}$$

where

$$r_{12}^2 = r_{31}^2 + r_{32}^2 - 2r_{31}r_{32}\cos\phi \tag{3}$$

is the square of the distance 1,2 and $C = \cos\phi$, $S = \sin\phi$.

To simplify the example, a valence-type potential function without interaction terms will be used; *i.e.*,

$$2V = F_{11}S_1^2 + F_{22}S_2^2 + F_{33}S_3^2 \tag{4}$$

where $S_1 = \Delta r_{31}$, $S_2 = \Delta r_{32}$, and $S_3 = \Delta\phi$. The determinant of the force constants is therefore

$$|\mathbf{F}| = \begin{vmatrix} F_{11} & 0 & 0 \\ 0 & F_{22} & 0 \\ 0 & 0 & F_{33} \end{vmatrix} \tag{5}$$

and because of the zeros many of its minors will be zero. By taking each of the nonzero minors of $|\mathbf{F}|$ and constructing the appropriate minors of $|\mathbf{G}|$, Eq. (3), Sec. 4-5, gives the expanded secular equation. The result is given in tabular form in Table 4-1.

In the table the entries are the coefficients of the mass factors above the entry and of the force constant factor on the left for the λ factor indicated. The sum of all these terms equated to zero is the final equation. If a more general type of potential function were used, additional minors of $|\mathbf{F}|$ would have been nonvanishing and would therefore have brought in additional terms. Furthermore, the minors of $|\mathbf{F}|$ used above would have had additional terms.

This example illustrates the need for great care and system in writing down the terms to be sure that all are correctly included. Thus the coefficient of $\lambda^0 F_{11}F_{22}F_{33}\mu_1\mu_2\mu_3$ is the sum of three contributions, corresponding to the six possible terms of Eq. (7), Sec. 4-5, namely, 123, 213, 132, 231, 312, and 321. Three of these (213, 231, and 312) give zero contributions because of the columns of zeros in $|\mathbf{H}^{(1)}|$ and $|\mathbf{H}^{(2)}|$.

By setting up the secular equation in cartesian coordinates, one sees that in general a reciprocal mass μ_α for a given atom could never occur to a power higher than 3 and that in the present case where the atoms are limited to motions in a plane μ_α^2 is the highest power for any atom. With these general rules in mind the zeros under μ_1^3, μ_2^3, and μ_3^3 in Table 4-1 would have been obvious without calculation.

TABLE 4-1. EXPANDED SECULAR EQUATION

Term in λ^3:	1

Terms in $-\lambda^2$:

	μ_1	μ_2	μ_3
F_{11}	1	0	1
F_{22}	0	1	1
F_{33}	$\frac{1}{r_{31}^2}$	$\frac{1}{r_{32}^2}$	$\frac{r_{12}^2}{r_{31}^2 r_{32}^2}$

Terms in λ:

	μ_1^2	μ_2^2	μ_3^2	$\mu_1\mu_2$	$\mu_2\mu_3$	$\mu_3\mu_1$
$F_{11}F_{22}$	0	0	S^2	1	1	1
$F_{22}F_{33}$	0	$\frac{1}{r_{32}^2}$	$\frac{r_{12}^2 - S^2 r_{32}^2}{r_{31}^2 r_{32}^2}$	$\frac{1}{r_{31}^2}$	$\frac{r_{12}^2 + r_{31}^2}{r_{31}^2 r_{32}^2}$	$\frac{1}{r_{31}^2}$
$F_{33}F_{11}$	$\frac{1}{r_{31}^2}$	0	$\frac{r_{12}^2 - S^2 r_{31}^2}{r_{31}^2 r_{32}^2}$	$\frac{1}{r_{32}^2}$	$\frac{1}{r_{32}^2}$	$\frac{r_{12}^2 + r_{32}^2}{r_{31}^2 r_{32}^2}$

Terms in $-\lambda^0$:

	μ_1^3	μ_2^3	μ_3^3	$\mu_1^2\mu_2$	$\mu_2^2\mu_1$	$\mu_1^2\mu_3$	$\mu_2^2\mu_3$	$\mu_1\mu_3^2$	$\mu_2\mu_3^2$	$\mu_1\mu_2\mu_3$
$F_{11}F_{22}F_{33}$	0	0	0	$\frac{1}{r_{31}^2}$	$\frac{1}{r_{32}^2}$	$\frac{1}{r_{31}^2}$	$\frac{1}{r_{32}^2}$	$\frac{r_{12}^2}{r_{31}^2 r_{32}^2}$	$\frac{r_{12}^2}{r_{31}^2 r_{32}^2}$	$\frac{(r_{12}^2 + r_{31}^2 + r_{32}^2)}{r_{31}^2 r_{32}^2}$

In Chap. 9, further illustrations of the expansion of secular equation are given.

4-7. Determination of Normal Coordinates

So far emphasis has been placed on the calculation of the normal frequencies of vibration rather than the normal modes or normal coordinates. In some problems, for example in the study of infrared and Raman intensities (Sec. 7-9), it is necessary to know the coefficients of the transformation from some set of known coordinates, such as internal coordinates, S_t, to normal coordinates Q_k. These can be found with any of the schemes described in Sec. 4-3 or 4-4, as will be shown below.

The secular equation always arises as the condition of compatibility for a set of simultaneous, homogeneous, linear algebraic equations (see, for example, Sec. 2-2). The coefficients of the unknowns in these equations form the elements of the secular equation. The unknowns themselves are either the desired amplitudes of the internal coordinates in the normal modes or are related to them.

Consider first the case which leads to the secular equation [Eq. (6), Sec. 4-3],

$$|\mathbf{F} - \mathbf{G}^{-1}\lambda| = 0$$

The kinetic and potential energies which gave rise to this equation were shown in Sec. 4-3 to have the forms

$$2T = \sum_{tt'} (G^{-1})_{tt'}\dot{S}_t\dot{S}_{t'} \tag{1}$$

$$2V = \sum_{tt'} F_{tt'}S_tS_{t'} \tag{2}$$

Application of Newton's equations of motion in the Lagrange form and substitution of a trial periodic solution of the form

$$S_t = A_t \cos (\lambda^{\frac{1}{2}}t + \epsilon) \tag{3}$$

leads, as in Sec. 2-2, to the following set of conditions on the A_t's:

$$\begin{aligned}
&[F_{11} - (G^{-1})_{11}\lambda]A_1 + [F_{12} - (G^{-1})_{12}\lambda]A_2 + \cdots \\
&\qquad\qquad + [F_{1n} - (G^{-1})_{1n}\lambda]A_n = 0 \\
&[F_{21} - (G^{-1})_{21}\lambda]A_1 + [F_{22} - (G^{-1})_{22}\lambda]A_2 + \cdots \\
&\qquad\qquad + [F_{2n} - (G^{-1})_{2n}\lambda]A_n = 0 \\
&\cdots\cdots\cdots\cdots\cdots\cdots\cdots\cdots\cdots\cdots \\
&[F_{n1} - (G^{-1})_{n1}\lambda]A_1 + [F_{n2} - (G^{-1})_{n2}\lambda]A_2 + \cdots \\
&\qquad\qquad + [F_{nn} - (G^{-1})_{nn}\lambda]A_n = 0
\end{aligned} \tag{4}$$

If a value of λ has been obtained, these equations may be solved for the amplitudes A_t, or rather for their ratios, since the equations are homogeneous. As shown in Sec. 2-4, these amplitudes are also, except for a constant undetermined factor K_k, the coefficients connecting the internal coordinates S_t and the normal coordinates Q_k. For each root λ_k of the secular equation, there will be a set of solutions for the amplitudes A_t. Therefore, a second subscript k can be added to A_t to distinguish these solutions. Consequently, one can write

$$S_t = \sum_{k=1}^{n} \mathfrak{N}_k A_{tk} Q_k = \sum_{k=1}^{n} L_{tk} Q_k \tag{5}$$

The normalizing factor $\mathfrak{N}_k$ can be found by using the fact that this expression for S_t must lead to a potential energy, in terms of the normal coordinates Q_k, of the form

$$2V = \sum_k \lambda_k Q_k^2 \tag{6}$$

so that from the substitution of (5) in (2),

$$\mathfrak{N}_k^2 \sum_{tt'} F_{tt'} A_{tk} A_{t'k} = \lambda_k \tag{7}$$

The solution of the simultaneous equations (4) and the normalization of the resulting amplitudes A_{tk} by means of (7) therefore yield both the normal modes of motion and the relation between the internal coordinates S_t and the normal coordinates Q_k. It must be emphasized that this relation is seldom orthogonal so that if the inverse transformation is desired, giving Q_k in terms of the S_t's, it is necessary to solve[1] the equations (5) as n equations in the n unknowns Q_k.

Equations in **GF** *or* **FG** *form.* It is more usual not to evaluate the elements $(G^{-1})_{tt'}$, but to use the secular equation in some other form, for example, the form given in Eq. (8), Sec. 4-3:

$$|\mathbf{GF} - \mathbf{E}\lambda| = 0 \tag{8}$$

The homogeneous equations (4) for the amplitudes A_t, or for the normalized amplitudes L_t, can be transformed, by multiplication of the t'th equation by $G_{tt'}$, addition of all the resulting equations, and use of the property that [see Eq. (3), Sec. 4-3]

$$\sum_{t'} G_{tt'}(G^{-1})_{t't''} = \delta_{tt''} \tag{9}$$

into the form

$$\begin{aligned}
(\Sigma G_{1t'}F_{t'1} - \lambda)L_1 + (\Sigma G_{1t'}F_{t'2})L_2 + \cdots + (\Sigma G_{1t'}F_{t'n})L_n &= 0 \\
(\Sigma G_{2t'}F_{t'1})L_1 + (\Sigma G_{2t'}F_{t'2} - \lambda)L_2 + \cdots + (\Sigma G_{2t'}F_{t'n})L_n &= 0 \\
\cdots\cdots\cdots\cdots\cdots\cdots\cdots\cdots\cdots\cdots\cdots\cdots \\
(\Sigma G_{nt'}F_{t'1})L_1 + (\Sigma G_{nt'}F_{t'2})L_2 + \cdots + (\Sigma G_{nt'}F_{t'n} - \lambda)L_n &= 0
\end{aligned} \tag{10}$$

These equations, for a given value of λ which satisfies (8), say λ_k, can be compactly written in symbolic or matrix notation as

$$(\mathbf{GF} - \mathbf{E}\lambda_k)\mathbf{L}_k = \mathbf{0} \tag{11}$$

where $\mathbf{L}_k$ is the kth column of the matrix $||L_{tk}|| = \mathbf{L}$ and $\mathbf{0}$ is a zero matrix of n rows and one column. Solution of these equations for the normalized

[1] This is equivalent to finding the reciprocal of the matrix formed by the coefficients in (5).

L_{tk} can then be carried out as before. The results should be the same, but by using (11) instead of (5), the quantities $G_{tt'}$, instead of $(G^{-1})_{tt'}$, are employed.

If the inverse transformation from S_t to Q_k is desired, it is possible to get it directly by solving the equations

$$(\mathbf{FG} - \mathbf{E}\lambda_k)(\mathbf{L}^{-1})_k^\dagger = \mathbf{0} \tag{12}$$

or, written out,

$$\begin{aligned}
(\Sigma F_{1t'}G_{t'1} - \lambda_k)(L^{-1})_{k1} + (\Sigma F_{1t'}G_{t'2})(L^{-1})_{k2} + \cdots \\
+ (\Sigma F_{1t'}G_{t'n})(L^{-1})_{kn} = 0 \\
(\Sigma F_{2t'}G_{t'1})(L^{-1})_{k1} + (\Sigma F_{2t'}G_{t'2} - \lambda_k)(L^{-1})_{k2} + \cdots \\
+ (\Sigma F_{2t'}G_{t'n})(L^{-1})_{kn} = 0 \\
\cdots\cdots\cdots\cdots\cdots\cdots\cdots\cdots\cdots\cdots \\
(\Sigma F_{nt'}G_{t'1})(L^{-1})_{k1} + (\Sigma F_{nt'}G_{t'2})(L^{-1})_{k2} + \cdots \\
+ (\Sigma F_{nt'}G_{t'n} - \lambda_k)(L^{-1})_{kn} = 0
\end{aligned} \tag{13}$$

The proof of this result is most easily given in matrix language and is found in Appendix VIII. The resulting transformation coefficients can be normalized by the procedure shown in Appendix VIII. It should be noted that, because of the symmetry of **F** and **G**, individually the coefficients of the unknowns $(L^{-1})_{kt}$ in (13) are the same numbers as appear in (10), if one simply interchanges rows for columns in (10).

The **F** *and* **G** *Form.* The form of the secular equation described in Sec. 4-4 also arises from a set of equations [Eqs. (12) and (13), Sec. 4-4] whose solution gives the normal modes. The coefficients A_t are the same quantities defined above in (4). It will be noted that there are $2n$ equations [Eqs. (12) and (13), Sec. 4-4] instead of n and that there are $2n$ amplitudes, the A_t's and the A_t''s. As stated in the footnote to Sec. 4-4, these quantities are connected. The amplitudes A_t' are proportional to the coefficients in the inverse transformation from internal coordinates to normal coordinates (see Appendix VIII). If the resistance-type machine (Sec. 9-10), is used to solve the secular equation, the amplitudes A_t (and also A_t') will be obtained as by-products.

4-8. Approximate Separation of High and Low Frequencies.[1]

In many molecules there will be some fundamental frequencies of vibration much higher than the rest. Hydrocarbons provide an illustration since these will have modes of vibration in which the hydrogen atoms vibrate along the direction of the CH bonds without much other motion in the molecule. Because of the small mass of the hydrogen atom these modes will have frequencies considerably higher than the other modes.

[1] B. L. Crawford, Jr., and J. T. Edsall, *J. Chem. Phys.*, **7**: 223 (1939).
E. B. Wilson, Jr., *J. Chem. Phys.*, **7**: 1047 (1939); **9**: 76 (1941).

Under these circumstances, it is possible to reduce the degree of the secular equation by the number of such frequencies and have left an equation of lower degree which gives good approximations for the lower frequencies.

One way of carrying out this approximation is to set the force constants mainly responsible for the high frequencies equal to infinity. (In the example of the hydrocarbons these would be the force constants for the stretching of the CH bonds.) This is equivalent to preventing any change in the corresponding coordinate. Therefore one could merely drop these coordinates out of the expressions for the kinetic and potential energies, Eqs. (1) and (4), Sec. 4-3. If the secular equation in the form of Eq. (6), Sec. 4-3,

$$|\mathbf{F} - \mathbf{G}^{-1}\lambda| = 0 \tag{1}$$

is used, one simply drops out the appropriate rows and columns and solves the equation which is left for the low frequencies. It is similarly easy to make this approximation when the secular equation is written in the expanded form of Sec. 4-5.

When the forms of the secular equation which use **G** instead of $\mathbf{G}^{-1}$ are desired (Sec. 4-3), further considerations enter. What is needed is the reciprocal of that part of $\mathbf{G}^{-1}$ which remains after dropping out the appropriate rows and columns. Appendix IX shows that one should drop the appropriate rows and columns from **F**, and use, instead of the elements $G_{tt'}$, the smaller set $G^0_{tt'}$ where

$$G^0_{tt'} = G_{tt'} - \sum_{ss'} G_{ts}X_{ss'}G_{s't'} \tag{2}$$

where the subscript s covers the coordinates S_s to be held rigid while the elements $X_{ss'}$ satisfy the equation[1]

$$\sum_{s'} X_{ss'}G_{s's''} = \delta_{ss''} \tag{3}$$

If there is only one coordinate S_s to be dropped out, $X_{ss} = 1/G_{ss}$ and the sum in (2) reduces to a single term.

This approximation is an excellent one when carbon-hydrogen stretching vibrations (which occur at about 3,000 cm^{-1}) are the ones eliminated and all the other frequencies are below 2,000 cm^{-1}. Since this treatment is equivalent to applying a constraint to the vibrations, it can be shown that the frequencies calculated in this way will always be higher than the exact solutions.[2] In some cases, this separation of the high frequencies

[1] In other words **X** is the matrix reciprocal to the matrix $||G_{s's''}||$, containing only the rows and columns for the coordinates to be held fixed.

[2] E. T. Whittaker, "Analytical Dynamics," Cambridge, New York, 1937.

can be used to break the problem down to individual frequencies as a good first approximation.

Similarly, the higher frequencies can be approximately calculated by setting certain force constants equal to zero, namely, those which govern the low frequencies. If the form

$$|\mathbf{FG} - \mathbf{E}\lambda| = 0 \tag{4}$$

of the secular equation is employed, the effect of setting certain force constants equal to zero can be achieved by merely leaving out the rows and columns of **F** and **G** corresponding to the coordinates whose diagonal force constants vanish. This follows from the expanded form of the secular equation discussed in Sec. 4-5. If, say, two rows and columns of **F** vanish, the constant term in λ will vanish in the expanded equation, by the rule of Eq. (3), Sec. 4-5. Therefore two zero roots can be factored out and the remaining terms will be just those which would have been obtained if the given rows and columns had never been used in **F** and **G**.

CHAPTER 5

SYMMETRY CONSIDERATIONS

Many common molecules, for example, water, ammonia, methane, etc., possess some symmetry. In calculating the normal modes and frequencies of vibration, symmetry considerations can reduce enormously the labor of the calculations. In addition, without any other information whatsoever, the symmetry and geometry of a molecular model can be used to determine the number of fundamental frequencies, their degeneracies, the selection rules for the infrared and Raman spectra, the degrees of the factors of the secular equation, the number of independent constants in the quadratic part of the potential energy function, the splitting of overtone levels, the possibility of perturbations due to resonance, the nature of the rotational structure of the infrared bands, the polarization properties of the Raman lines, and other useful information. It is, therefore, of great importance to master the simple technique which enables this information to be obtained most easily. The theorems on which this method depends are not as easy as the method itself, so that many of them have been put in Appendixes X and XI. The reader who desires to study the subject more thoroughly should consult some of the standard treatments of group theory.[1]

[1] A few of these are:

E. Wigner, "Gruppentheorie," Vieweg, Brunswick, 1931. (Reprinted Edwards, Ann Arbor, Mich., 1944.)

S. Bhagavantam and T. Venkatarayudu, "Theory of Groups and Its Application to Physical Problems," 2d ed., Andhra University, Waltair, India, 1951.

A. Speiser, "Die Theorie der Gruppen von endlicher Ordnung," Springer, Berlin, 1927. (Reprinted Dover, New York, 1945.)

F. D. Murnaghan, "The Theory of Group Representations," Johns Hopkins Press, Baltimore, 1938.

I. Schur, "Die algebraischen Grundlagen der Darstellungstheorie der Gruppen," Frey and Kratz, Zurich, 1936.

D. E. Littlewood, "The Theory of Group Characters and Matrix Representations of Groups," Oxford, New York and London, 1940.

H. Weyl, "The Classical Groups; Their Invariants and Representations," University Press, Princeton, 1939.

Discussions of the application of group theory to the treatment of molecular vibrations may be found in the following:

J. E. Rosenthal and G. M. Murphy, *Revs. Mod. Phys.*, **8**: 317 (1936).

E. Wigner, *Nachr. Ges. Wiss. Göttingen*, p. 133 (1930).

5-1. The Symmetry of Molecules

It is first necessary to have a definite method of classifying and describing the symmetry of a given molecule. The method which is used will be introduced by the discussion of an example—ammonia. This molecule has the shape of a regular triangular pyramid, the nitrogen atom occupying the apex and the three equivalent hydrogen atoms being located at the corners of the equilateral triangle which forms the base. If the molecule is rotated[1] by 120° about the axis passing through the apex and the center of the base, it assumes a position indistinguishable from its original position, unless some kind of distinguishing label is attached to the hydrogen atoms. This illustrates the concept of *equivalent configurations;* two or more configurations of a molecule are said to be *equivalent* if they are indistinguishable when atoms of the same kind are considered to be indistinguishable. It is evident that the rotation described above carries the NH_3 molecule into a configuration equivalent to its initial configuration, since an observer who had seen the molecule only before and after the rotation could not tell whether or not the molecule had been moved.

A second rotation of 120° brings the molecule to another position equivalent to the other two, so that there are three equivalent configurations related by rotations about the axis through the apex and the center of the base. Such an axis is one illustration of a *symmetry element.* As will be shown, the symmetry of a molecule is best described by listing the symmetry elements it possesses.

The particular symmetry element described above is called a *threefold proper axis of symmetry*, or merely a *threefold axis.* Molecules may also possess twofold axes (example H_2O), fourfold axes, etc. In general, a molecule with an n-fold axis possesses n equivalent positions obtainable one from the other by rotations about this axis. Crystals can possess only 2-, 3-, 4-, and 6-fold axes, but in molecules the value of n is not limited to these values; 5-fold, 7-fold, and higher axes are presumably rare.

The operation of rotating a molecule about an axis from one equivalent position to another is an illustration of a *symmetry operation*, which is to be carefully distinguished from a *symmetry element*, the latter in this case being the axis of rotation.

Another possible symmetry element is the *plane of symmetry*, the corresponding symmetry operation being reflection of the molecule through the plane. The H_2O molecule possesses two planes of symmetry (Fig. 5-1), one the plane of the molecule itself, the other the plane per-

[1] Three-dimensional models are very helpful in following many of the arguments of this chapter.

pendicular to the plane of the molecule and passing through the oxygen and the mid-point of the line joining the hydrogens.

A third possible symmetry element is the *center of symmetry*. If there exists a point within the molecule such that for every atom a straight line can be drawn through the point connecting this atom with an equivalent atom at an equal distance from the point, then this point is called a center of symmetry. Carbon dioxide (OCO, linear), benzene (plane hexagon), ethylene, and SF_6 (octahedral) each possesses a center of symmetry on the basis of the present view of its structure. Methane, although it has a carbon atom at the center of gravity of the molecule, does not possess a center of symmetry. Molecules which do possess a center of symmetry may have an atom at the center (SF_6) or they may not (benzene).

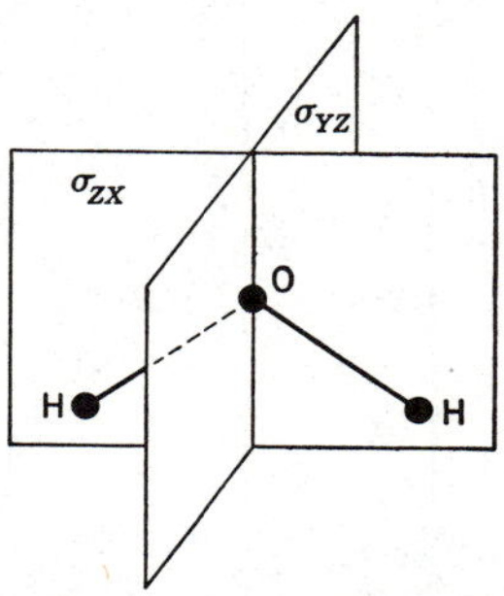

FIG. 5-1. The planes of symmetry possessed by the water molecule.

There remains one more type of symmetry element, called an *alternating axis of symmetry*,[1] best illustrated by an example. If a methane molecule (Fig. 5-2) is oriented so that two hydrogen atoms are in one horizontal plane and the other two hydrogen atoms are in another, lower, horizontal plane, then there exists a vertical fourfold alternating axis of symmetry through the carbon atom. If the molecule is rotated one quarter of a revolution about the axis and then reflected through a plane passing through the carbon atom and perpendicular to the axis, it will attain a position equivalent to the original position. In general, an n-fold alternating axis is an axis with which are associated n equivalent positions of the molecule so related that the molecule can be shifted from one position to the next by a rotation of $2\pi/n$ about the axis, followed by a reflection through a plane perpendicular to the axis. An alternating axis of even order implies the presence of an ordinary rotation axis of half that order, and one of odd order implies the presence of an ordinary rotation axis of the same order. A molecule with an alternating axis does not necessarily possess a plane of symmetry perpendicular to the alternating axis, an illustration being methane (Fig. 5-2). Another molecule which possesses a 4-fold alternating axis is allene ($CH_2{=}C{=}CH_2$), if the model based on the tetrahedral carbon atom is accepted. Benzene possesses a 6-fold alternating axis, which illustrates the fact

[1] This is also called a rotary-reflection axis, or an improper axis. In some applications, notably E. Wigner, *Ges. Wiss. Göttingen*, p. 133 (1930), a different definition is used in which the molecule is first rotated and then inverted through a center, rather than reflected in a plane. The two definitions are closely related.

that a molecule possessing a plane of symmetry perpendicular to an n-fold alternating axis also possesses an ordinary n-fold axis.

These symmetry elements are given symbols, C_n in referring to an n-fold rotation axis, σ to a plane of symmetry, i to a center of symmetry, and S_n to an n-fold alternating axis. Closely related symbols are used for the symmetry operations as distinct from the symmetry elements. Thus C_n^k means a rotation by $2\pi k/n$ about an n-fold rotation axis. σ is the symbol for a reflection through a plane of symmetry as well as for the plane itself. Likewise i is used for both the operation of inversion through a center of symmetry and for the center itself. S_n^k refers to the operation in which k successive rotations of $2\pi/n$ about an alternating axis are carried out, each rotation being followed by a reflection in the plane perpendicular to the axis. If k is even, S_n^k is equivalent to C_n^k. Another symbol, E, is used for the trivial operation of leaving a molecule in its original position. It is called the *identity* operation.

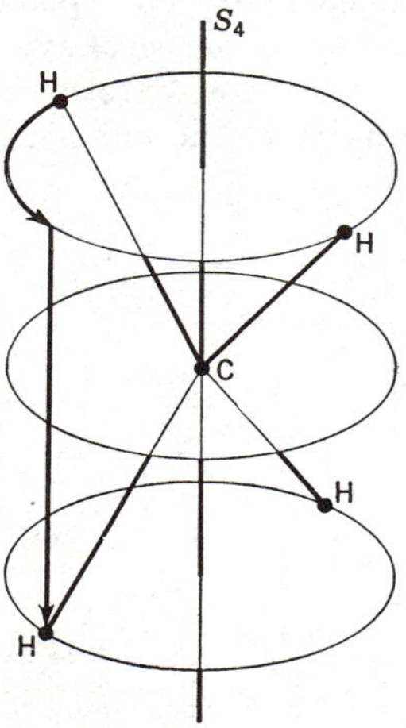

FIG. 5-2. Alternating axis of symmetry: illustrated by an S_4 axis of methane. The arrows show the path of one of the hydrogen atoms for the operation $S_4{}^1$, described as rotating about the vertical axis, S_4, by $2\pi/4$, then reflecting through the horizontal plane containing the carbon atom.

Methane is a good molecule to use as an illustration of the concepts of this section. It possesses the following elements of symmetry: four threefold axes of rotation along the CH bonds, three twofold rotation axes bisecting the pairs of CH bonds, six planes of symmetry and three four-fold alternating axes. These elements generate twenty-four different symmetry operations: E, C_3^{I}, $(C_3^{\mathrm{I}})^2$, C_3^{II}, $(C_3^{\mathrm{II}})^2$, C_3^{III}, $(C_3^{\mathrm{III}})^2$, C_3^{IV}, $(C_3^{\mathrm{IV}})^2$, C_2, C_2', C_2'', σ^{I}, σ^{II}, σ^{III}, σ^{IV}, σ^{V}, σ^{VI}, S_4, S_4^3, S_4', $(S_4')^3$, S_4'', $(S_4'')^3$. Primes and Roman superscripts distinguish operations generated by different elements of the same kind. Operations such as S_4^2 which equal other operations included in the list (in this case C_2) are omitted. Other molecules which may be used as illustrations by the reader are SF_6, C_2H_4, C_6H_6, C_3H_6 (cyclopropane).

5-2. Groups of Symmetry Operations

When a molecule is subjected to two symmetry operations in succession, it reaches a position which could have been obtained by the use of some one symmetry operation. For example, if a carbonate ion (plane model) is reflected through the plane of symmetry σ_v'' (the subscript v indicating that the plane is vertical, *i.e.*, parallel to C_n) and then rotated counterclockwise a third of a revolution (C_3^1) it reaches a position obtainable in one step by reflection in the plane of symmetry σ_v' (see Fig. 5-3). This statement can be written symbolically by the relation $C_3^1\sigma_v'' = \sigma_v'$. The operation which is equivalent to the successive application of two or more symmetry operations is called the *product* of those operations. The analogy between algebraic products and products of symmetry oper-

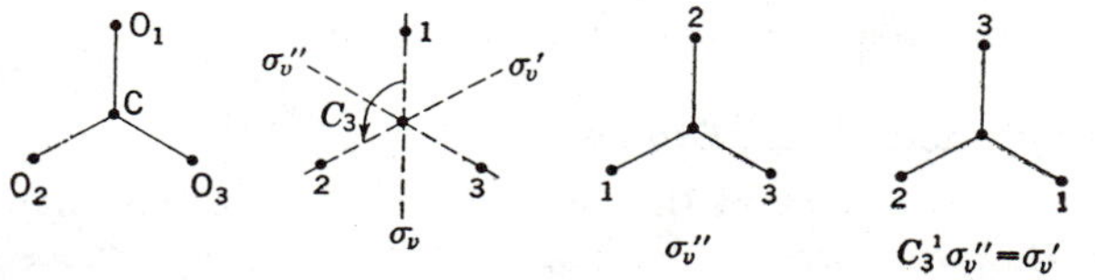

FIG. 5-3. Illustration of a product of symmetry operations, $C_3{}^1\sigma_v''$, for the carbonate ion.

ations is complete except for one point—the order in which symmetry operations are applied is usually important. Thus, in the example just given $C_3^1\sigma_v'' = \sigma_v'$, but $\sigma_v''C_3^1 = \sigma_v$ (see Fig. 5-4). The convention will be adopted that the *last operation to be applied is written to the left* in the expression for a product. Another convention employed in this chapter is that in labeling *symmetry operations* and *symmetry elements*, they will be *regarded as fixed in space*. Thus, in Fig. 5-3, σ_v'' interchanges the atoms 1 and 2 in the first application but if applied to the molecule after C_3^1 it would exchange 3 and 1, as in Fig. 5-4.

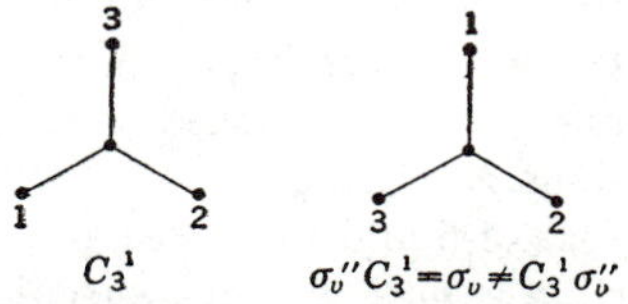

FIG. 5-4. Example showing that the products of symmetry operations may depend upon the order of application of the operations, $\sigma_v''C_3{}^1 \neq C_3{}^1\sigma_v''$, for the carbonate ion (see Fig. 5-3).

It is evident that for every symmetry operation possessed by a molecule there is some operation, either the same one or a different one, which undoes the work of the first. That is, if an operation R changes a molecule from the position I to the position II, then there is an operation which shifts the molecule from II to I. Such an operation is called the *inverse* of the original operation and is written as R^{-1}. Symbolically, $R^{-1}R = E$, since the symbol E represents the identity operation, which leaves the molecule unaltered. Since $RR^{-1} = E$, also, $RR^{-1} = R^{-1}R$. In the example of the carbonate ion, the inverse of C_3^1 is C_3^2, while the inverse of σ_v is σ_v itself.

The set of symmetry operations possessed by a molecule forms an example of what is known to mathematicians as a *group*. A group of operations is defined as a set of operations with the following properties: (*a*) The product of two or more operations is defined in such a way as to obey the associative law of multiplication. (*b*) The product of two or more operations of the set is equivalent to some member of the set. (*c*) The set possesses an identity operation. (*d*) Every operation possesses an inverse which is a member of the set. It is evident that the set of all the symmetry operations of a molecule, including the identity operation E, satisfies the above requirements. The operations of a group are not restricted, however, to symmetry operations. Thus, in Chap. 6 groups will occur which consist of permutations.

5-3. The Symmetry Point Groups

One of the advantages of the method described in this chapter is that it enables the symmetry of the hundreds of thousands of possible molecules to be classified in a small number of groups, according to the number and nature of the symmetry elements. These groups are called the symmetry point groups because they are composed of symmetry operations which form a group in the mathematical sense and because some point in the molecule is not altered by any of the operations.[1]

The simplest point group[2] is that to which molecules with no symmetry whatever belong. It consists of but one operation, the identity E. The symbol for this trivial group is $\mathcal{C}_1$. It will be noted that the symbols for the point groups are closely similar to those for the symmetry operations. To prevent confusion, distinctive script type will be used for the group symbols. A more important group is one made up of the n operations associated with an n-fold rotational axis C_n; that is, the operations $E, C_n^1, C_n^2, \ldots, C_n^{n-1}$. Such a group is given the symbol $\mathcal{C}_n$ (see Figs. 5-5 and 5-6). As previously remarked, there is no limit to the possible values of n for molecules, in contradistinction to the case of crystals, where only $n = 1, 2, 3, 4,$ and 6 are possible. However, there are very few molecules with $n = 5, 7,$ or higher.

Another group is one differing from $\mathcal{C}_n$ only in the presence of n twofold axes at right angles to the principal symmetry axis C_n. Such a group is called the group $\mathcal{D}_n$. $\mathcal{D}_2$, which has three equivalent twofold axes at right angles to each other, is usually called $\mathcal{V}$. If in addition there is added a plane σ_h at right angles to C_n, a new group $\mathcal{D}_{nh}$ is produced

[1] In studying crystals, another kind of symmetry group is often used, called a space group. This includes operations which translate the whole crystal in one direction. Point groups are also important in crystal studies, the thirty-two crystal classes being the thirty-two possible point groups for crystals.

[2] For longer discussions of the material in this section, see A. M. Schoenflies, "Theorie der Kristallstruktur," Borntraeger, Berlin, 1923; P. Niggli, "Geometrische Kristallographie des Diskontinuums," Borntraeger, Berlin, 1919; the article by P. P. Ewald in the "Handbuch der Physik," Vol. 24, p. 191; F. Seitz, *Z. Krist.*, (*A*) **88:** 433 (1934); J. E. Rosenthal and G. M. Murphy, *Revs. Mod. Phys.*, **8:** 317 (1936).

(the subscript h on σ indicates that the plane is horizontal, *i.e.*, perpendicular to C_n). However, σ_h cannot be introduced without bringing the alternating axis S_n with it, since $\sigma_h C_n^k = S_n^k$ ($\sigma_h C_n^k = C_n^k \sigma_h$ in this case) for k odd; n vertical planes of symmetry σ_v through C_n also appear. The special case when $n = 2$, $\mathfrak{D}_{2h}$, is called $\mathfrak{V}_h$. The symbols $\mathfrak{D}$ and $\mathfrak{V}$ come from the German *Diedergruppe* and *Vierergruppe*, respectively.

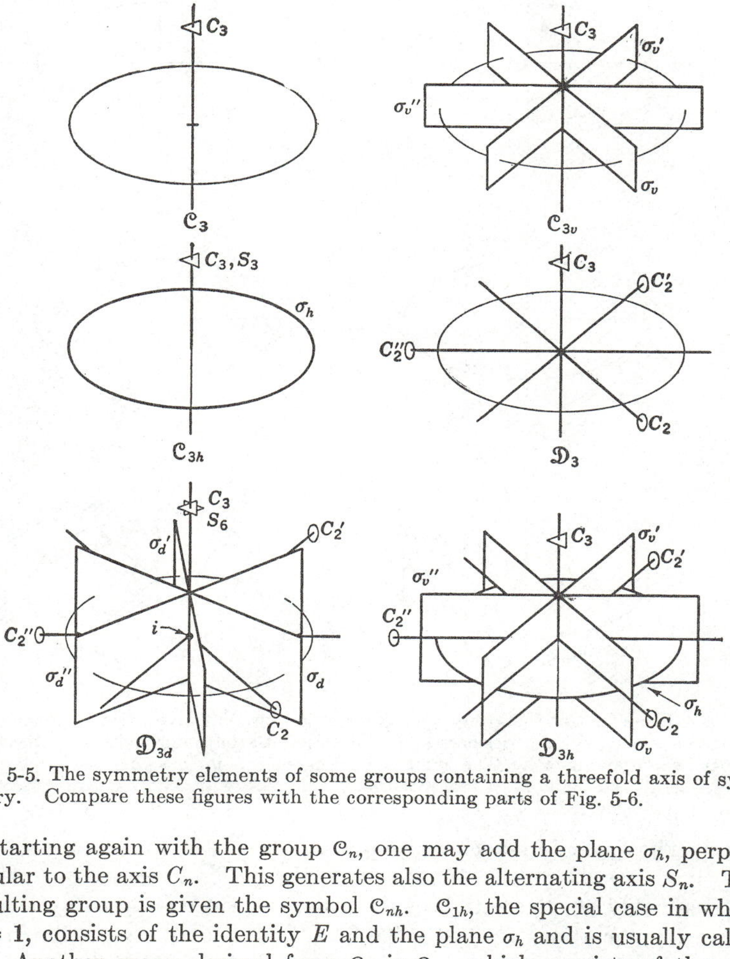

FIG. 5-5. The symmetry elements of some groups containing a threefold axis of symmetry. Compare these figures with the corresponding parts of Fig. 5-6.

Starting again with the group $\mathcal{C}_n$, one may add the plane σ_h, perpendicular to the axis C_n. This generates also the alternating axis S_n. The resulting group is given the symbol $\mathcal{C}_{nh}$. $\mathcal{C}_{1h}$, the special case in which $n = 1$, consists of the identity E and the plane σ_h and is usually called $\mathcal{C}_s$. Another group derived from $\mathcal{C}_n$ is $\mathcal{C}_{nv}$, which consists of the axis C_n plus n vertical planes of symmetry through the axis. The ammonia molecule belongs to the group $\mathcal{C}_{3v}$.

$\mathcal{S}_n$ denotes the group with the elements E, S_n^1, S_n^2, . . . , S_n^{n-1}. Only when n is even are these groups new, since $\mathcal{S}_n = \mathcal{C}_{nh}$ if n is odd. When $n = 2$, $\mathcal{S}_2 = \mathcal{C}_i$, a group with the two operations E and $S_2^1 = i$.

Another series of groups arises when to $\mathcal{D}_n$ is added a set of vertical planes σ_d through the axis C_n bisecting the angles between the twofold

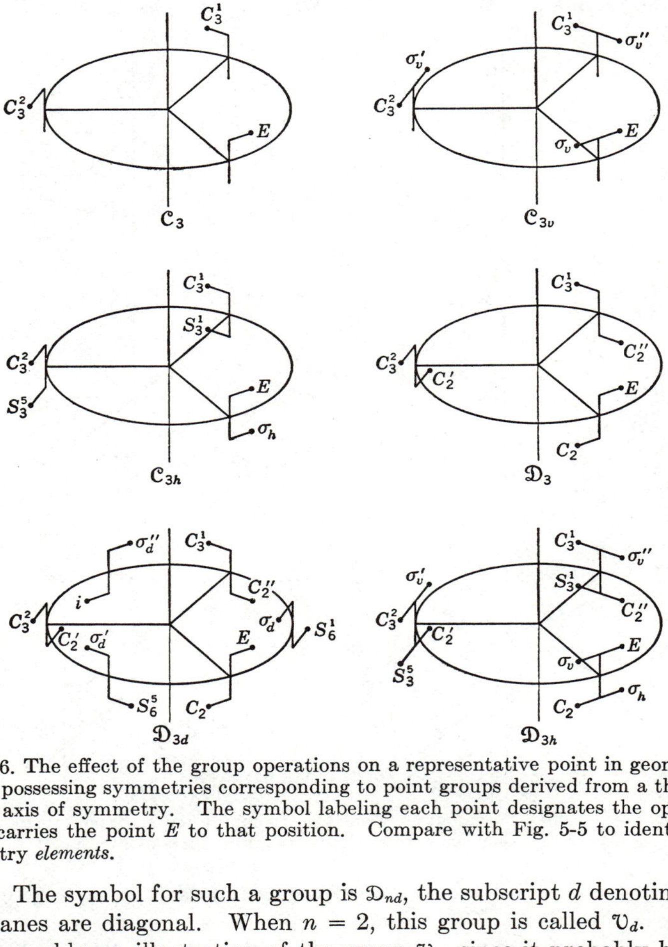

FIG. 5-6. The effect of the group operations on a representative point in geometrical figures possessing symmetries corresponding to point groups derived from a threefold proper axis of symmetry. The symbol labeling each point designates the operation which carries the point E to that position. Compare with Fig. 5-5 to identify the symmetry *elements*.

axes. The symbol for such a group is $\mathcal{D}_{nd}$, the subscript d denoting that the planes are diagonal. When $n = 2$, this group is called $\mathcal{V}_d$. Allene is presumably an illustration of the group $\mathcal{V}_d$, since it probably has the symmetry operations E, S_4, C_2, S_4^3, C_2', C_2'', σ_d, and σ_d'. $\mathcal{D}_{3d}$ is illustrated by the model of ethane in which the two methyl groups are turned about the carbon-carbon bond so that they make an angle of 60° with one another; *i.e.*, the "staggered" configuration.

The remaining seven possible point groups do not fit into a series. They are $\mathcal{T}$, $\mathcal{T}_d$, $\mathcal{T}_h$, $\mathcal{O}$, $\mathcal{O}_h$, $\mathcal{I}$, and $\mathcal{I}_h$. The group $\mathcal{T}$ contains all the rotation axes which are possessed by a regular tetrahedron, while $\mathcal{T}_d$ contains the planes, etc., of a regular tetrahedron as well. The operations of $\mathcal{T}_d$ have already been listed in connection with methane. In a similar manner $\mathcal{O}$ contains the rotations and $\mathcal{O}_h$ all the elements possessed by a cube. The other groups $\mathcal{T}_h$, $\mathcal{I}$, and $\mathcal{I}_h$ are not very important but are described in Appendix X. Detailed information on all the point groups is also contained in Appendix X.

To summarize this section: Every molecule may be assigned to one of the point groups $\mathcal{C}_n$, $\mathcal{D}_n$, $\mathcal{D}_{nh}$, $\mathcal{C}_{nh}$, $\mathcal{C}_{nv}$, $\mathcal{S}_n$, $\mathcal{D}_{nd}$, $\mathcal{T}$, $\mathcal{T}_d$, $\mathcal{T}_h$, $\mathcal{O}$, $\mathcal{O}_h$, $\mathcal{I}$ or $\mathcal{I}_h$. These point groups consist of certain definite symmetry operations (having a certain definite arrangement in space) and no other symmetry point groups are possible using the operations described.

5-4. Symmetrically Equivalent Atoms and Subgroups

Once the over-all symmetry of a given molecule is determined, it is easy to classify the atoms constituting the molecule into symmetrically equivalent sets. The atoms equivalent to any arbitrarily selected atom are all the atoms into which the given atom can be sent by operations of the symmetry group. All atoms of a symmetrically equivalent set are of course of the same chemical species, but it does not follow that all atoms of a given species fall into a single set. In ethane, the two carbon atoms constitute one set, the six hydrogen atoms a second set; on the other hand, in propane, assuming a bonded chain of the form H—C—C—C—H to be in a plane ($\mathcal{C}_{2v}$ symmetry) there are five sets of symmetrically equivalent atoms composed of one carbon (the central one), two carbons (the end chain atoms), two hydrogens (lying in the same plane as the carbons), four hydrogens (bonded to the end chain carbon atoms), and two hydrogens (bonded to the central carbon).

It is clear that the number of atoms in a symmetrically equivalent set cannot exceed the order (*i.e.*, the number of operations) of the group of over-all symmetry; the number may, however, be less than the order of the group, since it is possible that more than one symmetry operation may send a given initial atom into the same final atom. A trivial example of such a set is the single carbon atom of methane which is sent into itself by all twenty-four operations of the group $\mathcal{T}_d$. For any symmetrically equivalent set, the collection of operations from the over-all symmetry group $\mathcal{G}$ which send a given atom into itself constitute a subgroup of $\mathcal{G}$ which will be designated by the symbol $\mathcal{H}$ in the following discussion; this is easily seen in terms of the four group postulates, which are satisfied by any collection of operations having in common the property that they leave a given atom invariant. Suppose now that $\mathcal{H}$ does not contain every operation of $\mathcal{G}$; then there necessarily exists another atom equivalent to the arbitrarily chosen one. Let R_2 be some element of $\mathcal{G}$ which sends atom 1 into atom 2. Then the set of group operations, $R_2\mathcal{H}$, will also send atom 1 into atom 2, since each operation of $\mathcal{H}$ leaves atom 1 fixed, and the subsequent transformation by R_2 sends 1 into 2. Every member of the set $R_2\mathcal{H}$ is distinct, *i.e.*, there are as many operations in $R_2\mathcal{H}$ as in $\mathcal{H} = R_1\mathcal{H}$, for if one supposes the contrary, then

$$R_2R_1' = R_2R_1'' \tag{1}$$

with

$$R_1' \neq R_1'' \tag{2}$$

(R_1' and R_1'' distinct members of $\mathcal{H}$) leads immediately to a contradiction, since multiplication of (1) on the left by R_2^{-1} gives $R_1' = R_1''$. Moreover, no group operation outside the set $R_2\mathcal{H}$ can send atom 1 into atom 2. Suppose, on the contrary, that R_2' is such an operation. Then the set $(R_2')^{-1}(R_2\mathcal{H})$ sends atom 1 into itself and must, by definition, be equivalent to $\mathcal{H}$:

$$(R_2')^{-1}R_2\mathcal{H} = \mathcal{H} \tag{3}$$

so that, multiplying on the left by R_2',

$$R_2\mathcal{H} = R_2'\mathcal{H} \tag{4}$$

which establishes a contradiction.

This sort of argument may be extended to include the operations which send atom 1 into each of its other equivalent atoms, *i.e.*, until the operations of $\mathcal{G}$ are exhausted. One may then formally express the operations of $\mathcal{G}$ by the sum

$$\mathcal{G} = R_1\mathcal{H} + R_2\mathcal{H} + \cdots + R_n\mathcal{H} \tag{5}$$

Since each term in the sum represents a collection of h group operations,[1] it is clear that

$$nh = g \tag{6}$$

i.e., that the number of atoms in a symmetrically equivalent set is always a divisor of the order of the group, since g and h are integers.

Several examples will now be given to illustrate the subgroup associated with each symmetrically equivalent set of atoms in a molecule. For ethane (the staggered configuration), $\mathcal{G} = \mathfrak{D}_{3d}$; the subgroup associated with the carbon atoms is $\mathcal{H} = \mathfrak{C}_{3v}$ since the following operations of

[1] Small italic letters corresponding to the group symbols will be used to designate the order of the group.

$\mathfrak{D}_{3d}$ send one of the carbons into itself: E, C_3^1, C_3^2, σ_v, $\sigma_{v'}$, $\sigma_{v''}$. The values $n = 2$, $h = 6$, and $g = 12$ clearly satisfy (6). For the hydrogen atoms of ethane on the other hand, the appropriate $\mathfrak{H}$ is $\mathfrak{C}_s$, since a given member of the set is sent into itself by E and σ_v. For methane, $\mathfrak{G} = \mathfrak{T}_d$ ($g = 24$); the $\mathfrak{H}$ associated with the single carbon atom is $\mathfrak{T}_d$, whereas the $\mathfrak{H}$ corresponding to the symmetrically equivalent set of four hydrogen atoms is $\mathfrak{C}_{3v}$.

In group theoretical language, the collections of operations $R_i\mathfrak{H}$ of (5), for $i > 1$, are called the *cosets* of $\mathfrak{G}$ determined by the subgroup $\mathfrak{H} = R_1\mathfrak{H}$. In summary of this section, it is to be noted that the atoms of any molecule may be classified in symmetrically equivalent sets, each of which is associated with some subgroup $\mathfrak{H}$ (which may in extreme cases be either $\mathfrak{G}$ itself or consist merely of the identity element) which is determined by those symmetry operations which send an arbitrarily chosen atom of the given set into itself.[1]

Figure 5-6, Sec. 5-3, provides further illustrations of these concepts. Taking the group $\mathfrak{D}_{3h}$ as an example, the largest possible set of equivalent atoms, for which $\mathfrak{H}$ consists of the identity only, is seen to be one distributed exactly like the labels giving the names of the twelve symmetry operations. On the other hand, another possible set, consisting of three atoms lying on the twofold axes, is seen to have a subgroup comprising E, C_2, σ_v, σ_h with cosets C_3^1, C_2'', σ_v'', S_3^1 and C_3^2, C_2', σ_v', S_3^5.

5-5. Symmetry of the Potential and Kinetic Energies

In classifying a molecule according to its symmetry, the equilibrium configuration is used, whereas in the study of vibration it is the distorted molecule which is of most interest. A given distortion of a molecule may be described by giving the vectors which represent the displacements of the atoms from their individual equilibrium positions. The components of the displacement vector for the atom α are the displacement coordinates Δx_α, Δy_α, and Δz_α previously introduced (Sec. 2-1). If a molecule which is in a given distorted configuration is acted upon by a symmetry operation belonging to the undistorted molecule, a new configuration will result which may be different from the old one but which will always be equivalent to the old one in the sense that the same interatomic distances and angles will occur in both configurations.

In dealing with such symmetry operations, it is most convenient to regard the operation as one which interchanges and transforms the dis-

[1] There are some subgroups which never occur in this connection for any molecule, for example, $\mathfrak{C}_3$ as a subgroup of $\mathfrak{C}_{3v}$, since an atom left invariant by the threefold rotations is necessarily also left invariant by the planes of reflection, so that $\mathfrak{H}$ becomes equal to $\mathfrak{G}$.

placement vectors of equivalent atoms without permuting the atoms themselves. That is to say, the atom 1 may exchange displacements with atom 2, but the atoms themselves do not exchange positions. This distinction is illustrated in Fig. 5-7, in which the central molecule represents a distorted molecule of carbon dioxide. The molecule on the right shows the result of a symmetry operation (reflection in a plane perpendicular to the axis), while the molecule on the left shows the result of permuting the oxygen atoms 1 and 2. The effect of permutations is important in connection with the rotational intensities,[1] but not in the vibrational problem.

From the assumption that the potential energy V of the molecular model is a function only of the distances between the atoms, it follows that it must be unchanged by the symmetry operations of the equilibrium configuration. This means that a molecule in a given distorted

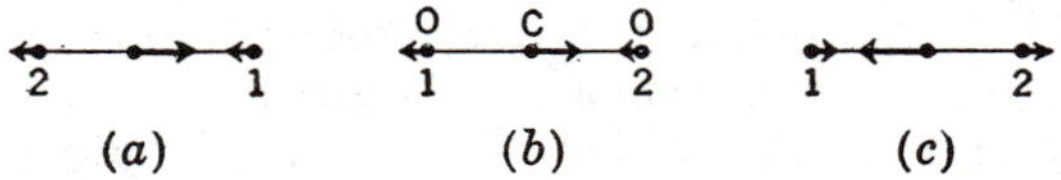

FIG. 5-7. The effect upon a distorted carbon dioxide molecule (*b*) of reflecting the displacements in a plane perpendicular to the axis (*c*) and of permuting the oxygen atoms (*a*).

configuration will possess the same numerical value of the potential energy in each of the equivalent configurations which result from the application of the symmetry operations of the molecule, acting in the sense just described.

The kinetic energy of a molecule possesses similar properties, except that it is a function of the velocities instead of the positions of the particles. At any instant the kinetic energy is determined by the values of $d(\Delta x_\alpha)/dt$, $d(\Delta y_\alpha)/dt$, $d(\Delta z_\alpha)/dt$. These quantities may be thought of as components of velocity vectors, one for each atom. The effect of a symmetry operation is then to transform these velocity vectors in just the same way as the displacement vectors are transformed. Therefore, the kinetic energy will have the same numerical value in any two states of motion which are symmetrically related.

5-6. Representations

The effect of a symmetry operation on a distorted molecule may be represented analytically by a linear transformation connecting the new values $\Delta x'_\alpha$, $\Delta y'_\alpha$, $\Delta z'_\alpha$ of the displacement coordinates with the old values Δx_α, Δy_α, and Δz_α. For example, the operation σ'_v of the carbonate ion, shown in Fig. 5-8, is represented by the linear transformation

[1] E. B. Wilson, Jr., *J. Chem. Phys.*, **3**: 276 (1935).

$$
\begin{aligned}
\Delta x_1 &\to \Delta x_1' = \tfrac{1}{2}\Delta x_3 + \tfrac{1}{2}\,3^{\frac{1}{2}}\Delta y_3 \\
\Delta y_1 &\to \Delta y_1' = \tfrac{1}{2}\,3^{\frac{1}{2}}\Delta x_3 - \tfrac{1}{2}\Delta y_3 \\
\Delta x_2 &\to \Delta x_2' = \tfrac{1}{2}\Delta x_2 + \tfrac{1}{2}\,3^{\frac{1}{2}}\Delta y_2 \\
\Delta y_2 &\to \Delta y_2' = \tfrac{1}{2}\,3^{\frac{1}{2}}\Delta x_2 - \tfrac{1}{2}\Delta y_2 \\
\Delta x_3 &\to \Delta x_3' = \tfrac{1}{2}\Delta x_1 + \tfrac{1}{2}\,3^{\frac{1}{2}}\Delta y_1 \\
\Delta y_3 &\to \Delta y_3' = \tfrac{1}{2}\,3^{\frac{1}{2}}\Delta x_1 - \tfrac{1}{2}\Delta y_1 \\
\Delta x_4 &\to \Delta x_4' = \tfrac{1}{2}\Delta x_4 + \tfrac{1}{2}\,3^{\frac{1}{2}}\Delta y_4 \\
\Delta y_4 &\to \Delta y_4' = \tfrac{1}{2}\,3^{\frac{1}{2}}\Delta x_4 - \tfrac{1}{2}\Delta y_4 \\
\Delta z_1 &\to \Delta z_1' = \Delta z_3 \\
\Delta z_2 &\to \Delta z_2' = \Delta z_2 \\
\Delta z_3 &\to \Delta z_3' = \Delta z_1 \\
\Delta z_4 &\to \Delta z_4' = \Delta z_4
\end{aligned}
\qquad (1)
$$

These equations enable the numerical values of the coordinates for the final configuration to be obtained from the numerical values (Δx_α, etc.)

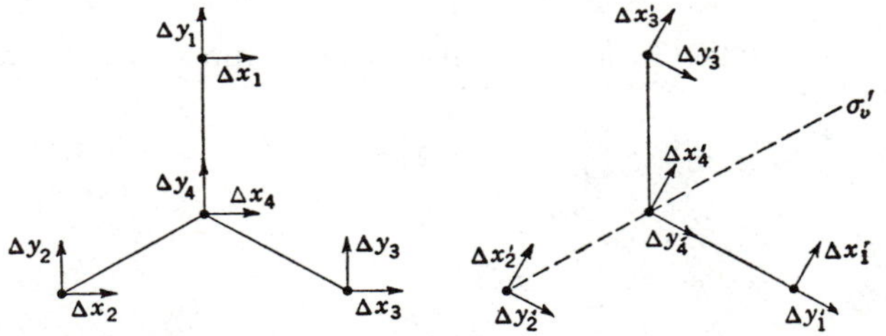

FIG. 5-8. The effect of the symmetry operation σ_v' on a set of cartesian displacement coordinates of the carbonate ion.

for the initial configuration. The reader will find it helpful to write out the transformation which corresponds to the operation C_3^1.

The potential energy of the initial configuration is $V(\Delta x_1, \ldots, \Delta z_N)$, whereas that of the final state is $V(\Delta x_1', \ldots, \Delta z_N')$. If the expressions for $\Delta x_\alpha'$, etc., given in (1) are substituted in $V(\Delta x_1', \ldots, \Delta z_N')$, the resulting function of Δx_α, etc., will be found to be identical with $V(\Delta x_1, \ldots, \Delta z_N)$, the initial potential energy. This is the analytic expression of the idea that the potential energies of the two configurations are identical. For illustrative purposes in testing the above statements, the reader may use the following approximate form of the potential energy for the carbonate ion:

$$
\begin{aligned}
V = \tfrac{1}{2}f_1[(\Delta y_1 - \Delta y_4)^2 &+ \tfrac{3}{4}(\Delta x_2 - \Delta x_4)^2 + \tfrac{1}{4}(\Delta y_2 - \Delta y_4)^2 \\
&+ \tfrac{3}{4}(\Delta x_3 - \Delta x_4)^2 + \tfrac{1}{4}(\Delta y_3 - \Delta y_4)^2] + \tfrac{1}{2}f_2[(\Delta x_3 - \Delta x_2)^2 \\
&+ \tfrac{1}{4}(\Delta x_3 - \Delta x_1)^2 + \tfrac{3}{4}(\Delta y_3 - \Delta y_1)^2 + \tfrac{1}{4}(\Delta x_2 - \Delta x_1)^2 \\
&\qquad + \tfrac{3}{4}(\Delta y_2 - \Delta y_1)^2]
\end{aligned}
$$

in which f_1 and f_2 are force constants. A function such as V which is unchanged in form by a given transformation is said to be *invariant* under that transformation. The kinetic energy T is also invariant under the transformation of velocities having the same coefficients as (1). As previously stated, T and V are assumed to be invariant under all the transformations representing the symmetry operations of the equilibrium configuration of the molecule; *i.e.*, the operations of the point group of the molecule.

The result of applying two linear transformations such as (1) in succession may be described by a third linear transformation, which is called the *product* of the first two. Thus, let

$$x'_j = \sum_i a_{ji} x_i \qquad j = 1, 2, 3, \ldots \tag{2}$$

be the first transformation and

$$x''_k = \sum_j b_{kj} x'_j \qquad k = 1, 2, 3, \ldots \tag{3}$$

be the second. Then by substituting the first expression into the second it is seen that

$$x''_k = \sum_i c_{ki} x_i \qquad k = 1, 2, 3, \ldots \tag{4}$$

with

$$c_{ki} = \sum_j b_{kj} a_{ji} \tag{5}$$

The transformation with the coefficients c_{ki} is the product transformation.[1] If the first two transformations represent symmetry operations of the molecule, the product transformation must also represent a symmetry operation of the molecule. Therefore, the linear transformations which represent the symmetry operations of a point group have the same multiplication properties as the corresponding operations themselves, and multiplication is evidently associative. The product of two or more members of this set of transformations is therefore also a member of the set. Furthermore, there exists an identity transformation

$$\Delta x'_\alpha = \Delta x_\alpha \qquad \Delta y'_\alpha = \Delta y_\alpha \qquad \Delta z'_\alpha = \Delta z_\alpha$$

correlated with the operation E. Finally, each transformation possesses an inverse transformation (see Sec. 5-2). Consequently, since this set of linear transformations possesses all the required properties, the transformations, as well as the corresponding symmetry operations, may be said to constitute a *group*.

The group formed by the physical symmetry operations themselves and the group formed by the linear transformations correlated with these

[1] Note that $\mathbf{c} = \mathbf{ba}$, using the rules of matrix multiplication.

symmetry operations are evidently closely related. Each of the members (known as *elements*) of one group is correlated to one of the members of the other, and the multiplication properties of the two groups are the same. Two groups which are related in this manner are said to be *isomorphic* and in particular the group of linear transformations is said to be a *representation* of the other group. The coordinates $\Delta x_1, \Delta y_1, \ldots, \Delta z_N$ in terms of which the transformations are written are said to form the *basis* of the representation.

The set of coordinates in terms of which the linear transformation of (1) is described was arbitrarily chosen, and it is evident that similar results would be obtained if, for example, the orientation of the x, y, z axes were selected differently. The coefficients in (1) would then be different, but all the general remarks which followed would be equally true for the new transformations. Since it is frequently helpful to use several coordinate systems for a single molecule, it is important to study the effect of a change of coordinates. Suppose that the new coordinate system is obtained from the old one by a linear transformation

$$\eta_i = \sum_{j=1}^{3N} a_{ij}\xi_j \qquad i = 1, 2, \ldots, 3N \tag{6}$$

This change of coordinates is more general than a mere rotation of axes, which is a special case. The inverse of (6) may be written as

$$\xi_j = \sum_{i=1}^{3N} (a^{-1})_{ji}\eta_i \tag{7}$$

while the effect of any symmetry operation R upon the coordinates ξ_j is described by the transformation

$$\xi'_{j'} = \sum_{j=1}^{3N} R_{j'j}\xi_j \tag{8}$$

The three equations (6), (7), and (8) may now be combined to give the new transformation which represents the effect of R upon the new coordinates η_i. The result is

$$\begin{aligned}\eta'_{i'} &= \sum_{j'} a_{i'j'}\xi'_{j'} = \sum_{j'j} a_{i'j'}R_{j'j}\xi_j \\ &= \sum_i \Big(\sum_{j'j} a_{i'j'}R_{j'j}(a^{-1})_{ji}\Big)\eta_i\end{aligned} \tag{9}$$

When two representations of a group differ only in that the basis coordinates of one are linear combinations, as in (6), of the basis coordinates of the other, the two representations are said to be *equivalent*. Equivalent representations may be recognized by the fact that corresponding transformations in the two representations have the same *characters*, the character of a transformation being defined as follows. If the transformation [for example, the one in (8)] is written out in full,

$$\begin{aligned}
\xi_1' &= R_{11}\xi_1 + R_{12}\xi_2 + \cdots + R_{1,3N}\xi_{3N} \\
\xi_2' &= R_{21}\xi_1 + R_{22}\xi_2 + \cdots + R_{2,3N}\xi_{3N} \\
&\cdots\cdots\cdots\cdots\cdots\cdots \\
\xi_{3N}' &= R_{3N,1}\xi_1 + R_{3N,2}\xi_2 + \cdots + R_{3N,3N}\xi_{3N}
\end{aligned} \tag{10}$$

the character of the transformation is by definition the quantity

$$\chi_R = \sum_{i=1}^{3N} R_{ii} = R_{11} + R_{22} + \cdots + R_{3N,3N} \tag{11}$$

The subscript R on χ denotes the transformation to which χ belongs. As an illustration, the character of the transformation representing the effect of σ_v' on the displacement coordinates of the carbonate ion is 2, which is obtained by inspection of (1). The only equations in that set which do not have $R_{ii} = 0$ are the 3d, 4th, 7th, 8th, 10th, and 12th, for which the diagonal coefficients are $\frac{1}{2}$, $-\frac{1}{2}$, $\frac{1}{2}$, $-\frac{1}{2}$, 1, 1, which add up to 2, as stated above.

To prove that corresponding transformations of equivalent representations have the same characters, use is made of (9), which shows that the character of the new transformation of the η's is

$$\begin{aligned}
\chi_R(\eta) &= \sum_{i'} \Big(\sum_{j',j} a_{i'j'} R_{j'j} (a^{-1})_{ji'} \Big) = \sum_{j',j} R_{j'j} \Big(\sum_{i'} (a^{-1})_{ji'} a_{i'j'} \Big) \\
&= \sum_{j',j} R_{j'j} \delta_{jj'} = \sum_j R_{jj} = \chi_R(\xi)
\end{aligned} \tag{12}$$

in which the equation $\sum_i (a^{-1})_{ji} a_{ij'} = \delta_{jj'}$, Eq. (4), Sec. 2-4, has been employed. Use has also been made of the fact that the quantities $a_{i'j'}$, etc., being ordinary numbers, can be written in any order.

A suggested exercise for the reader is to obtain the transformations and their characters for some of the symmetry operations of H_2O. Then, by orienting the x, y, z axes differently, setting up the new transformations, and determining their characters, one may test the above theorem.

5-7. Symmetry of Normal Coordinates

The whole method of applying group theory to vibrational problems depends upon the fact that the normal coordinates and normal modes of vibration of symmetrical molecules have certain special symmetry properties. For example, Fig. 5-9 shows the normal modes of the carbonate ion, point group $\mathfrak{D}_{3h}$, determined by the methods of Chap. 2. The first mode has as much symmetry as the undistorted molecule itself; that is, the figure is unchanged by all the rotations, reflections, etc., which make up the point group $\mathfrak{D}_{3h}$ $(E,C_3^1,C_3^2,\sigma_v,\sigma_v',\sigma_v'',\sigma_h,S_3^1,S_3^5,C_2,C_2',C_2'')$. The second mode is unaltered by the threefold rotations C_3^1 and C_3^2 and by reflection in the vertical planes σ_v, σ_v', and σ_v''. However, each of the other operations, for example the horizontal plane σ_h, reverses each of the arrows, thus changing the sign of the corresponding displacements. In general, it will be shown that *nondegenerate normal modes of vibration are always either symmetrical (unaltered) or antisymmetrical (changed in sign) with respect to a given symmetry operation of the point group of the undistorted molecule.*

The behavior of degenerate vibrations is illustrated by the modes 3 and 4 of Fig. 5-9, since these have the same frequency. Both are symmetrical with respect to σ_h but C_3^1 changes 3 into a new mode which is found to be a combination of the original modes 3 and 4. Likewise 4 is changed into a combination of 3 and 4. This is the general result: *A symmetry operation of the molecule will transform a member of a degenerate set of vibrations into a linear combination of the members of the degenerate set.*

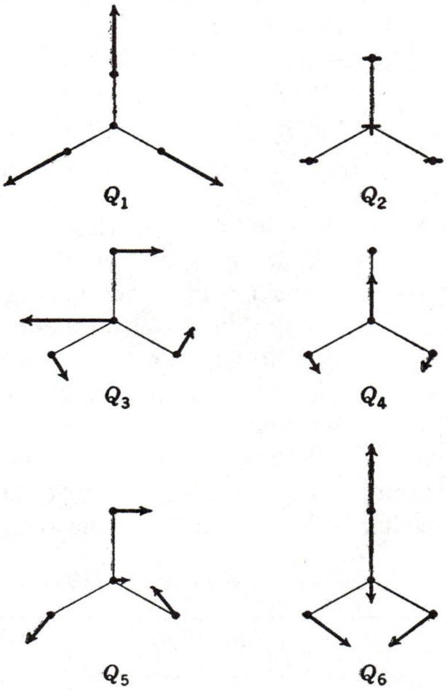

FIG. 5-9. Normal modes of $CO_3^{=}$.

The above conclusions can be analytically expressed if normal coordinates are used instead of normal modes. The components along various coordinate directions of the arrows in the diagrams of Fig. 5-9 are measures of the coefficients l_{ik} in the equation

$$Q_k = \sum_i l_{ik} q_i \tag{1}$$

which defines the normal coordinate Q_k in terms of the displacement coordinates q_i (see Sec. 2-4). Therefore, what has been said about the symmetry properties of the normal modes applies equally well to the Q's. Thus

$$Q_1 \xrightarrow{R} +Q_1 \tag{2}$$

under the influence of every symmetry operation R of the molecule, whereas

$$Q_2 \xrightarrow{R'} +Q_2 \tag{3}$$

for $R' = E, C_3^1, C_3^2, \sigma_v, \sigma_v', \sigma_v''$, but

$$Q_2 \xrightarrow{R''} -Q_2 \tag{4}$$

for $R'' = \sigma_h, S_3^1, S_3^5, C_2, C_2', C_2''$. The degenerate coordinates Q_3 and Q_4 transform as follows under the influence of C_3^2

$$\begin{aligned} Q_3 &\xrightarrow{C_3^2} -\tfrac{1}{2}Q_3 - \tfrac{1}{2}3^{\frac{1}{2}}Q_4 \\ Q_4 &\xrightarrow{C_3^2} \tfrac{1}{2}3^{\frac{1}{2}}Q_3 - \tfrac{1}{2}Q_4 \end{aligned} \tag{5}$$

It will be evident later that the symmetry properties of the normal coordinates are of great practical importance; it is therefore useful to have a compact way of tabulating them. For the *nondegenerate* coordinates this is simple as it is only necessary to make a table showing whether a given symmetry operation changes Q into $+1$ times itself or into -1 times itself. The rows of Table 5-1 labeled Q_1 and Q_2 show the results for the nondegenerate vibrations of the carbonate ion. It is not as easy to tabulate the complete transformations involving the *degenerate* normal coordinates, but it is not necessary to do so. All that is ordinarily needed is the character of the transformation.

TABLE 5-1. SYMMETRY PROPERTIES OF THE NORMAL COORDINATES OF $CO_3^=$

	E	C_3^1	C_3^2	σ_v	σ_v'	σ_v''	σ_h	S_3^1	S_3^5	C_2	C_2'	C_2''
Q_1	1	1	1	1	1	1	1	1	1	1	1	1
Q_2	1	1	1	1	1	1	−1	−1	−1	−1	−1	−1
Q_3, Q_4	2	−1	−1	0	0	0	2	−1	−1	0	0	0
Q_5, Q_6	2	−1	−1	0	0	0	2	−1	−1	0	0	0

Evidently, the character of the transformation (3) is 1, that of (4) is -1, and that of (5) is -1. Table 5-1 lists the characters for the transformations of the normal coordinates of $CO_3^=$.

If the transformations which represent the effect of the symmetry operations upon the whole set of normal coordinates of the molecule are examined, it is found that they form a *representation* of the symmetry group in the same sense that the transformations of the displacement coordinates ξ_i form a representation. This statement is true whether the normal coordinates of translation and rotation are included or excluded. An illustration is provided by the transformations, corresponding to the two operations C_3^1 and σ_v (vertical plane), of the $CO_3^=$ vibrational normal coordinates:

$$
\begin{array}{llllll}
Q_1 \xrightarrow{C_3^1} Q_1 & & & & & \\
Q_2 \xrightarrow{C_3^1} & +Q_2 & & & & \\
Q_3 \xrightarrow{C_3^1} & & -\frac{1}{2}Q_3 & -\frac{1}{2}3^{\frac{1}{2}}Q_4 & & \\
Q_4 \xrightarrow{C_3^1} & & +\frac{1}{2}3^{\frac{1}{2}}Q_3 & -\frac{1}{2}Q_4 & & \\
Q_5 \xrightarrow{C_3^1} & & & & -\frac{1}{2}Q_5 & -\frac{1}{2}3^{\frac{1}{2}}Q_6 \\
Q_6 \xrightarrow{C_3^1} & & & & +\frac{1}{2}3^{\frac{1}{2}}Q_5 & -\frac{1}{2}Q_6
\end{array}
\qquad (6)
$$

$$
\begin{array}{lllllll}
Q_1 \xrightarrow{\sigma_v} +Q_1 & & & & & & \\
Q_2 \xrightarrow{\sigma_v} & +Q_2 & & & & & \\
Q_3 \xrightarrow{\sigma_v} & & -Q_3 & & & & \\
Q_4 \xrightarrow{\sigma_v} & & & +Q_4 & & & \\
Q_5 \xrightarrow{\sigma_v} & & & & -Q_5 & & \\
Q_6 \xrightarrow{\sigma_v} & & & & & +Q_6 &
\end{array}
\qquad (7)
$$

An important difference exists between the representation of the symmetry group formed by transformations of the displacement coordinates ξ_i and that formed by the normal coordinates Q_k. If the general form of one of the transformations involving the ξ's is written out in full, it is

$$
\begin{aligned}
\xi_1' &= R_{11}\xi_1 + R_{12}\xi_2 + \cdots + R_{1,3N}\xi_{3N} \\
\xi_2' &= R_{21}\xi_1 + R_{22}\xi_2 + \cdots + R_{2,3N}\xi_{3N} \\
&\cdots\cdots\cdots\cdots\cdots\cdots\cdots\cdots \\
\xi_{3N}' &= R_{3N,1}\xi_1 + R_{3N,2}\xi_2 + \cdots + R_{3N,3N}\xi_{3N}
\end{aligned}
\qquad (8)
$$

By changing to a new set of variables η_i in the manner described in the last section, it is possible, if the η's are chosen properly, to simplify the form of the transformation which represents R. In fact, it can be shown that the new coordinates can be chosen in such a manner that the transformation representing any single symmetry operation R takes on the extremely simple form, called the *diagonal* form, given below:

$$
\begin{aligned}
\eta_1' &= R_{11}\eta_1 \\
\eta_2' &= \qquad R_{22}\eta_2 \\
\eta_3' &= \qquad\qquad R_{33}\eta_3 \\
&\qquad \text{etc.}
\end{aligned}
\qquad (9)
$$

In other words, every coordinate η_i is transformed into some multiple of itself; different coordinates η_i and η_j do not mix.

But it is not always possible to find a single set of coordinates η_i which will *simultaneously* reduce *every* transformation of the group (that is, the transformations representing all of the symmetry operations) to the simple form of (9). *Each* transformation can be reduced to diagonal form by *some* set of coordinates, but it may not be possible to reduce *all* by the *same* change of coordinates. However, it is usually possible to find a single change of coordinates which will greatly simplify the transformations; furthermore, there is a certain maximum simplicity which may be obtained in many ways but never exceeded. By this reduction the transformations of the group are simplified in that the variables η_1, η_2, η_3, etc., can be separated into sets which do not mix with one another in *any* of the transformations. When the coordinate system has been found such that it is impossible to break the coordinates down into any smaller nonmixing sets, the representation for which these coordinates form the basis is said to be *completely reduced.*

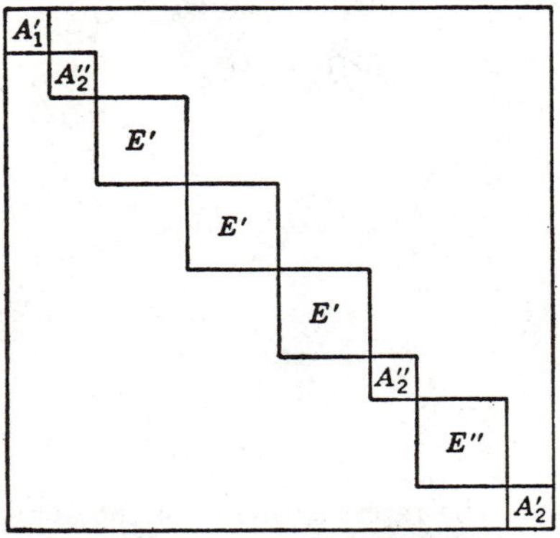

FIG. 5-10. Form of the completely reduced representation for the plane equilateral triangular model of the carbonate ion, CO_3^-, an example of the group $\mathfrak{D}_{3h}$. The symbols indicate the irreducible representation corresponding to each nonmixing block.

It will be shown (Sec. 6-2) that when the transformations representing the symmetry operations of a molecule are written in terms of the normal coordinates [see (6) and (7)] the representation will be completely reduced.

When the representation of $\mathfrak{D}_{3h}$ formed by the transformations of the displacement coordinates of $CO_3^=$ is completely reduced in one of the many possible ways, the transformation representing σ_v' has the form below [compare Eq. (1), Sec. 5-6]

$$\begin{aligned}
Q_1' &= Q_1 \\
Q_2' &= Q_2 \\
Q_3' &= \tfrac{1}{2}Q_3 - \tfrac{1}{2}3^{\frac{1}{2}}Q_4 \\
Q_4' &= -\tfrac{1}{2}3^{\frac{1}{2}}Q_3 - \tfrac{1}{2}Q_4 \\
Q_5' &= \tfrac{1}{2}Q_5 - \tfrac{1}{2}3^{\frac{1}{2}}Q_6 \\
Q_6' &= -\tfrac{1}{2}3^{\frac{1}{2}}Q_5 - \tfrac{1}{2}Q_6 \\
Q_7' &= \tfrac{1}{2}Q_7 - \tfrac{1}{2}3^{\frac{1}{2}}Q_8 \\
Q_8' &= -\tfrac{1}{2}3^{\frac{1}{2}}Q_7 - \tfrac{1}{2}Q_8 \\
Q_9' &= Q_9 \\
Q_{10}' &= \tfrac{1}{2}Q_{10} - \tfrac{1}{2}3^{\frac{1}{2}}Q_{11} \\
Q_{11}' &= -\tfrac{1}{2}3^{\frac{1}{2}}Q_{10} - \tfrac{1}{2}Q_{11} \\
Q_{12}' &= -Q_{12}
\end{aligned} \tag{10}$$

The structure of this transformation (and of all other transformations of the reduced representation) is shown in Fig. 5-10. There are twelve coordinates because the translational and rotational coordinates are still included ($Q_7 - Q_{12}$).

5-8. Irreducible Representations

One further and very remarkable theorem needs to be introduced before the method of applying group theory to molecular vibrations can be stated. Since the coordinates which form the basis for a completely reduced representation separate into sets which do not mix with one another, the equations involving the members of any one set can be considered by themselves as making up transformations which form a representation of the group. As an illustration, consider the coordinates Q_3 and Q_4 of the carbonate ion, Sec. 5-7. As seen in Eqs. (6) and (7), Sec. 5-7, they form a nonmixing set in so far as the operations C_3^1 and σ_v are concerned, and it will be found that this is still true when all the symmetry operations of the group $\mathfrak{D}_{3h}$ are considered. A few of these other transformations are given below.

$$\begin{array}{lll} Q_3 \xrightarrow{\sigma_h} Q_3 & Q_3 \xrightarrow{C_2} -Q_3 & Q_3 \xrightarrow{C_2''} \tfrac{1}{2}Q_3 + \tfrac{1}{2}3^{\frac{1}{2}}Q_4 \\ Q_4 \xrightarrow{\sigma_h} Q_4 & Q_4 \xrightarrow{C_2} Q_4 & Q_4 \xrightarrow{C_2''} \tfrac{1}{2}3^{\frac{1}{2}}Q_3 - \tfrac{1}{2}Q_4 \end{array} \qquad (1)$$

No change of coordinates can reduce such a representation any further; it is therefore called an *irreducible representation.*[1] A completely reduced representation is evidently made up of a number of irreducible representations, each of its noncombining sets of coordinates forming the basis for one irreducible representation.

In order to give the concept of an irreducible representation a more concrete reality for the reader, the transformations of normal coordinates have been used throughout as examples, but the concept itself is quite independent of the idea of normal coordinates or the problem of molecular vibrations. It arises whenever a set of linear transformations has the properties of a group, no matter what the meaning of the transformation variables. Later, it will be necessary to regard the transformations of a set of wave functions as forming a group, and the present theory will be applied.

The fundamental theorem concerning irreducible representations states that for *each point group there are only a definite small number of nonequivalent irreducible representations possible.* As previously mentioned, a representation is characterized by the characters of its transformations, so that the important properties, for our purposes, of the irreducible representations of any point group can be presented in tabular form by giving the characters corresponding to each symmetry operation for each

[1] In the molecular application the term *symmetry species* is used.

irreducible representation. These tables have been obtained for all of the symmetry point groups,[1] by methods described in treatments of group theory.[2] Such a table for the group $\mathfrak{D}_{3h}$ is given below.

Appendix X contains tables similar to Table 5-2 for each of the point groups. These tables are derived by methods which have nothing to do with the problem of molecular vibrations; as stated earlier, they are much more general than this application.

The numbers in Table 5-2 lead to the following conclusion: Any representation of the symmetry group $\mathfrak{D}_{3h}$ can be reduced by a change of

TABLE 5-2. CHARACTERS OF THE IRREDUCIBLE REPRESENTATIONS OF $\mathfrak{D}_{3h}$

$\mathfrak{D}_{3h}$	E	C_3^1	C_3^2	C_2	C_2'	C_2''	σ_h	S_3^1	S_3^5	σ_v	σ_v'	σ_v''
$\Gamma^{(1)} = A_1'$	1	1	1	1	1	1	1	1	1	1	1	1
$\Gamma^{(2)} = A_1''$	1	1	1	1	1	1	−1	−1	−1	−1	−1	−1
$\Gamma^{(3)} = A_2'$	1	1	1	−1	−1	−1	1	1	1	−1	−1	−1
$\Gamma^{(4)} = A_2''$	1	1	1	−1	−1	−1	−1	−1	−1	1	1	1
$\Gamma^{(5)} = E'$	2	−1	−1	0	0	0	2	−1	−1	0	0	0
$\Gamma^{(6)} = E''$	2	−1	−1	0	0	0	−2	1	1	0	0	0

coordinates to a completely reduced form in which none of the nonmixing sets of basis functions contains more than two functions, since each of these nonmixing sets forms the basis of one of the irreducible representations in the table, all of which correspond to one- or two-coordinate transformations. (*The number of basis functions for any transformation is evidently equal to the character of the transformation representing E.*) Furthermore, it is easy to find the number of times a given irreducible representation appears in the completely reduced form of a given reducible representation. Suppose that the character for the transformation representing the symmetry operation R is χ_R for the original reducible representation. Since the process of reduction consists in a mere change of coordinates, the completely reduced representation, taken as a whole, is equivalent to the original representation and will therefore have the same characters (see Sec. 5-6). The character χ_R for the completely reduced representation will, however, be the sum of the characters of the various irreducible representations of which it is composed. Equation (10), Sec. 5-7, and Fig. 5-10, Sec. 5-7, illustrate this. If the representation

[1] H. Bethe, *Ann. Physik*, **3**: 133 (1929).
E. Wigner, *Nachr. Ges. Wiss. Göttingen*, p. 133 (1930).
L. Tisza, *Z. Physik*, **82**: 48 (1933).
R. S. Mulliken, *Phys. Rev.*, **43**: 279 (1933).
G. Placzek, "Marx Handbuch der Radiologie," Vol. VI, Part II, p. 209, 1934.
J. E. Rosenthal and G. M. Murphy, *Revs. Mod. Phys.*, **8**: 317 (1936).

[2] See the references given on p. 77.

$\Gamma^{(\gamma)}$ occurs $n^{(\gamma)}$ times in the completely reduced form of the original representation, then

$$\chi_R = n^{(1)}\chi_R^{(1)} + n^{(2)}\chi_R^{(2)} + \cdots = \sum_\gamma n^{(\gamma)}\chi_R^{(\gamma)} \tag{2}$$

for each operation R, $\chi_R^{(\gamma)}$ being the character of the irreducible representation $\Gamma^{(\gamma)}$, for the operation R. Since the quantities χ_R can easily be obtained by inspection of the original reducible representation, while the $\chi_R^{(\gamma)}$'s are given in the tables of which Table 5-2 is an example, it is therefore possible to solve the set of g equations (2) for the unknowns $n^{(1)}$, $n^{(2)}$, etc. In fact, it is possible to show that the solutions have the form

$$n^{(\gamma)} = \frac{1}{g}\sum_R \chi_R^{(\gamma)*}\chi_R \tag{3}$$

in which g is the number of operations in the symmetry group and the sum is over all these operations. The asterisk denotes the complex conjugate.

Most of the applications of this simple formula will be reserved for the next chapter, but one will be given here. The characters χ_R for the transformations of the displacement coordinates of $CO_3^=$ are given in Table 5-3.

TABLE 5-3. CHARACTERS FOR THE REPRESENTATION OF $\mathfrak{D}_{3h}$ FORMED BY TRANSFORMATIONS OF THE DISPLACEMENT COORDINATES OF $CO_3^=$

$R =$	E	C_3^1	C_3^2	C_2	C_2'	C_2''	σ_h	S_3^1	S_3^5	σ_v	σ_v'	σ_v''
$\chi_R =$	12	0	0	−2	−2	−2	4	−2	−2	2	2	2

One of these ($R = \sigma_v$) has been found already, and the others can be determined in the same way, or by far simpler methods given in Chap. 6. By inserting these values of χ_R in (3), together with values of $\chi_R^{(\gamma)}$ taken from Table 5-2, the result is obtained (as may readily be verified) that the reduced representation contains the irreducible representation A_1' once, A_2' once, A_2'' twice, E' three times, and E'' once (see Fig. 5-10, Sec. 5-7). But since it has already been stated that the normal coordinates form the basis of a completely reduced representation, there must be one normal coordinate which transforms like A_1', one like A_2', three *pairs* like E', etc. Referring to Fig. 5-9 and Table 5-1, one sees that Q_1 is of symmetry A_1', Q_2 of symmetry A_2'', and the pairs Q_3, Q_4, and Q_5, Q_6 are of symmetry E'. The other six coordinates represent translation and rotation.

5-9. Classes of Symmetry Operations

There is one more concept which simplifies still further the application of group theory. This is the observation that in each point group, the symmetry operations can be divided up into *classes* with the property that the members of any one class always have the same character χ. Therefore, it is not necessary to find the character for each operation but merely for a sample operation in each class. Furthermore, in the tables of irreducible representations in Appendix X, the characters for each class, rather than for each operation, are given. If g_j is the number of opera-

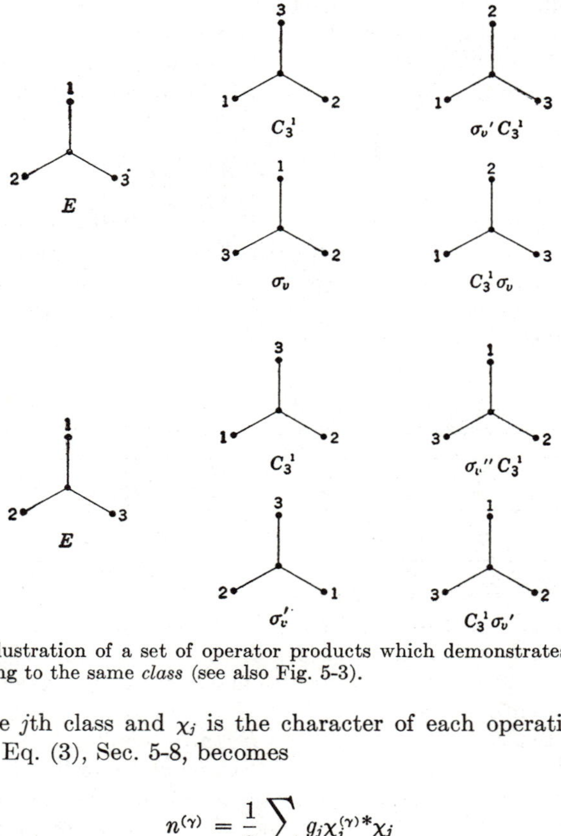

FIG. 5-11. Illustration of a set of operator products which demonstrates that σ_v, σ_v', and σ_v'' belong to the same *class* (see also Fig. 5-3).

tions in the jth class and χ_j is the character of each operation in that class, then Eq. (3), Sec. 5-8, becomes

$$n^{(\gamma)} = \frac{1}{g} \sum_j g_j \chi_j^{(\gamma)*} \chi_j \tag{1}$$

in which the sum is over all the classes of the group. As before, g is the number of operations in the group, χ_j refers to the *reducible* representation, $\chi_j^{(\gamma)}$ to the γth *irreducible* representation, and $n^{(\gamma)}$ is the number of times the irreducible representation $\Gamma^{(\gamma)}$ appears in the *reduced* representation.

The fundamental definition of a class is as follows. If A, B, and X are operations of the group, and if

$$X^{-1}AX = \mathrm{B} \tag{2}$$

or, by multiplying on the left, by X,

$$AX = XB \tag{3}$$

for *some* operation X of the group, then A and B are said to belong to the same class. As an illustration consider the three vertical planes of symmetry σ_v, σ_v', σ_v'' of the group $\mathfrak{D}_{3h}$. It is seen that (Fig. 5-11)

$$\sigma_v' C_3^1 = C_3^1 \sigma_v \tag{4}$$

and

$$\sigma_v'' C_3^1 = C_3^1 \sigma_v' \tag{5}$$

so that σ_v, σ_v', and σ_v'' belong to one class. The plane σ_h, however, does not belong to this class; in fact, it forms a class by itself, since $\sigma_h X = X\sigma_h$ for every operation X of the group.

If in a given representation of the group, the coefficients of the transformations A, B, X, and X^{-1} are A_{ij}, B_{ij}, X_{ij}, and $(X^{-1})_{ij}$, respectively, then the character of B is

$$\chi_B = \sum_i B_{ii} \tag{6}$$

By applying the rule for the product of transformations given in Eq. (5), Sec. 5-6, it is seen that

$$\begin{aligned} \chi_{X^{-1}AX} &= \sum_{i,j,k} (X^{-1})_{ij} A_{jk} X_{ki} \\ &= \sum_{j,k} A_{jk} \Big(\sum_i X_{ki} (X^{-1})_{ij} \Big) \\ &= \sum_j A_{jj} = \chi_A \end{aligned} \tag{7}$$

since by Eq. (4), Sec. 2-4, $\sum_i X_{ki}(X^{-1})_{ij} = \delta_{kj}$. Equation (2) shows that $\chi_{X^{-1}AX} = \chi_B$ so that, finally,

$$\chi_A = \chi_B \tag{8}$$

Therefore, operations in the same class have the same character. The classes for all the point groups have been worked out and are included in the tables of irreducible representations given in Appendix X.

CHAPTER 6

APPLICATIONS OF GROUP THEORY TO THE ANALYSIS OF MOLECULAR VIBRATIONS

In the previous chapter certain concepts and theorems of group theory were introduced in an intuitive and descriptive manner, while the reader was referred to Appendix XI for proofs of the theorems. In the present chapter, these ideas will be applied to the problem of the vibrations of symmetrical molecules. It will be possible in this chapter to be somewhat more rigorous and complete than in the preceding one.

The material of this chapter falls into two main categories. Sections 6-1 through 6-3 treat the vibration problem qualitatively in the sense that they provide the reader with a rapid method of deducing the number and degeneracies of the normal modes of vibration and the size of the factors of the secular determinant. The remaining sections (6-4 through 6-8) are devoted to detailed methods of carrying out the factoring of the secular determinant in order to make quantitative evaluations of the normal frequencies. These latter sections may be omitted by the reader who is interested only in the general inferences concerning molecular structure and symmetry which may be drawn from a qualitative study of the vibration spectrum.

6-1. Determination of the Characters χ_R

In Chap. 5 it was indicated that the number of normal coordinates with given symmetry properties can be obtained when the characters of the transformations of the displacement coordinates are known. These may be obtained directly by writing out the transformations, but much easier methods can be used. By definition

$$\chi_R = \sum_i R_{ii} \tag{1}$$

in which R_{ii} is a diagonal coefficient in the transformation

$$\xi_i' = \sum_{j=1}^{3N} R_{ij}\xi_j \tag{2}$$

which corresponds to the symmetry operation R. $\xi_1, \xi_2, \ldots, \xi_{3N}$ are the displacement coordinates $\Delta x_1, \Delta y_1, \ldots, \Delta z_N$ before application of the symmetry operation, while the primed letters represent the coordinates after the transformation. If a given operation R replaces atom 1 by atom 2 when applied to the molecule in its equilibrium configuration, then, when applied to the distorted molecule, it will replace the displacement of atom 1 by that of atom 2. Consequently, in the transformation (2) the new coordinate ξ_i' for atom 1 will involve the old coordinates ξ_j for atom 2 only, so that $R_{ii} = 0$ for all i's which refer to atom 1 (see Fig. 6-1).

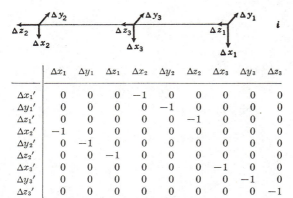

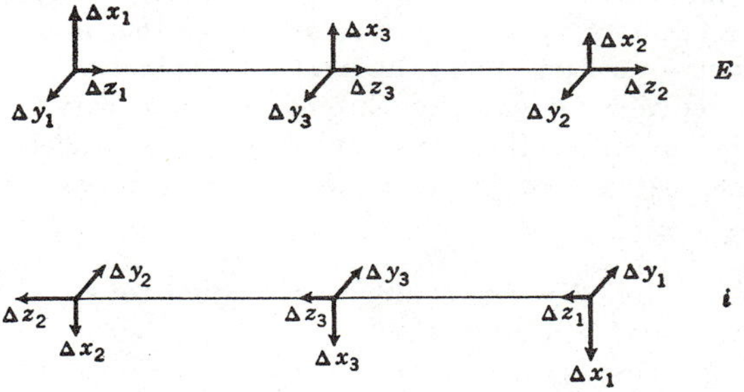

	Δx_1	Δy_1	Δz_1	Δx_2	Δy_2	Δz_2	Δx_3	Δy_3	Δz_3
$\Delta x_1'$	0	0	0	−1	0	0	0	0	0
$\Delta y_1'$	0	0	0	0	−1	0	0	0	0
$\Delta z_1'$	0	0	0	0	0	−1	0	0	0
$\Delta x_2'$	−1	0	0	0	0	0	0	0	0
$\Delta y_2'$	0	−1	0	0	0	0	0	0	0
$\Delta z_2'$	0	0	−1	0	0	0	0	0	0
$\Delta x_3'$	0	0	0	0	0	0	−1	0	0
$\Delta y_3'$	0	0	0	0	0	0	0	−1	0
$\Delta z_3'$	0	0	0	0	0	0	0	0	−1

FIG. 6-1. Transformation of displacements in linear symmetric triatomic molecule by the inversion operation, i. Note that shifted atoms (1 and 2) make no contribution to $\chi_i = -3$.

Therefore in finding the character of a transformation, one need consider only those atoms whose equilibrium positions are not altered; *i.e.*, all atoms for the identity operation, and those atoms which lie on a plane or on an axis of symmetry, etc., as the case may be, for the other group operations. Thus in the example of the carbonate ion (Fig. 5-8, Sec. 5-6) one need consider only Δx_4, Δy_4, Δz_4, the coordinates of the carbon atom, when obtaining the character for C_3^1, since C_3^1 moves the displacement of 1 to 2, 2 to 3, and 3 to 1. Furthermore, for σ_v, the plane through atoms 1 and 4, only the coordinates of 1 and 4 need to be considered.

Another very useful simplification results from the theorem, proved in Sec. 5-6, that the character of a transformation is independent of the choice of coordinate orientation. Consequently, in each case one may use the most convenient set of coordinates for the operation under consideration. Thus if reflection in a plane of symmetry is being treated, let the x and y axes lie in the plane and the z axis be perpendicular to it. Then the effect of the reflection on the coordinates of any atom lying in the plane (only such atoms need be considered) is given by the equations

$$\Delta x'_\alpha = \Delta x_\alpha \qquad \Delta y'_\alpha = \Delta y_\alpha \qquad \Delta z'_\alpha = -\Delta z_\alpha \tag{3}$$

The values of R_{ii} for these three equations are $+1$, $+1$ and -1, respectively, their sum being $+1$. Consequently, the character χ_R, when R is a reflection in a plane of symmetry σ, is $+1$ times the number of atoms whose equilibrium positions lie in the plane of symmetry.

Similarly when R is an inversion through a center of symmetry i, the character χ_R is zero unless there is an atom located at the center of symmetry. If there is, its coordinates evidently transform as follows

$$\Delta x'_\alpha = -\Delta x_\alpha \qquad \Delta y'_\alpha = -\Delta y_\alpha \qquad \Delta z'_\alpha = -\Delta z_\alpha \tag{4}$$

The character χ_R for $R = i$ is consequently equal to -3 if there is an atom at i, and is zero otherwise.

Another simple case is the twofold axis C_2. Let z be along C_2, x and y perpendicular to C_2. Then, for each atom located on C_2,

$$\Delta x'_\alpha = -\Delta x_\alpha \qquad \Delta y'_\alpha = -\Delta y_\alpha \qquad \Delta z'_\alpha = \Delta z_\alpha \tag{5}$$

so that such an atom contributes -1 to χ_R. Atoms not on C_2 contribute zero.

The same choice of axes for the threefold rotation yields the equations

$$\begin{aligned} \Delta x'_\alpha &= -\frac{1}{2}\,\Delta x_\alpha - \frac{3^{\frac{1}{2}}}{2}\,\Delta y_\alpha \\ \Delta y'_\alpha &= +\frac{3^{\frac{1}{2}}}{2}\,\Delta x_\alpha - \frac{1}{2}\,\Delta y_\alpha \\ \Delta z'_\alpha &= \qquad\qquad\qquad\qquad \Delta z_\alpha \end{aligned} \tag{6}$$

The contribution to the character is $-\frac{1}{2} - \frac{1}{2} + 1 = 0$, so that the character for C_3^1 for any molecule is always zero. The same result is obtained for C_3^2. In general for the rotation C_n^k the transformation is (see Fig. 6-2)

$$\begin{aligned} \Delta x'_\alpha &= \cos\frac{2\pi k}{n}\,\Delta x_\alpha - \sin\frac{2\pi k}{n}\,\Delta y_\alpha \\ \Delta y'_\alpha &= \sin\frac{2\pi k}{n}\,\Delta x_\alpha + \cos\frac{2\pi k}{n}\,\Delta y_\alpha \\ \Delta z'_\alpha &= \qquad\qquad\qquad\qquad\qquad \Delta z_\alpha \end{aligned} \tag{7}$$

so that each atom on the axis contributes $1 + 2\cos(2\pi k/n)$ to the character.

The only remaining type of operation is S_n^k, the rotary reflection. Since this operation first rotates the system about an axis, then reflects it through a plane perpendicular to the axis of rotation, χ_R will be zero unless there is an atom located at the intersection of the axis and the plane. For such an atom, the equations of transformation for Δx and Δy are the same as in (7), but $\Delta z' = -\Delta z$. Therefore the contribution to

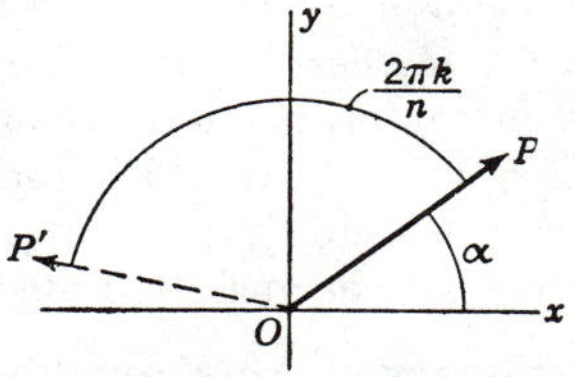

FIG. 6-2. Effect of rotation of a displacement by $2\pi k/n$.

$$\Delta x = OP \cos \alpha$$
$$\Delta y = OP \sin \alpha$$
$$\begin{aligned}\Delta x' &= OP' \cos\left(\alpha + \frac{2\pi k}{n}\right) \\ &= OP \cos \alpha \cos \frac{2\pi k}{n} - OP \sin \alpha \sin \frac{2\pi k}{n} \\ &= \Delta x \cos \frac{2\pi k}{n} - \Delta y \sin \frac{2\pi k}{n} \\ \Delta y' &= OP' \sin\left(\alpha + \frac{2\pi k}{n}\right) \\ &= OP \cos \alpha \sin \frac{2\pi k}{n} + OP \sin \alpha \cos \frac{2\pi k}{n} \\ &= \Delta x \sin \frac{2\pi k}{n} + \Delta y \cos \frac{2\pi k}{n}\end{aligned}$$

the character is $-1 + 2 \cos (2\pi k/n)$. At this point, the reader may observe that the character of E is a special case of that for C_n^k obtained by putting $n = 1$, and that the characters of σ and i are special cases of those for S_n^k obtained by putting $n = k = 1$ and $k = 1$, $n = 2$, respectively. The general types of operations C_n^k and S_n^k are sometimes referred to as "proper" and "improper" operations, respectively.

TABLE 6-1. CONTRIBUTION TO CHARACTER PER UNSHIFTED ATOM

Proper operations		Improper operations	
R	χ_R	R	χ_R
C_n^k	$1 + 2 \cos (2\pi k/n)$	S_n^k	$-1 + 2 \cos (2\pi k/n)$
$E = C_1^k$	3	$\sigma = S_1^1$	1
C_2^1	-1	$i = S_2^1$	-3
C_3^1, C_3^2	0	S_3^1, S_3^5	-2
C_4^1, C_4^3	1	S_4^1, S_4^3	-1
C_6^1, C_6^5	2	S_6^1, S_6^5	0

The character for the identity operation E is of course equal to the number of coordinates. This quantity is known as the *dimension* of the representation.

To summarize this section: *To find the character, χ_R, for the transformation R of the displacement coordinates of a molecule, first count the number of atoms which are not shifted when the symmetry operation R acts on the atoms of the molecule. Then multiply this number by the appropriate factor characteristic of the operation R, these factors being listed in Table* 6-1.

To obtain directly the character, χ_R, in terms of the internal coordinate representation of dimension $3N - 6$, treat improper operations exactly as above but for proper operations multiply the appropriate factor of Table 6-1 by two less than the number of unshifted atoms.[1]

6-2. Symmetry and Degeneracy of the Normal Modes

In Sec. 5-8, it has already been indicated how the number of normal modes of motion of each possible symmetry can be obtained. This section will treat the problem more thoroughly.[2] The fact that the transformations of the displacement coordinates ξ_i of the atoms of the molecule form a reducible representation of the symmetry point group of the molecule has been discussed in Sec. 5-6. It follows directly from this that the $3N$ normal coordinates (including translation and rotation) also form a basis of a representation of the group, since the normal coordinates Q_k are linear combinations of the displacement coordinates ξ_i (see Sec. 2-4).

This normal coordinate representation is at least partially, and usually completely, reduced, in that the transformations do not mix normal coordinates corresponding to different frequencies. Suppose that this is not true so that, for example, the transformation representing the operation R mixes the coordinates $Q_{k'}$ and $Q_{k''}$ corresponding to two different frequencies $\lambda_{k'}$ and $\lambda_{k''}$; that is,

$$Q_{k'} \xrightarrow{R} Q'_{k'} = aQ_{k'} + bQ_{k''} \qquad Q_{k''} \xrightarrow{R} Q'_{k''} = cQ_{k'} + dQ_{k''} \tag{1}$$

in which a, b, c, d are the transformation coefficients.

By definition of the normal coordinates

$$2T = \sum_k \dot{Q}_k^2 \qquad 2V = \sum_k \lambda_k Q_k^2 \tag{2}$$

If a particular state of motion of the molecule is specified by $Q_k = \dot{Q}_k = 0$ for $k \neq k'$ and $Q_{k'} = \dot{Q}_{k'} = 1$, (2) gives

$$T = \tfrac{1}{2} \qquad V = \tfrac{1}{2}\lambda_{k'} \tag{3}$$

[1] J. E. Rosenthal and G. M. Murphy, *Revs. Mod. Phys.*, **8**: 317 (1936).
A. G. Meister, F. F. Cleveland, and M. J. Murray, *Am. J. Phys.*, **11**: 239 (1943).

[2] The original application of group theory to the vibrational problem was made by E. Wigner, *Nachr. Ges. Wiss. Göttingen*, p. 133 (1930).

After the operation of R, the kinetic and potential energies become

$$\begin{aligned} T &= \tfrac{1}{2}(a^2 + c^2) \\ V &= \tfrac{1}{2}(a^2\lambda_{k'} + c^2\lambda_{k''}) \end{aligned} \tag{4}$$

in which (1) has been used; note that the transformation by (1) should be applied first, followed by substitution of $Q_{k'} = \dot{Q}_{k'} = 1$. Therefore, $a^2 + c^2 = 1$, by comparison of the kinetic energy in (3) and (4). The corresponding comparison of the potential energy expressions shows that $a^2\lambda_{k'} + c^2\lambda_{k''} = \lambda_{k'}$. For these conditions to be compatible, either $\lambda_{k'} = \lambda_{k''}$ or $c^2 = 0$, $a^2 = 1$. The latter alternative is equivalent to no mixing of the coordinates $Q_{k'}$ and $Q_{k''}$, when $\lambda_{k'} \neq \lambda_{k''}$.

Consequently, the normal coordinates corresponding to nondegenerate frequencies will form the basis of one-dimensional (and hence irreducible) representations of the group of symmetry operations while the d coordinates corresponding to a d-fold degenerate frequency will form the basis of a d-dimensional representation. There are two possible types of degeneracy leading to equal or degenerate frequencies. The first type arises because of the presence of threefold or higher axes of symmetry in the molecule and is quite independent of the numerical values of the force constants or masses. The second type is called "accidental" degeneracy and arises only when the force constants and masses of the molecule have certain special numerical values which cause two or more of the normal frequencies to be approximately equal. The higher dimensional representations formed by the degenerate normal coordinates are irreducible if the degeneracy is due to molecular symmetry, but reducible if the degeneracy is accidental.

Therefore, barring cases of accidental degeneracy,[1] the representation formed by the normal coordinates is a completely reduced one. It is consequently legitimate to apply the equation [Eq. (1), Sec. 5-9]

$$n^{(\gamma)} = \frac{1}{g}\sum_j g_j \chi_j^{(\gamma)*}\chi_j \tag{5}$$

to find the number of normal coordinates of each symmetry species, that is, in each irreducible representation. The methods given in Sec. 6-1 may be used to determine χ_j, while the values of g_j and $\chi_j^{(\gamma)*}$ are taken from the appropriate table of Appendix X.

As an illustration consider the ammonia molecule (NH_3). The character table for its point group, $\mathcal{C}_{3v}$, is given in Appendix X. There are three classes: E (which is always a class by itself, since $ER = RE$), C_3^1 and C_3^2 (written as $2C_3$ in the tables), and σ_v, σ_v', σ_v'' (written as $3\sigma_v$). The

[1] This exclusion is not really necessary since even in this exceptional case the normal coordinates can be chosen so as to form a completely reduced representation.

character for E is $3N = 12$, the number of coordinates. The character for C_3 is zero (see Table 6-1) while that for σ_v is 2 (the number of atoms in the plane σ_v times 1, the character for each such atom, from Table 6-1). Therefore $\chi_E = 12$, $\chi_{C_3} = 0$, $\chi_{\sigma_v} = 2$. Also $g_E = 1$, $g_{C_3} = 2$, $g_{\sigma_v} = 3$. Substitution of these in (5) together with appropriate values of $\chi_j^{(\gamma)*}$ taken from Appendix X leads to the result that

$$\begin{aligned} n^{(A_1)} &= (\tfrac{1}{6})(1 \cdot 1 \cdot 12 + 2 \cdot 1 \cdot 0 + 3 \cdot 1 \cdot 2) = 3 \\ n^{(A_2)} &= (\tfrac{1}{6})(1 \cdot 1 \cdot 12 + 2 \cdot 1 \cdot 0 + 3 \cdot -1 \cdot 2) = 1 \\ n^{(E)} &= (\tfrac{1}{6})(1 \cdot 2 \cdot 12 + 2 \cdot -1 \cdot 0 + 3 \cdot 0 \cdot 2) = 4 \end{aligned} \tag{6}$$

There are thus three totally symmetrical normal coordinates (A_1), one coordinate with symmetry species A_2, and four *pairs* of coordinates

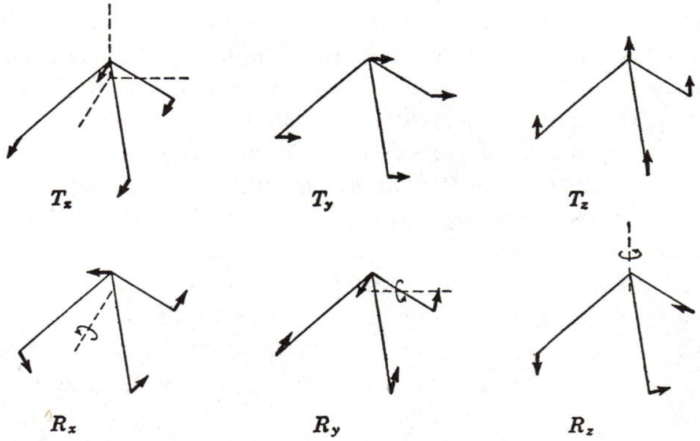

FIG. 6-3. Translational and rotational normal modes for NH_3.

with the symmetry species E. In other words, the *structure* of the representation formed by the cartesian coordinates of ammonia is $\Gamma = 3A_1 + A_2 + 4E$.

These twelve coordinates include, of course, the three coordinates representing translation and the three representing rotation (see Sec. 2-5). If it is desired to find the number of vibrational normal coordinates with each symmetry, the symmetry of the translational and rotational coordinates may be obtained and the proper numbers subtracted from the total values of $n^{(\gamma)}$. Figure 6-3 shows these motions for NH_3. By inspection it is seen that the translations along the x, y, and z axes will transform like x, y, and z and therefore the first two will belong to E and the third to A_1. Likewise the rotations about the x and y axes form a degenerate pair and belong to E while the rotation about the z axis is symmetrical to C_3 but antisymmetrical to σ_v, and therefore belongs to

A_2.‡ There are left two normal coordinates in A_1, none in A_2, and two pairs in E, these being now vibration coordinates.

It is not necessary to carry out this process for each molecule inasmuch as the symmetry of the translations and rotations is always the same for all molecules with the same point group, so that the number of such zero frequency motions for each symmetry may be readily determined from the symmetry species of the translations (T_x,T_y,T_z) and rotations (R_x,R_y,R_z) which are included in the tables of Appendix X.

The vibrations of A_1 are nondegenerate unless they accidentally coincide, but any two vibrations forming a pair of species E must have the same frequency because they are "mixed" on transformation. There are two such pairs in ammonia. Consequently, the group-theoretical treatment predicts that NH_3 will have four normal frequencies, two single and two doubly degenerate.

Use of Internal Coordinates. The reader should now have no difficulty in carrying out such an analysis for any molecule using the methods described above. There is, however, an alternative approach with certain advantages. This uses internal coordinates, such as a suitable set of changes in valence bond lengths and interbond angles (Sec. 4-1), instead of cartesian displacements. First assume that $3N - 6$ ($3N - 5$ for linear cases) independent internal coordinates have been chosen. In NH_3, for example, these could be the extensions of the three NH bonds (r_s) and the increases in the three interbond angles (α_t). The increases in the HH nonbonded distances could be used instead of the angles. Any distorted configuration of the molecule can be described by a set of values of these internal coordinates (except for the position and orientation of the molecule as a whole) so that if a symmetry operation is applied to a generally distorted configuration, a new configuration will result which can be described by new values of the internal coordinates. The transformations of these coordinates, therefore, also form a representation of the point group of the molecule, which is reducible, and can thus be reduced by the construction of suitable linear combinations. Since for infinitesimal displacements the internal coordinates are linearly related to the cartesian displacements and to the normal coordinates, and since they do not involve translation or rotation [*i.e.*, are orthogonal to $\mathcal{R}_1, \ldots, \mathcal{R}_6$ of Eq. (1), Sec. 2-5] the structure of the representation formed by $3N$-6 independent internal coordinates must be the same as that from the $3N$-6 normal coordinates. The internal coordinates can

‡ The strict analytical method of carrying out this process is to examine the expressions for $\mathcal{R}_1$ to $\mathcal{R}_6$ given in Eq. (1), Sec. 2-5, and to find their characters, the $\mathcal{R}_1$, etc., being considered as the basis of a reducible representation. Substitution of these in (5) will give the same results as obtained above by inspection.

therefore be used as the basis of an alternative method of finding this structure.

The procedure is, as before, to find the characters for the reducible representation and then to substitute them in the basic equation [Eq. (5)] for $n^{(\gamma)}$. It is immediately seen that the internal coordinates can be chosen so that they divide into sets such that the members of each set transform only among themselves. In many cases the situation is even simpler in that the internal coordinates simply permute among their own kind, *i.e.*, without needing to be expressed as general linear combinations of the original values. In NH_3 this is clearly so: a threefold rotation may replace r_1 by r_2, r_2 by r_3, r_3 by r_1, α_1 by α_2, α_2 by α_3, and α_3 by α_1. Under these conditions, the characters are obtainable separately for each set of equivalent[1] internal coordinates, and the character for the whole transformation is the sum of these separate contributions. Further, if a given internal coordinate S_t is transformed into some other member $S_{t'}(t' \neq t)$ of the set under an operation R, S_t will contribute nothing to the character χ_R, since $R_{tt} = 0$. Only those coordinates which are transformed into themselves (sometimes with reversal of sign) will contribute. The process of determining the character thus consists, for each equivalent set of coordinates, of selecting one operation R from each class in the group, counting the number of internal coordinates of the set which are not permuted into others by R, and multiplying by -1 if R reverses the sign of these unpermuted coordinates. Each set of equivalent coordinates forms a representation which can be reduced separately, and the structure of the normal coordinate representation can be found by summing the contributions from the different sets.

Ammonia provides a simple illustration of this method. Table 6-2 shows the steps involved.

TABLE 6-2. SYMMETRY SPECIES FOR THE VIBRATIONS OF AMMONIA

$\mathcal{C}_{3v}$	E	$2C_3$	$3\sigma_v$	$n_r^{(\gamma)}$	$n_\alpha^{(\gamma)}$	$n_Q^{(\gamma)}$
A_1	1	1	1	1	1	2
A_2	1	1	-1	0	0	0
E	2	-1	0	1	1	2
$\chi_j(r)$	3	0	1			
$\chi_j(\alpha)$	3	0	1			

The first four rows and columns form the standard table of characters for the irreducible representations of $\mathcal{C}_{3v}$, taken from Appendix X. The row labeled $\chi_j(r)$ gives the characters for the reducible representation

[1] Equivalent coordinates are defined as coordinates which are permuted by one or more symmetry operations of the point group.

formed by the NH extensions r_1, r_2, and r_3. Since there are three of them, $\chi_E(r) = 3$. C_3 permutes them all; thus $\chi_{C_3}(r) = 0$. Each plane of symmetry, σ_v, contains one bond, but the reflection interchanges the other two r's; therefore $\chi_{\sigma_v}(r) = 1$. Similar considerations apply to the angles α, whose characters are given in the last row. These happen in this case to be identical with those for the r's. The column labeled $n_r^{(\gamma)}$ gives the structure of the representation for the r's obtained by substituting the $\chi_j(r)$ in (5)

$$n^{(\gamma)} = \frac{1}{g} \sum_j g_j \chi_j^{(\gamma)*} \chi_j$$

in which as usual g is the number of group elements (six), g_j the number of elements in the jth class (1,2,3), $\chi_j^{(\gamma)}$ the character for class j and irreducible representation $\Gamma^{(\gamma)}$ as given in the upper left part of the table, and χ_j is the character for the reducible representation for any operation in class j (3,0,1). These values of $n_r^{(\gamma)}$ (1,0,1) mean that the representation formed by r_1, r_2, r_3 has the structure

$$\Gamma(r) = A_1 + E$$

as does also that formed by α_1, α_2, α_3.

The last column gives the structure of the normal coordinate representation, and is the sum of the previous two. In this case there are two normal frequencies of species A_1 and two of species E, the latter species being doubly degenerate and therefore having two normal coordinates each.

One advantage of this approach arises from the fact that the normal frequencies are sometimes largely determined by the type of internal coordinate which is most strongly involved in the given mode. Thus, in hydrocarbons certain modes are mainly CH stretching vibrations, and their frequencies are little changed from molecule to molecule. The number of such frequencies of each symmetry is readily determined by the above method. Even where one type of internal coordinate does not dominate all others in a given normal mode, it is often convenient to label a mode in terms of its principal component, so that one might speak of a degenerate "bending" mode for NH_3, even though no mode consists exclusively of the angle changes α_1, α_2, α_3. Another important advantage will appear when the factoring of the secular equation is treated.

The reader is warned that the method just described requires that the internal coordinates employed form complete sets of independent symmetrically equivalent coordinates. This is not always possible because redundancies may occur.

Treatment of Redundancy. In CH_4, the complete set of symmetrically equivalent angular coordinates consists of six HCH angle increments.

Geometrical considerations show that these are not all independent; their sum always equals zero. These coordinates are said to form a *redundant* set, with one *redundancy* condition connecting them. If the four obviously independent CH extensions are also used as internal coordinates, the total is ten, or one more than $3N - 6 = 9$, thus indicating the presence of redundancy.

It is possible to exclude one of the angles arbitrarily and to modify the treatment of the previous subsection to accommodate situations in which not all the members of a symmetrically equivalent set are used, but it is usually more convenient to retain complete sets and cope with the redundancy later. The procedure then follows exactly that of the previous subsection, except that the total number of combinations of internal coordinates will exceed the number of normal coordinates by the number of redundancy conditions. These redundancy conditions, or suitable linear combinations of them, will fall into the various symmetry species, and when these species are identified, the number of combinations of internal coordinates of each species can be suitably diminished to give the number of normal coordinates of each species. More formal methods of treating this problem will be given later (Sec. 6-8), but the problem can often be handled by inspection.

Table 6-3. Symmetry Species for Normal Coordinates of Methane

$\mathfrak{T}_d$	E	$8C_3$	$6\sigma_d$	$6S_4$	$3C_2$	$n_r^{(\gamma)}$	$n_\alpha^{(\gamma)}$	$n_Q^{(\gamma)}$
A_1	1	1	1	1	1	1	1[a]	1
A_2	1	1	−1	−1	1	0	0	0
E	2	−1	0	0	2	0	1	1
F_1	3	0	−1	1	−1	0	0	0
F_2	3	0	1	−1	−1	1	1	2
$\chi_j(r)$	4	1	2	0	0			
$\chi_j(\alpha)$	6	0	2	0	2			

[a] Redundancy condition.

The case of methane is tabulated in Table 6-3. The arrangement is exactly the same as that of Table 6-2, which should be consulted for an explanation. Now, however, $n_Q^{(\gamma)}$ is not always the sum of $n_r^{(\gamma)}$ and $n_\alpha^{(\gamma)}$, but the sum minus the number of sets of redundancy conditions of the given species. Since here the redundancy condition, *i.e.*, the sum of the angle increments equals zero, is by inspection completely symmetrical (species A_1), the unity under $n_\alpha^{(\gamma)}$ opposite A_1 represents this condition and is rejected.

In general, it is best to determine the symmetry species for the normal coordinates by the cartesian method, since, in more complicated cases, it is not always easy to recognize the redundancies in advance. However, the analysis of the internal coordinates just described is a necessary preliminary step in setting up the factored secular equation in terms of internal coordinates, and it is therefore convenient to carry out symmetry analyses of both cartesian and internal coordinates at an early stage in the consideration of a symmetric molecule.

A Special Case: The Regular Representation. The twelve CCH angles in cyclopropane provide an example of a set of equivalent coordinates of special interest. If every operation R of the group (except the identity) transforms each coordinate of an equivalent set into a different member of the set, the representation is called a *regular* representation. There is then no operation except the identity which transforms a coordinate into itself. The characters of a regular representation are clearly all zero except χ_E, which equals the order of the group, which must be the number of coordinates in the set. Then (5) yields

$$n^{(\gamma)} = \frac{1}{g} \sum_R \chi_R^{(\gamma)*} \chi_R = \chi_E^{(\gamma)} = d_\gamma$$

so that each species is represented by d_γ sets, where d_γ is the degeneracy or dimension of the species. Thus there will be two *pairs* for a species E and three *triplets* (nine coordinates) for a species F.

No equivalent set of coordinates can give any larger values of $n^{(\gamma)}$ than the above type. For example, suppose that for each coordinate there is one operation (besides identity) which sends the coordinate into itself so that the set contains $g/2$ coordinates. A set with the same transformation properties as this could be obtained by taking g coordinates forming a regular representation and constructing the sums of pairs of coordinates so that each pair was symmetrical to the appropriate symmetry operation. Since this process essentially has selected a subset of the original set of g coordinates, or of linear combinations of them, the subset cannot have any $n^{(\gamma)}$ greater than that yielded by the regular representation. Thus the CH stretch coordinates in cyclopropane transform identically with appropriate combinations of the CCH angles (see Fig. 6-9, Sec. 6-7).

Therefore in the reduction of the representation formed by a single complete set of equivalent coordinates, a nondegenerate species (A) can occur at most once, a two-dimensional species (E) can occur no more than twice, and a three-dimensional species (F) can occur at most three times (see also Sec. 6-8).

6-3. Factoring of the Secular Equation

As shown in Chap. 2, the determination of the normal frequencies, normal modes, and normal coordinates depends upon the solution of a secular equation, *e.g.*, Eq. (11), Sec. 2-2. In Sec. 2-8 it was shown that for linear and planar molecules, the secular equation can be factored if the proper choice of coordinates is made. In this section it will be shown that this phenomenon is not limited to such molecules, but occurs whenever the molecule possesses any symmetry.

For example, consider the simple case of the water molecule. If r_1, r_2, and α are, respectively, the extensions of the first and second OH bonds and the increase in the bond angle, a simple approximate potential energy expression is

$$2V = F_r(r_1^2 + r_2^2) + F_\alpha \alpha^2 \tag{1}$$

where F_r and F_α are force constants, while the kinetic energy is found to have the form

$$2T = M_r(\dot{r}_1^2 + \dot{r}_2^2) + M_\alpha \dot{\alpha}^2 + 2M_{rr}\dot{r}_1\dot{r}_2 + 2M_{r\alpha}(\dot{r}_1 + \dot{r}_2)\dot{\alpha} \tag{2}$$

in which the M's are certain reduced masses which constitute the inverse **G** matrix[1] and $\dot{r}_1$ is the time derivative of r_1, etc. The reader can readily verify that the coordinates[2]

$$\begin{aligned} \mathsf{S}_1 &= \alpha \\ \mathsf{S}_2 &= 2^{-\frac{1}{2}}(r_1 + r_2) \\ \mathsf{S}_3 &= 2^{-\frac{1}{2}}(r_1 - r_2) \end{aligned} \tag{3}$$

on substitution of the inverse relations in (1) and (2), lead to

$$2V = \mathsf{F}_\alpha \mathsf{S}_1^2 + \mathsf{F}_r \mathsf{S}_2^2 + \mathsf{F}_r \mathsf{S}_3^2 \tag{4}$$

and

$$2T = M_\alpha \dot{\mathsf{S}}_1^2 + 2\sqrt{2}\, M_{\alpha r}\dot{\mathsf{S}}_1\dot{\mathsf{S}}_2 + (M_r + M_{rr})\dot{\mathsf{S}}_2^2 + (M_r - M_{rr})\dot{\mathsf{S}}_3^2 \tag{5}$$

The secular equation, Eq. (3), Sec. 2-6, in terms of the S's then becomes

$$\begin{vmatrix} F_\alpha - M_\alpha\lambda & -\sqrt{2}\, M_{\alpha r}\lambda & 0 \\ -\sqrt{2}\, M_{\alpha r}\lambda & F_r - (M_r + M_{rr})\lambda & 0 \\ 0 & 0 & F_r - (M_r - M_{rr})\lambda \end{vmatrix} = 0 \tag{6}$$

This clearly factors into one linear and one quadratic factor because of the absence in the potential and kinetic energies of cross terms connecting S_3 with either S_1 or S_2.

[1] The M's may be evaluated by using the inverse of Eq. (1), Sec. 4-6, with $\mu_1 = \mu_2$, $r_{31} = r_{32}$.

[2] Sans serif type is used to indicate coordinates, potential energy constants, etc., when referred to the basis which accomplishes the maximum factoring of the secular equation possible by symmetry.

The coordinates S_1, etc., used above are illustrations of *symmetry coordinates*,[1] *i.e.*, coordinates in terms of which the secular equation is factored to the maximum extent made possible by the symmetry. In Appendix XII it is proved formally that coordinates are symmetry coordinates if (*a*) they form the basis of a completely reduced unitary[2] representation of the point group of the molecule; (*b*) sets of coordinates of the same degenerate symmetry species have identical transformation coefficients.

In other words, in order to factor the secular equation, the coordinates are formed into linear combinations such that each combination (or new coordinate) belongs to one of the symmetry species of the molecular point group. In the water illustration, S_1 and S_2 are of species A_1 while S_3 is B_1. When only real one-dimensional species occur, the proof is immediate that no cross terms will occur in either the kinetic or potential energies between two coordinates, $S^{(\gamma)}$ and $S^{(\gamma')}$ say, of different symmetry species $\Gamma^{(\gamma)}$ and $\Gamma^{(\gamma')}$. There will always be some operation R of the group for which

$$S^{(\gamma)} \xrightarrow{R} +S^{(\gamma)}$$

and

$$S^{(\gamma')} \xrightarrow{R} -S^{(\gamma')}$$

so that

$$S^{(\gamma)}S^{(\gamma')} \xrightarrow{R} -S^{(\gamma)}S^{(\gamma')}$$

Since this reversal of sign is incompatible with the fundamental property that the kinetic and potential energies must be unchanged by an operation of the group, such cross terms must have zero coefficients. The vanishing of all cross terms coupling coordinates of two different species clearly puts zeros in the secular equation which break it up into separate factors, one for each species.

In cases of degeneracy, where two- or three-dimensional species occur, the above argument may often still suffice, but in any event the general

[1] The term *symmetry coordinate* has been used differently by different authors. See, for example, the following:

J. B. Howard and E. B. Wilson, Jr., *J. Chem. Phys.*, **2**: 630 (1934).

C. Manneback, *Ann. soc. sci. Bruxelles*, **55**: 129 (1935).

J. E. Rosenthal and G. M. Murphy, *Revs. Mod. Phys.*, **8**: 317 (1936).

O. Redlich and H. Tompa, *J. Chem. Phys.*, **5**: 529 (1937).

It seems desirable to use this term for all the coordinates which factor the secular equation fully and to add special designations for the special types described later in this section.

[2] A unitary representation is a set of unitary matrices which have the same multiplication table as the group operations. In many cases, the representation will be real, which means that the elements of the matrices are real numbers, so that the matrices are orthogonal matrices (see Appendix V).

proof of Appendix XII covers all cases. But it is then necessary to observe requirement (*b*) above. The symmetry coordinates of a two-dimensional species will occur in degenerate pairs such that an operation *R* of the group will transform one coordinate into a linear combination of the two. If there are two or more pairs of the same species, unless all pairs transform exactly alike under all operations, the full symmetry factoring will not be achieved. This requirement will not be automatically satisfied because each set can be chosen in an infinite variety of ways. If all sets of degenerate symmetry coordinates are to have the same transformation coefficients, the choice for the first set may be made arbitrarily, but once it is made, the proper choice for all other sets is completely fixed. When correctly chosen, symmetry coordinates of a degenerate species will not only show no cross terms in T or V with coordinates of other species, but also no cross terms between different components of the degenerate sets, that is, if S_a, S_b, S_c and S'_a, S'_b, S'_c are two triply degenerate sets identically "oriented," there will be no terms of the type S_aS_b, $S_aS'_b$, etc., but only terms of type S_a^2, $S_aS'_a$, S_b^2, etc. See Table 6-4. Furthermore, if d_γ is the dimension of the species $\Gamma^{(\gamma)}$, the

TABLE 6-4. FORCE CONSTANTS IN SYMMETRY COORDINATES FOR THE E FACTOR OF NH_3[a]

	S_{ra}	$S_{\alpha a}$	S_{rb}	$S_{\alpha b}$
S_{ra}	F_{rr}	$F_{r\alpha}$	0	0
$S_{\alpha a}$	$F_{r\alpha}$	$F_{\alpha\alpha}$	0	0
S_{rb}	0	0	F_{rr}	$F_{r\alpha}$
$S_{\alpha b}$	0	0	$F_{r\alpha}$	$F_{\alpha\alpha}$

[a] The F_{rr}, etc., are appropriate linear combinations of the potential energy constants in terms of the valence coordinates r_i and α_i.

secular equation in terms of symmetry coordinates will not only show d_γ factors corresponding to this species, but also these d_γ factors will be identical (see Appendix XII).

Finally, therefore, if symmetry coordinates are used, the secular equation will factor, there being d_γ *equal factors for each species* $\Gamma^{(\gamma)}$ *(irreducible representation) which occurs in the structure of the representation formed by the coordinates. The degree of a given factor is the number of symmetry coordinates involved in the factor, i.e., the number* $n^{(\gamma)}$ *determined in Sec.* 6-2.

As an example, consider again the case of ammonia, NH_3. In the previous section it was found that the representation formed by the cartesian displacement coordinates could be reduced to give the structure

$$3A_1 + A_2 + 4E \tag{7}$$

where translational and rotational modes were included. Accordingly, the secular equation, if expressed in terms of symmetry coordinates, will have one cubic factor ($3A_1$), one linear factor (A_2), and two identical quartic factors ($4E$).

Some of these factors will contain zero roots. When these are taken out, as was done in Sec. 6-2, the representation has the structure

$$2A_1 + 2E \tag{8}$$

which corresponds to one quadratic factor ($2A_1$), involving the two totally symmetric normal coordinates, and two identical quadratic factors ($2E$), each involving one member of each of the two pairs of degenerate normal vibrations.

6-4. Construction of Internal Symmetry Coordinates

In order to carry out in practice the factoring of the secular equation described in the previous section, it is first necessary to construct the symmetry coordinates.[1]

There are several useful types of symmetry coordinates; one of the simplest consists of linear combinations of internal coordinates, *i.e.*, changes in interatomic distances and angles within the molecule. These combinations are chosen so as to conform to the requirements of Sec. 6-3 and therefore serve to factor the secular equation to the maximum extent possible from symmetry. The coordinates used in the water example of Sec. 6-3 are internal symmetry coordinates.

Symmetrically Complete Sets. It is easiest to construct each internal symmetry coordinate out of *equivalent* internal coordinates, *i.e.*, internal coordinates which are exchanged by the symmetry operations of the molecules, such as the four CH bond extensions in CH_4. Then the construction of the symmetry coordinates for a given molecule breaks down into separate problems for the different equivalent sets. Further, it is very desirable to utilize *symmetrically complete* sets; *i.e.*, sets containing all the coordinates resulting from the application of the symmetry operations of the molecule to an arbitrarily chosen coordinate. Thus the six HCH bond angles in CH_4 form such a set and should all be used, even though only five are independent. This use of redundant coordinates will introduce zero roots into the secular equation, but they are most

[1] Alternative methods of constructing symmetry coordinates have been described. See the following sources:

B. L. Crawford, Jr., *J. Chem. Phys.*, **21**: 1108 (1953).

J. E. Kilpatrick, *J. Chem. Phys.*, **16**: 749 (1948).

J. R. Nielsen and L. H. Berryman, *J. Chem. Phys.*, **17**: 659 (1949).

T. Venkatarayudu, *Proc. Indian Acad. Sci.*, (A)**17**: 75 (1943).

E. Wigner, *Nachr. Ges. Wiss. Göttingen*, p. 133 (1930).

E. Wigner, "Gruppentheorie," p. 123, paragraph 2, Edwards, Ann Arbor, Mich., 1944. (Reprint.)

easily removed after the factoring of the secular equation has been accomplished.

It should be noted that there are two kinds of internal coordinates, those which are permuted with unchanged sign and those which change sign under some operations of the group. The first kind includes changes in interatomic distances and simple bond angles while the second includes the angle between a bond and plane initially containing the bond (*e.g.*, between CO and the plane O′CO″ in $CO_3^=$) and the angle between two planes (twist angles, such as between H_2C and CH_2 in ethylene). The distinction between these two types should be kept in mind during the following sections.

A Simple Rule for Nondegenerate Coordinates. Nondegenerate internal symmetry coordinates are very easily written down by the use of a rule[1] involving the characters $\chi_R^{(\gamma)}$ for the given species $\Gamma^{(\gamma)}$. What is desired is a method for determining the coefficients $U_i^{(\gamma)}$ in the equation

$$\mathsf{S}^{(\gamma)} = \sum_i U_i^{(\gamma)} S_i \tag{1}$$

For the nondegenerate symmetry coordinates, the effect of a symmetry operator R is

$$\mathsf{S}^{(\gamma)} \xrightarrow{R} \mathsf{S}^{(\gamma)\prime} = \chi_R^{(\gamma)} \mathsf{S}^{(\gamma)} \tag{2}$$

But R also transforms the internal coordinates S_i:

$$S_i \xrightarrow{R} S_i' = \sum_{i'} R_{ii'} S_{i'} = R_{ij} S_j \tag{3}$$

in which the elements $R_{ii'}$ vanish except for the one element R_{ij} because the S's simply permute (with possibly a change of sign). Therefore $R_{ij} = \pm 1$ when the operator R moves a displacement at S_j to the position S_i. Comparison of these ways of expressing the effect of R leads to the relation

$$\begin{aligned} \mathsf{S}^{(\gamma)\prime} &= \chi_R^{(\gamma)} \mathsf{S}^{(\gamma)} = \chi_R^{(\gamma)} \sum_j U_j^{(\gamma)} S_j \\ &= \sum_i U_i^{(\gamma)} S_i' = \sum_{i,j} U_i^{(\gamma)} R_{ij} S_j \end{aligned} \tag{4}$$

The coefficients of S_j must be the same in the two expressions; thus, for the particular value, $j = 1$,

$$\chi_R^{(\gamma)} U_1^{(\gamma)} = \sum_i U_i^{(\gamma)} R_{i1} \tag{5}$$

[1] E. Wigner, "Gruppentheorie," Chap. 12; see also *Nachr. Ges. Wiss. Göttingen*, p. 133 (1930).

H. Eyring, J. Walter, and G. E. Kimball, "Quantum Chemistry," Chap. 10, Wiley, New York, 1944.

If the internal coordinates are of the first kind, R_{i1} is $+1$ for the coordinate S'_i to which R moves the displacement S_1 and zero otherwise; then

$$U_i^{(\gamma)} = \chi_R^{(\gamma)} U_1^{(\gamma)} \propto \chi_R^{(\gamma)} \qquad \text{when } S_i \overset{R}{\rightarrow} S'_i = S_1 \tag{6}$$

since $U_1^{(\gamma)}$ is arbitrary except for normalization. A compact and easily remembered form of this equation is

$$\mathsf{S}^{(\gamma)} = \mathfrak{N} \sum_R \chi_R^{(\gamma)} R S_1 \tag{7}$$

in which the notation RS_1 stands for the coordinate to which the displacement of S_1 is transferred by the operation R. The sum need be taken only over those operations R required to generate all the S_i, although a correct result will be obtained if all the group operations are used. $\mathfrak{N}$ is a normalizing factor.

This equation will also apply when S_1 is a coordinate of the second kind, provided that RS_1 is taken with the appropriate sign. Here R_{i1} will sometimes be -1, but in just these cases RS_1 gives a minus sign.

Modified Rule for Degenerate Cases; Orientation. A very similar rule permits the construction of degenerate symmetry coordinates. In cases of degeneracy, pairs or triplets of symmetry coordinates of the same species and from the same internal set will occur, but since the different degenerate components give rise to identical factors of the secular determinant (see Appendix XII), it is unnecessary to consider more than one representative of each degenerate set. It may also happen that a given internal coordinate set may contribute more than one (but never more than d_γ) degenerate sets to a given species, as described in Sec. 6-2. However, a consideration of such cases ($n^{(\gamma)} > 1$) will be postponed until Sec. 6-7, since, as will be shown there, they can always be reduced to several separate problems in each of which $n^{(\gamma)} \leq 1$.

In case only one internal coordinate set contributes to a given degenerate species, (7) needs no modification whatsoever. If, however, several internal coordinate sets contribute to the same degenerate species, (7) can be used with some single internal coordinate set, S_t, but in general must be modified for the other sets, $S'_{t'}$, $S''_{t''}$, etc. This modification consists in replacing a single S'_1 or S''_1 in (7) by a properly "oriented" linear combination. Proper orientation has the following meaning: the linear combination is to be chosen so as to exhibit exactly the same symmetry properties as the S_1 used for the generation of the first set. A general proof of the validity of this method will be given later; it is first illustrated with some examples.

Consider the doubly degenerate symmetry coordinates of species E in ammonia. According to Table 6-2 the three NH stretches, r_1, r_2, and r_3

and the three H—N—H bendings, α_{12}, α_{23}, α_{31} each contribute one degenerate set to E. Suppose that (7) is applied to r_1:

$$\mathsf{S}_r^{(E)} = \mathfrak{N} \sum_R \chi_R^{(E)} R r_1 = 6^{-\frac{1}{2}}(2r_1 - r_2 - r_3)$$

In order to get the bending symmetry coordinate, one must operate upon a linear combination of the α's which has symmetry properties identical with r_1. Perhaps the simplest such combination is the single angle opposite r_1, namely, α_{23}. This is easily grasped by geometric intuition, but is demonstrated explicitly by noting that identical sets of group operations, namely, E and σ_v, send r_1 into itself and also send α_{23} into itself. The desired representative symmetry coordinate is therefore:

$$\mathsf{S}_\alpha^{(E)} = \mathfrak{N} \sum_R \chi_R^{(E)} R \alpha_{23} = 6^{-\frac{1}{2}}(2\alpha_{23} - \alpha_{31} - \alpha_{12}) \quad (8)$$

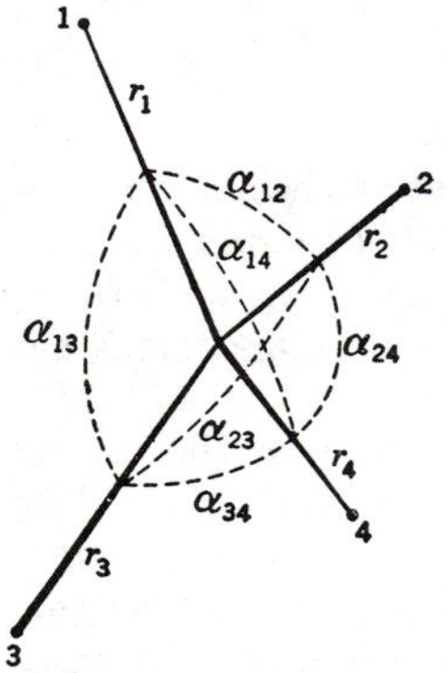

FIG. 6-4. The valence coordinates r_i and α_{ij} for CH_4.

As a second, slightly more complicated example, consider the four stretches and six bending coordinates of methane (Fig. 6-4). Only the α set is involved in the doubly degenerate species, so that it is permissible to construct the symmetry coordinate according to the formula

$$\begin{aligned}\mathsf{S}_\alpha^{(E)} &= \mathfrak{N} \sum_R \chi_R^{(E)} R \alpha_{12} \\ &= 12^{-\frac{1}{2}}(2\alpha_{12} - \alpha_{13} - \alpha_{14} - \alpha_{23} - \alpha_{24} + 2\alpha_{34}) \quad (9)\end{aligned}$$

The triply degenerate (F_2) coordinate can also be constructed by operating upon α_{12}:

$$\mathsf{S}_\alpha^{(F_2)} = \mathfrak{N} \sum_R \chi_R^{(F_2)} R \alpha_{12} = 2^{-\frac{1}{2}}(\alpha_{12} - \alpha_{34}) \quad (10)$$

A symmetry coordinate of species F_2 constructed from the bond stretches is also required. Evidently one cannot simply operate upon r_1, since r_1 is symmetric with respect to E, C_3, C_3^2, σ_{12}, σ_{13}, and σ_{14} ($\mathcal{C}_{3v}$), which are not identical with the operations which leave α_{12} invariant ($\mathcal{C}_{2v}$). However, a simple linear combination, namely, $r_1 + r_2$, has the desired property: it is sent into itself by E, C_2, σ_{12}, and σ_{34}, just as is α_{12}. Therefore

$$\begin{aligned}\mathsf{S}_r^{(F_2)} &= \mathfrak{N} \sum_R \chi_R^{(F_2)} R(r_1 + r_2) \\ &= \tfrac{1}{2}(r_1 + r_2 - r_3 - r_4) \quad (11)\end{aligned}$$

It was not necessary to choose α_{12} in (10). An equally satisfactory procedure would employ

$$\mathsf{S}_r^{(F_2)} = \mathfrak{N} \sum_R \chi_R^{(F_2)} R r_1 \tag{12}$$

and

$$\mathsf{S}_\alpha^{(F_2)} = \mathfrak{N} \sum_R \chi_R^{(F_2)} R(\alpha_{12} + \alpha_{13} + \alpha_{14}) \tag{13}$$

since each of these combinations is invariant under the same group operations: E, C_3, C_3^2, σ_{12}, σ_{13}, σ_{14}.

Note that the summation over R need not be carried out literally in evaluating (11) or (13). All that is necessary, say, in (11), is to write down the same function of $r_1 + r_2$, $r_1 + r_3$, etc., as (10) is of α_{12}, α_{13}, etc. This point is of great importance in factoring the secular determinant (Sec. 6-6) and is also a considerable shortcut in large molecules, since it reduces all the symmetry coordinates of a given species to a standard form. The normalization constant, of course, may be different.

The Correlation Theorem.[1] A remarkable theorem can be proved concerning the number of times, $n^{(\gamma)}$, a given species γ will occur in the reduction of the representation generated by a complete equivalent set of coordinates, $S_1, S_2, \ldots, S_i, \ldots$. The operations of the complete point group $\mathcal{G}$ which are such that

$$S_1 \xrightarrow{R} \pm S_1 \tag{14}$$

for any one particular coordinate S_1 of the set form a subgroup $\mathcal{H}$ of $\mathcal{G}$. Furthermore, S_1 will belong to some nondegenerate species η' of $\mathcal{H}$. Any irreducible representation of $\mathcal{G}$ will also be a representation of $\mathcal{H}$ but not necessarily an irreducible one. It can, however, always be reduced. Suppose that $\Gamma^{(\gamma)}$ of $\mathcal{G}$, when reduced on $\mathcal{H}$, has the structure

$$\Gamma^{(\gamma)} = \sum_\eta n_\gamma^{(\eta)} \Gamma^{(\eta)}$$

in which $\Gamma^{(\gamma)}$ is an irreducible representation of $\mathcal{G}$, $\Gamma^{(\eta)}$ one of $\mathcal{H}$. Then it will be proved that

$$n^{(\gamma)} = n_\gamma^{(\eta')} \tag{15}$$

in which η' is the special, nondegenerate species of S_1 described above.

The proof is as follows: from Eq. (3), Sec. 5-8,

$$n^{(\gamma)} = \frac{1}{g} \sum_R \chi_R^{(\gamma)*} \chi_R$$

[1] See A. Speiser, "Die Theorie der Gruppen von endlicher Ordnung," Springer, Berlin, 1927. The application to the molecular problem was made by R. Burgess.

where

$$\chi_R = \sum_i R_{ii}$$

R_{ii} being a diagonal element in the transformation of the equivalent coordinates. But these coordinates simply permute, possibly with change of sign. Thus

$$R_{ii} = 0, \text{ or } \pm 1$$

Then

$$n^{(\gamma)} = \frac{1}{g} \sum_i \sum_R \chi_R^{(\gamma)*} R_{ii}$$

and, for a given i, the terms will vanish except for operations R which do not permute S_i with other coordinates. These operations form a group $\mathcal{H}_i$ which is the same as $\mathcal{H}$, except for orientation. For such operations

$$R_{ii} = \chi_R^{(\eta')}$$

the character of the special species of $\mathcal{H}$. Thus

$$n^{(\gamma)} = \frac{1}{g} \sum_i \sum_{R \text{ in } \mathcal{H}} \chi_R^{(\gamma)*} \chi_R^{(\eta')}$$

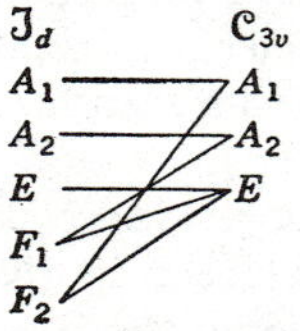

Fig. 6-5. Correlation table between $\mathcal{G} = \mathcal{T}_d$ and $\mathcal{H} = \mathcal{C}_{3v}$.

The inner sum will be the same for all values of i because all the coordinates S_i are equivalent and each is left unpermuted by a group $\mathcal{H}_i$ differing from $\mathcal{H}$ only in orientation in space. There are g/h coordinates S_i, where h is the number of operations in $\mathcal{H}$. Therefore

$$n^{(\gamma)} = \frac{1}{g}\frac{g}{h} \sum_{R \text{ in } \mathcal{H}} \chi_R^{(\gamma)*} \chi_R^{(\eta')}$$

$$= \frac{1}{h} \sum_{R \text{ in } \mathcal{H}} \chi_R^{(\eta')*} \chi_R^{(\gamma)} = n_\gamma^{(\eta')}$$

where the last result is simply the application of the basic formula, Eq. (5), Sec. 6-2, to the representation of $\mathcal{H}$ formed by the coordinates of species γ of $\mathcal{G}$. Since $n^{(\gamma)}$ is certainly a real quantity it is legitimate to replace $\chi_R^{(\gamma)*}\chi_R^{(\eta')}$ by its complex conjugate $\chi_R^{(\eta')*}\chi_R^{(\gamma)}$.

This theorem will now be illustrated and its implications discussed. Consider first the CH stretches r_1, r_2, r_3, and r_4 of CH_4. Here $\mathcal{G}$ is $\mathcal{T}_d$; $\mathcal{H}$ is $\mathcal{C}_{3v}$. The special species of $\mathcal{H}$ is A_1. The correlation table is shown in Fig. 6-5. From this it is seen, for example, that

$$\Gamma^{(A_1)} \text{ of } \mathcal{G} = \Gamma^{(A_1)} \text{ of } \mathcal{H}$$

and

$$\Gamma^{(F_2)} \text{ of } \mathcal{G} = \Gamma^{(A_1)} + \Gamma^{(E)} \text{ of } \mathcal{H}$$

Then the theorem above states that there is one combination of r's in A_1 of $\mathcal{T}_d$ and one degenerate set of three in F_2, since A_1 and F_2 are connected with the special species A_1 of $\mathcal{H}$. If properly oriented, one combination of the F_2 set will be in A_1 of $\mathcal{C}_{3v}$, the other two in E of $\mathcal{C}_{3v}$.

Proof of the Modified Rule for Degenerate Cases. Degenerate symmetry coordinates S_a, S_b will transform according to the formula

$$\mathsf{S}_a^{(\gamma)} \xrightarrow{R} \mathsf{S}_a^{(\gamma)\prime} = R_{aa}^{(\gamma)}\mathsf{S}_a^{(\gamma)} + R_{ab}^{(\gamma)}\mathsf{S}_b^{(\gamma)} = \sum_{a'} R_{aa'}^{(\gamma)}\mathsf{S}_{a'}^{(\gamma)} \tag{16}$$

If γ is a triply degenerate species, an additional term $R_{ac}^{(\gamma)}\mathsf{S}_c^{(\gamma)}$ must be included. As in the nondegenerate case:

$$\begin{aligned} \mathsf{S}_a^{(\gamma)\prime} &= \sum_i U_{ai}^{(\gamma)} S_i' = \sum_{i,j} U_{ai}^{(\gamma)} R_{ij} S_j \\ &= \sum_{a'} R_{aa'}^{(\gamma)}\mathsf{S}_{a'}^{(\gamma)} = \sum_{a',j} R_{aa'}^{(\gamma)} U_{a'j}^{(\gamma)} S_j \end{aligned} \tag{17}$$

Here R_{ij} is ± 1 when S_i is the coordinate which receives the displacement of S_j when R is applied. R_{ij} is zero for other values of j. Since the coefficient of a given S_j must be the same in the two expressions above for $\mathsf{S}_a^{(\gamma)\prime}$, it follows that

$$\sum_i U_{ai}^{(\gamma)} R_{ij} = \sum_{a'} R_{aa'}^{(\gamma)} U_{a'j}^{(\gamma)} \tag{18}$$

Now put $j = 1$ and let i designate the index of the *single* coordinate which receives the displacement S_1 when R is applied ($S_i \xrightarrow{R} S_i' = S_1$), so that

$$U_{ai}^{(\gamma)} = \sum_{a'} R_{aa'}^{(\gamma)} U_{a'1}^{(\gamma)} \tag{19}$$

In Appendix XIII it is proved that

$$\sum_{\text{coset } i} R_{aa'}^{(\gamma)} = \delta_{aa'} \sum_{\text{coset } i} \chi_R^{(\gamma)} \tag{20}$$

when the summation is over all the group operations which transfer the displacement S_1 to S_i (this set of operations is called coset i; see Sec. 5-4). Since there are h such operations, upon summing over these (19) becomes

$$\begin{aligned} \sum_{\text{coset } i} U_{ai}^{(\gamma)} = hU_{ai}^{(\gamma)} &= \sum_{\text{coset } i} \sum_{a'} R_{aa'}^{(\gamma)} U_{a'1}^{(\gamma)} \\ &= \sum_{a'} \Big(\sum_{\text{coset } i} R_{aa'}^{(\gamma)} \Big) U_{a'1}^{(\gamma)} = \sum_{a'} \delta_{aa'} \sum_{\text{coset } i} \chi_R^{(\gamma)} U_{a'1}^{(\gamma)} \\ &= U_{a1}^{(\gamma)} \sum_{\text{coset } i} \chi_R^{(\gamma)} \end{aligned} \tag{21}$$

This last equation shows that the desired coefficients $U_{ai}^{(\gamma)}$ are proportional to the sums of the characters of the group operations which transfer the displacement S_1 to S_i. Therefore the final result is the same as for the nondegenerate cases, namely,

$$\mathsf{S}_a^{(\gamma)} = \sum_i U_{ai}^{(\gamma)} S_i = \mathfrak{N} \sum_R \chi_R^{(\gamma)} R S_1 \tag{22}$$

To be sure, (22) gives only one member of a degenerate set of symmetry coordinates, but this is all that is needed for any ordinary calculations.

A slight modification of the proof is necessary in case the special species $\Gamma^{(\eta')}$ is not totally symmetric. However, the final result (22) is the same, with the understanding that

$$RS_1 = -S_i$$

whenever the displacement changes sign under the operation R.

In the above derivation no consideration was given to the question of "orientation." If more than one set of internal coordinates give rise to symmetry coordinates in the same degenerate species γ, proper orientation is necessary, as was illustrated in the CH_4 example. In (22) there is leeway for varying the orientation in that the choice of S_1 as the coordinate from which the others were generated was an arbitrary choice. Use of S_2, say, might have given a differently oriented coordinate, still of species γ. Furthermore, any nonvanishing linear combination

$$H_1 = \sum_i a_i S_i$$

will also generate a symmetry coordinate of species γ, that is,

$$\mathsf{S}^{(\gamma)} = \mathfrak{N} \Sigma \chi_R^{(\gamma)} R H_1 \tag{23}$$

The choice of H_1, that is, of the coefficients a_i, determines the orientation of the symmetry coordinate constructed from H_1.

Suppose that one coordinate, say S_1, of one of the equivalent sets of internal coordinates is left invariant, or turned into $-S_1$, by the operations of a particular subgroup $\mathcal{H}$ of $\mathcal{G}$. Then S_1 will be of species η' of $\mathcal{H}$, where η' is nondegenerate.

Further, any set $\mathsf{S}_a^{(\gamma)}$, $\mathsf{S}_b^{(\gamma)}$, $\mathsf{S}_c^{(\gamma)}$ can be considered as the basis of a representation of $\mathcal{H}$ and, by forming proper linear combinations, this representation can be reduced. Then, by (15), since $n^{(\gamma)} = 1$, only one of $\mathsf{S}_a^{(\gamma)}$, $\mathsf{S}_b^{(\gamma)}$, $\mathsf{S}_c^{(\gamma)}$ can be of species η' of $\mathcal{H}$. Let that one be $\mathsf{S}_a^{(\gamma)}$.

If (23) is used to construct a symmetry coordinate from each of the equivalent sets, in each case starting with a coordinate S_1 or combination H_1 of species η' of one particular subgroup $\mathcal{H}$ as defined above, then in each case an a-type symmetry coordinate will be generated, that is, one of species η' under $\mathcal{H}$. This cannot be otherwise, because such symmetry coordinates will contain S_1 (or H_1) and the effect of an operation of $\mathcal{H}$ on the symmetry coordinate will be the same as its effect on H_1.

The other symmetry coordinates $\mathsf{S}_b^{(\gamma)}$, etc., cannot belong to η' of $\mathcal{H}$ and therefore must be of some different species, say η'' of $\mathcal{H}$. This is all that is needed to prove the correctness of the orientation, because two coordinates such as $\mathsf{S}_{1a}^{(\gamma)}$ and $\mathsf{S}_{2b}^{(\gamma)}$, say, from different equivalent sets and different degeneracy indices a and b cannot enter any cross terms in the potential or kinetic energy because they are of different species under $\mathcal{H}$. The purpose of the orientation requirement was to ensure that no such cross terms occurred.

6-5. External Symmetry Coordinates[1]

Another type of symmetry coordinate that is often useful consists of linear combinations of cartesian displacement coordinates chosen to satisfy the requirements laid down for all symmetry coordinates in Sec. 6-3. These will be called *external symmetry coordinates*, and have the same advantages and disadvantages as cartesian coordinates compared with internal coordinates, Sec. 4-1.

In constructing this type, it is first clear that the different sets of equivalent atoms can be treated separately, except that all degenerate sets of symmetry coordinates belonging to one species must be identically "oriented," *i.e.*, must have the same transformation coefficients. In many cases a further breakdown is possible which separates the z coordinates, say, from the x and y coordinates. This will be the case for all noncubic point groups. By properly orienting the x and y axes separately for each atom, it is sometimes convenient to separate the x from the y coordinates initially. All these remarks represent simple methods of partially reducing the representation formed by the cartesian coordinates.

In constructing external symmetry coordinates, Eq. (7), Sec. 6-4, can also be applied, *i.e.*,

$$\mathsf{S}_k^{(\gamma)} = \mathfrak{N} \sum_R \chi_R^{(\gamma)} R \xi_i \tag{1}$$

where ξ_i is some one of the cartesian displacement coordinates of a given set, or a simple linear combination of several such coordinates. $R\xi_i$ is the result of operating on ξ_i with the symmetry operation R of the group, $\chi_R^{(\gamma)}$ is the character for R in the irreducible representation of species γ, the sum is over all R, and $\mathfrak{N}$ is a normalizing factor.

[1] This section is not essential to the understanding of the rest of the chapter, but it describes a useful alternative type of coordinates.

If the species is degenerate, the problem of "orientation" can be solved by the same method as used in Sec. 6-4 for internal symmetry coordinates; that is, select ξ_i to be a coordinate, or simple combination of coordinates, symmetrical (or antisymmetrical) to some chosen set of group operations constituting a subgroup.

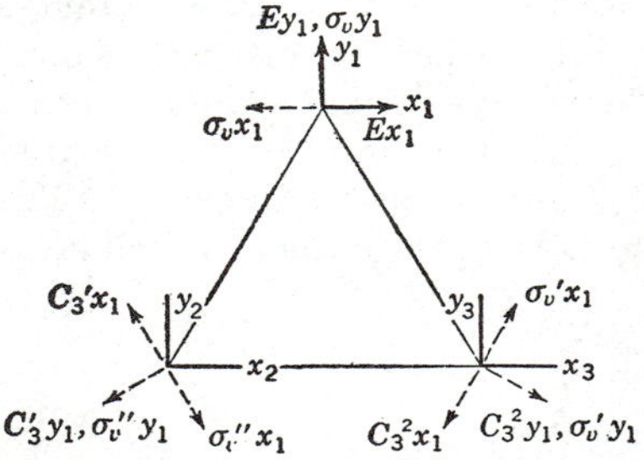

FIG. 6-6. Effect of symmetry operations on x_1 and y_1 in NH_3, for use in generating symmetry coordinates.

As a simple example, consider the x and y coordinates of the three equivalent hydrogen atoms of the pyramidal molecule NH_3, Fig. 6-6 ($\mathcal{C}_{3v}$). The structure of the representation of $\mathcal{C}_{3v}$ is, by the methods of Sec. 6-2,

$$A_1 + A_2 + 2E$$

The choice of y_1 as ξ_1 leads to

$$\begin{aligned} \mathsf{S}^{(A_1)} = \tfrac{1}{3}\sqrt{3}\,(y_1 - \tfrac{1}{2}y_2 - \tfrac{1}{2}\sqrt{3}\,x_2 \\ - \tfrac{1}{2}y_3 + \tfrac{1}{2}\sqrt{3}\,x_3) \end{aligned} \quad (2)$$

Since A_2 is odd to the planes, it is necessary to start with x_1 instead of y_1 in this case, whence

$$\mathsf{S}^{(A_2)} = \tfrac{1}{3}\sqrt{3}\,(x_1 - \tfrac{1}{2}x_2 + \tfrac{1}{2}\sqrt{3}\,y_2 - \tfrac{1}{2}x_3 - \tfrac{1}{2}\sqrt{3}\,y_3) \quad (3)$$

For the representative of one degenerate pair, start with y_1;

$$\mathsf{S}_1^{(E)} = \frac{1}{\sqrt{6}}\left(2y_1 + \frac{1}{2}\,y_2 + \frac{1}{2}\sqrt{3}\,x_2 + \frac{1}{2}\,y_3 - \frac{1}{2}\sqrt{3}\,x_3\right) \quad (4)$$

Now since y_1 is totally symmetric under E and σ_v, the representative of the second degenerate pair should use for ξ_i either $y_2 + y_3$ or $x_2 - x_3$, since these combinations are symmetric with respect to σ_v.

If $y_2 + y_3$ is used for ξ_i, a coordinate S' is obtained which is independent of $\mathsf{S}_1^{(E)}$ but not orthogonal to it.

$$\begin{aligned} \mathsf{S}' &= 2(y_2 + y_3) - (-\tfrac{1}{2}y_3 - \tfrac{1}{2}\sqrt{3}\,x_3 - \tfrac{1}{2}y_1 - \tfrac{1}{2}\sqrt{3}\,x_1) \\ &\qquad - (-\tfrac{1}{2}y_1 + \tfrac{1}{2}\sqrt{3}\,x_1 - \tfrac{1}{2}y_2 + \tfrac{1}{2}\sqrt{3}\,x_2) \\ &= y_1 + \tfrac{5}{2}y_2 + \tfrac{5}{2}y_3 - \tfrac{1}{2}\sqrt{3}\,x_2 + \tfrac{1}{2}\sqrt{3}\,x_3 \end{aligned} \quad (5)$$

To obtain a coordinate orthogonal to $\mathsf{S}_1^{(E)}$ and therefore suitable as the representative of the other degenerate pair, use may be made of an orthogonalization method similar to that described in Sec. 9-2. It is thus found that $\mathsf{S}' - (\sqrt{6}/2)\mathsf{S}_1^{(E)}$ is the desired combination. When normalized, this gives

$$\mathsf{S}_2^{(E)} = \frac{1}{2\sqrt{6}}\,(3y_2 + 3y_3 - \sqrt{3}\,x_2 + \sqrt{3}\,x_3) \quad (6)$$

A Geometrical Approach. It is possible to apply (1) in a geometrical manner. Instead of ξ_i use an arbitrary (except as limited below) vector ϱ_α attached to some atom α of an equivalent set. Then $R\varrho_\alpha$ represents the vector resulting from the operation of R on ϱ_α. Multiply $R\varrho_\alpha$ by $\chi_R^{(\gamma)}$ and sum over all operations of the group as before. In case of degeneracy, choose ϱ_α so that it has the appropriate symmetry under some fixed subgroup as before. The resulting expression is to be interpreted in the sense that the coefficient of each cartesian coordinate is the component in the cartesian direction of the vector attached to the atom in question.

The hydrogen coordinates of NH_3 may be constructed by this vector method. For example, consider the A_1 combination. Start with an arbitrary vector on H_1 (see Fig. 6-7). It actually does not need to lie in the plane of the three hydrogens, but since no symmetry operations in this group mix components in the plane with those perpendicular to it, it is sensible to start in one case with an initial vector in the plane, in another case with one perpendicular to it. The operations R which do not shift ϱ_α to another atom are E and σ_v, each with character unity in A_1. Clearly $E\varrho_\alpha + \sigma_v\varrho_\alpha$ is a vector lying in σ_v. A slight amount of intuition would have warned that the final ϱ on each atom must lie in the corresponding plane and thus led to the choice of such a vector for ϱ_α, but if this is not done, that is, if ϱ_α is arbitrary, (1) will take care of the situation anyway. The result is that the final symmetry coordinate is represented by vectors on each H atom, each in the plane through that atom and all of equal length, as shown in the figure.

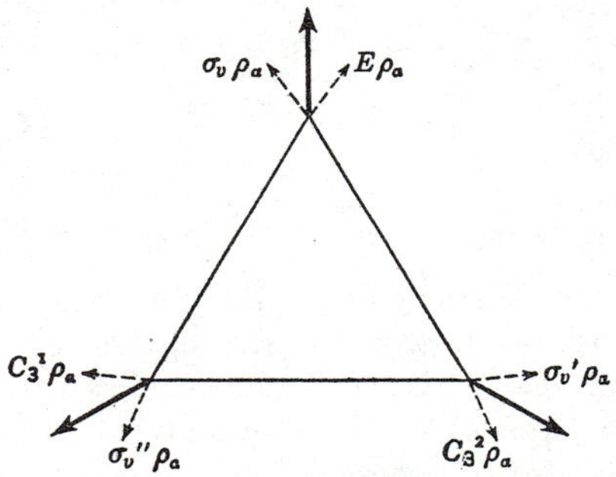

FIG. 6-7. Transformation of an arbitrary vector ϱ_α in xy plane of NH_3 by operations of $\mathcal{C}_{3v}$. The composition of these to form $S^{(A_1)}$ is shown by the heavy arrows.

6-6. The Potential and Kinetic Energies in Terms of Symmetry Coordinates

In Sec. 6-3 the potential and kinetic energy expressions for the water molecule were obtained in terms of internal symmetry coordinates, and it was found that no cross terms occurred between symmetry coordinates of different species. The method employed there was the obvious one of solving for the internal coordinates in terms of the symmetry coordinates and then substituting in the original expressions for T and V. This can always be done, even in complicated cases, although short cuts are available.

The first step, that of inverting the equations defining the symmetry coordinates, is greatly simplified if these equations form a unitary (orthogonal if real) transformation, and it is therefore important to construct them so that this property holds. Then the unitary property ensures that the coefficients of the inverse transformation may be obtained by reading down the columns of the forward coefficients and taking the complex conjugate (usually all coefficients are real).

An Illustration with the CH Stretches of Ethylene. It will be assumed that ethylene has $\mathfrak{D}_{2h} = \mathfrak{V}_h$ symmetry, and that the molecule lies in the xy plane. The species of the four CH stretches (Fig. 6-8) are determined to be

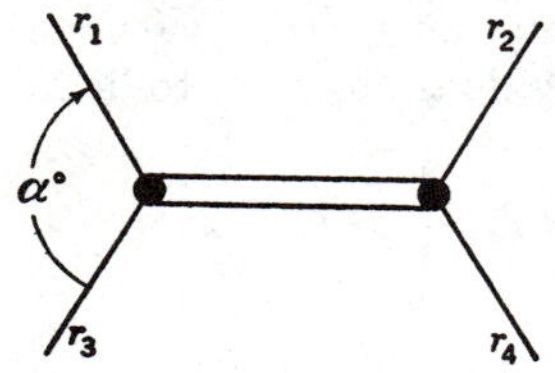

FIG. 6-8. The CH stretching coordinates of C_2H_4.

$$\Gamma = A_g + B_{1g} + B_{2u} + B_{3u} \tag{1}$$

Then, employing Eq. (7), Sec. 6-4, it is easily found that

$$\begin{aligned} S^{(A_g)} &= \tfrac{1}{2}(r_1 + r_2 + r_3 + r_4) \\ S^{(B_{1g})} &= \tfrac{1}{2}(r_1 - r_2 - r_3 + r_4) \\ S^{(B_{2u})} &= \tfrac{1}{2}(r_1 + r_2 - r_3 - r_4) \\ S^{(B_{3u})} &= \tfrac{1}{2}(r_1 - r_2 + r_3 - r_4) \end{aligned} \tag{2}$$

By reading down the first column of coefficients (those of r_1), the coefficients of r_1 in terms of the S's are obtained, since all quantities here are real. Similarly, the coefficients of r_2 above give the expression for r_2 in terms of the S's. In matrix language, the inverse is equal to the (conjugate) transpose. The result is that

$$\begin{aligned} r_1 &= \tfrac{1}{2}(S^{(A_g)} + S^{(B_{1g})} + S^{(B_{2u})} + S^{(B_{3u})}) \\ r_2 &= \tfrac{1}{2}(S^{(A_g)} - S^{(B_{1g})} + S^{(B_{2u})} - S^{(B_{3u})}) \\ r_3 &= \tfrac{1}{2}(S^{(A_g)} - S^{(B_{1g})} - S^{(B_{2u})} + S^{(B_{3u})}) \\ r_4 &= \tfrac{1}{2}(S^{(A_g)} + S^{(B_{1g})} - S^{(B_{2u})} - S^{(B_{3u})}) \end{aligned} \tag{3}$$

The complete potential function for ethylene will involve coordinates other than the CH bond stretches, but that part of it depending upon the r's alone can be written,[1]

$$\begin{aligned} 2V = F(r_1^2 + r_2^2 + r_3^2 + r_4^2) &+ 2F'(r_1r_2 + r_3r_4) \\ &+ 2F''(r_1r_3 + r_2r_4) + 2F'''(r_1r_4 + r_2r_3) \end{aligned} \tag{4}$$

In order to obtain the potential energy in terms of symmetry coordinates, one must now substitute (3) in (4). There is an obvious short cut for replacing $r_1^2 + r_2^2 + r_3^2 + r_4^2$. Since the transformation (3) is orthogonal, it follows that

$$\sum_t r_t^2 = \sum_\gamma (S^{(\gamma)})^2 \tag{5}$$

[1] The fact that r_1r_2, r_3r_4 or r_1r_3, r_2r_4, etc., have the same coefficient should be clear intuitively, but will be discussed below.

Furthermore, no cross terms between symmetry coordinates of different species can occur, so that it saves effort if the several species are treated separately. For example, from (3)

$$r_t = \tfrac{1}{2}\mathsf{S}^{(A_g)} \tag{6}$$

if the other S's are zero; thus the coefficient of $(\mathsf{S}^{(A_g)})^2$ in $2V$ must be, from substitution of (6) in (4),

$$F + F' + F'' + F''' \tag{7}$$

In like manner, the coefficient of each term can be constructed, yielding

$$\begin{aligned} 2V = {} & (F + F' + F'' + F''')(\mathsf{S}^{(A_g)})^2 + (F - F' - F'' + F''')(\mathsf{S}^{(B_{1g})})^2 \\ & + (F + F' - F'' - F''')(\mathsf{S}^{(B_{2u})})^2 + (F - F' + F'' - F''')(\mathsf{S}^{(B_{3u})})^2 \end{aligned} \tag{8}$$

When the potential and kinetic energy have been transformed to symmetry coordinates, the secular equation may be written down in any of the forms previously described in Chap. 4. The transformation of **G** to symmetry coordinates may be obtained by inspection of (8). From the methods of Sec. 4-2 it is easily found that

$$\begin{aligned} G_{tt} &= \mu_C + \mu_H = G \\ G_{13} &= G_{24} = \mu_C \cos\alpha^0 = G'' \\ G_{12} &= G_{34} = G' = 0 \\ G_{14} &= G_{23} = G''' = 0 \end{aligned} \tag{9}$$

Therefore

$$\begin{aligned} \mathsf{G}^{(A_g)} &= G + G' + G'' + G''' = \mu_H + \mu_C(1 + \cos\alpha^0) \\ \mathsf{G}^{(B_{1g})} &= G - G' - G'' + G''' = \mu_H + \mu_C(1 - \cos\alpha^0) \\ \mathsf{G}^{(B_{2u})} &= G + G' - G'' - G''' = \mu_H + \mu_C(1 - \cos\alpha^0) \\ \mathsf{G}^{(B_{3u})} &= G - G' + G'' - G''' = \mu_H + \mu_C(1 + \cos\alpha^0) \end{aligned} \tag{10}$$

A More General Method.[1] The illustrative example just given lacked generality for two reasons: (*a*) only a single set of equivalent coordinates was treated and (*b*) no degenerate symmetry coordinates were involved. The method would fail if degeneracies occurred, since in order to solve for internal coordinates in terms of symmetry coordinates, expressions for all the degenerate components would be required, and the methods of Sec. 6-4 have not prescribed how more than one representative of each degenerate set may be obtained.

Actually, there is a much easier method of getting the potential energy in terms of symmetry coordinates, with the added advantage that it

[1] The more general method of this section is based on the unpublished work of Robert Burgess. The notation and presentation are somewhat different from his, but the basic principles were discovered by him.

requires only a single representative of each degenerate set. Using the previous example as a first illustration, suppose the force constants of (4) are written in tabular (or matrix) form.

	r_1	r_2	r_3	r_4			r_1	r_2	r_3	r_4	
r_1	F_{11}	F_{12}	F_{13}	F_{14}		r_1	F	F'	F''	F'''	
r_2	F_{21}	F_{22}	F_{23}	F_{24}	=	r_2	F'	F	F'''	F''	(11)
r_3	F_{31}	F_{32}	F_{33}	F_{34}		r_3	F''	F'''	F	F'	
r_4	F_{41}	F_{42}	F_{43}	F_{44}		r_4	F'''	F''	F'	F	

Consider the first row of the force constant table (11) above. The columns are labeled by the internal coordinates r_1, r_2, r_3, and r_4. For a given species (for example, B_{1g}) each of these coordinates will appear in the symmetry coordinate with a coefficient which has already been obtained (*e.g.*, $\frac{1}{2}$, $-\frac{1}{2}$, $-\frac{1}{2}$, $\frac{1}{2}$). Then the rule for getting the potential energy coefficient of the square of the symmetry coordinate is: *Multiply the force constant in the first row and in the column labeled by a given internal coordinate by the coefficient with which that internal coordinate appears in the symmetry coordinate. Then divide by the coefficient of the first internal coordinate (row label). Do this for each column and add the results.* Thus for B_{1g} in the present example, this rule yields

$$(\tfrac{1}{2}/\tfrac{1}{2})F + (-\tfrac{1}{2}/\tfrac{1}{2})F' + (-\tfrac{1}{2}/\tfrac{1}{2})F'' + (\tfrac{1}{2}/\tfrac{1}{2})F''' = F - F' - F'' + F'''$$

With the other species, this rule yields [see (2) for symmetry coordinate coefficients]

$$\begin{aligned} \mathsf{F}^{(A_g)} &= F + F' + F'' + F''' \\ \mathsf{F}^{(B_{2u})} &= F + F' - F'' - F''' \\ \mathsf{F}^{(B_{3u})} &= F - F' + F'' - F''' \end{aligned} \tag{12}$$

in agreement with (8).

Although the symmetry coordinates of (2) were all nondegenerate, the same rule works in degenerate cases as well. The force constant matrix for the bending coordinates of methane (Fig. 6-4) has the form

	α_{12}	α_{13}	α_{14}	α_{23}	α_{24}	α_{34}	
α_{12}	F	F'	F'	F'	F'	F''	
α_{13}	F'	F	F'	F'	F''	F'	
α_{14}	F'	F'	F	F''	F'	F'	
α_{23}	F'	F'	F''	F	F'	F'	(13)
α_{24}	F'	F''	F'	F'	F	F'	
α_{34}	F''	F'	F'	F'	F'	F	

The symmetry coordinates, by Eqs. (9) and (10), Sec. 6-4, are

$$S_\alpha^{(A_1)} = 6^{-\frac{1}{2}}(\alpha_{12} + \alpha_{13} + \alpha_{14} + \alpha_{23} + \alpha_{24} + \alpha_{34}) \tag{14}$$
$$S_\alpha^{(E)} = 12^{-\frac{1}{2}}(2\alpha_{12} - \alpha_{13} - \alpha_{14} - \alpha_{23} - \alpha_{24} + 2\alpha_{34}) \tag{15}$$
$$S_\alpha^{(F_2)} = 2^{-\frac{1}{2}}(\alpha_{12} - \alpha_{34}) \tag{16}$$

Therefore

$$F_\alpha^{(A_1)} = F_\alpha + F'_\alpha + F'_\alpha + F'_\alpha + F'_\alpha + F''_\alpha = F_\alpha + 4F'_\alpha + F''_\alpha \tag{17}$$
$$F_\alpha^{(E)} = F_\alpha - 4(\tfrac{1}{2})F'_\alpha + F''_\alpha = F_\alpha - 2F'_\alpha + F''_\alpha \tag{18}$$
$$F_\alpha^{(F_2)} = F_\alpha - F''_\alpha \tag{19}$$

The complete **F** matrix for methane also includes blocks corresponding to bond stretches and to interactions between stretches and bends. These blocks have the form:

	r_1	r_2	r_3	r_4
r_1	F_r	F'_r	F'_r	F'_r
r_2	F'_r	F_r	F'_r	F'_r
r_3	F'_r	F'_r	F_r	F'_r
r_4	F'_r	F'_r	F'_r	F_r

and

	α_{12}	α_{13}	α_{14}	α_{23}	α_{24}	α_{34}
r_1	$F_{r\alpha}$	$F_{r\alpha}$	$F_{r\alpha}$	$F'_{r\alpha}$	$F'_{r\alpha}$	$F'_{r\alpha}$
r_2	$F_{r\alpha}$	$F'_{r\alpha}$	$F'_{r\alpha}$	$F_{r\alpha}$	$F_{r\alpha}$	$F'_{r\alpha}$
r_3	$F'_{r\alpha}$	$F_{r\alpha}$	$F'_{r\alpha}$	$F_{r\alpha}$	$F'_{r\alpha}$	$F_{r\alpha}$
r_4	$F'_{r\alpha}$	$F'_{r\alpha}$	$F_{r\alpha}$	$F'_{r\alpha}$	$F_{r\alpha}$	$F_{r\alpha}$

(20)

The expressions for the symmetry coordinates constructed from the bond stretches are, by Eqs. (7) and (11), Sec. 6-4,

$$S_r^{(A_1)} = \tfrac{1}{2}(r_1 + r_2 + r_3 + r_4) \tag{21}$$
$$S_r^{(F_2)} = \tfrac{1}{2}[(r_1 + r_2) - (r_3 + r_4)] \tag{22}$$

The straightforward construction, with the rule above, of the diagonal force constants corresponding to (21) and (22) gives

$$F_r^{(A_1)} = F_r + 3F'_r \tag{23}$$
$$F_r^{(F_2)} = F_r - F'_r \tag{24}$$

There remains the problem of treating the off-diagonal constants, such as $F_{r\alpha}^{(F_2)}$, the coefficient of $S_r^{(F_2)}S_\alpha^{(F_2)}$. The rule for the off-diagonal constants is as follows: *Multiply the force constant in the first row and in the column labeled by a given internal coordinate by the coefficient with which that internal coordinate appears in the symmetry coordinate. Then divide by the coefficient of the first internal coordinate of the other set (row label). Do this for each column and add.* In the totally symmetric species of the methane example, this rule yields

$$F_{r\alpha}^{(A_1)} = \frac{1}{\frac{1}{2}}[6^{-\frac{1}{2}}(F_{r\alpha} + F_{r\alpha} + F_{r\alpha} + F'_{r\alpha} + F'_{r\alpha} + F'_{r\alpha})]$$
$$= 6^{\frac{1}{2}}(F_{r\alpha} + F'_{r\alpha}) \tag{25}$$

while for the triply degenerate species one finds

$$\begin{aligned}\mathsf{F}_{r\alpha}^{(F_2)} &= \frac{1}{\frac{1}{2}}\,[2^{-\frac{1}{2}}(F_{r\alpha} + 0 + 0 + 0 + 0 - F'_{r\alpha})] \\ &= 2^{\frac{1}{2}}(F_{r\alpha} - F'_{r\alpha}) \qquad (26)\end{aligned}$$

Proof of the General Method. The complete potential energy expression is of the form

$$2V = \sum_{tt'} F_{tt'} S_t^* S_{t'} \qquad (27)$$

(usually the complex conjugate sign is omitted since the S_t are real). This sum can be broken up into parts in which t and t' run, respectively, over single equivalent coordinate sets. Then the $F_{tt'}$ in a single such part would all come from a single block of the type indicated in (11), (13), or (20). Since the method of constructing symmetry coordinates does not mix nonequivalent internal coordinates, it follows that the introduction of symmetry coordinates will not mix the additive parts of the potential energy. In the methane example, the rr, $r\alpha$, and $\alpha\alpha$ parts retain their identity.

Consider a single such part, in particular, one whose force constants form a diagonal block, such as rr or $\alpha\alpha$, in the force constant matrix. Let the symmetry coordinates be given by

$$\mathsf{S}_k = \sum_t U_{kt} S_t \qquad (28)$$

whose inverse is

$$S_t = \sum_k (U^{-1})_{tk} \mathsf{S}_k = \sum_k U_{kt}^* \mathsf{S}_k \qquad (29)$$

since the U_{kt} are unitary. When this is inserted in (27), one finds

$$\begin{aligned}2V &= \sum_{tt'} F_{tt'} \Big(\sum_k U_{kt}^* \mathsf{S}_k\Big)^* \Big(\sum_{k'} U_{k't'}^* \mathsf{S}_{k'}\Big) \\ &= \sum_{kk'} \sum_{tt'} (U_{kt} U_{k't'}^* F_{tt'}) \mathsf{S}_k^* \mathsf{S}_{k'} \qquad (30)\end{aligned}$$

From (30), evidently the desired coefficients are

$$\mathsf{F}_{kk'} = \sum_{tt'} U_{kt} U_{k't'}^* F_{tt'} \qquad (31)$$

If this equation is multiplied on both sides by $U_{kt''}^*$ and summed over k, it follows from the unitary property of U that

$$\begin{aligned}\sum_k U_{kt''}^* \mathsf{F}_{kk'} &= \sum_{tt'} U_{k't'}^* F_{tt'} \sum_k U_{kt''}^* U_{kt} \\ &= \sum_{tt'} U_{k't'}^* F_{tt'} \delta_{tt''} = \sum_{t'} U_{k't'}^* F_{t''t'} \qquad (32)\end{aligned}$$

Now the indices k' and t'' can be chosen arbitrarily. Suppose first that $S_{t''}$ is selected as the first coordinate of an internal set which gives rise to the symmetry coordinate $\mathsf{S}_{k'}$; in such a case the sum over k in (32) gives rise to only a single term:

$$\sum_k U^*_{kt''}\mathsf{F}_{kk'} = U^*_{k't''}\mathsf{F}_{k'k'} \tag{33}$$

This is true because (*a*) the $U_{kt''}$ will vanish unless k refers to one of the symmetry coordinates generated from the internal coordinate $S_{t''}$ and (*b*) for those $U_{kt''}$ which do not vanish by (*a*) all the coefficients $\mathsf{F}_{kk'}$ vanish except for $k = k'$, since it is here assumed that $n^{(\gamma)} \leq 1$. Therefore it follows from (32) and (33) that

$$\mathsf{F}_{k'k'} = \frac{1}{U^*_{k't''}} \sum_{t'} U^*_{k't'} F_{t''t'} \tag{34}$$

which proves the rule given above for computing diagonal elements of F, except that the complex conjugate signs were dropped, since in the present example, as in most cases, the U_{kt} are real.

To prove the rule for an off-diagonal element, choose t'' as the first member of an internal coordinate set which yields a symmetry coordinate $\mathsf{S}_{k''}$ in the same species as $\mathsf{S}_{k'}$ (which must therefore be constructed from a different internal set, if $n^{(\gamma)} \leq 1$ for all sets). By an argument paralleling the one given above, it follows that

$$\sum_k U^*_{kt''}\mathsf{F}_{kk'} = U^*_{k''t''}\mathsf{F}_{k''k'}$$

so that

$$\mathsf{F}_{k''k'} = \frac{1}{U^*_{k''t''}} \sum_{t'} U^*_{k't'} F_{t''t'} \tag{35}$$

which completes the proof. In some cases $U_{k''t''}$ will vanish if t'' refers to the first member of a given internal set; when this happens it is merely necessary to choose t'' as the index of some member of the set for which $U_{k''t''}$ does not vanish, and then select the elements $F_{t''t'}$ from the t''th row instead of the first.

Abbreviation of the Factoring. The reader may verify by inspection of (13) through (16) that the coefficients of any two α's are equal in all symmetry coordinate expressions if the corresponding force constants in the first row of (13) are equal because of symmetry. If this fact can be proved to be true in general, the work of setting up the symmetry coordinate formulas can clearly be abbreviated. The proof that this is true follows. First, if there exists any group operation, R', which simultaneously sends S_1 into S_1 and S_t into $S_{t'}$, then $F_{1t} = F_{1t'}$, since otherwise the potential energy would not be invariant as it must be. The existence of such operations is the reason for the equality of the force constants F'_α under α_{13}, α_{14}, α_{23}, and α_{24} in the first row of (13). This fact will now be utilized in the main proof.

The coefficient of any internal coordinate, S_t, in a symmetry coordinate formula such as (14), (15), or (16) is, except for the normalization constant, just the sum of the characters of the group operations which send S_1 into S_t.‡

Let X_t be an operation which sends S_1 into S_t. Also let R' be an operation which sends S_1 into itself and which sends S_t into $S_{t'}$. Consequently, $F_{1t} = F_{1t'}$. But since the inverse of R' also sends S_1 into itself, the operation $[R'X_t(R')^{-1}]$ will send S_1 into itself and then into S_t, which will finally go into $S_{t'}$; in short, $R'X_t(R')^{-1}$ sends S_1 into $S_{t'}$. But since $R'X_t(R')^{-1}$ is, by definition, an operation in the same class as X_t, it follows that the operations which send S_1 into $S_{t'}$ will be from the same classes as those which send S_1 into S_t. Therefore the sum of the characters of all the group operations which send S_1 into S_t will equal the similar sum for the operations which send S_1 into $S_{t'}$. This completes the proof of the equality of the coefficients, $U_{kt} = U_{kt'}$ if $F_{1t} = F_{1t'}$.

It should now be evident that the bookkeeping involved in transforming the potential (and kinetic) energies to symmetry coordinates can be greatly abbreviated. Once a standard orientation is adopted, it is only necessary to write down a few rows of the **F** matrix. In the methane case, these would be:

	α_{12}	α_{13}	α_{14}	α_{23}	α_{24}	α_{34}	r_1	r_2	r_3	r_4
α_{12}	F_α	F'_α	F'_α	F'_α	F'_α	F''_α	$F_{\alpha r}$	$F_{\alpha r}$	$F'_{\alpha r}$	$F'_{\alpha r}$
r_1							F_r	F'_r	F'_r	F'_r

Since there are only three distinct potential constants in the $\alpha\alpha$ block, and since in the present example all the symmetry coordinates involving the bond stretches depend upon $r_i + r_j$ in the same way as a symmetry coordinate involving the angles depends upon α_{ij}, the character sums in Eq. (7), Sec. 6-4, need only be evaluated in three cases, namely, for those operations which send α_{12} into α_{12}, α_{13}, and α_{34}, respectively. Such calculations are summarized in Table 6-5.

The lower half of the table shows the symmetry coordinates which follow from the upper half immediately upon normalization, noting that the unnormalized r combination is the same function of $r_i + r_j$ as the α combination is of α_{ij}.

‡ If S_1 is not totally symmetric, the characters must be taken with negative sign if $S_1 \rightarrow -S_t$.

Once the properly normalized symmetry coordinates are tabulated, the elements of the factored potential energy matrix can be read by inspection of Table 6-6, which is constructed with the aid of the rules symbolized by (34) and (35).

The final form of the potential energy matrix is as shown in Table 6-7.

TABLE 6-5. ABBREVIATED CALCULATION OF SYMMETRY COORDINATES FOR METHANE

		$\alpha_{12} \to \alpha_{12}$	$\alpha_{12} \to \alpha_{13}$	$\alpha_{12} \to \alpha_{34}$
	R:	$E,\ C_2,\ 2\sigma_d$	$\sigma_d,\ S_4,\ 2C_3$	$2C_2,\ 2S_4$
$\sum_R \chi_R^{(\gamma)}$	A_1:	4	4	4
	E:	4	−2	4
	F_2:	4	0	−4

$$S_\alpha^{(A_1)} = 6^{-\frac{1}{2}}(\alpha_{12} + \alpha_{13} + \alpha_{14} + \alpha_{23} + \alpha_{24} + \alpha_{34})$$
$$S_r^{(A_1)} = \tfrac{1}{2}(r_1 + r_2 + r_3 + r_4)$$
$$S_\alpha^{(E)} = 12^{-\frac{1}{2}}(2\alpha_{12} - \alpha_{13} - \alpha_{14} - \alpha_{23} - \alpha_{24} + 2\alpha_{34})$$
$$S_\alpha^{(F_2)} = 2^{-\frac{1}{2}}(\alpha_{12} - \alpha_{34})$$
$$S_r^{(F_2)} = \tfrac{1}{2}(r_1 + r_2 - r_3 - r_4)$$

TABLE 6-6. SUMMARY OF CALCULATIONS FOR FACTORING SECULAR DETERMINANT OF METHANE

		F_α	$4F'_\alpha$	F''_α	F_r	F'_r	F'_r	F'_r	$2F_{\alpha r}$	$2F'_{\alpha r}$
A_1	U_{kt}	$6^{-\frac{1}{2}}$	$6^{-\frac{1}{2}}$	$6^{-\frac{1}{2}}$	$\frac{1}{2}$	$\frac{1}{2}$	$\frac{1}{2}$	$\frac{1}{2}$		
	$U_{kt}/U_{kt''}$	1	1	1	1	1	1	1	$6^{\frac{1}{2}}/2$	$6^{\frac{1}{2}}/2$
E	U_{kt}	$2(12^{-\frac{1}{2}})$	$-12^{-\frac{1}{2}}$	$2(12^{-\frac{1}{2}})$	0	0	0	0		
	$U_{kt}/U_{kt''}$	1	$-\frac{1}{2}$	1	0	0	0	0	0	0
F_2	U_{kt}	$2^{-\frac{1}{2}}$	0	$-2^{-\frac{1}{2}}$	$\frac{1}{2}$	$\frac{1}{2}$	$-\frac{1}{2}$	$-\frac{1}{2}$		
	$U_{kt}/U_{kt''}$	1	0	−1	1	1	−1	−1	$2^{-\frac{1}{2}}$	$-2^{-\frac{1}{2}}$

6-7. Correlation Tables and the Treatment of Internal Coordinate Sets in Which $n^{(\gamma)} > 1$

In Secs. 6-4 and 6-6 it was assumed that $n^{(\gamma)} \leq 1$ in all species for any internal coordinate set. This will not always be true, but in this section it will be shown that any internal coordinate set for which $n^{(\gamma)} > 1$ can be broken into smaller, nonmixing sets in each of which $n^{(\gamma)} \leq 1$, so that the methods of Secs. 6-4 and 6-6 can then be applied to any molecule.

In other words, when $n^{(\gamma)} > 1$, it is always possible to choose as new coordinates certain linear combinations of the members in the original set such that the new coordinates belong to two or more different, equivalent sets. These new combinations can often be chosen *ab initio* as group motions such as twists, wags, etc. The following paragraphs outline a formal method for obtaining the appropriate combinations if they are not intuitively obvious.

TABLE 6-7. COMPLETELY FACTORED POTENTIAL ENERGY MATRIX FOR METHANE

$S_{\alpha}^{(A_1)}$	$S_r^{(A_1)}$	$S_{\alpha a}^{(E)}$	$S_{\alpha b}^{(E)}$	$S_{\alpha a}^{(F_2)}$	$S_{ra}^{(F_2)}$	$S_{\alpha b}^{(F_2)}$	$S_{rb}^{(F_2)}$	$S_{\alpha c}^{(F_2)}$	$S_{rc}^{(F_2)}$
$F_{\alpha} + 4F'_{\alpha} + F''_{\alpha}$	$6^{1/2}(F_{\alpha r} + F'_{\alpha r})$	0	0	0	0	0	0	0	0
	$F_r + 3F'_r$	0	0	0	0	0	0	0	0
		$F_{\alpha} - 2F'_{\alpha} + F''_{\alpha}$	0	0	0	0	0	0	0
			$F_{\alpha} - 2F'_{\alpha} + F''_{\alpha}$	0	0	0	0	0	0
				$F_{\alpha} - F''_{\alpha}$	$2^{1/2}(F_{\alpha r} - F'_{\alpha r})$	0	0	0	0
					$F_r - F'_r$	0	0	0	0
						$F_{\alpha} - F''_{\alpha}$	$2^{1/2}(F_{\alpha r} - F'_{\alpha r})$	0	0
							$F_r - F'_r$	0	0
								$F_{\alpha} - F''_{\alpha}$	$2^{1/2}(F_{\alpha r} - F'_{\alpha r})$
									$F - F'_r$

Consider an example in which $n^{(\gamma)} > 1$, namely, that of the twelve HCC angles in cyclopropane ($\mathfrak{D}_{3h}$), Fig. 6-9. A single angle is invariant only under the subgroup $\mathcal{C}_1$ consisting merely of E. In such a case (the regular representation), $n^{(\gamma)} = d_\gamma$, or

$$\Gamma = A_1' + A_1'' + A_2' + A_2'' + 2E' + 2E'' \tag{1}$$

An inspection of Fig. 6-9 shows that the twelve angles, when arranged into three sets of four members each (having a common central carbon atom) have the property that each set of four is either merely permuted within itself, or sent as a rigid unit into another set of four, by all symmetry operations of the molecule. Call the collection of group operations

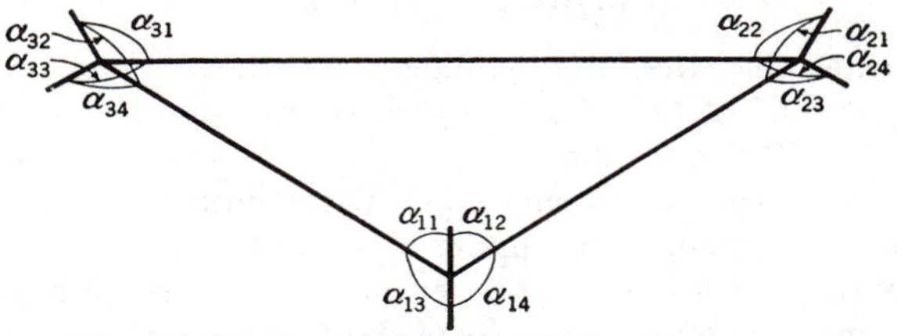

FIG. 6-9. The twelve HCC angles in cyclopropane.

(E, C_2, σ_h, and σ_v in the present example) which merely permute one of these sets $\mathcal{K}$. But $\mathcal{K}$, which is a subgroup of $\mathcal{G}$, itself contains $\mathcal{H}$. In all practical cases it will be found that such an intermediate subgroup can be found if $n^{(\gamma)} > 1$.

The utility of $\mathcal{K}$ lies in the following. The original set of internal coordinates can be reorganized into k/h nonmixing sets each containing g/k equivalent members; k is the order of $\mathcal{K}$. When the representation afforded by each equivalent set is analyzed, it will be found that $n^{(\gamma)} \leq 1$ for all $\Gamma^{(\gamma)}$.

Moreover, each independent set is distinguished by its species under $\mathcal{K}$. In fact, the reorganized coordinates can be constructed as if they were nondegenerate symmetry coordinates under $\mathcal{K}$, employing the usual formula for such cases,

$$K^{(\kappa)} = \mathfrak{N} \sum_V \chi_V^{(\kappa)} V S_t \tag{2}$$

in which V is an operation of $\mathcal{K}$, $\chi_V^{(\kappa)}$ is the character of a one-dimensional species of $\mathcal{K}$, and S_t is some one internal coordinate.

In the cyclopropane example, where $\mathcal{K} = \mathcal{C}_{2v}$, (2) gives four separate combinations when applied to $S_t = \alpha_{11}$:

$$\begin{aligned} K_1^{(A_1)} &= \tfrac{1}{2}(\alpha_{11} + \alpha_{12} + \alpha_{13} + \alpha_{14}) \\ K_1^{(A_2)} &= \tfrac{1}{2}(\alpha_{11} - \alpha_{12} - \alpha_{13} + \alpha_{14}) \\ K_1^{(B_1)} &= \tfrac{1}{2}(\alpha_{11} + \alpha_{12} - \alpha_{13} - \alpha_{14}) \\ K_1^{(B_2)} &= \tfrac{1}{2}(\alpha_{11} - \alpha_{12} + \alpha_{13} - \alpha_{14}) \end{aligned} \tag{3}$$

Similar expressions would be obtained by operating upon $S_t = \alpha_{21}$ or $S_t = \alpha_{31}$. Then the symmetry operations which send carbon atom 1 into carbon atom 2 will send $K_1^{(A_1)}$ into $K_2^{(A_1)}$, $K_1^{(A_2)}$ into $\pm K_2^{(A_2)}$, etc., but will never mix $K_t^{(A_1)}$ and $K_t^{(A_2)}$, etc.

The final symmetry coordinates under $\mathcal{G} = \mathfrak{D}_{3h}$ can be constructed by treating the $K^{(A_1)}$, $K^{(A_2)}$, $K^{(B_1)}$, and $K^{(B_2)}$ families separately. In the present example, use of the correlation table shows that

$$\begin{aligned} \Gamma(K^{(A_1)}) &= A_1' + E' \\ \Gamma(K^{(A_2)}) &= A_1'' + E'' \\ \Gamma(K^{(B_1)}) &= A_2'' + E'' \\ \Gamma(K^{(B_2)}) &= A_2' + E' \end{aligned} \tag{4}$$

which indicates that the final symmetry coordinates can be assembled using the methods of Sec. 6-4 in which it was assumed that $n^{(\gamma)} \leq 1$. Note that the $K^{(A_2)}$, $K^{(B_1)}$, and $K^{(B_2)}$ families behave somewhat like the twisting or out-of-plane internal coordinates described earlier, in that they are not totally symmetric under their respective subgroups.

One complication arises in the degenerate species. Consider E', in which there must be representatives of the $K^{(A_1)}$ and of the $K^{(B_2)}$ families. Since $K_1^{(B_2)}$, by its definition in (3), is of a different species under the subgroup $\mathcal{K}$ than is $K_1^{(A_1)}$, it will not be correctly oriented as is required for the construction of a degenerate symmetry coordinate. Furthermore, if one attempts to force it into the correct orientation, the result will be zero:

$$\begin{aligned} \sum_T \chi_T^{(A_1)} T K_1^{(B_2)} &= E K_1^{(B_2)} + C_2 K_1^{(B_2)} + \sigma(xz) K_1^{(B_2)} + \sigma(yz) K_1^{(B_2)} \\ &= K_1^{(B_2)} - K_1^{(B_2)} + K_1^{(B_2)} - K_1^{(B_2)} = 0 \end{aligned} \tag{5}$$

where the summation is carried over the subgroup $\mathcal{H} = \mathcal{K} = \mathcal{C}_{2v}$. However, if the above character operator formula is applied to $K_2^{(B_2)}$ instead of to $K_1^{(B_2)}$, the linear combination $K_2^{(B_2)} - K_3^{(B_2)}$ is obtained, and the representative symmetry coordinate of this type in E' may then be obtained by applying the character operator formula to $K_2^{(B_2)} - K_3^{(B_2)}$ over the whole group.

In order to obtain the potential energy in terms of these symmetry coordinates, it is necessary to revise the initial **F** matrix, which has the form indicated in Table 6-8 (for simplicity, the force constants are written $F_\alpha = a$, $F'_\alpha = b$, etc.)

TABLE 6-8. FORCE CONSTANT MATRIX FOR HCC ANGLES IN CYCLOPROPANE

	α_{11}	α_{12}	α_{13}	α_{14}	α_{21}	α_{22}	α_{23}	α_{24}	α_{31}	α_{32}	α_{33}	α_{34}
α_{11}	a	b	c	d	e	f	g	h	e	i	g	j
α_{12}	b	a	d	c	i	e	j	g	f	e	h	g
α_{13}	c	d	a	b	g	h	e	f	g	j	e	i
α_{14}	d	c	b	a	j	g	i	e	h	g	f	e

As the next step, consider the set of α's to have been broken up into the nonmixing sets $K^{(A_1)}$, $K^{(A_2)}$, etc. Then the first rows of the various blocks would have the form indicated in Table 6-9.

TABLE 6-9. REVISED FORCE CONSTANT MATRIX FOR CYCLOPROPANE

	$K_1^{(A_1)}$	$K_2^{(A_1)}$	$K_3^{(A_1)}$	$K_1^{(A_2)}$	$K_2^{(A_2)}$	$K_3^{(A_2)}$	$K_1^{(B_1)}$	$K_2^{(B_1)}$	$K_3^{(B_1)}$	$K_1^{(B_2)}$	$K_2^{(B_2)}$	$K_3^{(B_2)}$
$K_1^{(A_1)}$	F_{A_1}	F'_{A_1}	F'_{A_1}							$F_{A_1B_2}$	$F'_{A_1B_2}$	$-F'_{A_1B_2}$
$K_1^{(A_2)}$				F_{A_2}	F'_{A_2}	F'_{A_2}	$F_{A_2B_1}$	$F'_{A_2B_1}$	$-F'_{A_2B_1}$			
$K_1^{(B_1)}$							F_{B_1}	F'_{B_1}	F'_{B_1}			
$K_1^{(B_2)}$										F_{B_2}	F'_{B_2}	F'_{B_2}

where
$$\begin{aligned} F_{A_1} &= a + b + c + d & F'_{A_1} &= e + g + (\tfrac{1}{2})(f + h + i + j) \\ F_{A_2} &= a - b - c + d & F'_{A_2} &= e - g + (\tfrac{1}{2})(-f + h - i + j) \\ F_{B_1} &= a + b - c - d & F'_{B_1} &= e - g + (\tfrac{1}{2})(f - h + i - j) \\ F_{B_2} &= a - b + c - d & F'_{B_2} &= e + g - (\tfrac{1}{2})(f + h + i + j) \\ F_{A_1B_2} &= 0 & F'_{A_1B_2} &= \tfrac{1}{2}(-f - h + i + j) \\ F_{A_2B_1} &= 0 & F'_{A_2B_1} &= \tfrac{1}{2}(f - h - i + j) \end{aligned}$$

Note that the interaction terms of the types A_1A_2, A_1B_1, A_2B_2, and B_1B_2 are omitted in Table 6-9 since (4) indicates that no such interactions will arise among the final symmetry coordinates. Thus $K^{(A_1)}(A_1' + E')$ and $K^{(A_2)}(A_1'' + E'')$ will have no interaction because they do not enter into symmetry coordinates of any common species.

The calculations giving the form of the symmetry coordinates are summarized in Table 6-10. Note that the degenerate coordinates formed

TABLE 6-10. SUMMARY OF CALCULATION OF SYMMETRY COORDINATES FOR HCC ANGLES IN CYCLOPROPANE

	$K_1 \rightarrow \pm K_1$	$K_1 \rightarrow \pm K_2$
R:	$E, C_2, \sigma_h, \sigma_v$	C_3, C_2, S_3, σ_v
$\sum_R \chi_R^{(\gamma)} \chi_R^{(\kappa')}$:		
A_1'	4	4
A_1''	4	4
A_2'	4	4
A_2''	4	4
E'	4	-2
E''	4	-2

$$\begin{aligned} S^{(A_1')} &= 3^{-\frac{1}{2}}(K_1^{(A_1)} + K_2^{(A_1)} + K_3^{(A_1)}) \\ S^{(A_1'')} &= 3^{-\frac{1}{2}}(K_1^{(A_2)} + K_2^{(A_2)} + K_3^{(A_2)}) \\ S^{(A_2')} &= 3^{-\frac{1}{2}}(K_1^{(B_2)} + K_2^{(B_2)} + K_3^{(B_2)}) \\ S^{(A_2'')} &= 3^{-\frac{1}{2}}(K_1^{(B_1)} + K_2^{(B_1)} + K_3^{(B_1)}) \\ S_{A_1}^{(E')} &= 6^{-\frac{1}{2}}(2K_1^{(A_1)} - K_2^{(A_1)} - K_3^{(A_1)}) \\ S_{B_2}^{(E')} &= \mathfrak{N}[2(K_2^{(B_2)} - K_3^{(B_2)}) - (K_3^{(B_2)} - K_1^{(B_2)}) - (K_1^{(B_2)} - K_2^{(B_2)})] \\ &= 2^{-\frac{1}{2}}(K_2^{(B_2)} - K_3^{(B_2)}) \\ S_{A_2}^{(E'')} &= 6^{-\frac{1}{2}}(2K_1^{(A_2)} - K_2^{(A_2)} - K_3^{(A_2)}) \\ S_{B_1}^{(E'')} &= 2^{-\frac{1}{2}}(K_2^{(B_1)} - K_3^{(B_1)}) \end{aligned}$$

from the B types of intermediate coordinates are, respectively, the same functions (except for normalization) of $K_2 - K_3$, $K_3 - K_1$, and $K_1 - K_2$, as the A types are of K_1, K_2, and K_3. This procedure easily vanquishes the difficulties of orientation.

From the expressions given in Table 6-10, the coefficients U_{kt} needed in Eqs. (34) and (35), Sec. 6-6, to find the elements of F may readily be obtained. A little caution must be employed in dealing with the $\mathsf{S}_{B_2}^{(E')}$ and $\mathsf{S}_{B_1}^{(E'')}$ coordinates. Thus in applying Eq. (34), Sec. 6-6, to find $\mathsf{F}_{B2}^{(E')}$, $U_{k1} = 0$ ($K_1^{(B_2)}$ does not appear in the symmetry coordinate). Therefore it is necessary to divide by U_{k2} and hence to employ revised elements from the second row of the $B_2 \times B_2$ block. Although this row is not shown in Table 6-9, it obviously consists of the elements F'_{B_2}, F_{B_2}, and F'_{B_2}, and leads to the result shown in Table 6-11.

By combining the information in Tables 6-9 and 6-11, one finds, for example, that the E' factor of the potential energy has the form:

$$\begin{array}{ll} (a + b + c + d) - \\ \quad [e + g + \frac{1}{2}(f + h + i + j)] & \frac{3^{\frac{1}{2}}}{2}(-f - h + i + j) \\ & (a - b + c - d) - \\ \frac{3^{\frac{1}{2}}}{2}(-f - h + i + j) & \quad [e + g - \frac{1}{2}(f + h + i + j)] \end{array}$$

In conclusion, it should be pointed out that use of the correlation tables is very desirable in order to determine the correct intermediate subgroup. In the tetramethyl methane molecule, there exists a set of twelve equivalent CH stretches if the over-all symmetry is $\mathfrak{T}_d$. Chemical intuition, which suggests that this set might be broken down by constructing intermediate symmetry coordinates at each methyl group, would prove incorrect. For when the correlation of A' in $\mathfrak{C}_s$ (the group which leaves a single CH stretch invariant) is traced into F_2 of $\mathfrak{T}_d$ via $\mathfrak{K} = \mathfrak{C}_{3v}$ (the local symmetry of the methyl group), it will be found that the degenerate species, E, of $\mathfrak{C}_{3v}$ is involved. Since the method described in this section requires that only nondegenerate species be involved, $\mathfrak{C}_{3v}$ would not be an acceptable $\mathfrak{K}$. On the other hand, $\mathfrak{C}_{2v}$ is readily found to be a suitable $\mathfrak{K}$.

6-8. Removal of Redundant Coordinates[1]

The simplest type of redundancy is that in which only the coordinates of a single symmetrically equivalent set are involved. Thus the six bending coordinates for methane are connected by the relation

$$\alpha_{12} + \alpha_{13} + \alpha_{14} + \alpha_{23} + \alpha_{24} + \alpha_{34} = 0 \tag{1}$$

[1] E. B. Wilson, Jr., *J. Chem. Phys.*, **9**: 76 (1941).
C. E. Sun, R. G. Parr, and B. L. Crawford, Jr., *J. Chem. Phys.*, **17**: 840 (1949).

TABLE 6-11. SUMMARY OF FACTORIZATION OF POTENTIAL ENERGY OF HCC ANGLES IN CYCLOPROPANE

$U_{kt}/U_{kt''}$	F_{A_1}	$2F'_{A_1}$	F_{A_2}	$2F'_{A_2}$	F_{B_1}	F'_{B_1}	F'_{B_1}	F_{B_2}	F'_{B_2}	F'_{B_2}	$F_{A_1B_2}$	$F'_{A_1B_2}$	$-F'_{A_1B_2}$	$F_{A_2B_1}$	$F'_{A_2B_1}$	$-F'_{A_2B_1}$
A_1'	1	1	0	0	0	0	0	0	0	0	0	0	0	0	0	0
A_1''	0	0	1	1	0	0	0	0	0	0	0	0	0	0	0	0
A_2'	0	0	0	0	0	0	0	1	1	1	0	0	0	0	0	0
A_2''	0	0	0	0	1	1	1	0	0	0	0	0	0	0	0	0
E'	+1	$-\frac{1}{2}$	0	0	0	0	0	1	−1	0	0	$\frac{3^{\frac{1}{2}}}{2}$	$-\frac{3^{\frac{1}{2}}}{2}$	0	0	0
E''	0	0	1	$-\frac{1}{2}$	1	−1	0	0	0	0	0	0	0	0	$\frac{3^{\frac{1}{2}}}{2}$	$-\frac{3^{\frac{1}{2}}}{2}$

which holds in the approximation used throughout this chapter, namely, that the internal coordinates represent small displacements from equilibrium. In the methane problem, (1) is a symmetry coordinate in A_1 (except for a normalizing constant). The elimination of the redundancy in such a case is trivial, since one merely omits the row and column in the $\mathbf{F}^{(A_1)}$ and $\mathbf{G}^{(A_1)}$ matrices corresponding to this coordinate. However, even such a simple redundancy is of value as a check on the computation of the $\mathbf{G}$ matrix, since the elements of the corresponding row and column should automatically vanish. This may be seen as follows. The internal coordinates were defined in terms of the $3N$ cartesian displacements, ξ_i, by Eq. (1), Sec. 4-2,

$$S_t = \sum_{i=1}^{3N} B_{ti}\xi_i \tag{2}$$

If

$$\sum_t a_t S_t = 0 \tag{3}$$

is the redundancy condition, then

$$\sum_{i=1}^{3N}\left(\sum_t a_t B_{ti}\right)\xi_i = 0 \tag{4}$$

or

$$\sum_t a_t B_{ti} = 0 \qquad i = 1, 2, \ldots, 3N \tag{5}$$

since the ξ_i are independent (the B_{ti} themselves are so defined as to exclude translation and rotation). But use of the definition of $G_{tt'}$ by Eq. (1), Sec. 4-2,

$$G_{tt'} = \sum_{i=1}^{3N} \mu_i B_{ti} B_{t'i}$$

in which μ_i is the reciprocal of the mass of the atom one of whose cartesian displacement coordinates is ξ_i, then shows that

$$\sum_t a_t G_{tt'} = \sum_t a_t G_{t't} = \sum_{i=1}^{3N} \mu_i B_{t'i} \sum_t a_t B_{ti} = 0 \tag{6}$$

since $\mathbf{G}$ is symmetric. But (6) proves that if a linear combination of internal coordinates vanishes, so does exactly the same linear combination of elements in any row or column of the $\mathbf{G}$ matrix. Although the proof has been given in terms of the S_t, it clearly is equally applicable to the symmetry coordinates, S_k.

In the methane example, the single symmetry coordinate $\mathsf{S}^{(A_1)}$ vanishes; hence all the elements in its row or column in the $\mathbf{G}$ matrix must vanish. Because of the definition of symmetry coordinates (no interaction terms between coordinates of different species) this assertion reduces to the statement that all elements in the $\mathbf{G}^{(A_1)}$ factor associated with $\mathsf{S}^{(A_1)}$ vanish.

In a slightly more complicated case, the redundancy may involve coordinates from distinct symmetrically equivalent sets. As an example, let α be an XCX bending, β an XCY bending in an X_3CY molecule. If the angles are tetrahedral,

$$\alpha_1 + \alpha_2 + \alpha_3 + \beta_1 + \beta_2 + \beta_3 = 0 \tag{7}$$

analogous to (1). Normalized sums of the α and β constitute two symmetry coordinates which appear in the totally symmetric species A_1 of the group $\mathcal{C}_{3v}$. It would accordingly be expected that the sums of the elements in the rows or columns of $\mathbf{G}^{(A_1)}$ corresponding to these two symmetry coordinates would vanish. The $\mathbf{G}^{(A_1)}$ matrix for the X_3CY molecule has the form (S_r designates a symmetrical CX stretching coordinate and S_s the CY stretch):

$\mathbf{G}^{(A_1)}$	S_r	S_s	S_α	S_β
S_r	$\mu_x + \frac{1}{3}\mu_c$	$-3^{-\frac{1}{2}}\mu_c$	$-\frac{2^{\frac{3}{2}}}{3r_0}\mu_c$	$\frac{2^{\frac{3}{2}}}{3r_0}\mu_c$
S_s		$\mu_y + \mu_c$	$\frac{2 \cdot 6^{\frac{1}{2}}}{3r_0}\mu_c$	$-\frac{2 \cdot 6^{\frac{1}{2}}}{3r_0}\mu_c$
S_α			$\frac{1}{r_0^2}\left(\mu_x + \frac{8}{3}\mu_c\right)$	$-\frac{1}{r_0^2}\left(\mu_x + \frac{8}{3}\mu_c\right)$
S_β				$\frac{1}{r_0^2}\left(\mu_x + \frac{8}{3}\mu_c\right)$

indicating that this expectation is confirmed (use the symmetry of the matrix for $G_{\beta\alpha}^{(A_1)}$, etc.).

The redundancy may be removed in this case as follows. The potential energy corresponding to the A_1 factor may be written out as:

$$\begin{aligned} 2V = \mathsf{F}_{rr}\mathsf{S}_r^2 + \mathsf{F}_{ss}\mathsf{S}_s^2 + \mathsf{F}_{\alpha\alpha}\mathsf{S}_\alpha^2 + \mathsf{F}_{\beta\beta}\mathsf{S}_\beta^2 \\ + 2\mathsf{F}_{rs}\mathsf{S}_r\mathsf{S}_s + 2\mathsf{F}_{r\alpha}\mathsf{S}_r\mathsf{S}_\alpha + 2\mathsf{F}_{r\beta}\mathsf{S}_r\mathsf{S}_\beta \\ + 2\mathsf{F}_{s\alpha}\mathsf{S}_s\mathsf{S}_\alpha + 2\mathsf{F}_{s\beta}\mathsf{S}_s\mathsf{S}_\beta + 2\mathsf{F}_{\alpha\beta}\mathsf{S}_\alpha\mathsf{S}_\beta \end{aligned} \tag{8}$$

[the (A_1) superscripts have been dropped for simplicity]. Since

$$\mathsf{S}_\alpha = 3^{-\frac{1}{2}}(\alpha_1 + \alpha_2 + \alpha_3) \qquad \text{and} \qquad \mathsf{S}_\beta = 3^{-\frac{1}{2}}(\beta_1 + \beta_2 + \beta_3)$$

the redundancy condition amounts to

$$\mathsf{S}_\alpha + \mathsf{S}_\beta = 0 \tag{9}$$

One can therefore eliminate either S_α or S_β from (8) and thereby reduce the order of the $\mathbf{F}^{(A_1)}$ matrix. For example, if S_β is eliminated (by putting $S_\beta = -S_\alpha$), the $\mathbf{F}^{(A_1)}$ matrix assumes the form

F	S_r	S_s	S_α
S_r	F_{rr}	F_{rs}	$F_{r\alpha} - F_{r\beta}$
S_s		F_{ss}	$F_{s\alpha} - F_{s\beta}$
S_α			$F_{\alpha\alpha} + F_{\beta\beta} - 2F_{\alpha\beta}$

The corresponding appropriate $\mathbf{G}^{(A_1)}$ matrix is then obtained by *merely omitting its row and column corresponding to* S_β. In the present case it would have been equally correct to eliminate S_α from the potential energy expression and then ignore the corresponding row and column of $\mathbf{G}^{(A_1)}$.

Since there is no way of giving independent numerical values to $F_{r\alpha}$ and $F_{r\beta}$, but only to the combination $F_{r\alpha} - F_{r\beta}$, it is perfectly allowable (but quite arbitrary) to give $F_{r\beta}$ the value zero. Similarly one can put $F_{s\beta}$, $F_{\beta\beta}$, and $F_{\alpha\beta}$ all equal to zero by the same argument. If this is done, the result is the same as if the S_β row and column had been omitted from both the **F** and **G** matrices. Similarly it would have been equally correct to omit the rows and columns labeled with S_α.

The proof of the correctness of this procedure may now be given for a more general case, namely, one in which there are r independent redundancies,[1] expressed by the conditions:

$$\sum_{k=1}^{n} a_{ck}S_k = 0 \qquad c = 1, 2, \ldots, r \tag{10}$$

First, it may be assumed without loss of generality that (10) may be used to eliminate the last r symmetry coordinates, namely, S_{n-r+1}, S_{n-r+2}, . . . , S_n (the coordinates can merely be renumbered). The first step is the elimination of these coordinates from the potential energy. This can be done in a stepwise fashion, each step of which would be similar to the process illustrated for the X_3CY molecule above.

When this has been done, the result is an expression for the potential energy in terms of a set of $n - r$ independent coordinates. It is then required to construct a **G** matrix in terms of these same independent coordinates. But examination of the rules for the construction of the elements of the **G** matrix shows that any element such as G_{kl} depends on the nature of the coordinates S_k and S_l and not at all on what other coordinates there may be. Consequently the $n - r$ rows and columns corresponding to the independent coordinates are calculated by the same rules whether or not the redundant coordinates are utilized. The rule is

[1] The remarks to follow are phrased in terms of symmetry coordinates, but actually apply to any secular determinant.

therefore proved that the **G** matrix to use is obtained by omitting the rows and columns for the coordinates which have been declared redundant.

In closing this section, a few remarks on the occurrence of redundancies may be made. For complex molecules it may be difficult to recognize a redundancy in advance. If a sufficient number of internal coordinates is used, the number of redundancies can be found by subtracting $3n - 6$ (the number of internal degrees of freedom) from the number of internal coordinates. However, essential internal coordinates may have been inadvertently omitted. The final test is the rank of the **G** matrix, that is, the dimension of the largest nonvanishing determinant contained in it. This should equal the true number of internal degrees of freedom.[1]

If the molecule has any symmetry, these considerations can be applied separately to each species. The number of independent coordinates in each species can be obtained by reducing the representation formed by the cartesian coordinates and subtracting the translations and rotations appropriately. The number of internal symmetry coordinates is similarly obtained for each species, and any excess represents redundancy. Likewise the rank of $\mathbf{G}^{(\gamma)}$ should equal the number of independent coordinates of species γ.

One very common form of redundancy involves the angles between the bonds attached to a single atom. If there are s noncoplanar bonds, there will be only $3(s + 1) - 6 - s = 2s - 3$ independent bond angles, as illustrated in the CH_4 case. Redundancies are also likely to appear in rings and these may involve the bond stretching coordinates as well as the bond angles. Some of these points are illustrated in Chap. 10 for the case of benzene.

[1] A method of selecting a set of internal coordinates which includes all essential vibrational motions has been described by J. C. Decius, *J. Chem. Phys.*, **17**: 1315 (1949).

CHAPTER 7

VIBRATIONAL SELECTION RULES AND INTENSITIES

The fundamental principles upon which the calculation of selection rules are based have been given in Secs. 3-4, 3-5, and 3-6. In this chapter these principles will be applied to the problem of determining the vibrational selection rules for symmetrical molecules. It will be found that certain transitions are forbidden merely because of the symmetry properties of the molecule. Other transitions are found not to be forbidden by symmetry considerations; such transitions may nevertheless be missed experimentally because of low intensity due to other causes. On the other hand, transitions forbidden by symmetry sometimes seem to appear in the spectra of liquids, presumably due to the distortion of the symmetry by the neighboring molecules. However, in spite of the fact that so-called "forbidden" transitions may occur weakly in liquids and so-called "allowed" transitions are quite frequently not observed, the selection rules given by symmetry considerations are of very great importance as a guide in the interpretation of molecular spectra.

In the discussion which follows, the harmonic oscillator functions will be employed as vibrational wave functions. The results obtained, however, are equally applicable to anharmonic oscillators for the following reasons. The wave functions for a molecule whose vibrations are anharmonic can always be expressed as linear combinations of the harmonic oscillator functions.

$$\psi = \sum_v a_v \psi_v$$

Furthermore, all these functions, ψ_v, which appear together in a given linear combination must transform exactly alike under the molecular symmetry operations. Thus the wave functions, ψ, of the anharmonic molecule belong to the same symmetry species as do those of the harmonic oscillator, and the symmetry species for a state of a real molecule can be identified with the species of the harmonic oscillator function which most closely approximates it and therefore contributes the largest term $a_v\psi_v$ to the linear combination.

For this reason the group theoretical methods employed in this chapter are of great generality, *i.e.*, they are not restricted to the special case assumed in Chap. 3, in which the wave functions were those of a set of harmonic oscillators and the expansion of the electric moment in the normal coordinates contained no terms higher than the first power in such coordinates. Therefore, the conditions developed here for selection rules will be necessary but not sufficient for a transition to occur in such a model. For instance, no vibrational transitions are prohibited for the water molecule by virtue of the symmetry of its equilibrium configuration ($\mathcal{C}_{2v}$), but many transitions would be prohibited by the additional restrictions imposed by the type of wave equation and electric moment expansion used in Chap. 3.

7-1. The Symmetry of Wave Functions

Just as the normal coordinates of a symmetrical molecule show certain symmetry properties, so do the vibrational wave functions. For example, the wave function for the lowest vibrational state of a molecule is (see Sec. 3-3)

$$\psi = \mathfrak{N} \exp\left[-\left(\frac{\gamma_1 Q_1^2}{2}\right) - \left(\frac{\gamma_2 Q_2^2}{2}\right) - \cdots - \left(\frac{\gamma_{3N-6} Q_{3N-6}^2}{2}\right)\right] \tag{1}$$

in which[1] $\gamma_k = 4\pi^2\nu_k/h$, ν_k being the classical frequency of vibration associated with the normal coordinate Q_k. This wave function is unchanged by the application of any of the symmetry operations of the molecule, for the following reasons. The argument of the exponential in (1) is

$$\tfrac{1}{2}(\gamma_1 Q_1^2 + \gamma_2 Q_2^2 + \cdots)$$

The nondegenerate Q's are always transformed into $\pm Q$, so that for them Q^2 is unaltered. A degenerate pair Q_k, Q_l always transforms so that $Q_k^2 + Q_l^2$ is invariant, with similar results for a triply degenerate set Q_k, Q_l, Q_m, since otherwise the potential energy V in terms of the Q's would not be completely symmetrical. But these same combinations of Q's occur in the exponential in (1) because, if $\nu_k = \nu_l$, then $\gamma_k = \gamma_l$. Therefore, the exponential and consequently ψ is invariant.

The wave function which corresponds to the state in which one quantum of one of the normal vibrations Q_k is excited is

$$\psi = \mathfrak{N} U Q_k \tag{2}$$

where

$$U = \exp \sum_{k=1}^{3N-6} \left(-\frac{\gamma_k Q_k^2}{2}\right) \tag{3}$$

It is evident, therefore, that such a function has the same symmetry properties as Q_k itself, since the exponential is completely symmetrical.

[1] This symbol is not to be confused with γ employed as the index of an irreducible representation (symmetry species).

In particular, if Q_k and Q_l correspond to the same classical frequency $\nu_k = \nu_l$, then the two corresponding wave functions of the type given in (2) will correspond to the same energy and will form the basis of an irreducible representation of the point group of the molecule, just as do Q_k and Q_l themselves. In fact, the representation formed by the ψ's will be identical with that formed by the Q's.

In general it will be true that the set of wave functions belonging to any one energy level will be transformed into linear combinations of one another by the symmetry operations of the group, so that they form the basis of a representation of the symmetry point group of the molecule. This follows from the fact that the wave equation is unaltered by the symmetry transformations of the group so that the two sets of solutions, one obtained by solving the original equation, the other by solving the transformed equation, cannot be independent of one another.

The ground state will have the symmetry of the molecule itself, as has been seen, while the fundamental levels will have the same symmetry as the corresponding normal coordinates. Higher levels—overtone and combination levels—will have other symmetries which are not as immediately obvious, but which can be obtained by the methods of Secs. 7-2 and 7-3.

7-2. Symmetry of Combination Levels

A combination level has been defined in Sec. 3-2 as a level in which two or more quantum numbers have values greater than zero, these quantum numbers corresponding to normal coordinates of different frequency. Consider, for example, the binary combination $v_k = 1$, $v_l = 1$, all other v's $= 0$. The wave function for such a state has the form

$$\psi = \mathfrak{N} U Q_k Q_l \tag{1}$$

If ν_k and ν_l are each nondegenerate frequencies, then the energy level corresponding to the above ψ must be a nondegenerate level (except in cases of accidental degeneracy) and the representation of the point group of the molecule formed by the transformations of ψ has the dimension unity. Also, if $Q_k \xrightarrow{R} \chi_R^{(k)} Q_k$ and $Q_l \xrightarrow{R} \chi_R^{(l)} Q_l$, where $\chi_R^{(k)}$ and $\chi_R^{(l)}$ are the characters of the irreducible representations to which Q_k and Q_l belong,[1] then

$$\psi \xrightarrow{R} \chi_R^{(k)} \chi_R^{(l)} \psi \tag{2}$$

[1] In Chaps. 5 and 6, γ was usually used as an index for an irreducible representation where it was desired to emphasize the relations between the irreducible representations of the group and its subgroups. In the present chapter, γ will frequently be replaced by k, l, etc., corresponding to the normal coordinates Q_k, Q_l, etc., which are the bases of specific irreducible representations.

Consequently, the character for the operation R of the representation formed by ψ is $\chi_R^{(k)}\chi_R^{(l)}$, the product of the corresponding characters of the two normal coordinates.

If ν_k, say, is degenerate while ν_l is nondegenerate, then the energy level will be degenerate to the same extent that ν_k is. For illustration, suppose that ν_k is doubly degenerate, then there will be two wave functions, ψ_a and ψ_b, given by

$$\begin{aligned} \psi_a &= \mathfrak{N} U Q_{ka} Q_l \\ \psi_b &= \mathfrak{N} U Q_{kb} Q_l \end{aligned} \tag{3}$$

in which Q_{ka} and Q_{kb} represent the pair of normal coordinates associated with ν_k. Suppose that

$$\begin{aligned} Q_{ka} &\xrightarrow{R} R_{aa} Q_{ka} + R_{ab} Q_{kb} \\ Q_{kb} &\xrightarrow{R} R_{ba} Q_{ka} + R_{bb} Q_{kb} \end{aligned} \tag{4}$$

while

$$Q_l \xrightarrow{R} \chi_R^{(l)} Q_l$$

Then

$$\begin{aligned} \psi_a &\xrightarrow{R} \chi_R^{(l)} R_{aa} \psi_a + \chi_R^{(l)} R_{ab} \psi_b \\ \psi_b &\xrightarrow{R} \chi_R^{(l)} R_{ba} \psi_a + \chi_R^{(l)} R_{bb} \psi_b \end{aligned} \tag{5}$$

The character for this transformation is

$$\chi_R^{(l)}(R_{aa} + R_{bb}) = \chi_R^{(l)} \chi_R^{(k)} \tag{6}$$

where $\chi_R^{(k)} = R_{aa} + R_{bb}$ is the character of the transformation of Q_{ka} and Q_{kb}. Hence, in this case also, the representation formed by the wave functions for the combination level has characters equal to the products of the corresponding characters of the representations formed by Q_{ka}, Q_{kb}, and Q_l.

Finally, if both ν_k and ν_l are degenerate, the same rule applies. For, if

$$\begin{aligned} Q_{ka} &\xrightarrow{R} R_{aa}^{(k)} Q_{ka} + R_{ab}^{(k)} Q_{kb} \\ Q_{kb} &\xrightarrow{R} R_{ba}^{(k)} Q_{ka} + R_{bb}^{(k)} Q_{kb} \end{aligned}$$

with similar equations for Q_{la} and Q_{lb}, then

$$\begin{aligned} \psi_{aa} &\xrightarrow{R} R_{aa}^{(k)} R_{aa}^{(l)} \psi_{aa} + \text{other terms} \\ \psi_{ab} &\xrightarrow{R} R_{aa}^{(k)} R_{bb}^{(l)} \psi_{ab} + \text{other terms} \\ \psi_{ba} &\xrightarrow{R} R_{bb}^{(k)} R_{aa}^{(l)} \psi_{ba} + \text{other terms} \\ \psi_{bb} &\xrightarrow{R} R_{bb}^{(k)} R_{bb}^{(l)} \psi_{bb} + \text{other terms} \end{aligned} \tag{7}$$

so that the character for the transformation of ψ_{aa}, etc., is

$$(R_{aa}^{(k)} + R_{bb}^{(k)})(R_{aa}^{(l)} + R_{bb}^{(l)}) = \chi_R^{(k)} \chi_R^{(l)}$$

Here ψ_{aa} is the wave function containing $Q_{ka}Q_{la}$; ψ_{ab} contains $Q_{ka}Q_{lb}$, etc.

If ν_k, ν_l, or both are triply degenerate, the rule applies equally well, so that in general to find the symmetry species of the wave functions comprising a combination level, first form the products of the corresponding characters of the representations formed by the normal coordinates which are excited. This gives the characters for the new representation, so that Eq. (1), Sec. 5-9,

$$n^{(\gamma)} = \frac{1}{g} \sum g_j \chi_j^{(\gamma)*} \chi_j \tag{8}$$

can be applied to determine the structure of the new representation, that is, the symmetries of the states composing the degenerate excited energy level. The representation formed by the wave functions for a combination level is said to be the *direct product* of the representations formed by the two sets of normal coordinates. This is depicted symbolically by the equation

$$\Gamma = \Gamma^{(k)} \times \Gamma^{(l)}$$

where $\Gamma^{(k)}$ is the symbol for the representation formed by the normal coordinates associated with ν_k, $\Gamma^{(l)}$ is the representation associated with ν_l, and Γ is the symbol for the representation for the combination level which may involve more than one irreducible representation.

An example may clarify the method. Suppose that Q_{ka} and Q_{kb} are two degenerate normal coordinates of methane ($\mathfrak{T}_d$) of species E, while Q_{la}, Q_{lb}, and Q_{lc} are taken as of species F_2. The combination level with one quantum of each vibration excited will have a symmetry $\Gamma = E \times F_2$. To find the structure of Γ, use is made of the following character table (see Appendix X).

CHARACTER TABLE FOR $\mathfrak{T}_d$

$\mathfrak{T}_d$	E	$8C_3$	$3C_2$	$6\sigma_d$	$6S_4$
A_1	1	1	1	1	1
A_2	1	1	1	−1	−1
E	2	−1	2	0	0
F_1	3	0	−1	−1	1
F_2	3	0	−1	1	−1
$E \times F_2$	6	0	−2	0	0

The last line gives the characters for Γ which are the products of the characters for E and for F_2. By the use of (8), or by inspection of the table, the components of Γ can be found to be

$$\Gamma = F_1 + F_2 \tag{9}$$

The meaning of this statement is that the six functions ψ_{aa}, ψ_{ab}, ψ_{ac}, ψ_{ba}, ψ_{bb}, ψ_{bc} can be formed into six linear combinations such that three of them transform into one another so as to form a representation of species F_1, while the other three linear combinations form a representation of species F_2. It will be shown in Sec. 8-6 that the energy level under discussion, which is sixfold degenerate when only the quadratic part of the potential energy is considered, will split into two levels under the influence of the anharmonic portion of V (cubic and higher terms). One of these levels is of species F_1, the other of species F_2.

In this way direct product tables for the irreducible representations of the symmetry point groups have been constructed and are useful in determining the symmetries of combination levels. Table X-12 summarizes rules through which straightforward derivations of these direct products may be achieved.

Combination levels which involve more than two frequencies can be treated in exactly the same way. The symmetry of such a level is the direct product of the irreducible representations of the fundamentals involved; *i.e.*,

$$\Gamma = \Gamma^{(k)} \times \Gamma^{(l)} \times \Gamma^{(m)} \tag{10}$$

Although this discussion has assumed that no $v_k > 1$, the following sections will show how to determine the symmetry of an individual level when $v_k > 1$, after which the over-all symmetry can be computed by a simple extension of the concepts of the present section.

7-3. Symmetry Species of Overtone Levels[1]

An overtone level is one for which all quantum numbers v_k are zero except those associated with some one frequency ν_l. If only one quantum of this frequency is excited the level is a fundamental level, but if more than one quantum of the same frequency is excited, the level is an overtone level. If the frequency which is excited is nondegenerate, all the corresponding overtone levels are likewise nondegenerate, but, as shown in Sec. 3-2, the overtone levels of a doubly degenerate frequency have a degeneracy $v + 1$, while the overtones of a triply degenerate frequency have a degeneracy $\frac{1}{2}(v + 1)(v + 2)$. The quantum number v is the sum of the quantum numbers v_{la}, v_{lb}, etc., associated with the degenerate frequency.

Nondegenerate Overtones. Consider first nondegenerate overtone levels. The wave function ψ for such a level is

$$\psi = \mathfrak{N} U H_{v_l}(\gamma_l^{\frac{1}{2}} Q_l) \tag{1}$$

[1] L. Tisza, *Z. Physik*, **82**: 48 (1933).

H_v being the vth Hermite polynomial (see Sec. 3-3 and Appendix III). These polynomials are even functions of Q_l if v_l is even and odd functions if v_l is odd. From this it follows that for v_l even

$$\psi \xrightarrow{R} \psi \tag{2}$$

For v_l odd,

$$\psi \xrightarrow{R} \chi_R^{(l)}\psi \tag{3}$$

if $Q_l \xrightarrow{R} \chi_R^{(l)} Q_l$. Consequently, when the fundamental is nondegenerate, the even overtones (v_l even) are completely symmetrical (*i.e.*, of the same symmetry as the ground state) while odd overtones have the same symmetry as the corresponding fundamental level—which is in turn the same as the appropriate normal coordinate Q_l.

Degenerate Fundamentals. The overtones of degenerate frequencies provide a somewhat more difficult problem. As an illustration, consider the level with $v = 3$ of a doubly degenerate fundamental. The four states have the quantum numbers (3,0), (2,1), (1,2), and (0,3). The wave functions are (see Sec. 3-3)

$$\begin{aligned} \psi_{3,0} &= \mathfrak{N}_{3,0}U(8\gamma^{\frac{3}{2}}Q_a^3 - 12\gamma^{\frac{1}{2}}Q_a) \\ \psi_{2,1} &= \mathfrak{N}_{2,1}U(4\gamma Q_a^2 - 2)2\gamma^{\frac{1}{2}}Q_b \\ \psi_{1,2} &= \mathfrak{N}_{1,2}U2\gamma^{\frac{1}{2}}Q_a(4\gamma Q_b^2 - 2) \\ \psi_{0,3} &= \mathfrak{N}_{0,3}U(8\gamma^{\frac{3}{2}}Q_b^3 - 12\gamma^{\frac{1}{2}}Q_b) \end{aligned} \tag{4}$$

It is clear that each of these contains a term which is cubic in Q_a, Q_b and a term which is linear in these coordinates. If a transformation R is applied to these wave functions, where

$$\begin{aligned} Q_a &\xrightarrow{R} R_{aa}Q_a + R_{ab}Q_b \\ Q_b &\xrightarrow{R} R_{ba}Q_a + R_{bb}Q_b \end{aligned} \tag{5}$$

it is evident that the cubic terms will remain cubic terms and the linear terms will remain linear terms after the transformation. Since the ψ's belonging to this degenerate energy level must form a representation of the group, the effect of any operation R on any one of the wave functions must be to generate a linear combination of these same wave functions; *i.e.*,

$$\psi_{vv'} \xrightarrow{R} \sum_{v'',v'''} R_{vv',v''v'''}\psi_{v''v'''} \tag{6}$$

in which the sum is over the quantum number pairs associated with the one energy level in question. In this transformation the terms of each power in the Q's must transform in the same way. Therefore, to study the transformation of the ψ's it is only necessary to study the transformation of the highest terms in them, in this case the cubic terms. In determining the character of a transformation, it does not matter what coordinate system is used (Sec. 5-5). It is therefore convenient for each operation R to use a set of normal coordinates Q_a and Q_b such that[1]

$$\begin{aligned} Q_a &\xrightarrow{R} R_a Q_a \\ Q_b &\xrightarrow{R} R_b Q_b \end{aligned} \tag{7}$$

so that $\chi(R) = R_a + R_b$. Naturally the same choice of Q_a and Q_b will not diagonalize the transformations for every R, but for each R some Q_a and Q_b can be found which will have this property. Since Q_a and Q_b are a degenerate pair, any orthogonal linear combinations of them are equally good as normal coordinates (see Sec. 2-3). Then

$$\begin{aligned} Q_a^3 &\xrightarrow{R} R_a^3 Q_a^3 \\ Q_a^2 Q_b &\xrightarrow{R} R_a^2 R_b Q_a^2 Q_b \\ Q_a Q_b^2 &\xrightarrow{R} R_a R_b^2 Q_a Q_b^2 \\ Q_b^3 &\xrightarrow{R} R_b^3 Q_b^3 \end{aligned} \tag{8}$$

so that

$$\chi_3(R) = R_a^3 + R_a^2 R_b + R_a R_b^2 + R_b^3 \tag{9}$$

if the symbol $\chi_v(R)$ is used for the character of the transformation of the functions for the level with quantum number v. Similarly

$$\chi_2(R) = R_a^2 + R_a R_b + R_b^2 \tag{10}$$

Furthermore, the character $\chi(R^v)$ for the transformation of Q_a and Q_b under the operation which is the vth power of R is evidently, by repeated application of (7),

$$\chi(R^v) = R_a^v + R_b^v \tag{11}$$

Comparison of Eqs. (9) to (11) shows that

$$\chi_3(R) = \tfrac{1}{2}[\chi(R)\chi_2(R) + \chi(R^3)] \tag{12}$$

A little consideration indicates that this result can be generalized, with the result that

$$\chi_v(R) = \tfrac{1}{2}[\chi(R)\chi_{v-1}(R) + \chi(R^v)] \tag{13}$$

an equation which can be used to calculate the character for any overtone of a doubly degenerate fundamental.

For overtones of triply degenerate frequencies the characters are given by the relation

$$\chi_v(R) = \tfrac{1}{3}\left(2\chi(R)\chi_{v-1}(R) + \tfrac{1}{2}\left\{\chi(R^2) - [\chi(R)]^2\right\}\chi_{v-2}(R) + \chi(R^v)\right)$$

[1] For certain operations, R_a and R_b may be complex numbers.

which can be checked by the same methods as used for the doubly degenerate case.[1] Here $\chi_1(R) = \chi(R)$, $\chi_0(R) = 1$, and $\chi_{-k}(R) = 0$.

With these equations it is possible to calculate the characters and then by the use of Eq. (8), Sec. 7-2, to find the species of any overtone level. Table X-13 summarizes the results of such calculations. As an example consider the states with $v = 2$, 3, and 4 for the doubly degenerate frequency of methane. The fundamental has the species E. The necessary steps and the final results are included in Table 7-1. The first row lists the classes of the operations in $\mathfrak{T}_d$. The next three rows list the classes of the operations which are the square, cube, and fourth power, respectively, of operations of the classes in the first row. $\chi(R)$ is the character of the irreducible representation E for the operation R. The final species of the states with $v = 2$, 3, and 4 are given in the last column and are obtained by the use of Eq. (8), Sec. 7-2, or by inspection.

TABLE 7-1. SYMMETRIES OF THE OVERTONES OF E IN $\mathfrak{T}_d$

R	E	$8C_3$	$3C_2$	$6S_4$	$6\sigma_d$	Species
R^2	E	C_3	E	C_2	E	
R^3	E	E	C_2	S_4	σ_d	
R^4	E	C_3	E	E	E	
$\chi(R)$	2	−1	2	0	0	E
$\chi(R^2)$	2	−1	2	2	2	
$\chi(R^3)$	2	2	2	0	0	
$\chi(R^4)$	2	−1	2	2	2	
$\chi_2(R)$	3	0	3	1	1	$A_1 + E$
$\chi_3(R)$	4	1	4	0	0	$A_1 + A_2 + E$
$\chi_4(R)$	5	−1	5	1	1	$A_1 + 2E$

Degenerate Fundamentals; an Alternate Method.[2] Since the method for finding the character just described depends upon recursion formulas involving the total vibrational quantum number, v, it would lead to rather lengthy computations for the higher overtones. A method is described in Appendix XIV whereby the characters $\chi_v(R)$ can be expressed in a closed formula. This method may also prove advantageous in discussing problems in which the interaction energy between vibration and rotation is taken into consideration.

The formula for the character of an overtone of a doubly degenerate species as found in Appendix XIV is

$$\chi_v(R) = \sum_{|l|} \chi_{|l|}(R)$$

[1] For the so far unused cases of degeneracies 4 and 5, see Tisza, *Z. Physik*, **82**: 48 (1933).

[2] J. C. Decius, *J. Chem. Phys.*, **17**: 504 (1949).

where $|l| = 1, 3, 5, \ldots, v$ if v is odd or $|l| = 0, 2, 4, \ldots, v$ if v is even. The value of $\chi_{|l|}(R)$ depends upon the following conditions:

If $|l| = 0$, then $\chi_{|0|}(R) = 1$

If $|l| \neq 0$, $\chi_R^{(\gamma)} = 0$, $\chi_{R^2}^{(\gamma)} > 0$, then $\chi_{|l|}(R) = 0$

Otherwise $\chi_{|l|}(R) = 2 \cos l\alpha_R$

The angle, α_R, may be determined from the fact that

$$\chi_R^{(\gamma)} = 2 \cos \alpha_R$$

In the triply degenerate case,

$$\chi_v(R) = \sum_l \chi_l(R)$$

where $l = 0, 2, 4, \ldots, v$ if v is even or $l = 1, 3, 5, \ldots, v$ if v is odd. If $\chi_R^{(\gamma)} \neq 0$ and $\chi_R^{(\gamma)} = [\chi_R^{(\gamma)}]^2 - 2\chi_R^{(\gamma)}$ or if $\chi_R^{(\gamma)} = 0$ and $\chi_{R^3}^{(\gamma)} = +3$, the expression for $\chi_l(R)$ is

$$\chi_l(R) = \frac{\sin [(2l + 1)\alpha_R/2]}{\sin (\alpha_R/2)}$$

otherwise

$$\chi_l(R) = \frac{\cos [(2l + 1)\alpha_R/2]}{\cos (\alpha_R/2)}$$

The value of α_R is determined by the formula

$$\chi_R^{(\gamma)} = \pm 1 + 2 \cos \alpha_R$$

in the triply degenerate case, where, as usual, the plus sign is used for proper operations and the minus sign for improper operations.

7-4. Symmetry Species of a General Vibrational Level

The results of the two previous sections make it a simple matter to deduce the symmetry of the most general excited vibrational level. As an example, consider the vibrational state of methane in which the A_1 mode is doubly excited, the E mode singly excited, one of the two F_2 modes singly, the other doubly excited. This could be represented symbolically by stating that $v_1 = 2$, $v_2 = 1$, $v_3 = 1$, $v_4 = 2$, where $\Gamma^{(1)} = A_1$, $\Gamma^{(2)} = E$, $\Gamma^{(3)} = \Gamma^{(4)} = F_2$. Using the results of Sec. 7-3, the symmetry species of each multiply excited frequency could be computed with the result that

$$(\Gamma^{(1)})^2 = A_1 \tag{1}$$
$$(\Gamma^{(4)})^2 = (F_2)^2 = A_1 + E + F_2 \tag{2}$$

Now it is easily seen that the combination species of wave functions of distinct frequencies are obtained by exactly the methods of Sec. 7-2; in short the characters of the combination are obtained as products of the characters of the individual frequencies. In the present case the desired product can be expressed symbolically as

$$\Gamma = (\Gamma^{(1)})^2 \times \Gamma^{(2)} \times \Gamma^{(3)} \times (\Gamma^{(4)})^2 \tag{3}$$

The reader should note carefully that for the *degenerate* cases

$$(\Gamma^{(k)})^2 \neq \Gamma^{(k)} \times \Gamma^{(k)}$$

for example, $(F_2)^2 = A_1 + E + F_2$, but $F_2 \times F_2 = A_1 + E + F_1 + F_2$ the former expression representing the species of a doubly excited F_2 mode, of degeneracy 6, the latter the combination species of two distinct F_2 modes, each singly excited, of degeneracy 9.

In the present example, the desired expression is thus of the form

$$\begin{aligned} \Gamma &= A_1 \times E \times F_2 \times (A_1 + E + F_2) \\ &= E \times F_2 + E \times F_2 \times E + E \times F_2 \times F_2 \end{aligned} \tag{4}$$

making use of the fact that the totally symmetric species, A_1, behaves like unity in a direct product. The final reduction of an expression like (4) can be completed either by computing the products of characters $\chi^{(E)}\chi^{(F_2)}$, $(\chi^{(E)})^2\chi^{(F_2)}$, etc., and then using the standard reduction formula, or by repeated application of the rules for determining direct products as given in Table X-12. If the latter procedure is followed, use of

$$\begin{aligned} E \times F_1 = E \times F_2 &= F_1 + F_2 \\ F_1 \times F_2 &= A_2 + E + F_1 + F_2 \\ F_2 \times F_2 &= A_1 + E + F_1 + F_2 \end{aligned} \tag{5}$$

in (4) yields

$$\begin{aligned} \Gamma &= F_1 + F_2 + (F_1 + F_2) \times E + (F_1 + F_2) \times F_2 \\ &= F_1 + F_2 + 2(F_1 + F_2) + (A_2 + E + F_1 + F_2) \\ &\qquad + (A_1 + E + F_1 + F_2) \\ &= A_1 + A_2 + 2E + 5F_1 + 5F_2 \end{aligned} \tag{6}$$

Clearly, this procedure can be applied to the most general case, expressed symbolically by

$$\Gamma = (\Gamma^{(1)})^{v_1} \times (\Gamma^{(2)})^{v_2} \times \cdots \times (\Gamma^{(f)})^{v_f} \tag{7}$$

where f is the number of distinct fundamental frequencies.

7-5. Symmetry of the Components of the Electric Moment

The vibrational selection rules for the infrared spectra are determined by the values of the integrals

$$\int \psi_{v''}^* \mu_x \psi_{v'} \, d\tau_v \qquad \int \psi_{v''}^* \mu_y \psi_{v'} \, d\tau_v \qquad \int \psi_{v''}^* \mu_z \psi_{v'} \, d\tau_v$$

as was shown in Sec. 3-5. Here $\psi_{v'}$ is the vibrational wave function for the vibrational state v', $\psi_{v''}$ is the function for the other state v'', μ_x, μ_y, and μ_z are the components of the electric moment $\boldsymbol{\mu}$ in the rotating system of axes, and $d\tau_v$ is the volume element $dQ_1 \cdot dQ_2 \dots, dQ_{3N-6}$. It is the purpose of this chapter to find the conditions under which the above integrals vanish because of the symmetry of $\psi_{v''}$ and $\psi_{v'}$; consequently it is necessary to know the symmetry properties of μ_x, μ_y, μ_z. It will actually be found that these transform in the same way as x, y, and z.

By definition (Sec. 3-5) these components are

$$\begin{aligned} \mu_x &= \sum_{\alpha=1}^{N} e_\alpha x_\alpha \\ \mu_y &= \sum_{\alpha=1}^{N} e_\alpha y_\alpha \\ \mu_z &= \sum_{\alpha=1}^{N} e_\alpha z_\alpha \end{aligned} \tag{1}$$

where e_α is the effective charge on the αth atom. Since symmetry operations will only mix coordinates of equivalent atoms and since equivalent atoms have the same effective charge, the problem of determining the transformation properties of μ_x, μ_y, and μ_z reduces to that of calculating the effect of the symmetry operations upon Σx_α, Σy_α, and Σz_α, the summations being carried out over equivalent atoms only. But a symmetry operation R acting on x_α, for example, may produce two effects. In the first place, it may change the subscript α to some other subscript α', corresponding to another atom equivalent to α, and in the second place it may change x to some linear combination of x, y, z. Since Σx_α includes all atoms α which are equivalent, the first effect of R does not change Σx_α but merely changes the order of the terms. The second effect of R will affect Σx_α in the same way as it affects the coordinate axis x, so that the general conclusion is that μ_x, μ_y, and μ_z will transform in the same way as x, y, and z, themselves.

The translational coordinates T_x, T_y, T_z also transform in the same way as x, y, z (Sec. 6-2). The translational properties of T_x, T_y, T_z for the various point groups are given in Appendix X.

7-6. The Transformation Properties of the Components of the Polarizability

The intensities of Raman lines are determined by the components of the polarizability α, which are defined by the relations

$$\begin{aligned} \mu_x &= \alpha_{xx}\mathcal{E}_x + \alpha_{xy}\mathcal{E}_y + \alpha_{xz}\mathcal{E}_z \\ \mu_y &= \alpha_{yx}\mathcal{E}_x + \alpha_{yy}\mathcal{E}_y + \alpha_{yz}\mathcal{E}_z \\ \mu_z &= \alpha_{zx}\mathcal{E}_x + \alpha_{zy}\mathcal{E}_y + \alpha_{zz}\mathcal{E}_z \end{aligned} \tag{1}$$

Here μ_x, μ_y, and μ_z are the components of the electric moment induced in the molecule by the electric field with the components $\mathcal{E}_x$, $\mathcal{E}_y$, and $\mathcal{E}_z$. The α's are functions of the normal coordinates Q_k and therefore will be transformed by the symmetry transformations R of the molecule. In Appendix XV it is shown that α_{xx}, α_{xy}, etc., transform in the same way as do xx, xy, etc. Consequently, for each point group it is easy to find the structure of the representation formed by the transformations of the polarizability components. This may be done directly, as illustrated in Appendix XV, by writing out enough of the transformations of xx, xy, etc., to get the characters for the representation (remembering that xy and yx are equal as are α_{xy} and α_{yx}), or one may use the method already given for computing the characters for the overtone with $v = 2$ of a triply degenerate frequency. The latter applies to this problem because in both cases the transformation properties of six quantities—xx, yy, zz, xy, yz, and zx in one case, Q_a^2, Q_b^2, Q_c^2, Q_aQ_b, Q_bQ_c, and Q_cQ_a in the other—are being sought.

Using Eqs. (30) and (32), App. XIV, with $v = 2$,

$$\chi_2(R^+) = \frac{\sin (3\alpha_{R^+}/2) \sin 2\alpha_{R^+}}{\sin (\alpha_{R^+}/2) \sin \alpha_{R^+}} \tag{2}$$

$$\chi_2(R^-) = \frac{\cos (3\alpha_{R^-}/2) \sin 2\alpha_{R^-}}{\cos (\alpha_{R^-}/2) \sin \alpha_{R^-}} \tag{3}$$

Use of standard trigonometric expressions for the sine of

$$3\alpha_R/2 = \alpha_R + \alpha_R/2$$

and for $\sin 2\alpha_R$ allows (2) and (3) to be written in the simpler form

$$\chi_2(R^\pm) = 2 \cos \alpha_R^\pm(\pm 1 + 2 \cos \alpha_R^\pm) \tag{4}$$

The value of α_R and the sign option are determined by visualizing the effect of each group operation on a cartesian axis system (cf. Sec. 6-1; the sign is + for a proper, − for an improper rotation by α_R).

As an illustration consider the point group $\mathfrak{D}_{3h}$. Table 7-2 summarizes the calculation of the symmetry species of α. For the first three classes,

TABLE 7-2. SYMMETRY SPECIES OF THE POLARIZABILITY COMPONENTS FOR $\mathfrak{D}_{3h}$

R	E	$2C_3$	$3C_2$	σ_h	$2S_3$	$3\sigma_v$
α_R	0	$2\pi/3$	π	0	$2\pi/3$	0
$2 \cos \alpha_R(\pm 1 + 2 \cos \alpha_R)$	6	0	2	2	2	2

the sign in (4) is positive, for the last three, negative. By the standard reduction method, the structure of the representation having the characters given in the last row of Table 7-2 is found to be

$$\Gamma = 2A_1' + E' + E'' \tag{5}$$

The conclusion is therefore reached that the six polarizability components can be formed into six linear combinations, of which two are of species A_1', one pair of species E', and the remaining pair of species E''.

Examination of the actual transformations of α_{xx}, α_{xy}, etc., shows that the proper combinations are $\alpha_{xx} + \alpha_{yy}$, $\alpha_{zz}(2A_1')$, $\alpha_{xx} - \alpha_{yy}$, $\alpha_{xy}(E')$, α_{xz}, $\alpha_{yz}(E'')$.

7-7. Determination of Selection Rules for Infrared Absorption

It is now possible to outline the cases in which transitions are forbidden by the symmetry of the molecule. Consider first the integrals which determine the intensity of the infrared spectrum, namely,

$$\int \psi_{v'}^* \mu_x \psi_{v''} \, d\tau \qquad \int \psi_{v'}^* \mu_y \psi_{v''} \, d\tau \qquad \int \psi_{v'}^* \mu_z \psi_{v''} \, d\tau \tag{1}$$

Since these are definite integrals over the whole configuration space of the molecule, they should be unchanged by a symmetry operation R, inasmuch as such an operation merely produces a transformation of coordinates.

Suppose at first that $\psi_{v''}$ is the wave function for the ground state, which is completely symmetrical. Furthermore, consider that linear combinations of the dipole moment components, and of the wave functions of the upper state, have been chosen in such a way as to form completely reduced representations. Since these representations may be degenerate, the operation R will have the effect

$$\begin{aligned} \mu_a &\xrightarrow{R} \sum_{a'} R_{aa'}^{(k)} \mu_{a'} \\ \psi_{v'b}^* &\xrightarrow{R} \sum_{b'} R_{bb'}^{(l)*} \psi_{v'b'}^* \end{aligned} \tag{2}$$

where $R_{aa'}^{(k)}$ and $R_{bb'}^{(l)}$ are transformation coefficients of the kth and lth irreducible representations, respectively, while a and b are used as degeneracy indices. Consequently, the effect of R upon the integral $\int \psi_{v'b}^* \mu_a \psi_{v''} \, d\tau$ can be expressed as

$$\int \psi_{v'b}^* \mu_a \psi_{v''} \, d\tau \xrightarrow{R} \sum_{a'b'} R_{bb'}^{(l)*} R_{aa'}^{(k)} \int \psi_{v'b'}^* \mu_{a'} \psi_{v''} \, d\tau = \int \psi_{v'b}^* \mu_a \psi_{v''} \, d\tau \tag{3}$$

Averaging over the g operations of the group yields the result:

$$\int \psi_{v'b}^* \mu_a \psi_{v''} \, d\tau = \frac{1}{g} \sum_{a'b'} \left(\sum_R R_{bb'}^{(l)*} R_{aa'}^{(k)} \right) \int \psi_{v'b'}^* \mu_{a'} \psi_{v''} \, d\tau \tag{4}$$

But the orthogonality theorem in Appendix XI shows that the sum enclosed in parentheses in (4) vanishes unless $\Gamma^{(k)} = \Gamma^{(l)}$ and unless

$a = b$ and $a' = b'$. Consequently, unless some of the functions $\psi_{v'b}$ belong to the same species as μ, all the integrals (1) will vanish and the corresponding transition will be forbidden.

For example, consider the possibility of transitions from the ground state of methane ($\mathfrak{T}_d$) to the level with 3 quanta of the doubly degenerate fundamental frequency excited. This excited level was found in Sec. 7-3 to have the species $A_1 + A_2 + E$. Furthermore, Table X-10 shows that μ_x, μ_y, and μ_z belong to the species F_2. Consequently, the transition is forbidden, since F_2 does not occur in $A_1 + A_2 + E$. As a second example, consider the transition from the ground state to the first overtone level of the triply degenerate frequency of methane. This upper level has the species $A_1 + E + F_2$ and the transition to it from the ground state will therefore be allowed; or more properly, there is allowed a transition from the ground state to three degenerate states with species F_2 of the excited level. The components of this degenerate level which are of species A_1 and E cannot combine with the ground state.

If the wave function $\psi_{v''}$ is not completely symmetrical it is necessary to compare the species of μ, not with the species of the functions $\psi_{v'}$, but with the species of the functions $\psi^*_{v'}\psi_{v''}$. The structure of these products is evidently (from the discussion of Sec. 7-2) the direct product of the species of $\psi^*_{v'}$ and $\psi_{v''}$ separately. For example, consider the possibility of transitions from one to the other of the two excited levels of methane previously used as examples. These had the structures $A_1 + A_2 + E$ and $A_1 + E + F_2$, respectively. Their direct product is

$$\begin{aligned}(A_1 + A_2 + E) \times (A_1 + E + F_2) &= A_1 \times A_1 + A_1 \times E + A_1 \times F_2 \\ &+ A_2 \times A_1 + A_2 \times E + A_2 \times F_2 + E \times A_1 + E \times E + E \times F_2 \\ &= 2A_1 + 2A_2 + 4E + 2F_2 + 2F_1 \quad (5)\end{aligned}$$

Since the species of μ (F_2) occurs, this transition is allowed.

More generally stated, the rule for the existence (nonvanishing) of the integrals $\int \psi^*_{v'}\mu\psi_{v''}\,d\tau$ is that the triple direct product of the species of $\psi^*_{v'}$, μ, and $\psi_{v''}$ must contain the totally symmetric species. Thus if $\Gamma^{(k)}$ is the species of $\psi_{v'}$, $\Gamma^{(l)}$ that of $\psi_{v''}$, and $\Gamma^{(m)}$ that of μ, the direct product[1]

$$\Gamma^{(k)} \times \Gamma^{(l)} \times \Gamma^{(m)}$$

must be formed, and the integral will not necessarily vanish if $\Gamma^{(k)} \times \Gamma^{(l)} \times \Gamma^{(m)}$ contains the totally symmetric species.

It has been shown in Sec. 7-2 that the character of a direct product is the product of the characters of the factors. Therefore the number of

[1] If $\Gamma^{(k)}$ is the species of $\psi_{v'}$, the representation $\Gamma^{(k)*}$ should, strictly speaking, be used for $\psi^*_{v'}$. However, in the subsequent discussion it will be assumed that all the characters are real; therefore the complex conjugate signs may be omitted.

times the totally symmetric species occurs in the triple direct product is

$$n^{(1)} = \frac{1}{g} \sum_R \chi_R^{(k)}(\chi_R^{(l)}\chi_R^{(m)}) \tag{6}$$

since $\chi_R^{(1)} = 1$. If $(\chi_R^{(l)}\chi_R^{(m)})$ is considered to be the character of the reducible representation $\Gamma^{(l)} \times \Gamma^{(m)}$, then $n^{(1)}$ is the number of times $\Gamma^{(k)}$ occurs in the reduction of $\Gamma^{(l)} \times \Gamma^{(m)}$. Therefore if $\Gamma^{(k)}$ does occur in $\Gamma^{(l)} \times \Gamma^{(m)}$, $n^{(1)}$ will be greater than zero and there will be no symmetry requirement that the integral vanish. Clearly there was nothing unique in the choice of l and m from k, l, and m.

When only fundamentals are considered, that is, transitions from the ground level to levels with only one quantum of one vibration excited, the selection rules can be expressed very simply. These fundamentals will be active if they have the species of μ_x, μ_y, or μ_z; otherwise they are inactive. Thus in benzene μ_z has the species A_{2u} while μ_x, μ_y have the species E_{1u}. Therefore, the only infrared active fundamentals are those whose normal coordinates are of species A_{2u} or E_{1u}.

7-8. Selection Rules for the Raman Effect

In Sec. 3-7 it was shown that the selection rules for Raman transitions are determined by the integrals

$$\int \psi_{v'}^* \alpha_{gg'} \psi_{v''} \, d\tau \tag{1}$$

in which $g,g' = x, y$, or z and $\alpha_{gg'}$ is one of the components of the polarizability. These integrals can be treated in exactly the same way as the integrals of the electric moment; that is to say, they vanish unless the symmetry of the set of functions $\psi_{v'}^* \psi_{v''}$ for the two levels under consideration has some species in common with the species of α_{xx}, α_{xy}, etc. For fundamentals it is true here also that this requires that the normal vibrations for a given frequency fall into one of the species associated with the α's, if this frequency is to occur in the Raman effect.

For example, in $\mathfrak{D}_{3h}$ it was found in Sec. 7-6 that the α's have the species $2A_1' + E' + E''$. Consequently, for molecules of this symmetry, such as cyclopropane, only those vibration frequencies with species A_1', E', or E'' should occur as fundamentals in the Raman effect. Since the tables in Appendix X indicate the irreducible representations for μ‡ and for α, it is possible to tell very quickly for any molecule how many fundamental frequencies should be allowed in the infrared and how many in the Raman spectrum.

In Sec. 3-7 it was pointed out that a Raman line could have a depolarization ratio, ρ, less than $\frac{6}{7}$ only if the matrix element

$$\alpha_{v'v''} = \int \psi_{v'}^* \tfrac{1}{3}(\alpha_{xx} + \alpha_{yy} + \alpha_{zz}) \psi_{v''} \, d\tau_v$$

‡ The species of μ are identical with those of T.

did not vanish. Since the sum of the diagonal terms of the polarizability is an invariant under a similarity transformation by any orthogonal matrix (Appendix XV), it must belong to the totally symmetric species of every molecular symmetry group. Therefore only the totally symmetric fundamentals can have $\rho < \frac{6}{7}$.

7-9. Absolute Infrared Absorption Intensities of Fundamentals

In Chap. 3 expressions were given connecting the Einstein coefficients governing the probabilities of transitions to the matrix elements of the dipole moment. In this section a relation will be found between the Einstein coefficients and an experimentally observable quantity, the integrated absorption coefficient.

The Integrated Absorption Coefficient. Although experimental difficulties impede the straightforward evaluation of the absorption coefficient, methods have been devised[1] to overcome the experimental problems, and it is possible to measure the integrated absorption coefficient defined by

$$\int \kappa(\nu)\, d\nu \tag{1}$$

where

$$\kappa = \frac{1}{l} \ln \frac{I_0}{I} \tag{2}$$

I_0 being the initial and I the final intensity of a parallel beam of radiation traversing a length l of absorbing gas.

In differential form (2) becomes

$$-dI = \kappa I\, dl \tag{3}$$

for the energy absorbed per unit area in unit time by a layer of infinitesimal thickness dl. The decrease in intensity of the beam, $-dI$, can also be expressed in terms of the Einstein coefficients $B_{n'n''}$ and $B_{n''n'}$ which were introduced in Sec. 3-4; $B_{n''n'}\rho(\nu)$ is the probability (per molecule) of a transition from the state n'' to the state n' per unit time in the presence of radiation of frequency ν having a density $\rho(\nu)$. If there are $N_{n'}$ molecules per unit volume in the state n' and $N_{n''}$ in the state n'', then the net rate of transitions in unit volume from n'' to n' is

$$(B_{n''n'}N_{n''} - B_{n'n''}N_{n'})\rho(\nu) = B_{n'n''}\rho(\nu_{n'n''})(N_{n''} - N_{n'}) \tag{4}$$

since $B_{n'n''} = B_{n''n'}$.

[1] D. G. Bourgin, *Phys. Rev.*, **29**: 794 (1927).
D. G. Bourgin, *Phys. Rev.*, **32**: 237 (1928).
E. C. Kemble, *J. Chem. Phys.*, **3**: 316 (1935).
L. A. Matheson, *Phys. Rev.*, **40**: 813 (1932).
R. Rollefson and A. H. Rollefson, *Phys. Rev.*, **48**: 779 (1935).
E. B. Wilson, Jr., and A. J. Wells, *J. Chem. Phys.*, **14**: 578 (1946).
A. M. Thorndike, A. J. Wells, and E. B. Wilson, Jr., *J. Chem. Phys.*, **15**: 157 (1947).

Each such transition reduces the energy of the radiation beam by $h\nu_{n'n''}$, so that the decrease in intensity for a length dl with unit cross-sectional area is given by

$$-dI = h\nu_{n'n''}B_{n'n''}\rho(\nu_{n'n''})(N_{n''} - N_{n'})\,dl \tag{5}$$

But the radiation flux intensity and density are related by

$$I = c\rho \tag{6}$$

where c is the velocity of light, so that

$$-dI = \frac{h\nu_{n'n''}}{c} B_{n'n''}(N_{n''} - N_{n'})I\,dl \tag{7}$$

Comparison of (7) with (3) indicates that

$$\kappa = \frac{h\nu_{n'n''}}{c} B_{n'n''}(N_{n''} - N_{n'}) \tag{8}$$

Finally, by Eq. (2), Sec. 3-4,

$$B_{n'n''} = \frac{8\pi^3}{3h^2}[|(\mu_X)_{n'n''}|^2 + |(\mu_Y)_{n'n''}|^2 + |(\mu_Z)_{n'n''}|^2] \tag{9}$$

so that

$$\kappa = \frac{8\pi^3}{3ch} \nu_{n'n''}(N_{n''} - N_{n'})|(\mu)_{n'n''}|^2 \tag{10}$$

In this derivation it has been assumed that the spectral line is perfectly sharp, whereas it is known both from experiment and from a more refined theory[1] that an absorption line always has a finite width. Since this width, usually less than 1 cm^{-1}, is very much less than the center frequency, which will be of the order of 100 to 1,000 cm^{-1} in the infrared, it is reasonable to treat the Einstein coefficient as a constant for a given line, as has been done above, but in place of κ it may seem more realistic to write $\int\kappa(\nu)\,d\nu$ for the total line absorption coefficient.

$$\int_{\text{line } n'n''} \kappa(\nu)\,d\nu = \frac{8\pi^3}{3ch}(N_{n''} - N_{n'})\nu_{n'n''}|(\mu)_{n'n''}|^2 \tag{11}$$

Summation of Intensity Over a Band. A given vibration-rotation band, however, contains a large number of rotational lines over which (11) should be summed if the experimental quantity on the left refers to such a

[1] W. Heitler, "Quantum Theory of Radiation," Oxford, London, 1954. J. H. Van Vleck and V. Weiskopf, *Revs. Mod. Phys.*, **17**: 227 (1945).

band rather than an individual line. To the approximation in which vibrational and rotational energies are separable,

$$\nu_{n'n''} = \nu_{v'v''} + \nu_{R'R''} \tag{12}$$

From Sec. 3-5 it can be shown that

$$\begin{aligned} |(\mu)_{n'n''}|^2 &= \sum_F \left| \sum_g (\Phi_{Fg})_{R'R''}(\mu_g)_{v'v''} \right|^2 \\ &= \sum_{Fgg'} (\Phi_{Fg})^*_{R'R''}(\Phi_{Fg'})_{R'R''}(\mu_g)^*_{v'v''}(\mu_{g'})_{v'v''} \end{aligned} \tag{13}$$

where the $(\Phi_{Fg})_{R'R''}$ are the direction cosine matrix elements between the space-fixed (F) and molecule-fixed (g) axes; $(\mu_g)_{v'v''}$ is defined by

$$(\mu_g)_{v'v''} = \int \psi^*_{v'}\mu_g\psi_{v''}\,d\tau \qquad g = x, y, z \tag{14}$$

In order to calculate the total band intensity for the transition $v'' \rightarrow v'$, it would be necessary to sum expression (11) over all R' and R'', using (12) and (13). Although such calculations can be carried out rigorously, only an approximate treatment will be attempted here. Introducing Boltzmann's expression for the relative populations of the states n' and n'':

$$N_{n'} = N_{n''}e^{-(W_{n'}-W_{n''})/kT}$$

or

$$N_{v'R'} = N_{v''R''}e^{-h(\nu_{v'v''}+\nu_{R'R''})/kT} \tag{15}$$

If now the rotational quantization were neglected, $\nu_{R'R''}$, which assumes both positive and negative values, might be omitted from (12) and (15) and the direction cosine matrix elements replaced by their classical average over all directions.

$$\overline{\Phi^*_{Fg}\Phi_{Fg'}} = \delta_{gg'}\tfrac{1}{3} \tag{16}$$

(see Appendix IV). Then (11) would become

$$\int_{\text{band } v'v''} \kappa(\nu)\,d\nu = \frac{8\pi^3}{3ch}(N_{v''} - N_{v'})\nu_{v'v''} \sum_g |(\mu_g)_{v'v''}|^2 \tag{17}$$

An exact summation[1] over the rotational components, in the case of a parallel band of a symmetric rotor molecule,[2] leads to the expression on the right-hand side of (17) multiplied by the factor

[1] B. L. Crawford, Jr., and H. L. Dinsmore, *J. Chem. Phys.*, **18**: 1682 (1950).

[2] The symmetric rotor has equal principal moments of inertia with respect to the x and y axes; its parallel bands are those for which only $\mu_z^{(k)}$ is different from zero.

$$1 + \frac{2Bc}{\nu_0} \frac{1 + \exp(-h\nu_0/kT)}{1 - \exp(-h\nu_0/kT)} \tag{18}$$

where B is the rotational constant, $B = h/8\pi^2 c I_B$, c the velocity of light, and I_B the moment of inertia perpendicular to the symmetry axis. The magnitudes of ν_0 (the frequency of the band center or the pure vibrational transition frequency) and B are ordinarily such that $(2Bc/\nu_0) < 5$ per cent and at ordinary temperatures $(h\nu_0/kT) > 1$, so that replacement of (18) by unity would rarely result in a gross error exceeding 10 per cent. Since (17) is commonly used to solve for the dipole moment matrix element, the error in the latter quantity would not exceed 5 per cent.

It must still be noted, however, that in the harmonic oscillator approximation, the transitions $v'' = 0 \rightarrow v' = 1$, $v'' = 1 \rightarrow v' = 2$, etc., all coincide exactly (lead to the same $\nu_{v'v''}$), and even in the presence of anharmonicity there will be approximate coincidence. Therefore (17) must still be summed over v''. For a fundamental transition of the type $v'_k = v''_k + 1$, $v'_l = v''_l$ for $l \neq k$, suppose Q_k is nondegenerate. Then

$$\nu_{v'v''} = \nu_k \tag{19}$$

$$N_{v_k} = \frac{N}{\Theta_k} e^{-(h\nu_k/kT)v_k} \tag{20}$$

N is the total number of molecules per unit volume, and where Θ_k is the partition function for the vibration Q_k defined by

$$\Theta_k = (1 - e^{-h\nu_k/kT})^{-1} \tag{21}$$

Furthermore, from Eqs. (7) and (8), Sec. 3-5, and Appendix III, the dipole moment matrix element for the transition $v_k \rightarrow v_k + 1$ is found to be

$$(\mu_g)_{v_k+1,v_k} = \mu_g^{(k)} \left[\frac{h}{8\pi^2\nu_k}(v_k + 1)\right]^{\frac{1}{2}} \tag{22}$$

Introducing the abbreviation $u_k = h\nu_k/kT$ and summing (17) over v_k with the aid of (19), (20), and (22),

$$\int_{\text{band } \Delta v_k = 1} \kappa(\nu)\, d\nu$$

$$= \frac{N\pi}{3c\Theta_k}[(\mu_x^{(k)})^2 + (\mu_y^{(k)})^2 + (\mu_z^{(k)})^2] \sum_{v_k=0}^{\infty} (e^{-u_k v_k} - e^{-u_k(v_k+1)})(v_k + 1) \tag{23}$$

The summation in (23) is readily found to have the value $(1 - e^{-u_k})^{-1}$ which is identical with the partition function Θ_k, so that the final result is

$$\int_{\text{band } \Delta v_k = 1} \kappa(\nu)\, d\nu = \frac{N\pi}{3c}[(\mu_x^{(k)})^2 + (\mu_y^{(k)})^2 + (\mu_z^{(k)})^2] \tag{24}$$

If the molecule possesses sufficient symmetry, only one of the three quantities, $\mu_x^{(k)}$, $\mu_y^{(k)}$, $\mu_z^{(k)}$, will be different from zero,[1] so that (24) can be used to calculate one of these quantities from the integrated band intensity, although there will be an ambiguity of sign. This is true for normal coordinates, Q_k, which belong to nondegenerate species which involve only one component of a translation.

In case ν_k is a degenerate frequency, expressions similar to (24) apply to each of the components of the degeneracy, so that the band intensity should be written

$$\int_{\text{band}} \kappa(\nu)\, d\nu = \frac{N\pi}{3c} \sum_a [(\mu_x^{(ka)})^2 + (\mu_y^{(ka)})^2 + (\mu_z^{(ka)})^2] \tag{25}$$

However, if the axes x, y, z are chosen properly, only the quantities $\mu_x^{(ka)}$, $\mu_y^{(kb)}$, (and $\mu_z^{(kc)}$ if triply degenerate) will be nonvanishing; moreover, since they are equal, they can be determined from (25) except for sign.

Relation between Dipole Moment Derivatives and Bond Moments. If the normal coordinates are known in terms of internal coordinates, it is possible to express dipole moment derivatives with respect to normal coordinates in terms of derivatives with respect to internal coordinates. If it is assumed that the molecular dipole moment is the sum of bond moments, expressions for such bond moments and their derivatives can be obtained in terms of the intensities. Since the normal coordinates and internal coordinates are related by the linear transformations,[2]

$$Q_k = \sum_t (L^{-1})_{kt} S_t \tag{26}$$

it follows that

$$\frac{\partial \mu_g}{\partial S_t} = \sum_k \frac{\partial \mu_g}{\partial Q_k} \frac{\partial Q_k}{\partial S_t} = \sum_k \mu_g^{(k)} (L^{-1})_{kt} \tag{27}$$

The assumption of bond moment additivity mentioned above leads to expressions for $\partial \mu_g / \partial S_t$ in terms of a relatively small number of parameters. For suppose that

$$\boldsymbol{\mu} = \sum_\beta \boldsymbol{\mu}_\beta = \sum_\beta \mu_\beta \mathbf{e}_\beta \tag{28}$$

[1] In case two or three of the $\mu^{(k)}$ differ from zero, it is theoretically possible to evaluate them separately by a study of the individual rotational line intensities, since the rotational selection rules differ for the components of the dipole moment along the three inertial axes.

[2] In practice, the transformation would be expressed in two steps, *i.e.*, from internal to symmetry coordinates and then from symmetry coordinates to normal coordinates.

where $\boldsymbol{\mu}_\beta$ is the moment of the βth bond, μ_β is the *magnitude* of the dipole moment of the βth bond, regarded as a function only of the length r_β of the bond, and $\mathbf{e}_\beta$ is a unit vector along the bond. Then the components of $\boldsymbol{\mu}$ are of the form

$$\mu_g = \sum_\beta \mu_\beta(r_\beta)\Phi_{\beta g} \tag{29}$$

in which $\Phi_{\beta g}$, the direction cosine on the βth bond, with respect to the g axis of the molecule, can be expressed as a function of the angular internal coordinates.

As a simple example of the application of (29) in computing the partial derivatives, $\partial\mu_g/\partial S_t$, consider the ethylene molecule. By symmetry, the x, y, z axes can clearly be chosen so that the x axis lies along the carbon-carbon bond, the y axis is a perpendicular bisector of the carbon-carbon

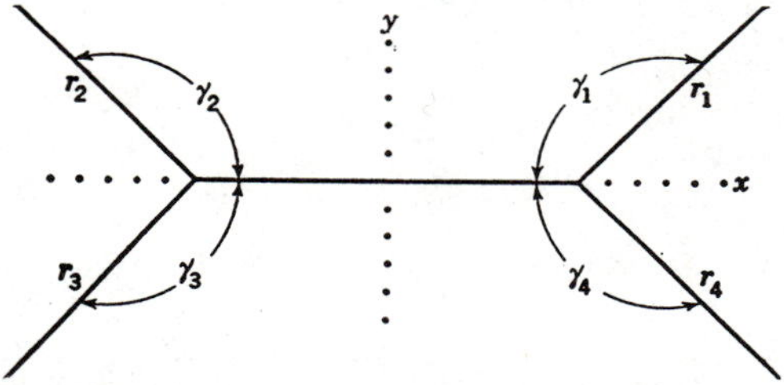

FIG. 7-1. Internal coordinates for ethylene used to express the x and y components of the dipole moment.

bond lying in the plane of the molecule, and the z axis is perpendicular to the plane of the molecule. Reference to Fig. 7-1 then shows that

$$\begin{aligned} \mu_x &= -\mu_1 \cos\gamma_1 + \mu_2 \cos\gamma_2 + \mu_3 \cos\gamma_3 - \mu_4 \cos\gamma_4 \\ \mu_y &= \mu_1 \sin\gamma_1 + \mu_2 \sin\gamma_2 - \mu_3 \sin\gamma_3 - \mu_4 \sin\gamma_4 \end{aligned} \tag{30}$$

Similarly the μ_z component could be expressed in terms of the μ_i's and suitable out-of-plane coordinates. From (30) one immediately obtains

$$\begin{aligned} \left(\frac{\partial\mu_x}{\partial r_1}\right)_0 &= -\left(\frac{d\mu}{dr}\right)_0 \cos\gamma_0 \\ \left(\frac{\partial\mu_y}{\partial r_1}\right)_0 &= \left(\frac{d\mu}{dr}\right)_0 \sin\gamma_0 \\ \left(\frac{\partial\mu_x}{\partial\gamma_1}\right)_0 &= \mu_0 \sin\gamma_0 \\ \left(\frac{\partial\mu_y}{\partial\gamma_1}\right)_0 &= \mu_0 \cos\gamma_0 \end{aligned} \tag{31}$$

in which μ_0 and $(d\mu/dr)_0$ are the equilibrium values of the magnitude and derivative of the magnitude with respect to bond length, respectively, for a single carbon-hydrogen bond; γ_0 is the equilibrium CCH angle. Inspection of the character table for the point group, $\mathcal{V}_h$, of the ethylene molecule shows that the normal coordinates which contribute to the μ_x and μ_y derivatives are of species B_{3u} and B_{2u}, respectively.

Actually the assumption of bond moment additivity does not hold even within 10 per cent,[1] but the procedure just described serves as a framework for the calculation of $\partial\mu_g/\partial S_t$ on the basis of more elaborate models, in which interaction effects between the bonds would presumably be taken into consideration.

The effects of isotopic substitutions on intensities are discussed in Sec. 8-5.

[1] A. M. Thorndike, A. J. Wells, and E. B. Wilson, Jr., *J. Chem. Phys.*, **15**: 157 (1947).

CHAPTER 8

POTENTIAL FUNCTIONS

In previous chapters, it has been assumed that the potential energy function could be expanded in a power series involving the displacement coordinates, and that only the quadratic terms need be considered. Moreover, it was implied that the coefficients in this expression were known constants and that the problem to be solved was the determination of the vibration frequencies as functions of these constants. Actually the force constants have been determined *a priori* for only a few diatomic and very simple polyatomic molecules,[1] so that in practice it is usually the reverse problem which is most important, that is, the determination of the force constants from the known vibrational frequencies. In order to accomplish this, it is usually necessary to make certain simplifying assumptions about the nature of the potential energy. These assumptions will be discussed in this chapter as part of a general treatment of the nature of the potential energy function. Some consideration will also be given to the influence of cubic and higher powers in the potential energy.

8-1. The General Quadratic Potential Function

If cubic and higher terms in V are neglected, then the potential energy can be written in the form

$$2V = \sum_{i,j=1}^{3N} f'_{ij}\xi_i\xi_j \tag{1}$$

in which ξ_i represents one of the $3N$ cartesian displacement coordinates and the f'_{ij} are the force constants. Since there are $3N$ coordinates ξ_i, it might at first sight seem that there were $(3N)^2$ quantities which required evaluation. But, as has been seen previously in Chaps. 2 and 6, these constants are not all independent. In the first place, it is evident that the coefficient of $\xi_i\xi_j = \xi_j\xi_i$ can be split into two parts such that $f'_{ij} = f'_{ji}$. In the second place, although there are $3N$ external coordinates, the potential energy actually depends only upon the internal coordinates,

[1] J. H. Van Vleck and P. C. Cross, *J. Chem. Phys.*, **1**: 357 (1933), have made the necessary quantum mechanical calculation for the H_2O molecule, for example.

which are $3N - 6$ in number for nonlinear molecules. Consequently, V can be written in the form

$$2V = \sum_{t,t'=1}^{3N-6} F_{tt'}S_tS_{t'} \tag{2}$$

where the S_t are internal coordinates. In the most general case there will then be a number of force constants equal to the number of elements on and above the diagonal of a square having $3N - 6$ elements on its side, which is

$$1 + 2 + 3 + \cdots + (3N - 7) + (3N - 6) = \frac{(3N - 6)(3N - 5)}{2} \tag{3}$$

This number applies only to unsymmetrical ($\mathcal{C}_1$) molecules, however, since the existence of symmetry implies further relations between constants and hence a still smaller number of independent parameters. As an illustration, consider the water molecule ($\mathcal{C}_{2v}$). A convenient set of internal coordinates is formed from r_1 and r_2, the stretches of the two OH bonds, and α, the distortion of the valence angle. The potential energy can be written as

$$2V = F_{11}r_1^2 + F_{22}r_2^2 + F_{33}\alpha^2 + 2F_{12}r_1r_2 + 2F_{13}r_1\alpha + 2F_{23}r_2\alpha \tag{4}$$

The symmetry of H_2O obviously requires that $F_{11} = F_{22}$ and that $F_{13} = F_{23}$ so that for this molecule there are only four independent force constants instead of six. The same reasoning may be applied to more complex molecules, the general principle being that whenever a symmetry operation exists which sends the coordinate product $S_tS_{t'}$ into $S_{t''}S_{t'''}$, then $F_{tt'}$ equals $F_{t''t'''}$. Cases also occur in which $S_tS_{t'}$ goes into $-S_{t''}S_{t'''}$; whence $F_{tt'} = -F_{t''t'''}$.

It is probably easier, however, to determine the number of independent force constants from a knowledge of the size of the factors of the secular determinant, made possible by the introduction of symmetry coordinates.[1]

The number of independent constants does not depend on the set of coordinates used to describe V, since a transformation of coordinates changes the original set of force constants $F_{tt'}$ into a new set, whose members $F_{kk'}$ are simply linear combinations of the original ones. Thus if there are $n^{(\gamma)}$ symmetry coordinates of species $\Gamma^{(\gamma)}$, there will be $n^{(\gamma)}(n^{(\gamma)} + 1)/2$ force constants F_{kl}^{γ} associated with this factor. If the symmetry coordinates of this species are degenerate, the factor is repeated identically d_γ times, so that there are still only $n^{(\gamma)}(n^{(\gamma)} + 1)/2$ different force constants. An example may assist in clarifying the method.

[1] J. B. Howard and E. B. Wilson, Jr., *J. Chem. Phys.*, **2**: 630 (1934), footnote 4.

For cyclopropane, it is found that the species of vibrations are

$$\Gamma = 3A_1' + A_2' + 4E' + A_1'' + 2A_2'' + 3E'' \tag{5}$$

Consequently, the total number of independent quadratic force constants is

$$(3 \times \tfrac{4}{2}) + (1 \times \tfrac{2}{2}) + (4 \times \tfrac{5}{2}) + (1 \times \tfrac{2}{2}) + (2 \times \tfrac{3}{2}) + (3 \times \tfrac{4}{2}) = 27$$

Note that $4E'$ contributes 10 and not 2×10 constants, since the two factors are identical.

It is apparent that the number of constants in this general type of potential function is almost always impractically large. It is always larger than the number of fundamental frequencies, except for symmetrical linear triatomic molecules and two or three other very symmetrical cases. Isotopic molecules, particularly those involving deuterium, are helpful in providing additional data for the determination of force constants, since it seems to be a valid assumption[1] that the potential function is unchanged by isotopic substitution. However, there are severe limitations to this method and unique solutions are not always possible.

In concluding this section, a few words of caution must be spoken in connection with the problem of redundancy. It was pointed out in Sec. 6-2 that in order to take advantage of symmetry, redundant coordinates should sometimes be employed. Naturally, the inclusion of such extra coordinates will result in the definition of redundant force constants. Thus when all six bending angles are used in the vibrational treatment of methane, three bending force constants can be defined, namely, the coefficients of α_{12}^2, $\alpha_{12}\alpha_{13}$, and $\alpha_{12}\alpha_{34}$ in the potential energy, designated as F_α, F'_α, and F''_α. When the secular determinant is factored it is found that the following combinations of force constants appear: $\mathsf{F}_{A_1} = F_\alpha + 4F'_\alpha + F''_\alpha$; $\mathsf{F}_E = F_\alpha - 2F'_\alpha + F''_\alpha$; and $\mathsf{F}_{F_2} = F_\alpha - F''_\alpha$. However, the symmetry coordinate corresponding to the A_1 factor is the redundancy condition (to the first order) so that no frequency exists which would yield the value of $F_\alpha + 4F'_\alpha + F''_\alpha$. Therefore it is impossible to determine more than two of the three parameters F_α, F'_α, and F''_α.

As a matter of fact, some authors have even introduced additional parameters in similar cases. To the first order in the small angular displacements, the redundancy condition for methane is

$$\mathsf{S}_{A_1} = \alpha_{12} + \alpha_{13} + \alpha_{14} + \alpha_{23} + \alpha_{24} + \alpha_{34} = 0 \tag{6}$$

[1] This assumption depends upon the accuracy of the Born-Oppenheimer separation of electronic and nuclear coordinates which is described in most standard textbooks on quantum mechanics. The force constants are theoretically determined by the equation for electronic motion, which involves the charges and configuration of the nuclei, but not their masses.

However, when the second-order terms are retained, the redundancy condition must be written in the form[1]:

$$\mathsf{S}_{A_1} + a\mathsf{S}_{A_1}^2 + b(\mathsf{S}_{Ea}^2 + \mathsf{S}_{Eb}^2) + c(\mathsf{S}_{F2a}^2 + \mathsf{S}_{F2b}^2 + \mathsf{S}_{F2c}^2) = 0 \tag{7}$$

in which S_{Ea}, etc., are symmetry coordinates formed from appropriate linear combinations of the six angular coordinates, and a, b, c are constants which depend upon the equilibrium geometry of the molecule.

It is now necessary to review the argument which was used to prove that no linear terms occurred in the expansion of the potential energy. Essentially, the argument was that at equilibrium the potential energy must be at a minimum; hence $(\partial V/\partial S_t)_0 = 0$ for $t = 1, 2, \ldots, 3N - 6$. But this argument is only valid if the coordinates are *independent*. If redundant coordinates are included, the argument must be revised as follows. Let $R(S_t) = 0$ be the redundancy condition; then the condition for a minimum in the potential energy, with unrestricted variation of all the coordinates, is that

$$\delta V - \rho\,\delta R = 0 \tag{8}$$

where ρ is an undetermined multiplier. When (8) is expressed in terms of the S_t, the result is

$$\sum_t \left(\frac{\partial V}{\partial S_t} - \rho\frac{\partial R}{\partial S_t}\right)_0 \delta S_t = 0 \tag{9}$$

whence

$$\left(\frac{\partial V}{\partial S_t}\right)_0 - \rho\left(\frac{\partial R}{\partial S_t}\right)_0 = 0 \tag{10}$$

since the variations in the S_t are now unrestricted. If a particular coordinate, say S_t', is not involved in a redundancy condition, then $(\partial V/\partial S_t')_0 = 0$; if, on the other hand, the coordinate is involved, the first derivative is in general not zero, but is equal to $\rho(\partial R/\partial S_t')_0$. In the methane example

$$\left(\frac{\partial V}{\partial \alpha_{12}}\right)_0 = \left(\frac{\partial V}{\partial \alpha_{13}}\right)_0 = \cdots = \left(\frac{\partial V}{\partial \alpha_{34}}\right)_0 = \rho \tag{11}$$

It is nevertheless still possible to formulate the potential energy as a pure quadratic form. For the linear term may be transformed to the quadratic terms with the aid of the explicit redundancy condition (7) given to the second order. Thus if the potential energy of bending of

[1] Since the redundancy condition in this case is known to belong to the totally symmetric species A_1, it follows that only the totally symmetric quadratic terms $\mathsf{S}_{A_1}^2$, $\mathsf{S}_{Ea}^2 + \mathsf{S}_{Eb}^2$, etc., appear.

methane is expressed in terms of symmetry coordinates, one has

$$\left(\frac{\partial V}{\partial S_{A_1}}\right)_0 = \rho \left(\frac{\partial R}{\partial S_{A_1}}\right)_0 = \rho \tag{12}$$

$$\left(\frac{\partial V}{\partial S_{E_a}}\right)_0 = 0, \text{ etc.} \tag{13}$$

so that

$$V = \rho S_{A_1} + (\tfrac{1}{2})[F_{A_1}S_{A_1}^2 + F_E(S_{E_a}^2 + S_{E_b}^2) + F_{F_2}(S_{F_{2a}}^2 + S_{F_{2b}}^2 + S_{F_{2c}}^2)] \tag{14}$$

the linear term can be replaced by quadratic terms with the aid of (7), the final expression for the potential energy thus becoming

$$2V = (F_{A_1} - 2a\rho)S_{A_1}^2 + (F_E - 2b\rho)(S_{E_a}^2 + S_{E_b}^2) + (F_{F_2} - 2c\rho)(S_{F_{2a}}^2 + S_{F_{2b}}^2 + S_{F_{2c}}^2) \tag{15}$$

From what has been said above, however, it is impossible to determine $F_{A_1} - 2a\rho$ from the observed frequencies, and only two experimental data are available with which to solve for three theoretical parameters, namely, F_E, F_{F_2}, and ρ. For this reason, it appears to be unnecessary ever to introduce linear terms into the discussion of the vibrational problem, even when redundancies are included, provided the most general number of quadratic terms, based on the true number of vibrational coordinates, are included in the problem.

8-2. The Approximation of Central Forces

One of the simplest approximations employed in reducing the number of independent force constants is the assumption of central forces.[1] It is assumed that the forces holding the atoms in their equilibrium positions act only along the lines joining pairs of atoms and that every pair of atoms is connected by such a force. This type of force function would result if the molecule were held together by purely ionic interactions. Also, this type of force yields only diagonal terms in the force constant matrix when the internal coordinate system is the complete set of interatomic distances (the central force coordinates).

In practice, this approximation has not been particularly successful and is now little used. As an illustration of one of its defects, the case of the bending motions of CO_2 might be mentioned. Since CO_2 is a linear molecule, a bending motion does not alter any of the interatomic distances for small displacements. Consequently, there would be no quadratic

[1] N. Bjerrum, *Verhandl. deut. physik. Ges.*, **16**: 737 (1914); D. M. Dennison, *Phil. Mag.*, **1**: 195 (1926). Bjerrum first applied the central force approximation to CO_2; Dennison gave a general discussion of nonlinear, symmetric XY_2.

For further applications with numerical data, see G. Herzberg, "Infrared and Raman Spectra of Polyatomic Molecules," Van Nostrand, New York, 1945.

terms in the central force potential energy which would resist this motion, so that zero frequency would be predicted for it.[1]

However, to illustrate this type of potential function, (1) gives $2V$ for the ammonia molecule.

$$2V = F_r(r_1^2 + r_2^2 + r_3^2) + F_s(s_{12}^2 + s_{23}^2 + s_{31}^2) \tag{1}$$

Here r_1, r_2, r_3 are the extensions of the three NH bonds and s_{12}, s_{23}, s_{31} are the changes in the three hydrogen-hydrogen distances, while F_r and F_s are the force constants.

If, however, one intends to use a general quadratic potential function, the applicability of which is as appropriate in any one coordinate system as in any other, the simplicity of the kinetic energy matrix elements in the central force coordinate system may favor the use of these coordinates for some molecules.

8-3. The Approximation of Valence Forces

An approximation which is at once free of the objection raised against central forces for linear molecules, and also more compatible with chemical ideas regarding interatomic forces, is the so-called valence force approximation.[2] Here the forces considered are those which resist the extension or compression of valence bonds, together with those which oppose the bending or torsion of bonds; forces between nonbonded atoms are not directly considered. For ammonia, the potential function would be

$$2V = F_r(r_1^2 + r_2^2 + r_3^2) + F_\alpha(\alpha_{12}^2 + \alpha_{23}^2 + \alpha_{31}^2) \tag{1}$$

where r_i is the extension of a bond, and α_{ij} is the distortion of the valence angle between bonds i and j. In comparing the central force and valence force functions, Eq. (1), Sec. 8-2, and Eq. (1) of this section, one must not expect that the force constants F_r will be numerically equal for the two cases. The force constants F_s and F_α even differ in their dimensions. Frequently $F_\alpha \alpha^2$ is replaced by $(r^0)^2 F_\alpha \alpha^2$ in the potential energy where r^0 is the equilibrium length of the bond defining the angle α. This is done in order to give the bending force constant the same dimensions as the stretching constants. In this book, however, F_α is consistently defined such that $F_\alpha \alpha^2$ has the dimensions of energy: in order to quote a force constant having the dimensions of dynes per centimeter, values of $F_\alpha / r_1^0 r_2^0$ or of $F_{r\alpha}/(r_1^0 r_2^0)^{1/2}$ will be quoted in the tables, where r_1^0 and r_2^0 are the equilibrium lengths of the bonds defining the sides of the angle. Since the magnitude of all force constants expressed in dynes per centimeter is

[1] A method of treating such motions in terms of a fourth-power potential function for small displacements has been developed by R. P. Bell, *Proc. Roy. Soc.* (*London*), (*A*), **183**: 328 (1945).

[2] The valence force function was first applied to CO_2 by N. Bjerrum, *Verhandl. deut. physik. Ges.*, **16**: 737 (1914).

of the order of 10^5, it is convenient to change the units to millidynes per angstrom (md/A).

Although better than the central force approximation, the simple valence force treatment often gives only a rough approximation. Inasmuch as the number of force constants is usually less than the number of frequencies (two force constants and four frequencies in the NH_3 example), the force constants can be computed from some of the frequencies, and the remaining frequencies used for checks. In such cases, calculated frequencies may sometimes deviate from the observed frequencies by as much as 10 per cent. Nevertheless, this simple picture is a very useful one in assigning observed spectral frequencies to modes of vibration. Furthermore, it is found that the force constants which occur are at least roughly characteristic of the type of bond, so that, for example, a carbon-carbon double bond will have a stretching force constant of about 9.7×10^5 dynes cm^{-1} (9.7 md/A) regardless of the molecule in which it occurs.

TABLE 8-1. APPROXIMATE BOND STRETCHING FORCE CONSTANTS

Bond	Molecule	*k, md/A*	Bond	Molecule	*k, md/A*
H—F	HF	9.67	F—B	BF_3	8.8
H—Cl	HCl	5.15	Cl—B	BCl_3	4.6
H—Br	HBr	4.11	Br—B	BBr_3	3.7
H—I	HI	3.16	P—P	P_4	2.1
H—O	H_2O	7.8	Si—Si	Si_2H_6	1.7
H—S	H_2S	4.3	S—S	S_2H_2	2.5
H—Se	H_2Se	3.3	B—N	$B_3N_3H_6$	6.3
H—N	NH_3	6.5	C$\overset{\cdots\cdot}{—}$C	C_6H_6	7.62
H—P	PH_3	3.1	N$\overset{\cdots\cdot}{—}$O	N_2O	11.5
H—As	AsH_3	2.6	C—C		4.5–5.6
H—C	CH_3X	4.7–5.0	C=C		9.5–9.9
H—C	C_2H_4	5.1	C≡C		15.6–17.0
H—C	C_6H_6	5.1	N—N		3.5–5.5
H—C	C_2H_2	5.9	N=N		13.0–13.5
H—Si	SiH_4	2.9	N≡N		22.9
F—O	F_2O	5.6	O—O		3.5–5.0
Cl—O	Cl_2O	4.9	C—N		4.9–5.6
F—C	CH_3F	5.6	C=N		10–11
Cl—C	CH_3Cl	3.4	C≡N		16.2–18.2
Br—C	CH_3Br	2.8	C—O		5.0–5.8
I—C	CH_3I	2.3	C=O		11.8–13.4

Owing to the large number of elements in the force constant matrix for most molecules, it is tempting to transfer force constant values from one molecule to a similar molecule in a given series to reduce the number of unknowns. Although this method has been applied with success in a few limited cases, it must be borne in mind that a force constant is defined only in terms of the complete potential function of which it is a member. Consequently, the success of transferring force constant values from one molecule to another is extremely sensitive to the differences between environments of the bond in the two molecules and also to the number of interaction constants employed in the potential functions.[1]

TABLE 8-2. APPROXIMATE BOND BENDING FORCE CONSTANTS

Angle	Molecule	$F_\alpha/r_1^0r_2^0$, md/A
HOH	H_2O	0.69
HSH	H_2S	0.43
HNH	NH_3	0.4–0.6
HPH	PH_3	0.33
HCH	CH_4	0.46
HCH	C_2H_4	0.30
FOF	F_2O	0.69
ClOCl	Cl_2O	0.41
FCF	CF_4	0.71
ClCCl	CCl_4	0.33
BrCBr	CBr_4	0.24
FBF	BF_3	0.37
ClBCl	BCl_3	0.16
BrBBr	BBr_3	0.13
HCF	CH_3F	0.57
HCCl	CH_3Cl	0.36
HCBr	CH_3Br	0.30
HCI	CH_3I	0.23
NNO	N_2O	0.49
OCO	CO_2	0.57
SCS	CS_2	0.23
HCC	C_2H_2	0.12
HCN	HCN	0.20

Some approximate numerical values for bond stretching constants are collected in Table 8-1. For the most part, these constants are based upon valence force functions, although a few of the values cited were obtained in the course of calculations which employed interaction constants (see Sec. 8-4).[2]

[1] A discussion of the variation of bond force constants may be found in T. Y. Wu, "Vibrational Spectra and Structure of Polyatomic Molecules," Supplement, Edwards, Ann Arbor, Mich., 1946.

[2] For other tabulations of force constants, see G. Herzberg, "Infrared and Raman Spectra of Polyatomic Molecules," Van Nostrand, New York, 1945; T. Y. Wu, "Vibrational Spectra and Structure of Polyatomic Molecules," Edwards, Ann Arbor, Mich., 1946; and W. Gordy, *J. Chem. Phys.*, **14**: 305 (1946).

The bending constants (when reduced to the units of millidynes per angstrom) have values of the order of one-tenth those of the stretching constants. Table 8-2 summarizes a few of the more important bending constants. Many of these constants were computed on the basis of potential functions including interaction terms.

8-4. Modification of the Simple Force Functions

The simple force functions described in Secs. 8-2 and 8-3 can be modified in several ways in order to give a more accurate description of the vibrational frequencies. Perhaps the most obvious procedure is the introduction of a few judiciously chosen interaction constants. Physically, this amounts to taking into account the change in the stiffness of a bond resulting from the distortion of other bonds.

As an example, CO_2 may be mentioned: the carbon-oxygen stretching constant is calculated to be 16.8 md/A from the symmetric vibration, but 14.2 md/A from the antisymmetric mode. Introduction of an interaction constant between the two CO bonds, however, would yield a stretching constant which is the mean of the above two values (15.5), and an interaction constant of 1.3, illustrating that the interaction constant is relatively small.

The positive sign of the interaction constant has been interpreted[1] in terms of the quantum mechanical resonance theory. If CO_2 is represented as a combination of the structures (*a*) O═C═O, (*b*) O^+≡C—O^-, (*c*) O^-—C≡O^+, and if the left-hand bond is lengthened, structure (*c*) will be favored over structure (*b*). But this would imply a stiffening of the right-hand bond, and therefore a positive interaction force constant.

Aside from the linear symmetric XY_2 molecule (of which CO_2 is an example), there are only a few other types of molecules for which all the force constants are known. These include the tetrahedral P_4 molecule, for which each symmetry species consists of a single vibration, and a few other simple molecules for which the additional data provided by isotopic frequencies (Sec. 8-5) yield enough equations to calculate all the force constants. The results are summarized in Table 8-3, which also shows parenthetically the valence force constants obtained by neglecting the interaction constants.

In the case of any but the simplest molecules, it becomes a matter of great difficulty to decide which of the numerous interaction constants to employ. One useful principle, however, may be borne in mind. If a given frequency is widely separated from all others of the same symmetry species, it will not be appreciably affected by small interaction constants. If, on the contrary, two frequencies of the same species are near one

[1] H. W. Thompson and J. W. Linnett, *J. Chem. Soc.*, p. 1384 (1937).

another, even a small interaction term, if it connects the two motions involved, will cause an appreciable displacement of the two frequencies. For a mathematical justification of this statement, the reader may refer to Chap. 9, especially Eqs. (3) and (4), Sec. 9-9.

TABLE 8-3. FORCE CONSTANTS INCLUDING INTERACTION CONSTANTS FOR SOME SIMPLE MOLECULES

	Bond stretching		*Bond bending*[a]		*Interaction constants*	
XY_2 (linear)	XY		YXY		XY,XY	
CO_2	15.5		0.57		1.3	
	(14.2,16.8)		(0.57)		(0)	
CS_2	7.5		0.23		0.6	
	(6.9,8.1)		(0.23)		(0)	
XYZ (linear)	XY	YZ	XYZ		XY, YZ	
HCN	5.7	18.6	0.20		−0.22	
	(5.8)	(17.9)	(0.20)		(0)	
NNO	17.88	11.39	0.49		1.36	
	(14.6)	(13.7)	(0.49)		(0)	
XY_2 (nonlinear)	XY		YXY		XY, XY	XY, YXY[a]
H_2O	8.43		0.77		−0.10	0.25
	(7.9)		(0.70)		(0)	(0)
H_2S	4.28		0.43		0.00	0.11
XY_3 (planar)	XY		YXY	⊥[a]	XY, XY	XY, YXY[a]
BF_3	7.27		0.52	0.87	0.78	−0.33
	(8.83)		(0.37)	(0.87)	(0)	(0)
BCl_3	4.02		0.18	0.42	0.31	−0.27
	(4.63)		(0.16)	(0.42)	(0)	(0)
X_4 (tetrahedral)	XX				XX, XX[b]	XX, XX[c]
P_4	2.06				−0.12	0.10

[a] Same dimensions as stretch, md/A.
[b] Interaction between adjacent bonds.
[c] Interaction between opposite bonds.

An interesting application of this criterion is the treatment of ethylene.[1] The simple valence force treatment is unable to fit the three frequencies, 1,342 cm^{-1}, 1,623 cm^{-1}, and 3,019 cm^{-1}, of the totally symmetric species. This is apparently due to neglect of the small interaction term which connects the bending of the CH bonds with the stretching of the C=C bond. Although relatively small, this constant is found to have an important effect on the two frequencies, 1,342 and 1,623, because they are

[1] H. W. Thompson and J. W. Linnett, *J. Chem. Soc.*, p. 1376 (1937).

fairly close together and their modes of vibration are principally bending of CH bonds and stretching of the C═C bond.

Another line of attack upon molecules which resist treatment by simple potential functions was opened by Urey and Bradley[1] who proposed the use of a mixed potential function, that is, a function which is basically of the valence force type, but which includes in addition some central force terms between nonbonded atoms. In physical terms, this sort of potential function seeks to take into account the van der Waals forces between the nonbonded Y atoms. Thus, for a tetrahedral XY_4 molecule, the potential function is assumed to have the form:

$$\begin{aligned} 2V = {} & 2f'(r_1 + r_2 + r_3 + r_4) + 2f''(q_{12} + q_{13} + q_{14} + q_{23} + q_{24} + q_{34}) \\ & + F_r(r_1^2 + r_2^2 + r_3^2 + r_4^2) + F_q(q_{12}^2 + q_{13}^2 + q_{14}^2 + q_{23}^2 + q_{24}^2 + q_{34}^2) \\ & + F_\alpha(\alpha_{12}^2 + \alpha_{13}^2 + \alpha_{14}^2 + \alpha_{23}^2 + \alpha_{24}^2 + \alpha_{34}^2) \qquad (1) \end{aligned}$$

Here f' and f'' are the strain constants which occur since the four XY stretches (r_i) are not independent of the six YY stretches (q_{ij}). In using this type of potential function, it is necessary eventually to eliminate the redundant coordinates, of which there are seven in the present example. Moreover, the redundancy conditions must be obtained to the second order to permit proper accounting for the linear potential energy constants f' and f''. When this is done, the potential energy may ultimately be written in the form

$$\begin{aligned} 2V = {} & (F_r + 4F_q)(r^{(A_1)})^2 \\ & + \left[F_\alpha + \frac{r_0^2}{3}\left(F_q + \frac{f''}{q_0}\right)\right](\alpha^{(E)})^2 + \left[F_r + \frac{4}{3}\left(F_q - \frac{f''}{q_0}\right)\right](r^{(F_2)})^2 \\ & + \left[F_\alpha + \frac{r_0^2}{3}\left(F_q + \frac{5f''}{q_0}\right)\right](\alpha^{(F_2)})^2 + \frac{4}{3}r_0\left(F_q - \frac{f''}{q_0}\right)r^{(F_2)}\alpha^{(F_2)} \qquad (2) \end{aligned}$$

in which r_0 and q_0 are the equilibrium XY and YY distances, respectively, and $r^{(A_1)}$, $\alpha^{(E)}$, etc., are appropriate symmetry coordinates. Note that only four parameters appear, namely, f'', F_r, F_q, and F_α. This is a consequence of the fact that f' and f'' are not independent. By the trigonometric law of cosines, it can be shown that, to the first order,

$$q_{ij} - \frac{6^{\frac{1}{2}}}{3}(r_i + r_j) - 3^{-\frac{1}{2}}r_0\,\alpha_{ij} = 0$$

If this expression is summed over all six q_{ij}, it becomes the redundancy relation

$$R = q_{12} + q_{13} + q_{14} + q_{23} + q_{24} + q_{34} - 6^{\frac{1}{2}}(r_1 + r_2 + r_3 + r_4) = 0$$

[1] H. C. Urey and C. A. Bradley, *Phys. Rev.*, **38:** 1969 (1931).

Then, by Eq. (10), Sec. 8-1,

$$\left(\frac{\partial R}{\partial \Sigma q}\right)_0 = 1 \quad \text{and} \quad \left(\frac{\partial R}{\partial \Sigma r}\right)_0 = -6^{\frac{1}{2}}$$

$$\left(\frac{\partial V}{\partial \Sigma q}\right)_0 - \rho \left(\frac{\partial R}{\partial \Sigma q}\right)_0 = f'' - \rho = 0$$

$$\left(\frac{\partial V}{\partial \Sigma r}\right)_0 - \rho \left(\frac{\partial R}{\partial \Sigma r}\right)_0 = f' + 6^{\frac{1}{2}}\rho = 0$$

from which, by elimination of ρ,

$$f' = -6^{\frac{1}{2}} f''$$

Results which are roughly consistent with known magnitudes of van der Waals' forces (between nonbonded atoms, such as the inert gases) are obtained, although the interpretation of the force constant parameters is subject to the qualifications noted at the end of Sec. 8-1.

This sort of a potential function has also been applied to molecules of the types CH_3Cl, CH_2Cl_2, $CHCl_3$, etc.[1]

In addition to the two factors whose qualitative influence on the interaction constants has already been mentioned, namely, the resonance effect (CO_2) and the forces between nonbonded atoms, the role of change of *sp* hybridization during vibration has been investigated.[2] From a detailed analysis of the spectra of H_2O and D_2O (see Sec. 8-6) it has been possible to obtain precise values, corrected for anharmonicity, of all four constants (in millidynes per angstrom) of the most general potential function, namely, $F_r = 8.425$, $F_{rr} = -0.100$, $F_\alpha/r^2 = 0.768$, and

$$\frac{F_{r\alpha}}{r} = 0.253$$

(see Table 8-3). Heath and Linnett[3] advance arguments based upon the sign of the interaction terms to show that neither a van der Waals nor an ionic repulsion between the two hydrogen atoms would account for the observed potential function. They rather believe that changes in *sp* hybridization with vibration will account for the observed force constants. Thus for pure *p* bond character, the oxygen atom would be expected to form two bonds at right angles, whereas complete sp^3 hybridization would lead to shorter OH bonds, forming a tetrahedral angle. Suppose, during vibration, that the HOH angle is increased over its equilibrium value (104°27′). The molecule would then naturally tend to contract and therefore stiffen its OH bonds, which accounts for the posi-

[1] T. Simanouti, *J. Chem. Phys.*, **17**: 245 (1949).
[2] C. A. Coulson, J. Duchesne, and C. Manneback, *Nature*, **160**: 793 (1947).
[3] D. F. Heath and J. W. Linnett, *Trans. Faraday Soc.*, **44**: 556 (1948).

tive sign of the constant, $F_{r\alpha}$. Moreover, since an extension of either OH bond would increase the pure p character of both bonds, it would tend to be accompanied by an extension of the other bond (decrease in the potential energy), and the interaction constant, F_{rr}, should be negative, as observed. The experimental value of the sign of F_{rr} is not always negative for nonlinear XY_2 molecules, however, since *inter alia*, F_2O apparently has a positive F_{rr}.‡

In the course of further development of these ideas, Heath and Linnett[1] have introduced the concept of the *orbital valency force field* in which the bending angles are essentially a measure of the decrease of electronic overlap. In the case of planar XY_3 molecules, for instance, the potential energy due to bending is proportional to the sum of squares of angles β_i, representing the deflections of the XY directions from an assumed rigid orbital framework (of planar, triangular, symmetry) which, however, is free to rotate so as to minimize $\sum_{i=1}^{3} \beta_i^2$. According to this definition, it is unnecessary to introduce a separate bending constant to represent the out-of-plane bending.

In XY_4 molecules, Linnett and Wheateley[2] have shown that the orbital valency force field should be modified to include the effect of change of hybridization described above for water. It is found, for example, that the simple orbital valence bending constant calculated from the E mode of methane is 1.35, while a value of 0.86 is found for the triply degenerate F_2 mode. Now in order to change the hybridization, it is necessary to adjust the coefficients of the s, p_x, p_y, and p_z orbitals in a wave mechanical calculation. Since s is of species A_1, and p_x, p_y, p_z are of species F_2, one would expect that changes of hybridization could be made for the F_2 vibration, but not for the doubly degenerate mode. Qualitatively this is in agreement with the fact that the F_2 distortion is less stiff than the E vibration.

Throughout this section thus far, it has been assumed that enough information has been at hand to assign the observed frequencies to their symmetry species. With more complex molecules, this is not so easy, and one of the principal aims in the evaluation of force constants for simple molecules has been the hope that the constants thus obtained could be employed in predicting the frequencies of larger molecules. Such a hope is, of course, based upon the reasonable assumption that interaction constants between widely separated parts of a large molecule

‡ J. Duchesne and L. Burnelle, *J. Chem. Phys.*, **19**: 1191 (1951).

[1] D. F. Heath and J. W. Linnett, *Trans. Faraday Soc.*, **44**: 873 (1948). See also J. B. Howard and E. B. Wilson, Jr., *J. Chem. Phys.*, **2**: 630 (1934).

[2] J. W. Linnett and P. J. Wheatley, *Trans. Faraday Soc.*, **45**: 33 (1949).

are small. It has been possible to transfer potential energy constants from one molecule to another, provided interaction terms are included.[1] Quite accurate reproduction of the observed carbon halogen stretching frequencies is possible for CX_4, CX_3H, CX_2H_2, and CXH_3 employing a CX stretching constant and interactions of the types CX with CX and CX with HCX which are assumed to be the same for all molecules of the series. Reasonable caution must be observed, however. The carbon-chlorine bond in cyanogen chloride, for example, has a stretching constant[2] which differs from the corresponding constant for the methane derivatives by about 35 per cent. This has been attributed to resonance between the electronic structures Cl—C≡N and Cl^+=C=N^-

The principal obstacle in the study of molecular potential functions more complicated than the simple valence force type is the difficulty of securing enough data with which to evaluate the numerous interaction constants. The isotope effect, discussed in Sec. 8-5, is particularly useful since it provides additional frequencies. Another property which has sometimes been used is the coupling of angular momenta of rotation and vibration, which affects the rotational fine structure[3] (see Appendix XVI). Centrifugal distortion as observed in the pure rotational spectrum in the microwave region can also provide additional data.[4]

8-5. The Isotope Effect

When an atom of a molecule is replaced by an isotopic atom of the same element, it is assumed that the potential energy function and configuration of the molecule are changed by negligible amounts.[5] The frequencies of vibration may, however, be appreciably altered because of the change in mass involved. This is especially true if hydrogen is the atom in question because of the large percentage change in mass. This shift or isotopic effect is very useful for several purposes. In the first place it may be used to help assign spectral lines to modes of vibration. Thus a normal mode of vibration in which the hydrogen atom in question is oscillating with a large relative amplitude will suffer a greater isotopic

[1] B. L. Crawford, Jr., and S. R. Brinkley, Jr., *J. Chem. Phys.*, **9**: 69 (1941).
J. C. Decius, *J. Chem. Phys.*, **16**: 214 (1948).
J. W. Linnett, *J. Chem. Phys.*, **8**: 91 (1940); *Trans. Faraday Soc.*, **37**: 469 (1941).
A. G. Meister, S. E. Rosser, and F. F. Cleveland, *J. Chem. Phys.*, **18**: 346 (1950).
B. Stepanov, *Acta Physicochim. U.S.S.R.*, **20**: 174 (1945).

[2] W. S. Richardson and E. B. Wilson, Jr., *J. Chem. Phys.*, **18**: 155 (1950).

[3] E. Teller, *Hand- und Jahrb. chem. Physik*, **9** (II): 43 (1934).
M. Johnston and D. M. Dennison, *Phys. Rev.*, **48**: 868 (1935).

[4] D. Kivelson, *J. Chem. Phys.*, **22**: 904 (1954).

[5] For evidence supporting this assumption, see G. Herzberg, "Spectra of Diatomic Molecules," 2d ed., Table 39, Van Nostrand, New York, 1950.
E. B. Wilson, Jr., *Ann. Rev. Phys. Chem.*, **2**: 151 (1951).

change in frequency than a normal mode in which this hydrogen is moving with a small relative amplitude. In the limiting case in which only hydrogen atoms are moving, replacement of all of them by deuterium atoms should decrease the corresponding fundamental frequency by the factor $1/\sqrt{2}$, this being the square root of the ratio of masses. The totally symmetric (A_1) vibration of methane is an example of this situation. For CH_4, the frequency is 2,914.2 cm^{-1} which decreases to 2,084.7 cm^{-1} in the case of CD_4. The ratio $\omega_{CD_4}/\omega_{CH_4}$ thus has an experimental value of 0.715, compared with a theoretically expected value of 0.707. The discrepancy is attributed to the fact that the observed frequencies are influenced by cubic and quartic terms in the potential energy, so that the vibration is not strictly harmonic as has been assumed in the theoretical development.

The Product Rule.[1] The vibration frequencies of isotopic molecules are related by a rule which is a generalization of the methane example given above. Suppose that the kinetic and potential energies are expressed in terms of external symmetry coordinates (Chap. 6)

$$\begin{aligned} 2V &= \sum_{kl} \mathsf{F}_{kl}\mathsf{S}_k\mathsf{S}_l \\ 2T &= \sum_{k} \mathsf{G}_{kk}\mathsf{P}_k^2 = \sum_{k} \mu_k\mathsf{P}_k^2 \end{aligned} \tag{1}$$

Here the kinetic energy expressed in terms of the momenta, P_k, conjugate to the coordinates S_k, contains no cross terms because the transformation from cartesian to external symmetry coordinates is orthogonal. Moreover, since each external symmetry coordinate is a linear combination of the cartesian displacement coordinates of a single equivalent set of atoms, the coefficient $\mathsf{G}_{kk} = \mu_k$ will be the reciprocal mass of an atom of the set from which S_k is constructed. Since the product of the characteristic values

$$\lambda_k = 4\pi^2\nu_k^2 = 4\pi^2c^2\omega_k^2 \tag{2}$$

is equal to the determinant of the coefficients of the secular equation,

$$|\mathsf{FG}| = |\mathsf{F}| \cdot \mu_1\mu_2 \cdots \mu_f = \lambda_1\lambda_2 \cdots \lambda_f \tag{3}$$

Moreover, since $|\mathsf{F}|$ will be the same for the isotopic molecule, it follows that

$$\frac{\lambda_1'\lambda_2' \cdots \lambda_f'}{\lambda_1\lambda_2 \cdots \lambda_f} = \frac{\mu_1'\mu_2' \cdots \mu_f'}{\mu_1\mu_2 \cdots \mu_f} \tag{4}$$

where the primes indicate the isotopic molecule; a more convenient form

[1] O. Redlich, *Z. physik. Chem.*, (*B*), **28**: 371 (1935).
Teller quoted by Angus, *et al.*, *J. Chem. Soc.*, p. 971 (1936).

of (4) is

$$\frac{\omega_1'\omega_2' \cdots \omega_f'}{\omega_1\omega_2 \cdots \omega_f} = \left(\frac{m_1m_2 \cdots m_f}{m_1'm_2' \cdots m_f'}\right)^{\frac{1}{2}} \tag{5}$$

This result is valid only if the symmetry species in question contains no translation or rotation. If such motions are involved, the corresponding ω_k will vanish, so that the ratios (4) or (5) become indeterminate. This difficulty can, however, be overcome by considering the result of applying weak forces which convert the motions of translation and rotation into oscillatory motions of low frequency. In the limit of vanishing forces (under which conditions the coupling with the internal vibrations vanishes), the ratio for the translational frequencies is

$$\frac{\omega_T'}{\omega_T} = \left(\frac{M}{M'}\right)^{\frac{1}{2}} \tag{6}$$

where M is the total mass of the molecule. By similar arguments, the ratio for rotational frequencies is

$$\frac{\omega_R'}{\omega_R} = \left(\frac{I}{I'}\right)^{\frac{1}{2}} \tag{7}$$

in which I is the moment of inertia with respect to the appropriate principal axis.

By using these results to eliminate the ratios of the vanishing frequencies, (5) can be used even for those factors of the secular equation which contain translations and/or rotations. In particular, it can be used for the whole secular equation, *i.e.*, for all the frequencies. It then has the form

$$\prod_{k=1}^{3N-6} \frac{\omega_k'}{\omega_k} = \prod_{i=1}^{3N} \left(\frac{m_i}{m_i'}\right)^{\frac{1}{2}} \left(\frac{M'}{M}\right)^{\frac{3}{2}} \left(\frac{I_x'I_y'I_z'}{I_xI_yI_z}\right)^{\frac{1}{2}} \tag{8}$$

When the product rule is applied to a single factor of the secular equation, the character tables may be used to indicate how many powers of $(M'/M)^{\frac{1}{2}}$ or which ratios $(I_x'/I_x)^{\frac{1}{2}}$, if any, will appear. Thus if T_z and R_z appear in the given symmetry species, $(M'/M)^{\frac{1}{2}}(I_z'/I_z)^{\frac{1}{2}}$ will appear in the product rule expression. If a doubly degenerate species is under consideration and R_x and R_y are involved, $(I_x'/I_x)^{\frac{1}{2}} = (I_y'/I_y)^{\frac{1}{2}}$ would be used in the product rule calculation, since the product of frequencies is extended over one of the degenerate components only.

It is also possible to determine which individual atomic mass ratios should appear as on the right-hand side of (5) in any factor with the aid of the character table. The character χ_R^s for each set, s, of equivalent atoms can be computed separately by the method of Sec. 6-2. Then if

$$\chi_R^s = \sum_\gamma n_s^{(\gamma)} \chi_R^\gamma \tag{9}$$

$n_s^{(\gamma)}$, which is the number of external symmetry coordinates of species γ constructed from the cartesian coordinates of set s, can be calculated by standard methods. With the aid of these quantities,

$$\left(\frac{m_1 m_2 \cdots m_f}{m_1' m_2' \cdots m_f'}\right)^{\frac{1}{2}} = \prod_s \left(\frac{m_s}{m_s'}\right)^{n_s^{(\gamma)}/2} \tag{10}$$

if the frequencies ω_1 through ω_f are of species γ.

In case isotopic substitution lowers the symmetry of the molecule, it is still possible to apply the product rule to the totality of frequencies. Some factoring is frequently possible, however. Suppose the isotopic substitution lowers the molecular symmetry from $\mathcal{G}$ to $\mathcal{H}$, the latter group naturally being a subgroup of $\mathcal{G}$. From Chap. 6 it is apparent that symmetry coordinates under $\mathcal{G}$ can always be so constructed as to be symmetry coordinates under $\mathcal{H}$ also. When the frequencies of a single species of $\mathcal{H}$ are multiplied together, however, it may happen that more than one species of $\mathcal{G}$ will be involved, so that the product may have to be formed over more than one factor of $\mathcal{G}$.

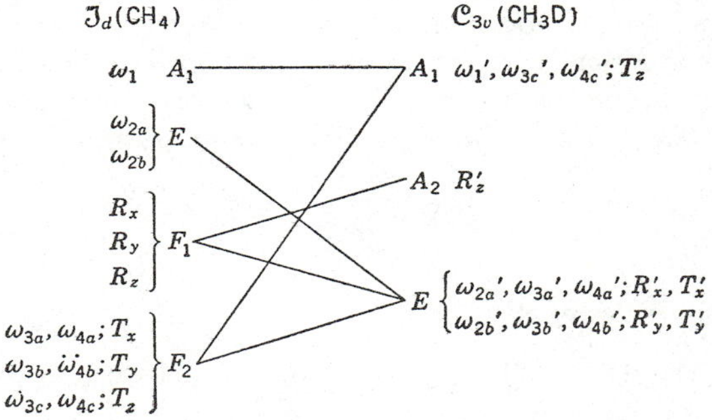

FIG. 8-1. Correlation scheme for CH_4 and CH_3D.

As an illustration of the procedure to be followed, consider the product rule for CH_4 and CH_3D. The respective symmetries are $\mathcal{T}_d$ and $\mathcal{C}_{3v}$. For CH_4, the vibrational coordinates form a representation of the structure

$$\Gamma_{CH_4} = A_1 + E + 2F_2 \tag{11}$$

whereas in CH_3D, the corresponding structure is

$$\Gamma_{CH_3D} = 3A_1 + 3E \tag{12}$$

The symmetry coordinates can be constructed in accordance with the correlation scheme of Fig. 8-1. If the coordinates are so numbered that in CH_4 ω_1 is in A_1, ω_2 is in E, and ω_3 and ω_4 are in F_2, then the frequencies of CH_3D may be classified as follows: ω_1, ω_{3c}, and ω_{4c} are in A_1 (of $\mathcal{C}_{3v}$) and the following degenerate pairs are in E, namely, ω_{2a}, ω_{2b}; ω_{3a}, ω_{3b}; ω_{4a}, ω_{4b}. Accordingly the product rule may be applied separately to the frequency combinations

$$\frac{\omega_1'\omega_{3c}'\omega_{4c}'}{\omega_1\omega_3\omega_4} = \left(\frac{m_H M'}{m_D M}\right)^{\frac{1}{2}} \tag{13}$$

and

$$\frac{\omega_2'\omega_{3a}'\omega_{4a}'}{\omega_2\omega_3\omega_4} = \left(\frac{m_H M' I_x'}{m_D M I_x}\right)^{\frac{1}{2}} \tag{14}$$

If the product rule is applied to two isotopic derivatives, such that both have less than the maximum symmetry, even less symmetry factoring may be possible. Thus for CH_3D and CH_2D_2, the only common symmetry elements are the identity and a single plane, constituting the group $\mathcal{C}_s$. Thus two factors are possible.

The Sum Rule.[1] In addition to the product rule, which relates the products of frequencies of any pair of isotopically related molecules, there exist certain sum rules which relate the sums of the squares of the frequencies of isotopic molecules. The basis of the rules is the fact that the sum of the squares of the frequencies is a linear function of the reciprocal masses of the atoms. If, therefore, several isotopic molecules can be found and geometrically superimposed with appropriate signs in such a way that the atoms vanish at all positions, the corresponding linear combination, *i.e.*, superposition, of the frequency sums should vanish. Thus if

$$\sigma = \sum_k \lambda_k = 4\pi^2 \sum_k \nu_k^2 \tag{15}$$

there exist sum rules of the forms

$$\sigma(\text{HOD}) + \sigma(\text{DOH}) - \sigma(\text{HOH}) - \sigma(\text{DOD}) = 0 \tag{16}$$

or

$$\sigma(\text{HOD}) + \sigma(\text{DOH}) = 2\sigma(\text{HOD}) = \sigma(\text{HOH}) + \sigma(\text{DOD}) \tag{17}$$

and

$$\sigma(\text{N}^{14}\text{H}_3) + \sigma(\text{N}^{15}\text{D}_3) - \sigma(\text{N}^{14}\text{D}_3) - \sigma(\text{N}^{15}\text{H}_3) = 0 \tag{18}$$

When every molecule has the same symmetry, as in (18), then the sum rule applies separately to each factor of the secular equation. But when the isotopic molecules possess different symmetry, the possibility of

[1] J. C. Decius and E. B. Wilson, Jr., *J. Chem. Phys.*, **19**: 1409 (1951). L. M. Sverdlov, *Doklady Akad. Nauk S.S.S.R.*, **78**: 1115 (1951).

factoring depends upon the following considerations (which lead, in general, to factoring different from that appropriate with the product rule). First, find those symmetry operations which are common to all molecules, as they appear in the superposition. In the water example (17), these operations are just the identity and the plane, constituting the group $\mathcal{C}_s$. In a more complicated example, this group may be smaller than that for any individual molecule involved. Thus, in the superposition involving para-dideuterobenzene,

$$3\sigma(p\text{-}C_6H_4D_2) = 2\sigma(C_6H_6) + \sigma(C_6D_6) \tag{19}$$

the deuteriums of the disubstituted molecule successively must occupy the 1,4, the 2,5, and the 3,6 positions. Although each molecule individually has the symmetry $\mathcal{V}_h$, the twofold axes in the plane are different for the 1,4 and the 2,5 positions, and the only common operations are, in fact, E, C_2 (perpendicular to the plane), σ_h, and i, constituting the group $\mathcal{C}_{2h}$.

Once this common denominator symmetry group has been identified, it is possible to express the potential and kinetic energies of each molecule in terms of symmetry coordinates under this group, and thus achieve a certain amount of factoring of the secular equation, which will certainly be the same for all molecules involved. Moreover, if F and G_i are the potential and kinetic energy matrices associated with some such factor, F will be the same for all isotopic molecules, and each element of G_i can be expressed in the form:

$$(\mathsf{G}_i)_{kl} = \sum_\alpha \mathsf{H}_{kl,\alpha}\mu_{i\alpha} \tag{20}$$

where the summation is over the atomic positions and $\mathsf{H}_{kl,\alpha}$ is the same for all isotopic molecules; $\mu_{i\alpha}$ is the reciprocal mass of the αth atom in the ith molecule (or molecular orientation) involved in the superposition. Therefore, since

$$\sigma = \sum_k \lambda_k = \sum_{kl} \mathsf{F}_{kl}\mathsf{G}_{lk} = \sum_{kl} \mathsf{F}_{kl}\mathsf{G}_{kl} = \sum_\alpha \Big(\sum_{kl} \mathsf{F}_{kl}\mathsf{H}_{kl,\alpha}\Big)\mu_\alpha \tag{21}$$

it follows that if n_i is the coefficient of the ith molecule in the superposition,

$$\begin{aligned}\sum_i n_i\sigma_i &= \sum_i n_i \sum_\alpha \Big(\sum_{kl} \mathsf{F}_{kl}\mathsf{H}_{kl,\alpha}\Big)\mu_{i\alpha} \\ &= \sum_\alpha \sum_{kl} (\mathsf{F}_{kl}\mathsf{H}_{kl,\alpha}) \sum_i n_i\mu_{i\alpha}\end{aligned} \tag{22}$$

But since by definition of the superposition,

$$\sum_i n_i\mu_{i\alpha} = 0 \qquad \text{all } \alpha \tag{23}$$

it follows that

$$\sum_i n_i \sigma_i = 0 \tag{24}$$

which proves the theorem.

It is of course necessary to employ a correlation diagram in order to select the frequencies of the more symmetrical molecules over which the summation is to be extended. Such a diagram, appropriate for the benzene superposition indicated in (19), is given in Fig. 8-2. This shows that four separate sum rules apply, corresponding to symmetry coordinates of species A_g, B_g, A_u, and B_u under the common group, $\mathfrak{C}_{2h}$. In particular, the first sums must be extended over the A_g and B_g species of $C_6H_4D_2$, and over the A_{1g}, A_{2g}, and E_{2g} species (the latter being counted

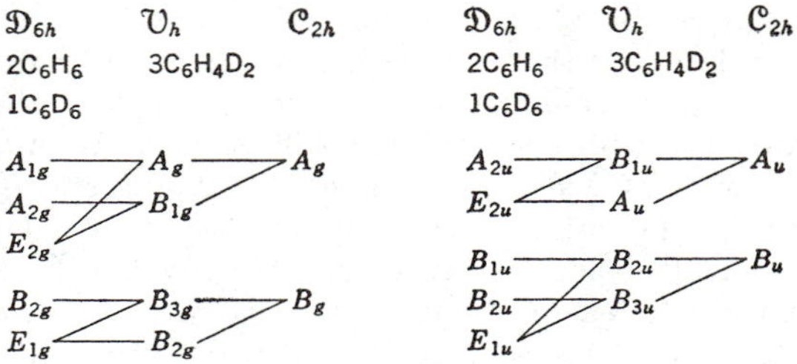

FIG. 8-2. Correlation of the species of $\mathfrak{D}_{6h}$, $\mathfrak{V}_h$, and $\mathfrak{C}_{2h}$ for application of the sum rule to benzene.

twice) of C_6H_6 and C_6D_6. In other cases, it may happen that only one of the members of a degenerate species is involved (connected with a single species of the common subgroup), and hence each squared frequency in such a species is to be counted only once.

In many of the elementary applications, no factoring is possible. This is true of the water example (17) due to the low symmetry of HOD, but it is also true of the superposition

$$4\sigma(CH_3D) = 3\sigma(CH_4) + 1\sigma(CD_4) \tag{25}$$

since the identity is the only symmetry operation common to all four orientations which must be assumed by CH_3D.

A Perturbation Treatment for Small Mass Changes.[1] Except in the case of hydrogen, substitution of isotopic atoms is accompanied by a fairly

[1] E. Teller, *Hand- und Jahrb. chem. Physik*, **9**(II)141: (1934).

W. Edgell, *J. Chem. Phys.*, **13**: 539 (1945).

A more cumbersome treatment was given by E. B. Wilson, Jr., *Phys. Rev.*, **45**: 427 (1934).

Special types of molecules had been treated earlier, for example, by A. Langseth, *Z. Physik*, **72**: 350 (1931); E. O. Salant and J. E. Rosenthal, *Phys. Rev.*, **42**: 812 (1932); **43**: 581 (1933).

small percentage change in mass, so that a perturbation method should give a reasonably accurate prediction of the corresponding change in frequency. To apply this method, it is necessary, however, to know the transformation to normal coordinates for the unsubstituted, or unperturbed molecule. Suppose that this transformation is designated by the coefficients $(L_0)_{tk}$ in Eq. (5), Sec. 4-7:

$$S_t = \sum_k (L_0)_{tk} Q_k^0 \tag{26}$$

Here the S_t may be considered to be symmetry coordinates. The coefficients for the inverse transformation are

$$Q_k^0 = \sum_t (L_0^{-1})_{kt} S_t \tag{27}$$

and, according to Eq. (13), Sec. 4-7, may be found by solving the simultaneous (secular) equations in the **FG**, rather than the **GF**, form.

In matrix language, Eq. (13), Sec. 4-7, becomes

$$\mathbf{F}\mathbf{G}^0(\mathbf{L}_0^{-1})^\dagger = (\mathbf{L}_0^{-1})^\dagger \mathbf{\Lambda}^0 \tag{28}$$

or

$$\mathbf{L}_0^\dagger \mathbf{F}\mathbf{G}^0(\mathbf{L}_0^{-1})^\dagger = \mathbf{L}_0^\dagger \mathbf{F}\mathbf{G}^0(\mathbf{L}_0^\dagger)^{-1} = \mathbf{\Lambda}^0 \tag{29}$$

Since the **F** matrix is unchanged upon isotopic substitution, it is unnecessary to give it distinctive symbols for the unperturbed and perturbed molecules; the **G** matrix for the perturbed molecule can be written

$$\mathbf{G} = \mathbf{G}^0 + \Delta\mathbf{G} \tag{30}$$

The matrix product $\mathbf{L}_0^\dagger \mathbf{F}\mathbf{G}(\mathbf{L}_0^\dagger)^{-1}$ can therefore be written

$$\begin{aligned} \mathbf{L}_0^\dagger \mathbf{F}\mathbf{G}(\mathbf{L}_0^\dagger)^{-1} &= \mathbf{L}_0^\dagger \mathbf{F}\mathbf{G}^0(\mathbf{L}_0^\dagger)^{-1} + \mathbf{L}_0^\dagger \mathbf{F}\,\Delta\mathbf{G}(\mathbf{L}_0^\dagger)^{-1} \\ &= \mathbf{\Lambda}^0 + \mathbf{L}_0^\dagger \mathbf{F}\,\Delta\mathbf{G}(\mathbf{L}_0^{-1})^\dagger \end{aligned} \tag{31}$$

The matrix represented in (31) is not diagonal, but the terms contributed by the perturbation, namely, $\mathbf{L}_0^\dagger \mathbf{F}\,\Delta\mathbf{G}(\mathbf{L}_0^{-1})^\dagger$, are small compared with the elements λ_k^0 of $\mathbf{\Lambda}^0$. Therefore, it is a good first-order approximation[1] to put

$$\lambda_k = \lambda_k^0 + [\mathbf{L}_0^\dagger \mathbf{F}\,\Delta\mathbf{G}(\mathbf{L}_0^{-1})^\dagger]_{kk} \tag{32}$$

where $\lambda_k = 4\pi^2\nu_k^2$ for the perturbed, that is, the isotopically substituted, molecule. It is fortunately unnecessary to know both $\mathbf{L}_0$ and $(\mathbf{L}_0^{-1})$ to employ (32). This follows from the fact that

$$\mathbf{L}_0^\dagger \mathbf{F}\mathbf{L}_0 = \mathbf{\Lambda}^0 \tag{33}$$

[1] This is the standard first-order approximation of perturbation theory. See V. Rojansky, "Introductory Quantum Mechanics," p. 368, Prentice-Hall, New York, 1938.

as shown in Appendix VIII. Solving (33) for $\mathbf{L}_0^\dagger \mathbf{F}$,

$$\mathbf{L}_0^\dagger \mathbf{F} = \mathbf{\Lambda}^0 \mathbf{L}_0^{-1} \tag{34}$$

and substituting in (32)

$$\lambda_k = \lambda_k^0 + [\mathbf{\Lambda}^0 \mathbf{L}_0^{-1}\, \Delta\mathbf{G}(\mathbf{L}_0^{-1})^\dagger]_{kk} \tag{35}$$

which upon expansion becomes

$$\begin{aligned} \lambda_k &= \lambda_k^0 + \lambda_k^0 \sum_{tt'} (L_0^{-1})_{kt}(L_0^{-1})_{kt'}\, \Delta G_{tt'} \\ &= \lambda_k^0(1 + \Delta_{kk}) \end{aligned} \tag{36}$$

where

$$\Delta_{kl} = \sum_{tt'} (L_0^{-1})_{kt}(L_0^{-1})_{lt'}\, \Delta G_{tt'} = (\mathbf{L}_0^{-1}\, \Delta\mathbf{G}\mathbf{L}_0^{-1})_{kl} \tag{37}$$

Although the above expression is only an approximate one for finite mass changes, it can be used to find an exact expression for the derivative of λ_k with respect to reciprocal mass. Since

$$G_{tt'} = \sum_{\alpha} \mathbf{s}_{t\alpha} \cdot \mathbf{s}_{t'\alpha}\mu_\alpha \tag{38}$$

then

$$\frac{\partial G_{tt'}}{\partial \mu_\beta} = \mathbf{s}_{t\beta} \cdot \mathbf{s}_{t'\beta} \tag{39}$$

But this expression is equal to the value which $G_{tt'}$ would assume if all $\mu_\alpha = 0$ except $\mu_\beta = 1$. Calling this value $G_{tt'}(\beta)$, with the aid of (36) and (37) one finds

$$\begin{aligned} \frac{\partial \lambda_k}{\partial \mu_\beta} &= \lambda_k^0 \sum_{tt'} (L_0^{-1})_{kt}(L_0^{-1})_{kt'} \frac{\partial G_{tt'}}{\partial \mu_\beta} \\ &= \lambda_k^0 \sum_{tt'} (L_0^{-1})_{kt}(L_0^{-1})_{kt'} G_{tt'}(\beta) \geq 0 \end{aligned} \tag{40}$$

The important inequality expressed by (40) shows for nondegenerate frequencies that no frequency can be decreased by the increase of any μ_β, that is, the decrease of the mass of any atom. The inequality follows from the fact that the kinetic energy can never be negative for any nonnegative values of the μ_β, since it is defined as

$$2T = \sum_i \mu_i p_i^2$$

But the expression $\sum_{tt'} (L_0^{-1})_{kt}(L_0^{-1})_{kt'} G_{tt'}(\beta)$ is equal to the kinetic energy if $(L_0^{-1})_{kt}$ is identified with the momentum conjugate to S_t, so that $\partial\lambda_k/\partial\mu_\beta$ is equal to λ_k^0 times a quantity which can never be negative, which proves the inequality.

The approximation made in (32) breaks down if two or more of the frequencies, and therefore λ's are nearly equal. Suppose $\lambda_k^0 \approx \lambda_l^0$. Then it is necessary to employ the off-diagonal perturbation terms coupling these frequencies and solve the secular determinant,

$$\begin{vmatrix} (\lambda_k^0 + \lambda_k^0 \Delta_{kk}) - \lambda & \lambda_k^0 \Delta_{kl} \\ \lambda_l^0 \Delta_{lk} & (\lambda_l^0 + \lambda_l^0 \Delta_{ll}) - \lambda \end{vmatrix} = 0$$

in order to find the perturbed λ_k and λ_l correct to the first order. This situation arises, in particular, if the molecular symmetry is lowered in such a way that two nearly equal frequencies in different factors of the unperturbed molecule are brought into the same factor (symmetry species) in the isotopic molecule, in accordance with the correlation diagram. This will certainly be true if the correlation of the degenerate species is of the form $E \rightarrow 2A$ or $\rightarrow 2B$, or $F \rightarrow 2A + B$ or $\rightarrow 3A$, etc., since in these cases the two or three degenerate modes of exactly equal frequency are brought together into the same determinant.

It can be shown that even in the case of degeneracy just considered, the quantities $\partial\lambda_k/\partial\mu_\beta$ are nonnegative, so that the theorem expressed in (40) is true in all cases.

Isotope Intensity Rules.[1] Although this chapter is concerned primarily with the potential energy function, it seems appropriate to follow the discussion of the effect of isotopic substitution on the frequencies with a brief account of the analogous effects on the intensities. In Sec. 7-9 it was demonstrated that the integrated intensity of a fundamental infrared band was proportional to the expression

$$I_k = \left(\frac{\partial \mu_x}{\partial Q_k}\right)_0^2 + \left(\frac{\partial \mu_y}{\partial Q_k}\right)_0^2 + \left(\frac{\partial \mu_z}{\partial Q_k}\right)_0^2 = \left(\frac{\partial \boldsymbol{\mu}}{\partial Q_k}\right)_0 \cdot \left(\frac{\partial \boldsymbol{\mu}}{\partial Q_k}\right)_0 \tag{41}$$

in which μ_x, μ_y, μ_z are components of the dipole moment along axes attached to the molecule as specified in Sec. 2-1 and Q_k is the normal coordinate corresponding to the fundamental mode under discussion. With the aid of the transformation connecting the normal coordinates, for example, with real internal symmetry coordinates

$$\mathsf{S}_{k'} = \Sigma L_{k'k} Q_k$$

the expression for I_k can be converted to one involving the derivatives with respect to the S, namely,

$$I_k = \left(\frac{\partial \boldsymbol{\mu}}{\partial Q_k}\right)_0 \cdot \left(\frac{\partial \boldsymbol{\mu}}{\partial Q_k}\right)_0 = \sum_{k'k''} \frac{\partial \boldsymbol{\mu}}{\partial \mathsf{S}_{k'}} \cdot \frac{\partial \boldsymbol{\mu}}{\partial \mathsf{S}_{k''}} L_{k'k} L_{k''k} \tag{42}$$

The transformation coefficients, $L_{k'k}$, however, are related to the **G** matrix as follows;

[1] B. L. Crawford, Jr., *J. Chem. Phys.*, **20**: 977 (1952).
J. C. Decius, *J. Chem. Phys.*, **20**: 1039 (1952).

$$\sum_k L_{k'k}L_{k''k} = \mathsf{G}_{k'k''} \tag{43}$$

and they also specify the relation between the $\mathbf{F}$ matrix and the λ_k according to the expression (see Appendix VIII)

$$\sum_{k'k''} L_{k'k}\mathsf{F}_{k'k''}L_{k''k} = \lambda_k \tag{44}$$

This latter relation can be converted to

$$\sum_k L_{k'k}\lambda_k^{-1}L_{k''k} = \mathsf{F}_{k'k''}^{-1} \tag{45}$$

where $\mathsf{F}_{k'k''}^{-1}$ is an element of the matrix inverse to the force constant matrix; $\mathbf{F}^{-1}$ as well as $\mathbf{F}$ will be the same for all isotopic molecules. Now in the light of (43) and (45) it is possible to form the following two sums over all coordinates of a given symmetry species:

$$\sum_k I_k = \sum_{k'k''} \frac{\partial \boldsymbol{\mu}}{\partial \mathsf{S}_{k'}} \cdot \frac{\partial \boldsymbol{\mu}}{\partial \mathsf{S}_{k''}} \sum_k L_{k'k}L_{k''k} = \sum_{k'k''} \frac{\partial \boldsymbol{\mu}}{\partial \mathsf{S}_{k'}} \cdot \frac{\partial \boldsymbol{\mu}}{\partial \mathsf{S}_{k''}} \mathsf{G}_{k'k''} \tag{46}$$

and

$$\sum_k \frac{I_k}{\lambda_k} = \sum_{k'k''} \frac{\partial \boldsymbol{\mu}}{\partial \mathsf{S}_{k'}} \cdot \frac{\partial \boldsymbol{\mu}}{\partial \mathsf{S}_{k''}} \sum_k L_{k'k}\lambda_k^{-1}L_{k''k} = \sum_{k'k''} \frac{\partial \boldsymbol{\mu}}{\partial \mathsf{S}_{k'}} \cdot \frac{\partial \boldsymbol{\mu}}{\partial \mathsf{S}_{k''}} \mathsf{F}_{k'k''}^{-1} \tag{47}$$

Under certain conditions, these two expressions will give rise to simple relations between the intensities of isotopic molecules, the latter analogous to the product rule, the former to the sum rule for the frequencies. The conditions are (*a*) that the molecule should have no dipole moment and/or (*b*) that the symmetry species of the vibrations over which the summation of intensities is carried out should not be the same as that of any rotation of the molecule which moves the permanent dipole moment. If one or both of these conditions are satisfied, $\partial\mu/\partial\mathsf{S}_{k'}$ is the same for all isotopic molecules. It then immediately follows that the intensity sum

$$\sum_k \frac{I_k}{\lambda_k}$$

is an isotopic invariant. This is the analogue of the frequency product rule.

From (46) it is also possible to obtain an isotopic intensity rule identical in form with the frequency sum rule (24). This follows from the fact that the right side of (46) varies with isotopic constitution in exactly the same manner as (21), provided $\partial\boldsymbol{\mu}/\partial\mathsf{S}_{k'}$ is an isotopic invariant.

The intensities of the Raman "lines" obey similar laws. The only change necessary in the statement of the Raman isotopic intensity rules is that the polarizability derivatives with respect to internal coordinates, such as $\frac{\partial \alpha_{xx}}{\partial S_{k'}}, \frac{\partial \alpha_{xy}}{\partial S_{k'}}$, etc., must be isotopic invariants. The sufficient conditions are that (*a*) the polarizability ellipsoid be a sphere and/or (*b*) the symmetry species over which the sum is extended must not be the same as that of any rotation which moves the polarizability ellipsoid into a distinguishable configuration.[1] Thus the intensity rules cannot be applied to the E_{1g} species of benzene, since this is the species of rotations R_x and R_y which (unlike the R_z rotation) would move the ellipsoid ($\alpha_x = \alpha_y \neq \alpha_z$) into a distinguishable configuration.

Finally, it should be mentioned that it is possible to revise these intensity rules so that they can be applied to symmetry species excluded by the above conditions, although the corrections are probably too cumbersome for all but the simplest molecules.

8-6. Anharmonic Terms in the Potential Energy

So far it has been assumed that only the quadratic terms in the potential energy need to be considered. In actual molecules the higher terms are of course not zero and may have to be taken into account for certain purposes. If the potential energy consisted of quadratic terms only, it would require infinite energy to break a valence bond. The quadratic, or harmonic, terms are consequently only a good approximation for small displacements from equilibrium.

The most obvious effect of the higher, or anharmonic terms, is upon the positions of overtone and combination levels. As pointed out in Sec. 3-2, a harmonic oscillator has equally spaced overtone levels so that the vibrational energy levels of a molecule with $3N - 6$ fundamental frequencies ν_k^0 would with neglect of anharmonic terms be given by the equation

$$W_{v_1, v_2, \ldots} = \sum_{k=1}^{3N-6} (v_k + \tfrac{1}{2}) h\nu_k^0 \tag{1}$$

where v_k is the quantum number of the kth normal mode and h is Planck's constant. In practice, however, it is found that a formula of the type

$$\frac{1}{hc} W_{v_1, v_2, \ldots} = X_0 + \sum_{k=1}^{3N-6} X_k \left(v_k + \frac{1}{2}\right) + \sum_{k=1}^{l} \sum_{l=1}^{3N-6} X_{kl} \left(v_k + \frac{1}{2}\right)\left(v_l + \frac{1}{2}\right) \tag{2}$$

[1] R. C. Lord and E. Teller, *J. Chem. Soc.*, p. 1728 (1937).
B. L. Crawford, Jr., *J. Chem. Phys.*, **20**: 977 (1952).

is needed to fit the observed frequencies;[1] c is the velocity of light, introduced so that the X's are in wave numbers.

Experimentalists usually use the form

$$\frac{1}{hc} W_{v_1, v_2, \cdots} = X'_0 + \sum_{k=1}^{3N-6} X'_k v_k + \sum_{k=1}^{l} \sum_{l=1}^{3N-6} X'_{kl} v_k v_l \tag{3}$$

instead of (2). The X's in the two equations are of course directly related, but (2) is the more useful form for theoretical discussions.

The terms of the type $X_{kl}(v_k + \frac{1}{2})(v_l + \frac{1}{2})$ arise from the effect of cubic, quartic, and other anharmonic terms in V. Except in the very simplest cases[2] (CO_2 and H_2O) no success has been attained in calculating the X_k's from the cubic and quartic coefficients in V, or vice versa. This is a formidable problem but one of considerable interest.

It is customary in applying normal coordinate theory to use the experimentally observed fundamental frequencies as the basis of the calculation of the quadratic force constants. From a strict viewpoint this is not justifiable inasmuch as the observed fundamental frequencies do not have the same values as they would if the anharmonic terms were zero. In order to calculate the quadratic force constants accurately, it is necessary to use the so-called *mechanical frequencies* of vibration, which are the frequencies which the molecule would exhibit if the anharmonic terms in V were all zero. When an empirical formula of the type given in (3) has been obtained, the mechanical frequencies ν_k^0 can be calculated from the relations

$$\frac{\nu_k^0}{c} = X_k = X'_k - \frac{1}{2} X'_{kk} - \frac{1}{2} \sum_{l=1}^{3N-6} X'_{kl} \tag{4}$$

as will now be shown. It is only necessary to prove that as the coefficients of the cubic and quartic terms in V are reduced to zero, the energy expression becomes

$$\frac{W}{hc} \to \sum_{k=1}^{3N-6} X_k \left(v_k + \frac{1}{2} \right) \tag{5}$$

thus permitting the identification made in (4). The cubic and quartic terms in V are of the general form

$$g_{klm} Q_k Q_l Q_m \qquad h_{klmn} Q_k Q_l Q_m Q_n$$

[1] There is no theoretical reason why higher terms, X_{klm}, should not be appreciable in some molecules. In fact, to account for the high overtones of HCN it has been found necessary to include such a term by E. Lindholm, *Z. Physik*, **108**: 454 (1937).

[2] A. Adel and D. M. Dennison, *Phys. Rev.*, **43**: 716 (1933).

B. T. Darling and D. M. Dennison, *Phys. Rev.*, **57**: 128 (1940).

These can be considered as a perturbation, the unperturbed problem being the harmonic oscillator. Since the average value[1] of $Q_kQ_lQ_m$ is zero in any state v, the first-order perturbation energy due to the cubic terms vanishes, but the second-order energy does not. The first-order energy from the quartic terms involves the mean value of $h_{klmn}Q_kQ_lQ_mQ_n$, which vanishes except for two classes of terms: $Q_k^2Q_l^2$ and Q_k^4. The mean values of terms of the first class are given by (see Appendix III)

$$h_{kkll}Q_k^2Q_l^2 \propto h_{kkll}(v_k + \tfrac{1}{2})(v_l + \tfrac{1}{2})$$

while for the second class

$$h_{kkkk}Q_k^4 \propto h_{kkkk}[(v_k + \tfrac{1}{2})^2 + \tfrac{1}{4}]$$

The second-order perturbation due to the cubic terms arises from quantities such as

$$\sum_{v'} g_{klm}g_{k'l'm'}(Q_kQ_lQ_m)_{vv'}\frac{(Q_{k'}Q_{l'}Q_{m'})_{v'v}}{W_v^0 - W_{v'}^0}$$

in which v represents all the quantum members of the state under consideration, while the sum is over all the other states v'. Several cases can arise. If $k \neq l \neq m$, then k', l', m' must be the same set as k, l, m or else the terms vanish. Also then v' has to be $v_k' = v_k \pm 1$, $v_l' = v_l \pm 1$, $v_m' = v_m \pm 1$. The eight states v' which give nonvanishing contributions to the sum can be grouped in pairs whose denominators differ only in sign. One such pair is proportional to

$$\begin{aligned} g_{klm}^2[(v_k + 1)(v_l + 1)(v_m + 1) - v_kv_lv_m] \\ = g_{klm}^2[(v_k + \tfrac{1}{2})(v_l + \tfrac{1}{2}) + (v_l + \tfrac{1}{2})(v_m + \tfrac{1}{2}) \\ + (v_m + \tfrac{1}{2})(v_k + \tfrac{1}{2}) + \tfrac{1}{4}] \end{aligned}$$

Similarly it is found that all the contributions of the cubic terms to the second-order energy involve the quantum numbers only as quadratic functions of $(v_k + \frac{1}{2})$, etc., with no terms linear in $(v_k + \frac{1}{2})$. This is also the result already obtained for the contributions of the quartic terms. Therefore the coefficients X_k of (2) will not involve either g_{klm} or h_{klmn}, whereas the coefficients X_{kl} will involve these constants and will vanish when they vanish. Consequently, as the constants g and h vanish, the energy does reduce to the form given in (5), thus completing the proof of (4).

In practice, there are extremely few molecules for which a sufficient number of overtone and combination bands have been accurately

[1] There are three cases: (1) $k \neq l \neq m$, (2) two subscripts the same, and (3) all three the same. In the first two cases, integrals of the form $(Q)_{v,v}$ enter as a factor of the complete expression, and in the third case, $(Q^3)_{v,v}$ is the relevant quantity. Both $(Q)_{v,v}$ and $(Q^3)_{v,v}$ vanish (see Appendix III).

measured for a complete determination of the constants X of (2), but there is no reason why such data should not eventually be obtained. When available, these data enable the true mechanical frequencies to be calculated by means of (4). These frequencies should give better results in normal coordinate treatments than the uncorrected fundamental frequencies. This is especially true for calculations involving the isotope effect, since the anharmonic corrections are different for the various isotopic species, being less for the heavier species. These effects are usually less than about 3 per cent.

Another effect of the anharmonic terms is to change the transition probabilities of vibrational transitions. If the electric moment were a linear function of the displacements from equilibrium and if the vibrational wave functions were accurately given by harmonic oscillator functions, no overtones or combinations should appear in infrared spectra. The fact that such bands do occur shows that one or the other of these conditions is not met; in fact, it is probable that neither condition is lived up to in actual molecules. It is evident from the convergence of overtone levels that the harmonic oscillator approximation is not exact, while considerations of intensities indicate that in addition the electric moment is not a strictly linear function of the displacements. For a further discussion of the effect of these factors on the intensities, the reader may refer to the work of Crawford and collaborators.[1]

Symmetry and the Number of Cubic and Quartic Constants. Since the terms $g_{klm}Q_kQ_lQ_m$ and $h_{klmn}Q_kQ_lQ_mQ_n$ are part of the potential energy, which is invariant under the symmetry operations, only those combinations of the normal coordinates whose products are totally symmetric can appear. The rules for determining the symmetry species of any binary product have already been given (Sec. 7-2), so that the symmetry species of $Q_kQ_lQ_m$ and $Q_kQ_lQ_mQ_n$ can easily be determined by a stepwise procedure. As an example, the possible cubic and quartic terms for H_2O will be deduced. Two of the normal coordinates, say Q_1 and Q_2, belong to the totally symmetric species A_1 of the group $\mathcal{C}_{2v}$; the third, Q_3, belongs to B_1. The following direct products must be considered:

$$A_1 \times A_1 \times A_1 = A_1;\ B_1 \times B_1 \times B_1 = B_1;\ A_1 \times A_1 \times B_1 = B_1;\ A_1 \times B_1 \times B_1 = A_1$$

Thus it can be said that among the cubic terms, all those which involve Q_3 an even number of times may occur. The following g_{klm} can therefore be different from zero; g_{111}, g_{222}, g_{112}, g_{122}, g_{133}, g_{233}; six cubic terms in all. An obvious extension yields the following nonvanishing quartic constants: h_{kkkk} (3); h_{1112}, h_{1222}; h_{kkll} (3); and h_{1233}, a total of nine quartic

[1] B. L. Crawford, Jr., and H. L. Dinsmore, *J. Chem. Phys.*, **18:** 893 (1950). D. F. Eggers, Jr., and B. L. Crawford, Jr., *J. Chem. Phys.* **19:** 1554 (1951).

terms. However, among the quartic terms only the constants of the type h_{kkkk} and h_{kkll} influence the energy according to the second-order perturbation theory outlined above. Since there are only six experimental constants, namely, X_{11}, X_{12}, X_{13}, X_{22}, X_{23}, and X_{33}, it is manifest that more data are required in order to determine the twelve g's and h's which appear in the energy expression. This has been accomplished[1] with the aid of the vibrational-rotational interactions which show that the changes of the moments of inertia with vibrational state depend upon the cubic constants. It has also been suggested[2] that the problem of the anharmonic terms can be made determinate by neglect of certain terms, just as in the treatment of the quadratic part of the potential function.

8-7. Quantum-mechanical Resonance

Whenever an approximate treatment of any problem is carried out using quantum mechanics, the possibility of *resonance* arises. If the approximate treatment yields two energy levels quite near to one another, then it will be shown that perturbation terms, neglected in the approximate treatment, may have an effect on the two nearby levels much greater than on isolated levels. The formula for the second-order correction to the energy level W_k^0, caused by a perturbation H', is[3]

$$\sum_l{}' \frac{H'_{kl}H'_{lk}}{W_k^0 - W_l^0} \tag{1}$$

in which

$$H'_{kl} = \int \psi_k^{0*} H' \psi_l^0 \, d\tau \tag{2}$$

and the sum is over all states l except the one in question, k. The superscript zero indicates the unperturbed quantities. If the perturbation is small, this correction will also usually be small. However, if there is some state l' with $W_k^0 - W_{l'}^0$ small for which $H'_{kl'} \neq 0$, then it is seen that one term in (1) may be quite large. If $W_{l'}^0 = W_k^0$, this term becomes infinite, indicating that the approximation involved in second-order perturbation theory has broken down. Under these conditions (when $W_{l'}^0 = W_k^0$ and the second-order terms due to other states can be neglected), it is necessary to solve the two-rowed secular equation

$$\begin{vmatrix} H'_{kk} + W_k^0 - W & H'_{kl'} \\ H'_{l'k} & H'_{l'l'} + W_{l'}^0 - W \end{vmatrix} = 0 \tag{3}$$

[1] B. T. Darling and D. M. Dennison, *Phys. Rev.*, **57**: 128 (1940).

[2] O. Redlich, *J. Chem. Phys.*, **9**: 298 (1941).

[3] See, for example, L. Pauling and E. B. Wilson, Jr., "Introduction to Quantum Mechanics," Sec. 25, McGraw-Hill, New York, 1935.

In the special case in which $W_k^0 + H'_{kk} = W_{l'}^0 + H'_{l'l'}$, the two roots of this equation are

$$W_k^0 + H'_{kk} + H'_{kl'} \qquad \text{and} \qquad W_k^0 + H'_{kk} - H'_{kl'} \tag{4}$$

so that the levels which are close together initially are spread apart by the perturbation. This is the phenomenon known as resonance.

There are several applications of resonance to molecular spectroscopy. In certain molecules when the anharmonic terms in the potential function are neglected, a calculation may show that two particular energy levels should lie close together. In such cases the anharmonic terms, which ordinarily might produce little effect, may split the two levels quite considerably. The actual extent of the splitting will depend upon how close the unperturbed levels are to one another and upon the magnitude of $H'_{kl'}$. If ψ_k^0 and $\psi_{l'}^0$ have different symmetries, that is if they belong to different species of the point group of the molecule, then the integral $H'_{kl'}$ will vanish, since H' itself is symmetrical, and under such conditions there will be no splitting of this kind. In general, if W_k^0 and $W_{l'}^0$ are degenerate levels so that the wave functions of the level W_k^0 and those for $W_{l'}^0$ have complex symmetries involving several species, there will be no interaction between the two levels unless they share at least one species in common.[1]

The first example of this type of resonance found[2] occurs in CO_2. This molecule has three fundamental frequencies, 667.5, 2,350, and about 1,300 cm^{-1}, of which 667.5 cm^{-1} is a doubly degenerate bending frequency. The overtone level 2×667.5 would lie quite near to the fundamental at about 1,300 and one of the components of this overtone is of the same symmetry as the fundamental in question. Consequently, resonance is to be expected between these nearby levels, pushing them further apart and causing each of the actual levels to be of mixed character; that is, each is part fundamental and part overtone. This mixing shows in the selection rules; both levels combine with the ground state to give Raman lines of about equal intensities, whereas without resonance one would expect one strong line or possibly one strong line and one very weak line. Furthermore, resonance also occurs between many sets of higher overtones of these same fundamentals. Other molecules in which resonance has been found include CCl_4,‡ C_6H_6,§ and CH_3Cl.‖ It is probably very widespread, especially in more complicated molecules.

[1] L. Tisza, *Z. Physik*, **82**: 48 (1933).

[2] E. Fermi, *Z. Physik*, **71**: 250 (1931).

D. M. Dennison, *Phys. Rev.*, **41**: 304 (1932).

A. Adel and D. M. Dennison, *Phys. Rev.*, **43**: 716 (1933); **44**: 99 (1933).

‡ G. Placzek, "Marx Handbuch der Radiologie," Vol. VI, Part II, p. 205, 1934.

§ E. B. Wilson, Jr., *Phys. Rev.*, **46**: 146 (1934).

‖ A. Adel and E. F. Barker, *J. Chem. Phys.*, **2**: 627 (1934).

In the case of H_2O, a second-order resonance is found to be of importance.[1] The two stretching modes, Q_1 and Q_3, have frequencies which are quite close to one another. Since they belong to different symmetry species, however, they cannot interact directly. Levels associated with the vibrational quantum numbers v_1, v_2, v_3, and $v_1 - 2$, v_2, $v_3 + 2$ will belong to identical symmetry species and should have approximately equal energies. The hypothesis of such resonance is supported by the fact that overtone bands with quantum numbers (2,0,1) and (0,0,3); (2,1,1) and (0,1,3), etc., occur with appreciable intensity and is confirmed by the detailed theoretical analysis of the observed frequencies.

Splitting of Overtone Levels.[2] The overtones of a degenerate frequency can have a very high degree of degeneracy if anharmonic terms are neglected. Thus the overtones of a triply degenerate fundamental frequency are successively 6-fold, 10-fold, etc., degenerate. The level with total quantum number equal to 2, for example, has the states (2,0,0), (0,2,0), (0,0,2), (1,1,0), (1,0,1), and (0,1,1). When the anharmonic effects are considered, the groups of states with two quantum numbers equal to zero will be somewhat differently affected compared with those having only one quantum number equal to zero, so that the level will split slightly into two levels, each triply degenerate.

The general procedure in finding the number of components involves the symmetry of the level. The methods which give the symmetry species of overtone levels have been described in Sec. 7-3. Suppose that the level has the structure

$$\sum_\gamma n^{(\gamma)}\Gamma^{(\gamma)}$$

in which the symbol $\Gamma^{(\gamma)}$ stands for the γth species and $n^{(\gamma)}$ is the number of times this species occurs in the reduction of the representation formed by the wave functions of the overtone level. Then the maximum number of components into which the level can be split by the anharmonic terms is

$$\sum_\gamma n^{(\gamma)} \tag{5}$$

The argument for this is as follows. The total number of wave functions for this level is $\sum_\gamma n^{(\gamma)}\chi_E^{(\gamma)}$, where $\chi_E^{(\gamma)}$ is the dimension of the irreducible representation (symmetry species) $\Gamma^{(\gamma)}$. In order to split the level into

[1] B. T. Darling and D. M. Dennison, *Phys. Rev.*, **57**: 128 (1940).

[2] L. Tisza, *Z. Physik*, **82**: 48 (1933).
D. M. Dennison, *Revs. Mod. Phys.*, **12**: 175 (1940).

more than $\sum_{\gamma} n^{(\gamma)}$ components it would be necessary for states belonging to the same degenerate representation to have different energies. The shift in energy caused by the anharmonic terms is determined by the values of the integrals

$$\int \psi_k^{0*} H' \psi_l^0 \, d\tau = H'_{kl}$$

All such integrals will vanish unless ψ_k^0 and ψ_l^0 are corresponding members of the same irreducible representation. Furthermore, if ψ_k^0 and $\psi_{k'}^0$ are degenerate functions of the same representation, then all integrals H'_{kl} can be grouped in equal pairs H'_{kl} and $H'_{k'l}$, so that no perturbation H' having the symmetry of the molecule can split a level whose wave functions belong to the same degenerate species. Consequently, the maximum possible splitting is that given by (5).

Usually the selection rules for transitions involving the various components of an overtone level are not the same for each component. This follows directly from the fact that the various components may have different symmetries. It is possible, for example, that certain components may combine with the ground level to give infrared bands while other components are forbidden to do so. Also the allowed bands may be of different kinds; both parallel and perpendicular type bands may occur together. These manifold possibilities add considerably to the complexity of overtone and combination bands.

8-8. Vibrations of Molecules with Several Equilibrium Positions

Ammonia (NH_3) is a molecule which possesses two distinct equilibrium configurations, distinguishable if the hydrogen atoms are given numbers. The equilibrium model is a pyramid, as in Fig. 8-3. It is not difficult for the nitrogen atom to pass through the plane of the hydrogens, bringing about the *inversion* of the molecule; *i.e.*, the molecule has turned inside out.

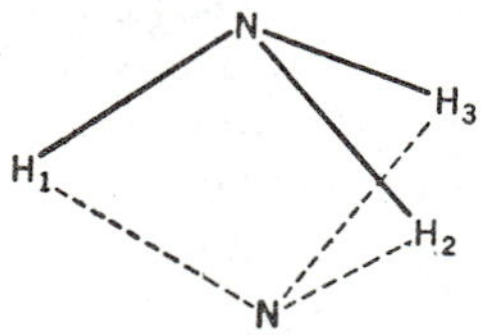

FIG. 8-3. Model for NH_3, N atom at apex of triangular pyramid. Dotted bonds show an inverted molecule.

A very large number of molecules possess more than one equilibrium configuration, and it is the aim of this section to describe qualitatively how they may be treated. Not only can molecules turn inside out, but one group, such as a methyl group, may turn about its axis from one to another position of minimum potential energy.

It is convenient for many purposes to use a different set of coordinates near each equilibrium configuration. The rotating axis system discussed in Sec. 2-1 and in more detail in Chap. 11 requires for its specification a knowledge of the equilibrium configuration. Therefore the

Eulerian angles, and also the vibrational coordinates, will be differently defined on the basis of the initial and inverted configurations. Nevertheless, any configuration can be described in terms of either set of coordinates, but then the displacements from equilibrium will not necessarily be small.

With the set of coordinates best adapted for one of the equilibrium configurations, the wave equation can be set up, approximately separated into translational, rotational, and vibrational parts, and then solved. The same procedure can be carried through for the other equilibrium configuration. The two wave functions thus obtained will be different, and general quantum mechanical arguments indicate that a better solution can be obtained by taking a linear combination of the two original ones.

If the energy barrier over which (or through which) the molecule must pass in order to get from one equilibrium position to another is great, compared with the molecular energy, the two wave functions described above will have little overlap and each can be obtained by ignoring the possibility of the other equilibrium configuration. Furthermore, if the two (or more) configurations are distinguishable only when identical atoms are distinguished, the energy levels (of rotation and vibration) will be the same for each configuration. In the high barrier case, the use of linear combinations will also yield the same energies (for levels with energy considerably less than the barrier height) because the cross term in the energy expression will vanish as the overlap vanishes. In other words, if the atoms always vibrate near the several equilibrium configurations and do not often venture into the regions in between, the existence of several equivalent configurations will not change the lower energy levels and can be ignored in this connection. Furthermore, the lower portion of the sinusoidal potential function often assumed can be closely approximated by a parabola, so that the standard methods for calculating vibration frequencies can be applied when the energy is small compared with the barrier height.

As the barrier is lowered, or as levels of higher energies are considered, the atoms begin to penetrate more often into the intermediate regions and the existence of several equilibrium configurations cannot be ignored. One effect will be the splitting of energy levels into two or more components, due to the different ways of making independent linear combinations of the wave functions for the several equilibrium configurations and the differing cross-term energies which occur. For example, if one equilibrium arrangement is labeled with A and the other with B, the appropriate linear combinations, when A and B are equivalent, are

$$2^{-\frac{1}{2}}(\psi_A + \psi_B) \qquad \text{and} \qquad 2^{-\frac{1}{2}}(\psi_A - \psi_B)$$

which lead to energies

$$W = \tfrac{1}{2}\int \psi_A^* H \psi_A \, d\tau + \tfrac{1}{2}\int \psi_B^* H \psi_B \, d\tau \pm \tfrac{1}{2}\int \psi_A^* H \psi_B \, d\tau \pm \tfrac{1}{2}\int \psi_B^* H \psi_A \, d\tau$$

or,

$$W = \int \psi_A^* H \psi_A \, d\tau \pm \int \psi_A^* H \psi_B \, d\tau$$

H is the Hamiltonian energy operator. The last term is the integral which depends on the overlap and which brings about a splitting of the energy into two levels. The lower the energy barrier and the higher the molecular energy, the greater is the splitting.

If there are three equivalent equilibrium configurations, as in ethane, the proper combinations are

$$3^{-\frac{1}{2}}(\psi_A + \psi_B + \psi_C)$$
$$6^{-\frac{1}{2}}(2\psi_A - \psi_B - \psi_C)$$
$$2^{-\frac{1}{2}}(\psi_B - \psi_C)$$

The latter two lead to identical energies; thus in this case also the original energy state is split into two states. However, coupling of over-all rotation can produce further splitting.

This treatment of the motion as that of a harmonic oscillator perturbed by the presence of several positions of minimum potential energy is quantitatively useful only when the splitting is small compared with the spacing of the groups of levels. Beyond this point, other approaches are desirable. For example, it may be possible to express the Eulerian angles from one set of coordinates in terms of those of another and with the aid of this relation the rotational parts of both ψ_A and ψ_B may be factored out of the combination, leaving

$$R(V_A \pm V_B)$$

where R is the rotational part and V the vibrational parts. The vibrational problem may then be solved separately, treating the problem as a whole and not as a superposition of equilibrium configurations.

8-9. Molecules with Internal Rotation

When a group such as the methyl group rotates about the bond which connects it to the rest of a molecule, the force acting on the group is a periodic function of the angle, ϕ, of rotation from an arbitrary zero position. This internal rotation is said to be "free" in the limiting case for which the force is independent of ϕ and "hindered" when the force is dependent upon ϕ.

In default of any a priori knowledge of the quantitative nature of the hindering forces,[1] it is customary to assume the potential barrier to be of a sinusoidal shape appropriate to the first term of a Fourier series

[1] See, however, E. N. Lassettre and L. B. Dean, *J. Chem. Phys.*, **17**: 317 (1949), for a theoretical discussion of the shape of potential barriers in several examples.

expansion of the periodic potential. Figure 8-4 illustrates the shape of $V(\phi)$ used in the case of three equivalent minima, or equilibrium configurations, *e.g.*, ethane.

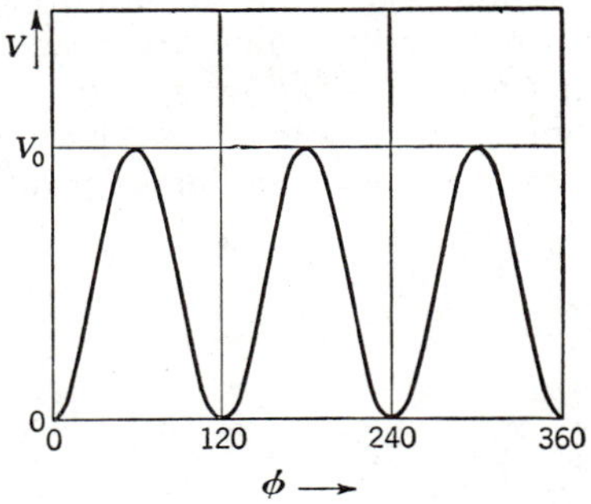

FIG. 8-4. Potential energy, V, versus angle of internal rotation, ϕ.

If the barrier height, V_0, is great compared with the molecular energy, the methods of the previous section can be employed, that is, the motions may be treated as harmonic oscillations in the potential wells with modifications resulting from quantum-mechanical tunneling due to the presence of several potential minima.

If, however, the barrier height is low or zero, a different approach is required. It then becomes desirable to introduce two or more rotating coordinate systems, one called the "framework" coordinate system describing the atomic system exclusive of rotating groups, to which are attached rotating coordinate systems which are fixed relative to each rotating group. The case which has been most thoroughly examined is that for which each attached group is "symmetric" or "balanced," *i.e.*, for which the moments of inertia would be independent of the orientation of an axis perpendicular to the connecting bond.[1] This follows from symmetry alone when the attached group has a threefold or higher symmetry about the bond joining it to the framework system.

For each extra rotating coordinate system introduced, one normal coordinate of vibration must be dropped and one new coordinate of rotational orientation introduced to describe the position of the group coordinate system or "top axes" relative to the molecular framework. One additional condition on the atomic displacements is therefore required to define each group coordinate system; these conditions are that there should be no angular momentum about each group axis of rotation due to vibrational motions of the atoms in the group—to the first order in the displacements from the equilibrium positions (see Sec. 2-1).

Expressed in the group coordinate systems the conditions for each group may be written as

$$\sum_{\beta} m_{g\beta}(\mathbf{a}_{g\beta} \times \boldsymbol{\rho}_{g\beta})_z = 0 \qquad (1)$$

[1] H. H. Nielsen, *Phys. Rev.*, **38**: 143 (1931).
L. J. B. La Coste, *Phys. Rev.*, **46**: 718 (1934).
J. B. Howard, *J. Chem. Phys.*, **5**: 442 (1937).
B. L. Crawford and E. B. Wilson, *J. Chem. Phys.*, **9**: 323 (1941).
K. S. Pitzer and W. D. Gwinn, *J. Chem. Phys.*, **10**: 428 (1942).

in which $m_{g\beta}$ is the mass of the βth atom in the gth group, $\mathbf{a}_{g\beta}$ is the vector from the origin of the group coordinates of the gth group to the equilibrium position of the βth atom, $\boldsymbol{\varrho}_{g\beta}$ is the vector displacement of this atom from its equilibrium position, and the parentheses enclose the component along the group axis of rotation, defined as the z axis, of the approximate angular momentum in the group coordinate system. The proper normal coordinate for internal rotation (Sec. 2-5) is

$$\begin{aligned} \mathcal{R}_\tau = \mathcal{R}_7 &= \mathfrak{N}_7 \sum_{\beta=1}^{N_g} m_{g\beta}(a_{g\beta}\,\Delta y_{g\beta} - b_{g\beta}\,\Delta x_{g\beta}) \\ &= \mathfrak{N}_7 \sum_{\beta=1}^{N_g} m_{g\beta}^{\frac{1}{2}}(a_{g\beta}q_{yg\beta} - b_{g\beta}q_{xg\beta}) \end{aligned} \tag{2}$$

To construct the normal coordinates so that they are orthogonal to these coordinates, suppose that one starts with internal coordinates which include a twist angle τ_g for each top. Then it can be shown that the proper procedure is simply to set the tops at any convenient angle of twist and to carry out a normal coordinate treatment in the usual way in terms of internal coordinates by simply ignoring the twist angles τ_g in both the **F** and **G** matrices.

The proof starts with a demonstration that the potential energy cannot depend upon the coordinate $\mathcal{R}_7$. For suppose the molecule is subjected to a small internal rotation. This means that the βth atom of the gth top is given a displacement

$$\boldsymbol{\varrho}_{g\beta} = \begin{pmatrix} -\tau b_{g\beta} \\ +\tau a_{g\beta} \\ 0 \end{pmatrix} = \begin{pmatrix} \Delta x_{g\beta} \\ \Delta y_{g\beta} \\ \Delta z_{g\beta} \end{pmatrix} \tag{3}$$

in which τ, which measures the twist, is small. Such a displacement yields the following value of $\mathcal{R}_7$, by substitution in (2):

$$\begin{aligned} \mathcal{R}_7 &= \mathfrak{N}_7 \sum_{\beta=1}^{N_g} m_{g\beta}(a_{g\beta}^2 + b_{g\beta}^2)\tau \\ &= \mathfrak{N}_7 I_{gz}\tau \end{aligned} \tag{4}$$

in which I_{gz} is the moment of inertia of the gth top about its z axis. But, by arguments identical with those given in (Sec. 2-5), it then follows that the potential energy cannot involve $\mathcal{R}_7$, since otherwise the potential energy would be changed by changing τ, which is contrary to the hypothesis of free internal rotation.

It then follows that the vibrational problem for the $3N - 7$ other normal coordinates can be solved by merely omitting the row and column of the **G** matrix (as well as that of the **F** matrix) which correspond to the twist angle.

It is also necessary that the frequencies obtained be independent of the angle of fixed twist employed. By ignoring interaction terms in the potential energy between internal coordinates in different axis systems within the molecule, or by taking their average values over all angles of twist, the potential energy can be put in a suitable form. The vibrational kinetic energy, for example in the form of the **G** matrix, will appear to depend on the top orientation (even if the dependence of the *velocity* of twist has been omitted as discussed above). However, it is found that, for symmetrical attached tops, linear combinations of internal coordinates within a given top can be formed so that the **G** matrix does not contain the twist angle explicitly. If S_a and S_b are a degenerate pair of internal coordinates for a given top, of the sort which have the transformation properties of a pair of translations perpendicular to the top axis and rotating with the top, then the combinations

$$\begin{aligned} S'_a &= S_a \cos \tau_0 + S_b \sin \tau_0 \\ S'_b &= S_a \sin \tau_0 - S_b \cos \tau_0 \end{aligned} \tag{5}$$

in which τ_0 is the arbitrarily fixed angle of twist, will transform like vectors fixed to the framework. In terms of these coordinates, τ_0 is eliminated from **G**. Furthermore, since S_a and S_b form a degenerate pair, they will always occur in pairs in the potential energy in such a way that τ_0 is not involved when V is expressed in terms of S'_a and S'_b. For example,

$$S_a^2 + S_b^2 \rightarrow S_a'^2 + S_b'^2$$

Consequently, it is possible to set up **F** and **G** in coordinates such that τ_0 does not enter explicitly. The frequencies will therefore not depend on τ_0, even if the above special coordinates are not used. The rule is thus proved that one should carry out the normal coordinate treatment in the usual way but ignore the twist angles. Set the tops at any convenient angles, and proceed as usual. All interaction constants between top and framework should be averaged over the twist angles.

Factoring of the Secular Equation. These considerations also lead to the conclusion that the secular equation of such molecules can often be factored further than would appear from the ordinary point group of the whole molecule, for any orientation of the tops. Thus in nitromethane (CH_3NO_2) the most favorable orientation leads to a point group $\mathcal{C}_s$, which would yield only two factors, of degrees 9 and 6. However, with the approximations used above it will be found that the secular equation actually factors into three parts, of degrees 5, 5, and 4.

The possible factoring can be established by the following considerations. First, construct internal symmetry coordinates appropriate to each top and to the framework, separately. For example, in nitromethane combine the coordinates of the methyl group into symmetry coordinates for the group $\mathcal{C}_{3v}$ and the nitro group coordinates into symmetry coordinates for $\mathcal{C}_{2v}$. Then there will be no cross terms in either the kinetic or potential energies between symmetry coordinates of different species as far as coordinates completely within one or the other part of the molecule are concerned. Now consider cross terms between a coordinate of a top (here CH_3) and a coordinate of the framework (here NO_2). Furthermore, concentrate attention on the kinetic energy, that is, the **G** matrix. A nondiagonal element of **G** can be written as (see Sec. 4-2)

$$\mathbf{G}_{kl} = \sum_{\alpha} \mu_\alpha \mathbf{s}_{k\alpha} \cdot \mathbf{s}_{l\alpha} \tag{6}$$

and the only atoms α which will be involved in the cross terms between top and framework are those at the ends of the single bond joining the top with the framework (here C—N). The vectors $\mathbf{s}_{k\alpha}$ for the top coordinates at the atoms in question will either be zero, or along the single bond, or perpendicular to the bond. In the last case there will always be a degenerate pair involved. These statements are a consequence of the fact that the top has a threefold or higher axis of symmetry. If the $\mathbf{s}$ vectors vanish at these atoms for all the coordinates of a given species under the top group, there will be a separate factor of the secular equation formed from the top coordinates of this species. The top coordinates of species which have $\mathbf{s}$ vectors along the single bond will go into the same factor of the secular equation as the framework coordinates which have $\mathbf{s}$ vectors with components along the single bond because there will be nonvanishing G elements according to (6) above.

Determination of Height of Potential Barriers. At the present time the heights of barriers inhibiting internal rotation must be obtained empirically. Thermodynamic data have been the most common source of information. The height of the barrier (assumed to be of sinusoidal shape) is adjusted until the heat capacity, entropy, or free energy calculated from statistical mechanics agrees with the measured values.[1] A second source of information has been measurements of the change of dipole moment with temperature. In a few cases observed Raman or infrared absorption lines have been assigned to transitions between torsional energy levels.

[1] E. Blade and G. E. Kimball, *J. Chem. Phys.*, **18**: 630 (1950).
K. S. Pitzer, *Chem. Rev.*, **27**: 39 (1940).
K. S. Pitzer, *Discussion Faraday Soc.*, **10**: 66 (1951).

Recently microwave spectroscopy has provided two new methods of estimating barrier heights.[1] In one technique the relative intensities of two absorption lines are measured. One line is due to a pure rotational transition for the molecule in its lowest state of vibration and internal torsional motion. The other line is due to the same rotational transition in molecules which are in an excited state of torsional motion. The ratio of these line intensities measures the population ratio of the two torsional states and therefore, via the Boltzmann factor, the frequency separation. This in turn is linked to the barrier height if the shape is assumed.

The other microwave technique involves measurement of frequencies of lines which involve "tunneling" of the molecule through the potential barrier. Methyl alcohol[2] is the outstanding example. "Tunneling" through the barrier can split the degeneracy which would otherwise be present because of the three equivalent minima in the potential barrier. This splitting combines with the over-all rotation in a rather complicated way, which has, however, been worked out. The extent of the splitting is related to the height of the barrier.

[1] B. P. Dailey, *Ann. N.Y. Acad. Sci.*, **55**: 915 (1952).

[2] E. V. Ivash and D. M. Dennison, *J. Chem. Phys.*, **21**: 1804 (1953).

CHAPTER 9

METHODS OF SOLVING THE SECULAR DETERMINANT

So far no attention has been given to practical methods of solving secular equations. When the determinant is small, it presents no problem, but when the number of rows exceeds about four, special methods are almost essential. These will be discussed in this chapter.

It might be pointed out here that secular equations are important in many other fields, such as quantum mechanics, electrical and mechanical engineering, astronomy, and statistics. The methods described below are applicable no matter what the source of the equation. For this reason, it will prove desirable in Sec. 9-1 to introduce some terminology and principles common to all secular equations. Section 9-2 gives a method for reducing the secular determinant to a symmetrical form which is frequently desirable, while Secs. 9-3 and 9-4 deal with methods of solution appropriate in case only the frequencies, and not the transformations to normal coordinates, are required. Following this, Secs. 9-5 to 9-7 deal with methods appropriate when both the frequencies and forms of the normal modes are desired. The remaining sections include discussions of machines which can be used to solve the problem.

The choice of the most suitable method for a given problem is not easy to make. A method which requires the least number of arithmetic operations for a given number of rows and columns in the secular determinant is highly desirable. This, however, is not the sole criterion, since such factors as ease of learning the routine and of applying ordinary computing machines or the question of the elimination of errors are also of importance. The reader will find discussions of these points in several of the articles cited in the footnotes.

The reader who does not wish to explore the various methods will find that the first scheme described in Sec. 9-7 is usually about as good as any when the secular equation is symmetrical and has more than four rows and columns. Smaller equations are probably best expanded into algebraic form.

9-1. Characteristic Values and Characteristic Vectors

One form in which secular equations often occur is

$$\begin{vmatrix} H_{11}-\lambda & H_{12} & H_{13} & \cdots & H_{1n} \\ H_{21} & H_{22}-\lambda & H_{23} & \cdots & H_{2n} \\ \cdots & \cdots & \cdots & \cdots & \cdots \\ H_{n1} & H_{n2} & H_{n3} & \cdots & H_{nn}-\lambda \end{vmatrix} = 0 \tag{1}$$

This can be abbreviated in matrix notation to

$$|\mathbf{H} - \lambda\mathbf{E}| \tag{2}$$

Examples of such equations are Eq. (11), Sec. 2-2, and Eqs. (7) and (8), Sec. 4-3. In the latter, **H** is **GF**. Usually the secular equation arises as the condition on λ required for the existence of nontrivial solutions of the set of simultaneous equations

$$(\mathbf{H} - \lambda_k\mathbf{E})\mathbf{A}_k = 0 \tag{3}$$

in which $\mathbf{A}_k$ is a one-column matrix, *i.e.*, a vector, whose components A_{tk} give the transformation from normal coordinates to internal coordinates. A given value of λ which satisfies (1) is related to the frequency in the kth normal mode by

$$\lambda_k = 4\pi^2\nu_k^2 \tag{4}$$

If (3) is rewritten in the form

$$\mathbf{H}\mathbf{A}_k = \lambda_k\mathbf{A}_k \tag{5}$$

it is apparent that multiplication of the vector $\mathbf{A}_k$ by the matrix **H** has the effect of multiplying all components of $\mathbf{A}_k$ by the same constant, λ_k. In this sense, $\mathbf{A}_k$ is called a *characteristic vector*[1] of the matrix, **H**, belonging to the *characteristic value*, λ_k. This fact in itself suggests a trial-and-error method of solution (not recommended without refinements described in Sec. 9-7): one could simply guess at a set of values of the A_{tk}, carry out the matrix multiplication on the left side of (5), and then see whether each component of the resulting vector were a constant multiple of its original value.

The characteristic vectors can be regarded as the columns of a square matrix, **A**. It will now be shown that the matrix **A** has the power to transform **H** into a diagonal matrix $\mathbf{\Lambda}$ whose diagonal values are the characteristic values λ. Multiply (5) on the left by $(\mathbf{A}^{-1})_l$, which is the lth *row* (not column) of the matrix reciprocal to **A**. Since, by the definition of reciprocal matrices,

$$(\mathbf{A}^{-1})_l\mathbf{A}_k = \sum_t (A^{-1})_{lt}A_{tk} = \delta_{lk} \tag{6}$$

[1] Also called an eigenvector, latent vector, modal column, etc.

it follows that

$$(\mathbf{A}^{-1})_l\mathbf{H}\mathbf{A}_k = (\mathbf{A}^{-1})_l\mathbf{A}_k\lambda_k = \delta_{lk}\lambda_k \tag{7}$$

The above equation is an equality between two single numbers which shows the truth of the matrix equation

$$\mathbf{A}^{-1}\mathbf{H}\mathbf{A} = \mathbf{\Lambda} = \begin{Vmatrix} \lambda_1 & 0 & \cdots & 0 \\ 0 & \lambda_2 & \cdots & 0 \\ \cdots & \cdots & \cdots & \cdots \\ 0 & 0 & \cdots & \lambda_n \end{Vmatrix} \tag{8}$$

which is the desired result.

Thus, another statement of the problem of solving the secular equation is: Find a matrix **A** which diagonalizes **H**.

If **H** is symmetric ($\mathbf{H}^\dagger = \mathbf{H}$), the transformation **A** has the special property that it can be made unitary ($\mathbf{A}^{-1} = \mathbf{A}^\dagger$) by a suitable normalization. For a proof, see Appendix V. In case **H** is not symmetric, it is often desirable to transform it to a symmetric form. A numerical method for doing this is given in Sec. 9-2, for the case $\mathbf{H} = \mathbf{GF}$.

Characteristic Values of Modified Matrix. This section will now be concluded with a theorem of considerable importance in connection with certain numerical methods for the determination of the characteristic values to be described in subsequent sections. Suppose a constant number, μ, is subtracted from each of the diagonal elements of **H**. Then the resultant matrix may be written

$$\mathbf{M} = \mathbf{H} - \mu\mathbf{E} \tag{9}$$

The characteristic vectors of **M** are, however, identical with those of **H** itself, while the characteristic values of **M** are $\lambda_1 - \mu$, $\lambda_2 - \mu$, etc. This is proved by multiplying **M** on the right by **A** (which, by definition, is composed of the characteristic vectors of **H**) and on the left by $\mathbf{A}^{-1}$. The result is

$$\begin{aligned} \mathbf{A}^{-1}\mathbf{M}\mathbf{A} &= \mathbf{A}^{-1}\mathbf{H}\mathbf{A} - \mu\mathbf{A}^{-1}\mathbf{A} = \mathbf{\Lambda} - \mu\mathbf{E} \\ &= \begin{Vmatrix} \lambda_1 - \mu & 0 & \cdots & 0 \\ 0 & \lambda_2 - \mu & \cdots & 0 \\ \cdots & \cdots & \cdots & \cdots \\ 0 & 0 & \cdots & \lambda_n - \mu \end{Vmatrix} \end{aligned} \tag{10}$$

This result can be generalized in the following manner. In place of (9), form

$$(\mathbf{H} - \mu_1\mathbf{E})(\mathbf{H} - \mu_2\mathbf{E}) \cdots (\mathbf{H} - \mu_p\mathbf{E}) \tag{11}$$

that is, multiply together the matrices formed by subtracting μ_1, μ_2, . . . , μ_p, respectively, from the diagonal terms of **H**. Then the characteristic vectors of (11) are again exactly those of **H**, while the characteristic values are $(\lambda_1 - \mu_1)(\lambda_1 - \mu_2) \cdots (\lambda_1 - \mu_p)$, $(\lambda_2 - \mu_1)(\lambda_2 - \mu_2) \cdots (\lambda_2 - \mu_p)$, etc. This is seen from the following equations:

$$\begin{aligned}\mathbf{A}^{-1}[(\mathbf{H} - \mu_1\mathbf{E})(\mathbf{H} - \mu_2\mathbf{E}) \cdots (\mathbf{H} - \mu_p\mathbf{E})]\mathbf{A} \\ = \mathbf{A}^{-1}(\mathbf{H} - \mu_1\mathbf{E})\mathbf{A}\mathbf{A}^{-1}(\mathbf{H} - \mu_2\mathbf{E})\mathbf{A} \cdots \mathbf{A}^{-1}(\mathbf{H} - \mu_p\mathbf{E})\mathbf{A} \\ = [\mathbf{A}^{-1}(\mathbf{H} - \mu_1\mathbf{E})\mathbf{A}][\mathbf{A}^{-1}(\mathbf{H} - \mu_2\mathbf{E})\mathbf{A}] \cdots [\mathbf{A}^{-1}(\mathbf{H} - \mu_p\mathbf{E})\mathbf{A}] \\ = (\mathbf{\Lambda} - \mu_1\mathbf{E})(\mathbf{\Lambda} - \mu_2\mathbf{E}) \cdots (\mathbf{\Lambda} - \mu_p\mathbf{E}) \qquad (12)\end{aligned}$$

in which $\mathbf{\Lambda}$ is the diagonal matrix with values $\lambda_1, \lambda_2, \ldots, \lambda_n$. Since each factor in the last form of (12) is diagonal, it is clear that the product will be diagonal and that the values of the elements are as asserted above.

9-2. Symmetrization of the Secular Determinant

The solution of the secular equation by several of the methods to be described below, Secs. 9-5 to 9-7, is appreciably simplified if **H** is symmetric. Moreover, in the case of one of the electric analogue devices, Sec. 9-10, which can be used to eliminate the numerical work, it is necessary that **H** be symmetric. Therefore a method which transforms $\mathbf{H} = \mathbf{GF}$ into a symmetrical matrix with the same characteristic values as **H** will now be described.

The method to be adopted consists in finding numerically the transformation to a new set of coordinates in terms of which **G** becomes simply a constant matrix, which, for convenience, can be taken as the unit matrix, **E**. Since **F** is always symmetric in any coordinate system, it is evident that in the new coordinate system

$$\bar{\mathbf{H}} = \bar{\mathbf{G}}\bar{\mathbf{F}} = \mathbf{E}\bar{\mathbf{F}} = \bar{\mathbf{F}} \qquad (1)$$

will then be symmetric. The desired transformation is carried out in two stages: first, a transformation which makes **G** diagonal, but not in general a unit matrix, is found. It is then not difficult to transform **G** into a unit matrix and to apply the corresponding transformation to **F**.

In order to see how **G** may be diagonalized, it is convenient (as in Appendix VII) to express its elements in terms of the transformation coefficients D_{ti} from mass-weighted coordinates (q_i) to internal coordinates (S_t), namely, as

$$G_{tt'} = \sum_{i=1}^{3N} D_{ti}D_{t'i} \qquad (2)$$

It will be convenient, moreover, to regard the elements of a given row of **D** as the components of a row vector, $\mathbf{D}_t$, whence (2) can be written in the form

$$G_{tt'} = \mathbf{D}_t\mathbf{D}_{t'}^\dagger \qquad (3)$$

The problem of diagonalizing **G** is now equivalent to constructing a new set of vectors $\bar{\mathbf{D}}_t$ defined as linear combinations of the $\mathbf{D}_t$ in such a way that the members of the new set are mutually orthogonal, *i.e.*,

$$\bar{\mathbf{D}}_t\bar{\mathbf{D}}_{t'}^\dagger = 0 \qquad \text{if } t \neq t' \tag{4}$$

This can be done by a standard method known as the Schmidt orthogonalization process for vectors. The new vectors are defined by equations

$$\begin{aligned} \bar{\mathbf{D}}_1 &= \mathbf{D}_1 \\ \bar{\mathbf{D}}_2 &= \mathbf{D}_2 + a\bar{\mathbf{D}}_1 \\ \bar{\mathbf{D}}_3 &= \mathbf{D}_3 + b\bar{\mathbf{D}}_1 + c\bar{\mathbf{D}}_2 \\ \bar{\mathbf{D}}_4 &= \mathbf{D}_4 + d\bar{\mathbf{D}}_1 + e\bar{\mathbf{D}}_2 + f\bar{\mathbf{D}}_3 \end{aligned} \tag{5}$$

in which the constants a, b, c, d, e, f, etc., are to be determined so that the orthogonality conditions (4) are satisfied. Thus a is determined by the condition

$$\bar{\mathbf{D}}_1\bar{\mathbf{D}}_2^\dagger = \bar{\mathbf{D}}_1\mathbf{D}_2^\dagger + a\bar{\mathbf{D}}_1\bar{\mathbf{D}}_1^\dagger = 0$$

or

$$a = -\frac{\bar{\mathbf{D}}_1\mathbf{D}_2^\dagger}{\bar{\mathbf{D}}_1\bar{\mathbf{D}}_1^\dagger} \tag{6}$$

similarly, b is found from the vanishing of $\bar{\mathbf{D}}_1\bar{\mathbf{D}}_3^\dagger$, c from the vanishing of $\bar{\mathbf{D}}_2\bar{\mathbf{D}}_3^\dagger$, etc., the results being

$$\begin{aligned} b &= -\frac{\bar{\mathbf{D}}_1\mathbf{D}_3^\dagger}{\bar{\mathbf{D}}_1\bar{\mathbf{D}}_1^\dagger} \\ c &= -\frac{\bar{\mathbf{D}}_2\mathbf{D}_3^\dagger}{\bar{\mathbf{D}}_2\bar{\mathbf{D}}_2^\dagger} \\ d &= -\frac{\bar{\mathbf{D}}_1\mathbf{D}_4^\dagger}{\bar{\mathbf{D}}_1\bar{\mathbf{D}}_1^\dagger} \\ e &= -\frac{\bar{\mathbf{D}}_2\mathbf{D}_4^\dagger}{\bar{\mathbf{D}}_2\bar{\mathbf{D}}_2^\dagger} \\ f &= -\frac{\bar{\mathbf{D}}_3\mathbf{D}_4^\dagger}{\bar{\mathbf{D}}_3\bar{\mathbf{D}}_3^\dagger} \end{aligned} \tag{7}$$

Now since $\bar{\mathbf{D}}_1 = \mathbf{D}_1$, it is seen from (3) that the coefficients a, b, d have the values of $a = -G_{12}/G_{11}$, $b = -G_{13}/G_{11}$, $d = -G_{14}/G_{11}$. The other coefficients can be found by the following process: first subtract $G_{1t}G_{1t'}/G_{11}$ from each element $(G_{tt'})$ of **G** except G_{11}, and designate the resulting matrix as $\mathbf{G}^{(1)}$. Note that according to this procedure, every off-diagonal element in the first row of $\mathbf{G}^{(1)}$ vanishes and that the calculation yields a $\mathbf{G}^{(1)}$ which is symmetric, so that the numerical work, *i.e.*, subtraction, is only carried out for the elements $G_{22}^{(1)}$, $G_{23}^{(1)}$, . . . , $G_{2n}^{(1)}$, $G_{33}^{(1)}$, $G_{34}^{(1)}$, etc.

It can now be shown that

$$\bar{\mathbf{D}}_2\bar{\mathbf{D}}_2^\dagger = G_{22}^{(1)} \tag{8}$$

and

$$\bar{\mathbf{D}}_2\mathbf{D}_{t'}^\dagger = G_{2t'}^{(1)} \qquad t' > 2 \tag{9}$$

For, from (5),

$$\begin{aligned}\bar{\mathbf{D}}_2\bar{\mathbf{D}}_2^\dagger &= \bar{\mathbf{D}}_2(\mathbf{D}_2^\dagger + a\bar{\mathbf{D}}_1^\dagger) = \bar{\mathbf{D}}_2\mathbf{D}_2^\dagger \\ &= (\mathbf{D}_2 + a\mathbf{D}_1)\mathbf{D}_2^\dagger = G_{22} - \frac{G_{12}^2}{G_{11}}\end{aligned} \tag{10}$$

by inserting the value of a found above. Moreover,

$$\bar{\mathbf{D}}_2\mathbf{D}_{t'}^\dagger = (\mathbf{D}_2 + a\mathbf{D}_1)\mathbf{D}_{t'}^\dagger = G_{2t'} - \frac{G_{12}G_{1t}}{G_{11}} \tag{11}$$

But these results are evidently equal to $G_{22}^{(1)}$ and $G_{2t'}^{(1)}$ by the definition of the process for obtaining $\mathbf{G}^{(1)}$.

If now a similar process is carried out upon $\mathbf{G}^{(1)}$, that is, if $G_{2t}^{(1)}G_{2t'}^{(1)}/G_{22}^{(1)}$ is subtracted from $G_{33}^{(1)}, G_{34}^{(1)}, \ldots, G_{44}^{(1)}$, etc., with the result being designated as $\mathbf{G}^{(2)}$, it can be shown by identical reasoning (although greater algebraic complexity will be encountered) that

$$\bar{\mathbf{D}}_3\mathbf{D}_{t'}^\dagger = G_{3t}^{(2)} \qquad t' > 3 \tag{12}$$

Therefore it is apparent that successive applications of the subtraction process will lead ultimately to a diagonal matrix $\mathbf{G}^{(n-1)}$ whose elements are G_{11}, $G_{22}^{(1)}$, $G_{33}^{(2)}$, etc., and that all the coefficients of the transformation (5) can be evaluated as ratios of the $\mathbf{G}^{(p)}$ elements, where $p = 0$ corresponds to $\mathbf{G}$ itself.

In order to subject $\mathbf{F}$ to the corresponding transformation, it is important to note a distinction in the transformations of $\mathbf{F}$ and $\mathbf{G}$ under a given coordinate transformation, *i.e.*, that $\mathbf{F}$ transforms as does $\mathbf{G}^{-1}$ (see Appendixes VII and VIII). If $\mathbf{S} \to \bar{\mathbf{S}} = \mathbf{CS}$ is the coordinate transformation $\mathbf{G} \to \bar{\mathbf{G}} = \mathbf{CGC}^\dagger$ and $\mathbf{F} \to \bar{\mathbf{F}} = (\mathbf{C}^{-1})^\dagger\mathbf{FC}^{-1}$. Therefore, in order to compute the $\mathbf{F}$ corresponding to the diagonal $\mathbf{G}$ obtained above, it is necessary to have the coefficients of the inverse transformation, $\mathbf{C}^{-1}$ which gives the old coordinates in terms of the new, *i.e.*,

$$\begin{aligned} S_1 &= (C^{-1})_{11}\bar{S}_1 + (C^{-1})_{12}\bar{S}_2 + \cdots \\ S_2 &= (C^{-1})_{21}\bar{S}_1 + (C^{-1})_{22}\bar{S}_2 + \cdots \\ &\text{etc.}\end{aligned} \tag{13}$$

Note, however, that the form of Eqs. (5) allows the inverse transformation to be written very simply by transposing $\bar{\mathbf{D}}_1$ and $\mathbf{D}_1$, $\bar{\mathbf{D}}_2$ and $\mathbf{D}_2$, etc.

$$\begin{aligned} \mathbf{D}_1 &= \bar{\mathbf{D}}_1 \\ \mathbf{D}_2 &= -a\bar{\mathbf{D}}_1 + \bar{\mathbf{D}}_2 \\ \mathbf{D}_3 &= -b\bar{\mathbf{D}}_1 - c\bar{\mathbf{D}}_2 + \bar{\mathbf{D}}_3 \\ \mathbf{D}_4 &= -d\bar{\mathbf{D}}_1 - e\bar{\mathbf{D}}_2 - f\bar{\mathbf{D}}_3 + \bar{\mathbf{D}}_4 \end{aligned} \tag{14}$$

Therefore, the matrix $\mathbf{C}^{-1}$ which is required in order to transform $\mathbf{F}$ is triangular, and has the form

$$\mathbf{C}^{-1} = \left\| \begin{matrix} 1 & 0 & 0 & 0 & \cdots \\ \dfrac{G_{12}}{G_{11}} & 1 & 0 & 0 & \cdots \\ \dfrac{G_{13}}{G_{11}} & \dfrac{G_{23}^{(1)}}{G_{22}^{(1)}} & 1 & 0 & \cdots \\ \dfrac{G_{14}}{G_{11}} & \dfrac{G_{24}^{(1)}}{G_{22}^{(1)}} & \dfrac{G_{34}^{(2)}}{G_{33}^{(2)}} & 1 & \cdots \\ \cdots & \cdots & \cdots & \cdots & 1 \end{matrix} \right\| \tag{15}$$

This does not quite complete the work, however, since $\mathbf{G}^{(n-1)}$ is diagonal, but not constant. It is made a unit matrix by a transformation equivalent to normalizing the vectors $\bar{\mathbf{D}}_t$ to unity, which amounts to defining

$$\mathbf{E}_t = \frac{\bar{\mathbf{D}}_t}{(\bar{\mathbf{D}}_t\bar{\mathbf{D}}_t^{\dagger})^{\frac{1}{2}}} \tag{16}$$

This is equivalent to a coordinate transformation by a matrix $(\mathbf{G}^{(n-1)})^{-\frac{1}{2}}$ which means that the desired final transform of $\mathbf{F}$ is

$$(\mathbf{G}^{(n-1)})^{\frac{1}{2}}\bar{\mathbf{F}}(\mathbf{G}^{(n-1)})^{\frac{1}{2}} \tag{17}$$

where $\bar{\mathbf{F}} = (\mathbf{C}^{-1})^{\dagger}\mathbf{F}\mathbf{C}^{-1}$. The general element of (17) may be computed from the simpler formula

$$\begin{aligned} [(\mathbf{G}^{(n-1)})^{\frac{1}{2}}\bar{\mathbf{F}}(\mathbf{G}^{(n-1)})^{\frac{1}{2}}]_{tt'} &= \sum_{ss'} (G^{(n-1)})^{\frac{1}{2}}_{ts}\bar{F}_{ss'}(G^{(n-1)})^{\frac{1}{2}}_{s't'} \\ &= (G_{tt}^{(n-1)}G_{t't'}^{(n-1)})^{\frac{1}{2}}\bar{F}_{tt'} \end{aligned} \tag{18}$$

9-3. Solution by Direct Expansion of the Secular Determinant

Suppose the secular equation is

$$\begin{vmatrix} H_{11}-\lambda & H_{12} & H_{13} & \cdots \\ H_{21} & H_{22}-\lambda & H_{23} & \cdots \\ H_{31} & H_{32} & H_{33}-\lambda & \cdots \\ \cdots & \cdots & \cdots & \cdots \end{vmatrix} = 0 \tag{1}$$

in determinantal form. Here $H_{kk'}$ is an element of the product $\mathbf{FG}$ or $\mathbf{GF}$. If the determinant is expanded, a polynomial equation in λ is obtained, with a highest power equal to n, the number of rows or columns in (1):

$$\lambda^n + c_1\lambda^{n-1} + c_2\lambda^{n-2} + \cdots + c_{n-1}\lambda + c_n = 0 \tag{2}$$

This equation is frequently called the *characteristic equation* of the matrix, $\mathbf{H}$. The coefficients, $c_1, c_2, \ldots, c_n$ can be seen to have the form:

$$
\begin{aligned}
c_1 &= (-1) \sum_{t=1}^{n} H_{tt} \\
c_2 &= (-1)^2 \sum_{t',t<t'} (H_{tt}H_{t't'} - H_{tt'}H_{t't}) \\
&\cdots\cdots\cdots\cdots\cdots\cdots \\
c_{n-1} &= (-1)^{n-1} \sum_{t=1}^{n} M_{tt} \\
c_n &= (-1)^n|\mathbf{H}|
\end{aligned} \tag{3}
$$

where M_{tt} is the determinant obtained by omitting the tth row and tth column of **H**. Similar, but more complicated, formulas could be given for the other c's. It is evident that it would be very laborious to expand a large determinant directly.

The reader should compare this development with the method described in Sec. 4-5, which was employed when **F** and **G** were treated separately. The present expansion is better suited to the case in which the product of **F** and **G** is available in numerical form.

Once a secular equation has been expanded, there are many standard methods for solving the resulting polynomial equation.[1]

As is shown in most standard textbooks on college algebra, the coefficients appearing in (2) are related to the roots $\lambda_1, \lambda_2, \lambda_3, \ldots, \lambda_n$ as follows:

$$
\begin{aligned}
c_1 &= -\sum_{k} \lambda_k \\
c_2 &= \sum_{k',k<k'} \lambda_k\lambda_{k'} \\
&\cdots\cdots\cdots\cdots \\
c_n &= (-1)^n\lambda_1\lambda_2\lambda_3 \cdots \lambda_n
\end{aligned} \tag{4}
$$

that is, c_p is $(-1)^p$ times the sum of all distinct products of p of the λ's. These relations may be combined with (3) to give convenient expressions for the determination of force constants as unknowns to fit an experimentally determined set of frequencies (λ_k's).

9-4. Indirect Expansion of the Secular Determinant[2]

There exist several indirect methods of obtaining the coefficients c_1 through c_n which appear in the expanded form of the secular equation

[1] H. Margenau and G. M. Murphy, "The Mathematics of Physics and Chemistry," p. 477, Van Nostrand, New York, 1943.

E. T. Whittaker and G. Robinson, "The Calculus of Observations," Chap. VI, Blackie, Glasgow, 1924.

[2] A review of the various methods under this heading together with comparisons of the relative efficiencies is given by H. Wayland, *Quart. Appl. Math.*, **2**: 277 (1945).

[Eq. (2), Sec. 9-3]. Their advantages are that they involve less arithmetic or a more routine procedure, although the former advantage is realized only when $n \geq 4$.

Trace Method. The arithmetic saving by this method is not as great as that accomplished by techniques to be described subsequently, but the procedure is quite simple and provides a good introduction to the other methods.

It will be noted that the simplest of the coefficients defined in terms of the elements $H_{kk'}$ by Eq. (3), Sec. 9-3, is c_1, which is just the negative of the sum of the diagonal elements of **H**. This sum, according to Eq. (4), Sec. 9-3, is equal to the sum of the roots

$$\sum_t H_{tt} = \sum_k \lambda_k \tag{1}$$

Suppose the square of the matrix **H** is computed numerically. From the arguments given in Sec. 9-1, it should be evident that the characteristic values of $\mathbf{H}^2$ are the squares of the characteristic values of **H**. Therefore

$$\sum_t (H^2)_{tt} = \sum_k \lambda_k^2 \tag{2}$$

This result can be generalized for any power, p, of **H**, in the form

$$\sum_t (H^p)_{tt} = \sum_k \lambda_k^p \tag{3}$$

Therefore, a computation of the first n powers of **H** makes possible the numerical evaluation of

$$S_p = \sum_k \lambda_k^p \tag{4}$$

for $p = 1, 2, \ldots, n$.

It is not difficult to evaluate the coefficients c_1 through c_n as defined in Eq. (4), Sec. 9-3, in terms of the S_p of (4). Clearly

$$c_1 = -S_1 \tag{5}$$

To find c_2, note that

$$\begin{aligned} c_2 &= \lambda_1\lambda_2 + \lambda_1\lambda_3 + \cdots + \lambda_{n-1}\lambda_n \\ &= \tfrac{1}{2}[(\lambda_1 + \lambda_2 + \cdots + \lambda_n)^2 - (\lambda_1^2 + \lambda_2^2 + \cdots + \lambda_n^2)] \\ &= \tfrac{1}{2}[(S_1)^2 - S_2] \end{aligned} \tag{6}$$

But (6) can also be written in the form

$$c_2 = -\tfrac{1}{2}(S_1c_1 + S_2) \tag{7}$$

upon substituting (5) for one power of S_1. By similar arguments, it is easily shown that

$$c_p = -\frac{1}{p}(S_1c_{p-1} + S_2c_{p-2} + \cdots + S_{p-1}c_1 + S_p) \tag{8}$$

Thus the computation of the c_p from the S_p is readily carried out in a stepwise manner, in which $c_1, c_2, \ldots, c_n$ are obtained successively.

Hamilton-Cayley Theorem. A second method of determining the coefficients c_p is based upon the following theorem: If the characteristic equation of **H** is

$$\lambda^n + c_1\lambda^{n-1} + \cdots + c_{n-1}\lambda + c_n = 0$$

then **H** satisfies the corresponding matrix equation:

$$\mathbf{H}^n + c_1\mathbf{H}^{n-1} + \cdots + c_{n-1}\mathbf{H} + c_n\mathbf{E} = 0 \tag{9}$$

The proof of this theorem follows directly from the remarks made at the end of Sec. 9-1. It is evident that the product matrix

$$\mathbf{P} = (\mathbf{H} - \lambda_1\mathbf{E})(\mathbf{H} - \lambda_2\mathbf{E}) \cdots (\mathbf{H} - \lambda_n\mathbf{E}) \tag{10}$$

has characteristic values which are *all zero*. This follows since the characteristic values of **P** are $(\lambda_k - \lambda_1)(\lambda_k - \lambda_2) \cdots (\lambda_k - \lambda_n)$, where λ_k is any characteristic value of **H**, so that one factor of the product will always vanish. But since $\mathbf{L}^{-1}\mathbf{PL}$ is the diagonal form of **P**, $\mathbf{L}^{-1}\mathbf{PL}$ must be a zero matrix

$$\mathbf{L}^{-1}\mathbf{PL} = \mathbf{O} \tag{11}$$

Therefore **P** is a zero matrix, since upon solving (11) for **P**,

$$\mathbf{P} = \mathbf{LOL}^{-1} = \mathbf{O} \tag{12}$$

Expansion of **P** and collection of terms in $\mathbf{H}$, $\mathbf{H}^2$, etc., will yield coefficients for the various powers $\mathbf{H}^p$ equal to c_p, making use of Eq. (4), Sec. 9-3.

In principle, n of the n^2 scalar equations

$$(H^n)_{tt'} + c_1(H^{n-1})_{tt'} + \cdots + c_{n-1}H_{tt'} + c_n = 0$$

embodied in (9) could be used as a set of simultaneous equations from which the unknown coefficients c_1 through c_n could be determined. However, it is possible to reduce the amount of arithmetic required by the following device: multiply an arbitrary column vector, **V**, by both sides of (9), thus obtaining, after rearrangement,

$$c_1\mathbf{H}^{n-1}\mathbf{V} + c_2\mathbf{H}^{n-2}\mathbf{V} + \cdots + c_n\mathbf{V} = -\mathbf{H}^n\mathbf{V} \tag{13}$$

In (13), there are only n simultaneous equations, since every term is a column vector with n rows. Moreover, it is unnecessary to compute the elements of $\mathbf{H}^2$, $\mathbf{H}^3$, etc., since only the columns $\mathbf{HV}$, $\mathbf{H}^2\mathbf{V} = \mathbf{H}(\mathbf{HV})$,

$\mathbf{H}^3\mathbf{V} = \mathbf{H}(\mathbf{H}^2\mathbf{V})$, etc., are required, and these can be computed successively, using only n^2 multiplications in each step rather than the n^3 multiplications necessary when the matrix multiplication **HH** is carried out. This method requires, of course, a solution of a set of simultaneous equations, but if an efficient method is used, there is a substantial saving as compared with the trace method when n exceeds a value of 4.

9-5. Solution by Evaluation of the Determinant

Another method for determining the characteristic values consists simply in computing the numerical value of the secular determinant for trial values of λ. When enough calculations have been performed to yield both positive and negative values of the determinant, interpolation may be used to approximate a root. A modification of this procedure involves a method of constructing the characteristic polynomial from the values of the determinant evaluated at $n + 1$ regularly spaced trial values of λ. The details of this procedure are given by Hicks.[1]

In either case considered above, the real problem is the computation of the determinant. One of the most efficient methods is described in Sec. 9-7 as "solution by successive elimination"; for further details and practical refinements of this method, the reader may refer to a paper by Dwyer.[2]

9-6. Rayleigh's Principle

At this point it is desirable to reconsider the problem of the secular equation from a somewhat different point of view. Such an equation has an interpretation as an *extreme value*, or *variational* problem which is very useful in connection with the approximate numerical methods of solving secular equations to be described in Secs. 9-8 and 9-9.

Suppose the secular determinant in question is

$$\begin{vmatrix} H_{11} - \lambda & H_{12} \\ H_{21} & H_{22} - \lambda \end{vmatrix} = 0 \tag{1}$$

and that **H** is symmetric ($H_{21} = H_{12}$). The characteristic values λ_1 and λ_2 ($\lambda_1 > \lambda_2$) which satisfy (1) have the following geometrical significance. Let x and y be a pair of numbers which satisfy the equation

$$H_{11}x^2 + (H_{12} + H_{21})xy + H_{22}y^2 = c \tag{2}$$

in which c is a constant not less than zero. It will now be shown that (2) is the equation of an ellipse in the xy plane.

Since **H** is symmetric, the transformation which diagonalizes **H** is an orthogonal one (Appendix V). But when **H** is diagonal, H_{11} becomes λ_1,

[1] B. L. Hicks, *J. Chem. Phys.*, **8**: 569 (1940).

[2] P. S. Dwyer, *Psychometrika*, **6**: 191 (1941).

H_{22} becomes λ_2, and the cross terms H_{12} and H_{21} disappear. Therefore, if ζ and η are the variables into which x and y, respectively, are sent by the transformation which diagonalizes $\mathbf{H}$, (2) becomes

$$\lambda_1\zeta^2 + \lambda_2\eta^2 = c \tag{3}$$

which is certainly the equation of an ellipse, since λ_1 and λ_2 are greater than zero. Since an orthogonal transformation is just a rotation of axes, it is true that (2) is likewise an equation of an ellipse. Furthermore, it is evident (see Fig. 9-1) that the maximum radius is $(c/\lambda_2)^{\frac{1}{2}}$ and the minimum radius is $(c/\lambda_1)^{\frac{1}{2}}$. Therefore any point on the ellipse in terms of the ζ, η coordinates satisfies

$$\frac{c}{\lambda_1} \leq \zeta^2 + \eta^2 \leq \frac{c}{\lambda_2} \tag{4}$$

since $\zeta^2 + \eta^2$ is the square of the radius.

Since the transformation is orthogonal,

$$\zeta^2 + \eta^2 = x^2 + y^2 \tag{5}$$

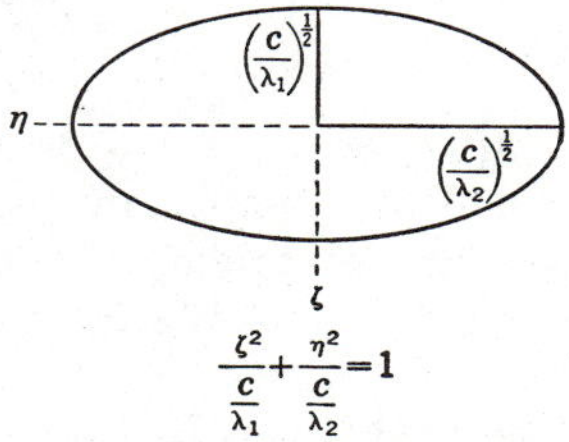

FIG. 9-1. Relation between the real roots λ_1 and λ_2 of a quadratic and the principal axes of an ellipse.

Suppose now that x and y are allowed to take on arbitrary values. This will have the effect of changing the scale of the ellipse, *i.e.*, the constant c. If c is eliminated from (4) by substitution of (2), and if (5) is used, the following inequality is found to hold for arbitrary x and y:

$$\lambda_1 \geq \frac{H_{11}x^2 + (H_{12} + H_{21})xy + H_{22}y^2}{x^2 + y^2} \geq \lambda_2 \tag{6}$$

This result is important in several ways. Since the $H_{tt'}$ are known numbers, one could in principle compute λ_1 and λ_2 by varying x and y in such a way as to yield the greatest and least values of the expression in the middle of (6). Moreover, the values of x and y at which this expression attains its extreme values define the characteristic vectors in the sense of Eq. (5), Sec. 9-1, which in the two-dimensional case amounts to the equations

$$\begin{aligned} H_{11}A_{1k} + H_{12}A_{2k} &= \lambda_k A_{1k} \\ H_{21}A_{1k} + H_{22}A_{2k} &= \lambda_k A_{2k} \end{aligned} \tag{7}$$

Multiplying the first equation of (7) by A_{1k}, the second by A_{2k} and adding,

$$H_{11}A_{1k}^2 + (H_{12} + H_{21})A_{1k}A_{2k} + H_{22}A_{2k}^2 = \lambda_k(A_{1k}^2 + A_{2k}^2) \tag{8}$$

or

$$\lambda_k = \frac{H_{11}A_{1k}^2 + (H_{12} + H_{21})A_{1k}A_{2k} + H_{22}A_{2k}^2}{A_{1k}^2 + A_{2k}^2} \tag{9}$$

The geometrical argument suggests another important application of (6). Suppose x and y are known to be approximately equal to A_{1k} and A_{2k}, respectively. Then on account of the fact that the derivative of the radius with respect to angle is zero by definition at the extreme points, one might expect that a small error in A_{1k} and A_{2k} (which amounts to a small error in the angle) would lead to only a very small error in the characteristic value as approximated by

$$\lambda_k = \frac{H_{11}x^2 + (H_{12} + H_{21})xy + H_{22}y^2}{x^2 + y^2} \tag{10}$$

Modification for $n > 2$ and **H** *Unsymmetric.* These considerations would be unimportant were it not for the fact that they generalize to a secular determinant of any size. Furthermore, the fact that the λ_k defined by (10) is a good approximation, even for rather poor approximations of x, y, etc., to the characteristic vector, can be generalized so as to hold even when **H** is not symmetric. This will now be done.

Let $U_1, U_2, \ldots, U_n$ and $V_1, V_2, \ldots, V_n$ be numbers which satisfy the condition

$$\sum_{t=1}^{n} U_t V_t = 1 \tag{11}$$

but which are otherwise arbitrary. Then define W as

$$W = \sum_{tt'} U_t H_{tt'} V_{t'} \tag{12}$$

It will now be proved that if $\delta W = 0$, the U's and V's satisfy the respective secular equations

$$\sum_{t} U_t H_{tt'} = \lambda U_{t'} \tag{13a}$$

$$\sum_{t'} H_{tt'} V_{t'} = \lambda V_t \tag{13b}$$

From (12), the variation of W expressed in terms of the variations of the U's and V's is

$$\delta W = \sum_{tt'} (\delta U_t\, H_{tt'} V_{t'} + U_t H_{tt'}\, \delta V_{t'}) \tag{14}$$

Also, from (11),

$$\delta \sum_{t} U_t V_t = \sum_{t} (U_t\, \delta V_t + \delta U_t\, V_t) = \sum_{t'} U_{t'}\, \delta V_{t'} + \sum_{t} V_t\, \delta U_t = 0 \tag{15}$$

Multiply (15) by the Lagrangian undetermined multiplier, λ, and subtract from (14), obtaining

$$\sum_{tt'} [(H_{tt'}V_{t'} - \lambda V_t)\,\delta U_t + (U_t H_{tt'} - \lambda U_{t'})\,\delta V_{t'}] = 0 \qquad (16)$$

Since the introduction of λ allows the variations δU_t and $\delta V_{t'}$ to be regarded as arbitrary, it follows that (16) can only be satisfied if

$$\sum_{t'} H_{tt'}V_{t'} = \lambda V_t$$

and

$$\sum_{t} U_t H_{tt'} = \lambda U_{t'}$$

as was to be proved.

Suppose (13b) is inserted in (12); then

$$W = \sum_{tt'} U_t H_{tt'} V_{t'} = \sum_t U_t \lambda V_t = \lambda \sum_t U_t V_t = \lambda$$

where the last step follows from the conditional equation, (11). This shows that the stationary values of W as defined by (12) are the characteristic values of $\mathbf{H}$.

Suppose now that $\mathbf{H}$ is symmetric and that its characteristic values are numbered in order of decreasing magnitude: $\lambda_1 > \lambda_2 > \cdots > \lambda_n$. Let the arbitrary $\mathbf{V}$, whose components are $V_{t'}$, be expressed as linear combinations of the characteristic vectors:

$$\mathbf{V} = \sum_k \alpha_k \mathbf{A}_k \qquad (17)$$

Then[1]

$$\begin{aligned} W = \mathbf{V}^\dagger\mathbf{H}\mathbf{V} &= \sum_{kl} \alpha_k^* \mathbf{A}_k^\dagger \mathbf{H} \alpha_l \mathbf{A}_l \\ &= \sum_{kl} \alpha_k^* \alpha_l (\mathbf{A})_k^\dagger \lambda_l \mathbf{A}_l = \sum_{kl} \alpha_k^* \alpha_l \lambda_l \delta_{kl} \\ &= \sum_k \alpha_k^* \alpha_k \lambda_k \end{aligned} \qquad (18)$$

Since

$$\mathbf{V}^\dagger\mathbf{V} = \sum_k \alpha_k^* (\mathbf{A})_k^\dagger \sum_l \alpha_l \mathbf{A}_l = \sum_k \alpha_k^* \alpha_k = 1$$

from (11), it follows that

$$\begin{aligned} \lambda_1 - W &= \lambda_1 \sum_k \alpha_k^* \alpha_k - \sum_k \alpha_k^* \alpha_k \lambda_k \\ &= \sum_k |\alpha_k|^2 (\lambda_1 - \lambda_k) \geq 0 \end{aligned} \qquad (19)$$

[1] When $\mathbf{H}$ is symmetric, $\mathbf{U}$ and $\mathbf{V}$ in (13a) and (13b) become identical.

Similarly, it can be shown that $W - \lambda_n \geq 0$. Therefore the value of W is bounded by the greatest and least characteristic values of $\mathbf{H}$,

$$\lambda_1 \geq W \geq \lambda_n \tag{20}$$

9-7. Solution of the Simultaneous Equations

We now turn to methods which do not attempt to obtain the characteristic values without the characteristic vectors. Of course, if the characteristic values have been obtained by the methods of Secs. 9-3 or 9-4, one may insert any given λ_k and solve the simultaneous equations. However, if one requires characteristic vectors as well as the λ's, for problems where $n > 3$, it is much better to use one of the methods to be described in this and the following sections. Furthermore, the subsequent methods can all be used in conjunction with the principle described in Sec. 9-6, namely, that an approximate solution of the simultaneous equations, *i.e.*, characteristic vector, $\mathbf{V}$, substituted in Eq. (18), Sec. 9-6, yields a relatively accurate estimate of the corresponding λ_k.

Solution by Successive Elimination.[1] A method of solving the simultaneous equations which is appropriate when $\mathbf{H}$ is a symmetric matrix proceeds as follows. A numerical value of $\lambda = \lambda'$ is guessed and inserted in Eq. (3), Sec. 9-1. The first component equation is left intact; the first and second equations are then combined to eliminate A_1 from the second equation; the first, second, and third equations are combined to eliminate A_1 and A_2 from the third equation. This process is continued until $A_1, A_2, \ldots, A_{n-1}$ have been eliminated from the last equation. The resulting set of equations will have the triangular form[2]

$$\begin{array}{rcl} H'_{11}A_1 + H'_{12}A_2 + H'_{13}A_3 + \cdots + H'_{1n}A_n &=& 0 \\ H'_{22}A_2 + H'_{23}A_3 + \cdots + H'_{2n}A_n &=& 0 \\ H'_{33}A_3 + \cdots + H'_{3n}A_n &=& 0 \\ \cdots\cdots\cdots\cdots\cdots\cdots\cdots\cdots && \\ H'_{nn}A_n &=& 0 \end{array} \tag{1}$$

Equations (1) possess a solution other than $A_t = 0$ for all t if and only if

$$\begin{vmatrix} H'_{11} & H'_{12} & H'_{13} & \cdots & H'_{1n} \\ 0 & H'_{22} & H'_{23} & \cdots & H'_{2n} \\ 0 & 0 & H'_{33} & \cdots & H'_{3n} \\ \cdots & \cdots & \cdots & \cdots & \cdots \\ 0 & 0 & 0 & \cdots & H'_{nn} \end{vmatrix} = 0 \tag{2}$$

[1] The elimination process, as applied to simultaneous equations, was described early by Gauss. It has been applied to the present problem, for example, by H. M. James and A. S. Coolidge, *J. Chem. Phys.*, **1**: 825 (1933).

[2] The actual numerical work is carried out in a stepwise manner similar to that employed in diagonalizing $\mathbf{G}$ (Sec. 9-2).

it being understood that the H''s depend upon the trial value of λ'. The value of a determinant of this form is simply the product of the elements on the principal diagonal

$$|\mathbf{H}'| = H'_{11}H'_{22}H'_{33} \cdot \cdot \cdot H'_{nn} \tag{3}$$

Unless this product vanishes, λ' is not an exact characteristic value. A second value of $\lambda = \lambda''$ can then be inserted and the equations again are triangularized by the successive elimination process. An improved λ can then be obtained by interpolation using λ', λ'', $|\mathbf{H}'|$, and $|\mathbf{H}''|$. The process is then continued until the determinant vanishes.

It may now appear to the reader that this method is simply a numerical procedure for evaluating a large determinant combined with interpolation. This is in fact true, the procedure being one of the best for the evaluation of large determinants, but it may be seen that the method really yields a characteristic vector as well as a characteristic value. To show this, it is first convenient to describe the practical procedure for triangularizing the equations.

The first equation is left unchanged, so that $H'_{11} = H_{11} - \lambda'$, $H'_{1t'} = H_{1t}$ for $t' = 2, 3, \ldots, n$. For the second row

$$H'_{2t'} = (H_{2t'} - \delta_{2t'}\lambda') - \frac{H'_{1t'}H'_{12}}{H'_{11}} \tag{4}$$

Since $H_{12} = H_{21} = H'_{12}$, it follows from (4) that $H'_{21} = 0$, that is, the first unknown, A_1, has been eliminated from the second equation as desired. The members of the third row are given by the equation

$$H'_{3t'} = (H_{3t'} - \delta_{3t'}\lambda') - \frac{H'_{1t'}H'_{13}}{H'_{11}} - \frac{H'_{2t'}H'_{23}}{H'_{22}} \tag{5}$$

In general, the elements of the tth row are given by

$$H'_{tt'} = (H_{tt'} - \delta_{tt'}\lambda') - \sum_{s=1}^{t-1} \frac{H'_{st'}H'_{st}}{H'_{ss}} \tag{6}$$

Now in order that the determinant should vanish, it is evident from (3) that one or more of the diagonal terms $H'_{11}, H'_{22}, \ldots, H'_{nn}$ must be zero, and all the nonzero terms must be finite. Suppose H'_{11} vanishes. Then according to (4), (5), and (6), some of the diagonal elements H'_{22}, H'_{33}, $\ldots$, H'_{nn} will therefore go to infinity as H'_{11} goes to zero unless all the elements $H'_{1t'} = H_{1t'}$ are zero. If the latter were true, numerical calculation would be unnecessary, however, since H_{11} would immediately be recognized as a characteristic value corresponding to a characteristic vector with components $A_1 \neq 0$, $A_2 = A_3 = \cdots = A_n = 0$. Therefore in the numerical calculation it may be assumed that $H'_{11} \neq 0$. A similar argument may be advanced to show that the determinant (3) will not vanish when $H'_{22}, H'_{33}, \ldots, H'_{n-1,n-1}$ are zero. One concludes that a necessary condition for $|\mathbf{H}| = 0$ is that the last factor should vanish

$$H'_{nn} = 0 \tag{7}$$

But when this is true, according to the last equation of (1), A_n is indeterminate. But the next to last equation of (1) may be solved for A_{n-1} in terms of A_n, the previous equation for A_{n-2} in terms of A_{n-1} and A_n, etc. On account of the triangular form, therefore, all the A's may be easily expressed in terms of A_n which, as has been seen previously, constitutes a complete solution (characteristic vector belonging to the λ' which makes $H'_{nn} = 0$) except for normalization.

Even when H'_{nn} is small, but not zero, an approximate set of A_t could be obtained by this procedure, and substitution in Eq. (18), Sec. 9-6, should then yield an improved characteristic value.

Still another method of obtaining an improved λ utilizes (6) with $t = t' = n$; instead of computing this final element with the value of λ' which has been used in the triangularization process up to this point, H'_{nn} is set equal to zero and the equation is used to solve for the next value of $\lambda = \lambda''$:

$$\lambda'' = H_{nn} - \sum_{s=1}^{n-1} \frac{(H'_{sn})^2}{H'_{ss}} \tag{8}$$

In some cases it may be found necessary to rearrange the order of the rows and columns of the original determinant in order to avoid situations where very high accuracy in numerical work is required to obtain reasonably accurate final results.

This method may also be used without modification for solving secular equations of the form $|F - \lambda G^{-1}| = 0$. By suitable modification it may be used with unsymmetrical secular equations, but only with somewhat greater labor.

Solution by Relaxation. The method which will next be described was first applied extensively by Southwell[1] in certain engineering problems dealing with static equilibrium. The displacements of a structure subjected to known loads are determined by assuming an arbitrary displacement, computing the forces of constraint necessary to maintain the structure in its arbitrary displacement, and then varying the assumed displacement so as to "relax" all the forces of constraint to negligible values.

[1] R. V. Southwell, "Relaxation Methods in Engineering Science," Oxford, New York, 1940.

Closely related methods had been described previously by Seidel, *Münch. Abh.*, **11** (3): 81 (1874), and by Gauss.

The method is applicable to any set of simultaneous equations; in the present example, the A_t are assigned arbitrary initial values, $A_t^{(0)}$, and λ is also assigned an initial trial value, $\lambda^{(0)}$. The quantities analogous to the forces of constraint are the e's defined by

$$\begin{aligned} e_1 &= (H_{11} - \lambda^{(0)})A_1^{(0)} + H_{12}A_2^{(0)} + \cdots + H_{1n}A_n^{(0)} \\ e_2 &= H_{21}A_1^{(0)} + (H_{22} - \lambda^{(0)})A_2^{(0)} + \cdots + H_{2n}A_n^{(0)} \\ &\cdots\cdots\cdots\cdots\cdots\cdots\cdots\cdots \\ e_n &= H_{n1}A_1^{(0)} + H_{n2}A_2^{(0)} + \cdots + (H_{nn} - \lambda^{(0)})A_n^{(0)} \end{aligned} \tag{9}$$

If the assumed A's and λ constituted a solution, all the e's would vanish. The relaxation procedure consists in altering the assumed $A^{(0)}$ so as to minimize $e_1, e_2, \ldots, e_n$ according to the following scheme: if e_t is the largest error and $H_{tt'}$ the largest coefficient in the tth equation, then $A_{t'}^{(0)}$ is altered by an amount sufficient to reduce e_t to approximately zero. Of course when this is done, $e_1, e_2, \ldots, e_{t-1}, e_{t+1}, \ldots, e_n$ are all altered. Next, the largest remaining error is relaxed by a similar procedure. The process is continued until all the e's are reasonably small. Naturally if $\lambda^{(0)}$ is not an exact characteristic value, the e's cannot all be made to vanish. When they are sufficiently small in the judgment of the computer, the modified values of $A_1^{(0)}, A_2^{(0)}, \ldots, A_n^{(0)}$ are then used to estimate an improved value of λ defined by

$$\lambda^{(1)} = \frac{\mathbf{A}^{(0)\dagger}\mathbf{H}\mathbf{A}^{(0)}}{\mathbf{A}^{(0)\dagger}\mathbf{A}^{(0)}} \tag{10}$$

The use of (10) takes advantage of the stationary property discussed in Sec. 9-6; division by $\mathbf{A}^{(0)\dagger}\mathbf{A}^{(0)}$ is equivalent to the normalization condition expressed by Eq. (11), Sec. 9-6.

The cycle may now be repeated using $\lambda = \lambda^{(1)}$ and starting the relaxation procedure with the $\mathbf{A}^{(1)}$ equal to the modified $\mathbf{A}^{(0)}$ which appeared at the end of the previous cycle. For practical details of the numerical relaxation procedure, the reader is referred to Southwell's book. It has been shown[1] that this procedure converges upon the greatest characteristic value λ_1, when $\lambda^{(0)} > H_{tt}$, all t, and upon the least characteristic value, λ_n, when $\lambda^{(0)} < H_{tt}$, all t. Actually, the convergence will be more rapid if the maximum e_t/H_{tt} rather than e_t is used as the criterion[2] for selecting the equation which determines the size of the correction to $A_{t'}^{(0)}$.

Once the characteristic vectors corresponding to the greatest and least values (λ_1 and λ_n) have been found, it is possible, in principle, to force convergence upon some other value by keeping the trial $\mathbf{A}$ orthogonal to $\mathbf{A}_1$ and $\mathbf{A}_n$. For another procedure which converges upon the intermediate characteristic values see Sec. 9-8.

[1] J. L. B. Cooper, *Quart. Appl. Math.*, **6**: 179 (1948).

[2] L. Fox, *Quart. J. Mechanics and Appl. Math.*, **1**: 253 (1948).

If **H** is unsymmetric, it can either be reduced to symmetric form by the method of Sec. 9-2, or the relaxation procedure can be applied to both sets of simultaneous equations (with coefficients $\mathbf{H} = \mathbf{GF}$ and $\mathbf{H}' = \mathbf{FG}$), using the successive approximate characteristic vectors as **V** and **U**, respectively, in Eq. (12), Sec. 9-6, to give an improved λ.

A distinctive feature of the relaxation method is the large freedom left to the judgment of the computer. This is particularly true in deciding how small to make the e's in (9) at any given stage (value of λ). For this reason, the speed of convergence of the method depends very largely on the experience of the computer.

9-8. Matrix Iteration Methods[1]

The methods of solution to be described under this heading have several advantages. Although they are not necessarily the most efficient in terms of number of arithmetic operations, they are in some ways the most systematic and therefore least liable to error. The typical arithmetic operation, namely, the addition of a set of products of the form $ab + cd +$ etc. lends itself more conveniently than some of the computations involved in the other methods to computation with ordinary calculating machines. Furthermore, if a numerical error is made at an early stage, it is eliminated as the calculation progresses. Also this iteration method is in many cases the most rapid for determining the value of the largest characteristic value if, for any reason, that value alone is required.

In its simplest form, the method consists in choosing an arbitrary set of numbers, $V_1^{(0)}, V_2^{(0)}, \ldots, V_n^{(0)}$, and computing from them a new set, $V_1^{(1)}$ to $V_n^{(1)}$, by means of the matrix multiplication

$$\mathbf{HV}^{(0)} = \mathbf{V}^{(1)} \tag{1}$$

in which $\mathbf{V}^{(0)}$ and $\mathbf{V}^{(1)}$ are the column matrices (vectors) whose components are the numbers described above and **H** is the matrix of the elements of the secular determinant which it is desired to solve.

This process may then be repeated to give

$$\mathbf{V}^{(2)} = \mathbf{HV}^{(1)} \tag{2}$$

and is continued until

$$\mathbf{V}^{(p+1)} \approx c\mathbf{V}^{(p)} \tag{3}$$

i.e., until the matrix multiplication yields a vector $\mathbf{V}^{(p+1)}$ all of whose components differ from those of the preceding vector $\mathbf{V}^{(p)}$ by approxi-

[1] See, for example, the following:
A. C. Aitken, *Proc. Roy. Soc. Edinburgh*, **57**: 172 (1937).
H. Hotelling, *Ann. Math. Stat.*, **14**: 1 (1943).
W. M. Kincaid, *Quart. Appl. Math.*, **5**: 320 (1947).

mately the same constant. It is evident that such a vector is approximately a characteristic vector, while the constant c is approximately a characteristic value.

It will now be proved that the process does indeed converge upon $c = \lambda_1$, the greatest characteristic value (and upon the corresponding characteristic vector, $\mathbf{A}_1$). The (unknown) characteristic vectors $\mathbf{A}_k$ are linearly independent (since they constitute the columns of a square matrix, $\mathbf{A}$, which has an inverse) so that a vector with arbitrary components, such as $\mathbf{V}^{(0)}$, can always be expressed as some linear combination of the $\mathbf{A}_k$:

$$\mathbf{V}^{(0)} = \sum_{k=1}^{n} C_k \mathbf{A}_k \tag{4}$$

Next, substitute (4) in (1), obtaining

$$\mathbf{V}^{(1)} = \mathbf{H}\left(\sum_{k=1}^{n} C_k \mathbf{A}_k\right) = \sum_{k=1}^{n} C_k \mathbf{H} \mathbf{A}_k \tag{5}$$

But since $\mathbf{A}_k$ is a characteristic vector of $\mathbf{H}$, $\mathbf{H}\mathbf{A}_k = \lambda_k \mathbf{A}_k$, so that (5) becomes

$$\mathbf{V}^{(1)} = \sum_{k=1}^{n} \lambda_k C_k \mathbf{A}_k \tag{6}$$

It is now easy to see the effect of repeated matrix multiplications of $\mathbf{V}^{(0)}$ by $\mathbf{H}$:

$$\mathbf{V}^{(p)} = \sum_{k=1}^{n} (\lambda_k)^p C_k \mathbf{A}_k \tag{7}$$

Assuming that $\lambda_1 > \lambda_2 > \cdots > \lambda_n > 0$, for large values of p it is apparent that the term $k = 1$ dominates all others in (7), provided only that $C_1 \neq 0$, so that in the limit of large p,

$$\mathbf{V}^{(p)} = (\lambda_1)^p C_1 \mathbf{A}_1 \tag{8}$$

In this same limit

$$\mathbf{V}^{(p+1)} = (\lambda_1)^{p+1} C_1 \mathbf{A}_1 = \lambda_1 \mathbf{V}^{(p)} \tag{9}$$

and since the characteristic vectors are determined only within a proportionality constant (unnormalized), $\mathbf{V}^{(p)}$ is just as good a characteristic vector as $\mathbf{A}_1$.

It should be clear from this argument that the rate of convergence is essentially determined by the ratio of the largest to next largest characteristic value, λ_1/λ_2. Naturally a good guess at $\mathbf{A}_1$, that is, a $\mathbf{V}^{(0)}$ such that C_1 is large while C_2, C_3, etc., are small, also speeds the process.

Use of the Rayleigh Principle. The present method can be combined with the principle developed in Sec. 9-6, namely, that an approximate characteristic vector will yield a better characteristic value when substituted in Eq. (18), Sec. 9-6. Normalization equivalent to Eq. (11), Sec. 9-6, is achieved by division by $\mathbf{V}^{(p)\dagger}\mathbf{V}^{(p)}$:

$$\begin{aligned}\lambda_1 &= \frac{\mathbf{V}^{(p)\dagger}\mathbf{H}\mathbf{V}^{(p)}}{\mathbf{V}^{(p)\dagger}\mathbf{V}^{(p)}} \\ &= \frac{\mathbf{V}^{(p)\dagger}\mathbf{V}^{(p+1)}}{\mathbf{V}^{(p)\dagger}\mathbf{V}^{(p)}}\end{aligned} \tag{10}$$

The last form of (10) is particularly simple and convenient, becoming upon expansion

$$\lambda_1 = \frac{\sum_{t=1}^{n} V_t^{(p)} V_t^{(p+1)}}{\sum_{t=1}^{n} (V_t^{(p)})^2} \tag{11}$$

In case $\mathbf{H}$ is not symmetric, the iterative procedure should be carried out using $\mathbf{H}^\dagger$ as well as $\mathbf{H}$; that is $\mathbf{U}^{(1)} = \mathbf{H}^\dagger\mathbf{U}^{(0)}$, $\mathbf{U}^{(2)} = \mathbf{H}^\dagger\mathbf{U}^{(1)}$, etc., should be computed. Then $\mathbf{U}^{(p)\dagger}$ is simply used in place of $\mathbf{V}^{(p)\dagger}$ in (10).

Procedure for Finding λ_2, λ_3, *etc.* The first of the two methods for this purpose which will be described here is a device which yields a new matrix which, in place of the characteristic values $\lambda_1, \lambda_2, \lambda_3, \ldots, \lambda_n$, has the characteristic values $0, \lambda_2, \lambda_3, \ldots, \lambda_n$. Thus, in the new matrix, λ_2 becomes dominant, so that iteration as above will converge upon $\mathbf{A}_2$ and λ_2.

The new matrix is obtained from $\mathbf{H}$ by subtracting $\lambda_1 L_{t1} L_{t'1}$ from $H_{tt'}$, if $\mathbf{H}$ is symmetric. Here L_{t1} is a component of the normalized characteristic vector corresponding to λ_1, which presumably has just been found by iteration with H. A suitable normalization procedure here consists in putting

$$L_{t1} = \frac{A_{t1}}{\left[\sum_{t'} (A_{t'1})^2\right]^{\frac{1}{2}}} \tag{12}$$

in which the A_{t1} are approximated by the $V_t^{(p)}$ obtained by iteration.

If $\mathbf{H}$ is unsymmetric, it is necessary instead to subtract $\lambda_1 U_t^{(p)} V_{t'}^{(p)} / \left(\sum_s U_s^{(p)} V_s^{(p)}\right)^{\frac{1}{2}}$, since this is an approximation to $\lambda_1 (L^{-1})^\dagger_{t1} L_{t'1}$ which replaces $\lambda_1 L_{t1} L_{t'1}$ when $\mathbf{L}$ is not orthogonal.

The third root is obtained in turn by subtracting $\lambda_2 L_{t2} L_{t'2}$

$$[\text{or } \lambda_2 (L^{-1})^\dagger_{t2} L_{t'2}]$$

from the matrix elements used in obtaining λ_2, etc. This procedure is called *deflation* and its validity is proved in the paper by Aitken cited in the footnote at the beginning of this section.

An alternate, and somewhat more versatile, procedure makes use of the theorem proved at the end of Sec. 9-1, namely, that a modified matrix, say $\lambda_1\mathbf{E} - \mathbf{H}$ (where λ_1 has just been determined by iteration with $\mathbf{H}$), has characteristic values $0, \lambda_1 - \lambda_2, \lambda_1 - \lambda_3, \ldots, \lambda_1 - \lambda_n$. Iteration with $\lambda_1\mathbf{E} - \mathbf{H}$ should therefore converge upon the largest of these, which is $\lambda_1 - \lambda_n$, the characteristic vector being identical with that associated with λ_n in $\mathbf{H}$.

Ordinarily, however, it is possible to use this method more effectively. If an estimate of the characteristic values is available, which is frequently the case in molecular vibrational problems, the approximate values $\lambda_1', \lambda_2', \ldots, \lambda_n'$ may be used very effectively to yield a matrix which will converge rapidly to any desired characteristic value in the iteration process. This may best be seen by a representative numerical example. Suppose the estimated values are $\lambda_1 = 5$, $\lambda_2 = 1$, $\lambda_3 = 0.5$, and $\lambda_4 = 0.4$. The matrix, $\mathbf{H} - 5\mathbf{E}$, would have approximate characteristic values of $0, -4, -4.5$, and -4.6. Similarly $\mathbf{H} - 1\mathbf{E}$ would have $4, 0, -0.5$, and -0.6, $\mathbf{H} - 0.5\mathbf{E}$ would have $4.5, 0.5, 0, -0.1$, and $\mathbf{H} - 0.4\mathbf{E}$ would have $4.6, 0.6, 0.1$, and 0 as characteristic values, respectively. As an example of a matrix which could be used in iteration to get λ_2, consider the product, $(\mathbf{H} - 5\mathbf{E})(\mathbf{H} - 0.5\mathbf{E})$. Its roots (characteristic values) would occur approximately at $0 \times 4.5 = 0$, $-4 \times 0.5 = -2.0$, $-4.5 \times 0 = 0$, and $-4.6 \times -0.1 = 0.46$, so that convergence should be reasonably rapid upon the value -2.0 which is $(\lambda_2 - 5)(\lambda_2 - 0.5)$. Obviously, a product of the form $(\mathbf{H} - 5\mathbf{E})(\mathbf{H} - 0.5\mathbf{E})(\mathbf{H} - 0.4\mathbf{E})$ would converge even more rapidly upon $(\lambda_2 - 5)(\lambda_2 - 0.5)(\lambda_2 - 0.4)$, but the computer would have to weigh the advantage of more rapid convergence against the greater labor of carrying out two matrix multiplications in place of one.

9-9. Perturbation Methods

In case the off-diagonal elements of $\mathbf{H} = \mathbf{GF}$ are small compared with the diagonal elements, the procedures of perturbation theory may be employed. The fact that the off-diagonal elements are small suggests that a fair approximation to the characteristic values would be $\lambda_t = H_{tt}$. Then a better approximation to λ_1 can be obtained through the following considerations. Replace the exact secular determinant by the approximate one,

$$\begin{vmatrix} H_{11} - \lambda & H_{12} & H_{13} & H_{14} & \cdots \\ H_{21} & H_{22} - H_{11} & 0 & 0 & \cdots \\ H_{31} & 0 & H_{33} - H_{11} & 0 & \cdots \\ H_{41} & 0 & 0 & H_{44} - H_{11} & \cdots \\ \cdots & \cdots & \cdots & \cdots & \cdots \end{vmatrix} = 0 \qquad (1)$$

The λ's which should appear in the second, third, etc., diagonal terms are replaced by the approximate value $\lambda_1 = H_{11}$. This will be justifiable, provided the differences $H_{tt} - H_{11}(t \neq 1)$ are large compared with the error in the initial approximation, $\lambda_1 = H_{11}$. Furthermore, the off-diagonal terms, other than those in the first row or column, are neglected. The latter approximation is appropriate inasmuch as the neglected off-diagonal terms can affect the desired λ only indirectly. But now the determinant in (1) can be evaluated by the method explained in Sec. 9-7 as "successive elimination." In the present case, multiply the second row by $H_{12}/(H_{22} - H_{11})$ and subtract from the first; then multiply the third row by $H_{13}/(H_{33} - H_{11})$ and subtract from the first, etc. This will eliminate the elements H_{12}, H_{13}, etc., so that the value of the determinant is

$$\left[(H_{11} - \lambda) - \sum_{t=2}^{n}\left(\frac{H_{1t}H_{t1}}{H_{tt} - H_{11}}\right)\right](H_{22} - H_{11})(H_{33} - H_{11}) \cdots (H_{nn} - H_{11}) = 0 \quad (2)$$

Then if H_{11} is not equal to any other diagonal term,

$$\lambda = H_{11} - \sum_{t=2}^{n}\left(\frac{H_{1t}H_{t1}}{H_{tt} - H_{11}}\right) \quad (3)$$

The same arguments can be employed to obtain a similar approximation to any other $\lambda_{t'}$ with the result

$$\lambda_{t'} = H_{t't'} - \sum_{t \neq t'}\left(\frac{H_{t't}H_{tt'}}{H_{tt} - H_{t't'}}\right) \quad (4)$$

In this same approximation, the characteristic vector corresponding to λ_1 is easily found to be (putting $A_{11} = 1$)

$$A_{t1} = \frac{H_{t1}}{H_{11} - H_{tt}}, \qquad t \neq 1 \quad (5)$$

or, corresponding to $\lambda_{t'}$ and putting $A_{t't'} = 1$

$$A_{tt'} = \frac{H_{tt'}}{H_{t't'} - H_{tt}}, \qquad t \neq t' \quad (6)$$

An approximation superior to (4) could be obtained by using the values for the $A_{tt'}$ given by (6) in connection with the Rayleigh principle (Sec. 9-6) to obtain

$$\lambda_{t'} = \frac{\sum_{tt''} A_{tt'}H_{tt''}A_{t''t'}}{\sum_{t} (A_{tt'})^2} \quad (7)$$

where the division by $\sum_t (A_{tt'})^2$ is equivalent to normalization of the $A_{tt'}$.

A somewhat different aspect of perturbation theory could be employed in the following problem: suppose the characteristic vectors of the matrix[1]

$$\mathbf{H}^0 = \mathbf{G}\mathbf{F}^0 \tag{8}$$

are known: this means that a matrix $\mathbf{L}^0$ is known which has the property that

$$\mathbf{L}^{0\dagger}\mathbf{F}^0\mathbf{L}^0 = \mathbf{\Lambda}^0 \qquad \text{and} \qquad \mathbf{L}^0\mathbf{L}^{0\dagger} = \mathbf{G} \tag{9}$$

(see Appendix VIII).

Now suppose it is desired to adjust $\mathbf{F}^0$ in such a way as to produce small changes in the values $\lambda_1^0, \lambda_2^0, \ldots$ presumably to secure better agreement with a set of observed λ's. Let $\mathbf{H}'$ be defined as

$$\mathbf{H}' = (\mathbf{L}^0)^{-1}(\mathbf{G}\,\Delta\mathbf{F})\mathbf{L}^0 = (\mathbf{L}^0)^{-1}\mathbf{L}^0\mathbf{L}^{0\dagger}\,\Delta\mathbf{F}\mathbf{L}^0 = \mathbf{L}^{0\dagger}\,\Delta\mathbf{F}\mathbf{L}^0 \tag{10}$$

Then the secular determinant for the perturbed problem would have the form

$$\begin{vmatrix} \lambda_1^0 + H'_{11} - \lambda & H'_{12} & H'_{13} & \cdots \\ H'_{21} & \lambda_2^0 + H'_{22} - \lambda & H'_{23} & \cdots \\ H'_{31} & H'_{32} & \lambda_3^0 + H'_{33} - \lambda & \cdots \\ \cdot & \cdot & \cdot & \cdots \\ \cdot & \cdot & \cdot & \cdots \\ \cdot & \cdot & \cdot & \cdots \end{vmatrix} = 0 \tag{11}$$

Since the elements $H'_{tt'}$ are presumably small, the first approximation would be:

$$\begin{aligned} \lambda_t = \lambda_t^0 + H'_{tt} &= \lambda_t^0 + \sum_{t't''} L^{0\dagger}_{tt'}\,\Delta F_{t't''}L^0_{t''t} \\ &= \lambda_t^0 + \sum_{t't''} L^0_{t't}L^0_{t''t}\,\Delta F_{t't''} \end{aligned} \tag{12}$$

A second-order approximation could be carried through employing the analogue of (4).

Just the opposite situation from the one just described may arise: the solution for $\mathbf{H}^0 = \mathbf{G}^0\mathbf{F}$ may be known, and that for

$$\mathbf{H} = \mathbf{G}\mathbf{F} = (\mathbf{G}^0 + \Delta\mathbf{G})\mathbf{F}$$

may be desired. This occurs in the case of isotopic substitution in a given molecule (see Sec. 8-5).

[1] In practice, $\mathbf{F}^0$ might be a diagonal force constant matrix.

Before applying perturbation methods it is important to be sure that the effect of the perturbation on a given root is small compared with the distance to the neighboring roots. Otherwise the method may not give a good result, even though the first- and second-order corrections (given by the above treatment) turn out to be small. For a method which overcomes this difficulty, the reader may refer to a procedure originated by Van Vleck.[1]

9-10. Use of Electric Circuit Analogues[2]

It is natural to consider the possibility of solving the secular equations by constructing physical models and observing their behavior. For instance, mechanical models have been used[3] to produce a system mechanically and geometrically similar to the molecule; the system is excited by a variable-speed motor and the resonant frequencies are noted. A more general type of mechanical model, which dispenses with the (unnecessary) geometrical similarity, but which is more versatile and accurate, has been described.[4]

In general, however, mechanical models are unwieldy in comparison with electrical models which form the main topic of this section. Two quite distinct types of electrical analogues have been employed for the solution of secular determinants. The first type to be described here solves the simultaneous equations from which the secular determinant arises by a method closely related to the relaxation procedure described in Sec. 9-6. The essential function of the circuit consists merely in carrying out potentiometrically the two fundamental arithmetic operations of multiplication and addition which appear in the simultaneous equations

$$\sum_{t'} (H_{tt'} - \delta_{tt'}\lambda)A_{t'} = 0 \qquad (1)$$

These basic operations are performed as follows: multiplication of $H_{tt'}$ by $A_{t'}$ is carried out by applying a voltage numerically equal to $A_{t'}$ across

[1] Described by E. C. Kemble, "The Fundamental Principles of Quantum Mechanics," McGraw-Hill, p. 394, New York, 1937.

[2] G. Kron, *J. Chem. Phys.*, **14**: 19 (1946).

G. K. Carter and G. Kron, *J. Chem. Phys.*, **14**: 32 (1946).

C. E. Berry, E. E. Wilcox, S. M. Rock, and H. W. Washburn, *J. Appl. Phys.*, **17**: 226 (1946).

R. H. Hughes and E. B. Wilson, Jr., *Rev. Sci. Instr.*, **18**: 103 (1947).

A. A. Frost and M. Tamres, *J. Chem. Phys.*, **15**: 383 (1947).

A. Many and S. Meiboom, *Rev. Sci. Instr.*, **18**: 831 (1947).

F. J. Murray, "The Theory of Mathematical Machines," King's Crown, New York, 1947.

W. A. Adcock, *Rev. Sci. Instr.*, **19**: 181 (1948).

[3] C. F. Kettering, L. W. Shutts, and D. H. Andrews, *Phys. Rev.*, **36**: 531 (1930).

[4] D. P. MacDougall and E. B. Wilson, Jr., *J. Chem. Phys.*, **5**: 940 (1937).

a potentiometer, whose slide wire is set at the value $H_{tt'}$. The voltage on the slide wire is then numerically equal to $H_{tt'}A_{t'}$ (see Fig. 9-2). Addition could obviously be accomplished by connecting the individual voltages in series; in practice, however, it is somewhat better to work with the average voltage, rather than the sum, the average being just as

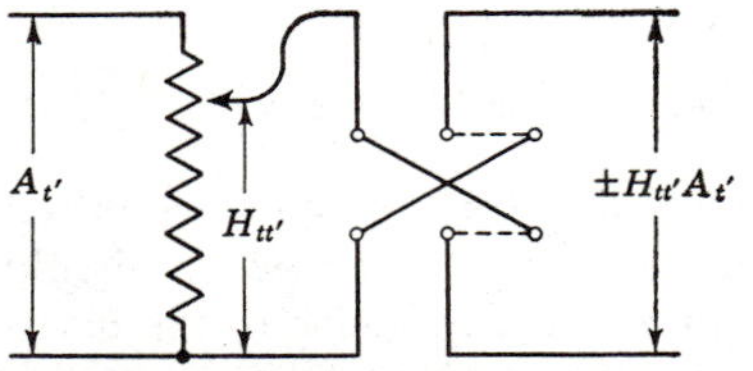

FIG. 9-2. Circuit yielding an output voltage which is the product of the input voltage $A_{t'}$ and a constant $H_{tt'}$. Actually $H_{tt'}$ is set on the potentiometer as a fraction of the total resistance. The double-pole double-throw switch may be used to reverse the polarity when $H_{tt'}$ is negative.

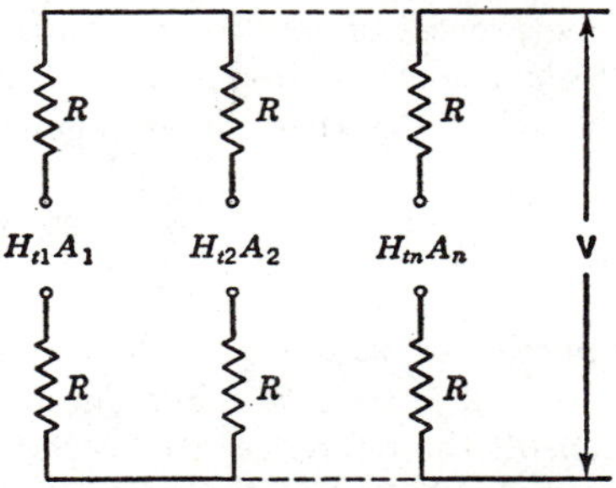

FIG. 9-3. Circuit for summing voltages $H_{tn}A_n$ through averaging. The equal resistances should be large compared to internal resistances in the $H_{tn}A_n$ and only one set of R's are necessary if all voltages are grounded on the same side.

useful for computational purposes. An averaging circuit for voltages is shown in Fig. 9-3. The currents through the (equal) large resistors R are $\frac{V - H_{t1}A_1}{2R}, \frac{V - H_{t2}A_2}{2R}, \ldots, \frac{V - H_{tn}A_n}{2R}$, respectively. By Kirchhoff's law, the sum of these currents vanishes so that

$$V - H_{t1}A_1 + V - H_{t2}A_2 \cdots + V - H_{tn}A_n = 0 \qquad (2)$$

or

$$nV = H_{t1}A_1 + H_{t2}A_2 + \cdots + H_{tn}A_n \qquad (3)$$

Frost and Tamres have described a circuit utilizing these basic principles which can be applied to the solution of the simultaneous equations corresponding to the secular determinant in the form

$$|\mathbf{F} - \lambda\mathbf{G}^{-1}| = 0 \qquad (4)$$

as well as

$$|\mathbf{FG} - \lambda\mathbf{E}| = 0 \qquad (5)$$

i.e., their device is not limited to the case where the λ's appear only on the diagonal. The sums of the F terms and G terms are formed separately, that is, the voltages corresponding to

$$\frac{1}{n}\sum_{t'=1}^{n} F_{tt'}A_{t'} \qquad (6)$$

and

$$\frac{1}{n}\sum_{t'=1}^{n} G_{tt'}^{-1}A_{t'} \tag{7}$$

are formed, the latter then being multiplied in turn by an estimated λ.[1] The process of solution is identical in principle with that described as a "relaxation" method under Sec. 9-7 above, the unknowns $A_{t'}$ being varied in such a way that the voltage corresponding to

$$\frac{1}{n}\left(\sum_{t'} F_{tt'}A_{t'} - \lambda \sum_{t'} G_{tt'}^{-1}A_{t'}\right)$$

approaches zero (as evidenced by a null detector) for each equation $t = 1, 2, \ldots, n$. For further practical details, the reader should consult the reference cited. Some difficulty in convergence to the intermediate characteristic values[2] may be encountered unless a good approximation is available to start with.

A modified procedure suggested by Kohn[3] accelerates convergence if the λ's occur only on the diagonal and if **H** is symmetric. According to this method a value of $\lambda = \lambda^{(0)}$ is guessed and the equation whose diagonal element, $H_{t''t''}$, lies closest to $\lambda^{(0)}$ is omitted. $A_{t''}$ is set equal to one and the $n - 1$ equations are solved with the circuit[4] using $\lambda = \lambda^{(0)}$. When the solution is obtained, the values $A_t^{(0)}$ thus found are inserted in the equation omitted earlier, which is then solved for $\lambda = \lambda^{(\frac{1}{2})}$, an intermediate approximation. Then the quantity $\sum_t (A_t^{(0)})^2$ is evaluated, and finally,

$$\lambda^{(1)} = \frac{\sum_{tt'} A_t^{(0)} H_{tt'} A_{t'}^{(0)}}{\sum_t (A_t^{(0)})^2} = \lambda^{(0)} + \frac{\lambda^{(\frac{1}{2})} - \lambda^{(0)}}{\sum_t (A_t^{(0)})^2}$$

is calculated. The process is then repeated, using $\lambda^{(1)}$ in place of $\lambda^{(0)}$.

[1] In practice, two potentiometers are used: if $\lambda < 1$, the voltage from (6) is compared with a certain fraction, λ, of (7), whereas if $\lambda > 1$, a certain fraction, $1/\lambda$, of (6) is compared with (7).

[2] It has been shown that no simple automatic system will converge in these cases. F. J. Murray, *Quart. Appl. Math.*, **7**: 263 (1948).

[3] W. Kohn, *J. Chem. Phys.*, **17**: 670 (1949).

[4] This method does not overcome the "instability" of the simultaneous equations problem when an intermediate λ is being sought.

To consider the possibility of automatic convergence, suppose first that it is desired to solve a set of inhomogeneous simultaneous equations

$$\sum_{t'} C_{tt'}A_{t'} = B_t \tag{8}$$

in which the $A_{t'}$ are unknowns. The errors

$$\epsilon_t = \sum_{t'} C_{tt'}V_{t'} - B_t \tag{9}$$

could be formed potentiometrically, the $V_{t'}$ being approximations to the $A_{t'}$. Now suppose

$$\frac{dV_t}{dt} = k\epsilon_t \tag{10}$$

where k is a negative constant. This last equation can be realized, for example, by driving an electric motor with the voltage $k\epsilon_t$, the motor being mechanically coupled to the potentiometer which produces the voltage V_t. The constant, k, can be taken to represent the gain of an amplifier whose input is ϵ_t. Such a system will converge to the solution of (8) subject only to certain restrictive conditions on the matrix of the coefficients $C_{tt'}$, namely, that the real parts of the characteristic values of C shall all have the opposite sign from that of the constant, k. When the simultaneous equations

$$\sum_{t'=1}^{n} (H_{tt'} - \lambda\delta_{tt'})A_{t'} = 0$$

are solved on a machine of the type now under discussion, the usual procedure is to set one of the unknowns, say A_n, equal to a constant, insert a trial value of λ, and then solve the first $n - 1$ inhomogeneous equations

$$\begin{aligned}
(H_{11} - \lambda)A_1 + H_{12}A_2 + \cdots + H_{1,n-1}A_{n-1} &= -H_{1n}A_n \\
H_{21}A_1 + (H_{22} - \lambda)A_2 + \cdots + H_{2,n-1}A_{n-1} &= -H_{2n}A_n \\
\cdots\cdots\cdots\cdots\cdots\cdots\cdots\cdots\cdots\cdots \\
H_{n-1,1}A_1 + H_{n-1,2}A_2 + \cdots + (H_{n-1,n-1} - \lambda)A_{n-1} &= -H_{n-1,n}A_n
\end{aligned} \tag{11}$$

The values of $A_1, A_2, \ldots, A_{n-1}$ may then be inserted in the last equation, which will be satisfied only if the trial λ is a true characteristic value. Thus, even though it is known that the characteristic values of the complete $\mathbf{H}$ matrix would satisfy the convergence criterion, the coefficients equivalent to the $C_{tt'}$ of (8) are the elements of $(\mathbf{H}' - \lambda\mathbf{E})$, where $\mathbf{H}'$ is the matrix obtained by omitting the last row and column of $\mathbf{H}$. The characteristic values of $(\mathbf{H}' - \lambda\mathbf{E})$ are $\lambda_1' - \lambda$, $\lambda_2' - \lambda$, . . . ,

$\lambda'_{n-1} - \lambda$, where λ'_1 through λ'_{n-1} are interleaved between the characteristic values λ_1 through λ_n of $\mathbf{H}$. Therefore, it follows from the convergence criterion mentioned above that the feedback device described above would only converge if $\lambda < \lambda'_{n-1}$, or if the sign of the constant k were changed and $\lambda > \lambda'_1$. It is therefore apparent that this method can be used only to find $\lambda_1 > \lambda'_1$ and $\lambda_n < \lambda'_{n-1}$.

A modification of this type of circuit, which has the distinct advantage of automatic convergence, has been described by Murray, and a fourth-order model of such a machine has been constructed by Adcock. The automatic operation is achieved through the use of a feedback device which is activated by the errors in the several equations and which then modifies the trial unknowns $V_{t'}$ in such a way that they approach the $A_{t'}$ as the errors approach zero.

The feedback arrangement, which is convergent for all the characteristic values, is based on the following principles. In place of solving (8), a machine is designed to solve the related equations

$$\sum_{t''} \left(\sum_{t'} C_{tt'} C_{t''t'} \right) X_{t''} = B_t \tag{12}$$

for the unknowns, $X_{t''}$. Since the coefficients in (12) are the elements of the matrix $\mathbf{CC}^\dagger$, whose characteristic values are positive for an arbitrary matrix, $\mathbf{C}$, the procedure is always convergent (for any trial value of λ).

If (9) and (12) are compared, evidently

$$V_{t'} = \sum_{t''} X_{t''} C_{t''t'} \tag{13}$$

when the ϵ_t approach zero.

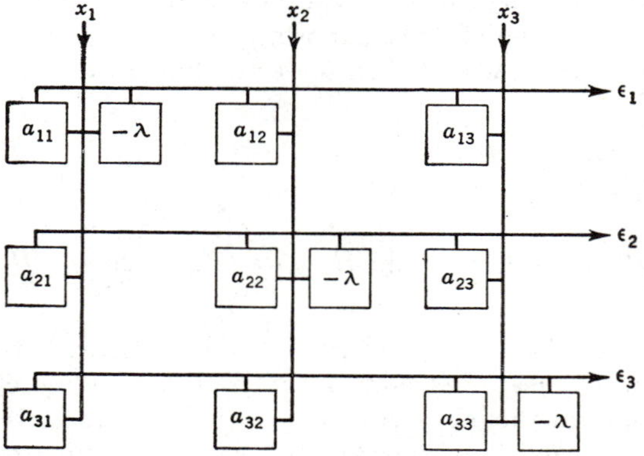

FIG. 9-4. Block diagram of the Adcock computer.

A block diagram of the machine described by Adcock is given in Fig. 9-4. Two networks are used, one which yields voltages corresponding to

(13), the $A_{t'}$, and the other which produces the error voltages, ϵ_t. The cycle is completed by the amplifiers and motors which adjust the voltages $X_{t''}$ according to

$$\frac{dX_{t''}}{dt} = k\epsilon_t \tag{14}$$

which replaces (10).

In principle, this last function could also be performed electronically by a feedback amplifier without mechanical parts.

The second type of electric circuit makes use of a well-known analogy between alternating-current networks containing inductances and capacitances and coupled mechanical systems. The single junction network shown in Fig. 9-5, for example, has a resonant frequency ν given by

$$4\pi^2\nu^2 = \lambda = \frac{1}{CL} \tag{15}$$

where L is the inductance of the coil and C the capacity of the condenser. The corresponding frequency expression for a mechanical system is

$$4\pi^2\nu^2 = \lambda = F_{11}G_{11} = \frac{F_{11}}{G_{11}^{-1}} \tag{16}$$

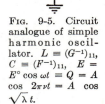

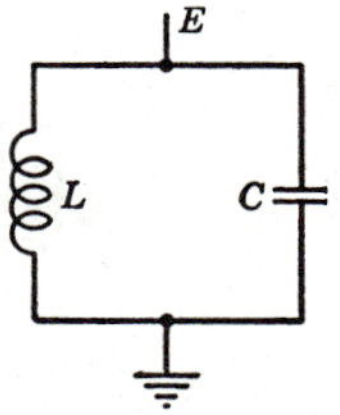

FIG. 9-5. Circuit analogue of simple harmonic oscillator. $L \equiv (G^{-1})_{11}$, $C \equiv (F^{-1})_{11}$, $E = E^\circ \cos \omega t \equiv Q = A \cos 2\pi\nu t = A \cos \sqrt{\lambda}\, t$.

Thus in order to solve the (trivial) one-dimensional secular equation, $F_{11}G_{11} - \lambda = 0$ or $F_{11} - \lambda G_{11}^{-1} = 0$, one could set $1/C$ numerically equal (in appropriate units) to the force constant, F_{11}, and L equal to G_{11}^{-1} and then determine the resonant frequency of the electrical network experimentally.

The analogy can be extended to the n-dimensional case, so that various condensers of the network stand for the elements $F_{tt'}$ and various inductance coils for the elements $G_{tt'}^{-1}$. But a serious practical limitation is then immediately evident, inasmuch as capacitances and inductances are limited to positive values, which is not in general true of $F_{tt'}$ and $G_{tt'}^{-1}$. One way out of this difficulty has been described by Kron. In place of setting the quantities analogous to the $F_{tt'}$ and the $G_{tt'}^{-1}$ at fixed values, and then searching for λ by applying an external current of varying frequency, the elements representing the terms $F_{tt'}$ are fixed, but the inductances, corresponding to terms of the type $-\lambda G_{tt'}^{-1}$, are varied until resonance with a fixed external frequency is found. As formulated by Kron and by Carter and Kron, such a circuit was applied to the case in which cartesian displacement coordinates were used, which means that $\mathbf{G}^{-1}$ was diagonal. Then the n inductances corresponding to the terms $G_{11}^{-1}, G_{22}^{-1}, \ldots, G_{nn}^{-1}$ were given values always in the ratio of the $G_{tt'}^{-1}$, but varied so as to correspond to different λ. It is evident that this procedure will become quite difficult if $\mathbf{G}^{-1}$ is not diagonal and n is large.

Hughes and Wilson have described an apparatus based on the principles used by Carter and Kron but meant to solve secular equations rather than to be a direct electrical analogue of a molecule. It is restricted to equations which are symmetric and have λ only on the diagonal. Consider the circuit shown in block form in Fig. 9-6 as an example for the case

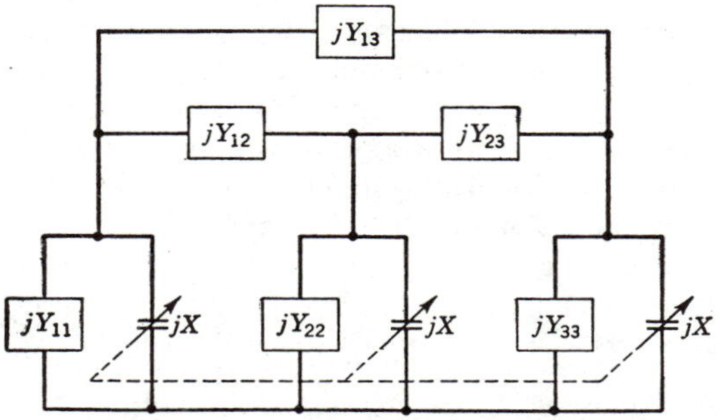

FIG. 9-6. Block diagram of the Hughes and Wilson circuit.

$n = 3$. Let the voltage amplitude at the tth junction be E_t. Then, by Kirchhoff's law, the current flowing out of the junction is

$$(Y_{tt} + X)E_t + \sum_{t' \neq t} Y_{tt'}(E_t - E_{t'}) = I_t \tag{17}$$

where the Y's are the admittances of the various fixed elements shown, and X is the admittance of the variable condensers. At frequency ν, the admittance of a condenser of capacity C connected in parallel with an inductance L is

$$Y = i\left(2\pi\nu C - \frac{1}{2\pi\nu L}\right) \tag{18}$$

Thus, by a suitable choice of C or L, Y can be given any desired positive or negative (imaginary) value. By collecting the coefficients of the E_t, (17) can be rewritten in the form

$$\left(-\sum_{t'} Y_{tt'} - X\right) E_t + \sum_{t' \neq t} Y_{tt'}E_{t'} = -I_t \tag{19}$$

Now, if the network is excited by applying a constant current generator (at the fixed frequency, ν) to one of the junctions, the I_t will be zero at all other junctions, and as resonance is approached, the E's will become very large so that the right-hand side of (19) can be neglected in comparison with the left-hand side. In other words, the condition for resonance is the vanishing of the determinant having diagonal terms $-\sum_{t'} Y_{tt'} - X$

and off-diagonal terms $Y_{tt'}$. The i's in (18) can be canceled from these equations, so that, in order to solve the secular determinant,

$$|\mathbf{H} - \lambda \mathbf{E}| = 0$$

it is merely necessary to put

$$\frac{Y_{tt'}}{i} = H_{tt'} = 2\pi\nu C_{tt'} - \frac{1}{2\pi\nu L_{tt'}} \tag{20}$$

and

$$H_{tt} = -\sum_{t'} \frac{Y_{tt'}}{i} \tag{21}$$

or, combining (21) and (20),

$$\frac{Y_{tt}}{i} = -\sum_{t'} H_{tt'} = 2\pi\nu C_{tt} - \frac{1}{2\pi\nu L_{tt}} \tag{22}$$

in order that the value of X/i at resonance shall be equal to λ.

A somewhat similar circuit has been described by Many and Meiboom. In their circuit, the network consists of n simple LC circuits, each one coupled to each of the others by two equal condensers. The sign of the coupling term is determined by the choice between "straight" and "crossed" coupling, as illustrated in Fig. 9-7. The inductances are

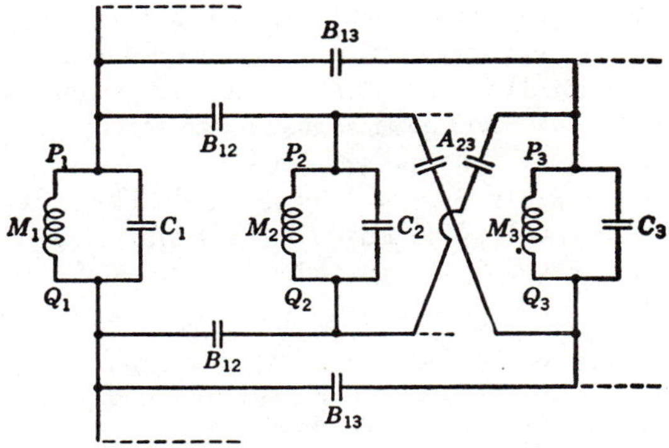

FIG. 9-7. An element of the Many and Meiboom circuit.

given a fixed constant value. The characteristic values are determined by varying an external frequency until resonance is detected. The characteristic values may also be determined. For practical details, the reader is referred to the papers cited.

CHAPTER 10

A SAMPLE VIBRATIONAL ANALYSIS: THE BENZENE MOLECULE

In this chapter a detailed analysis of the vibrational spectrum of the benzene molecule will be carried out in order to illustrate in a coordinated form the material from several previous chapters, in particular, Chap. 4 through 9. Several representative references to original papers on this subject are given in the footnote.[1]

10-1. Structure and Symmetry of the Benzene Molecule

The structure for benzene which will be assumed in this chapter is one in which all atoms are coplanar with the carbon atoms and the hydrogen atoms at the corners of concentric, regular hexagons (Fig. 10-1). The Kekulé structure, in which alternate carbon-carbon bonds, but not adjacent ones, are equivalent, would be somewhat less symmetrical, but modern theories of valence regard all six such bonds as equivalent. Moreover, there is good experimental evidence (aside from the vibrational spectrum) for the most symmetrical planar structure; for example, it is supported by electron diffraction experiments.[2]

The symmetry elements possessed by the model illustrated in Fig. 10-1 will now be enumerated. The axis of highest multiplicity is, of course, the sixfold axis, which will be regarded as the z axis of cartesian coordi-

[1] E. B. Wilson, Jr., *Phys. Rev.*, **45**: 706 (1934).
C. Manneback, *Ann. soc. sci. Bruxelles*, (*B*), **55**: 129 (1935).
C. Manneback, *Ann. soc. sci. Bruxelles*, (*B*), **55**: 237 (1935).
C. K. Ingold *et al.*, *J. Chem. Soc.*, 1936, pp. 912–987.
R. C. Lord, Jr., and D. H. Andrews, *J. Chem. Phys.*, **41**: 149 (1937).
A. Langseth and R. C. Lord, Jr., *Kgl. Danske Videnskab. Selskab. Mat.-fys. Medd.*, **16**: 6 (1938).
J. Duchesne and W. G. Penney, *Bull. soc. roy. sci. Liége*, **8**: 514 (1939).
E. Bernard, C. Manneback, and A. Verleysen, *Ann. soc. sci. Bruxelles*, **59**: 376 (1939).
R. P. Bell, *Trans. Faraday Soc.*, **41**: 293 (1945).
F. A. Miller and B. L. Crawford, Jr., *J. Chem. Phys.*, **14**: 282 (1946).
B. L. Crawford, Jr., and F. A. Miller, *J. Chem. Phys.*, **17**: 249 (1949).

[2] L. Pauling and L. O. Brockway, *J. Chem. Phys.*, **2**: 867 (1934).

nates. Lying in the plane of the molecule, perpendicular to the sixfold axis, are two sets of twofold axes, each consisting of three members. The first set is designated as C'_2, the second as C''_2. As indicated in Fig. 10-1, it is assumed that the cartesian y axis is coincident with one of the symmetry axes, C'_2, while the x axis is coincident with one of the set of C''_2 symmetry axes. The molecule also possesses a horizontal plane of symmetry, σ_h (coincident with the xy plane), three "vertical" planes, σ_v, one

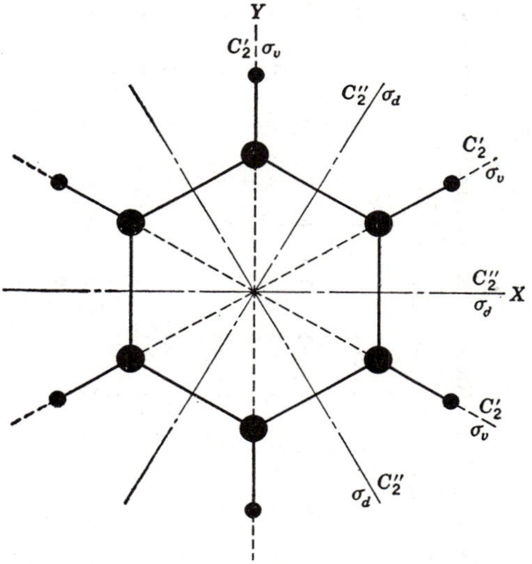

FIG. 10-1. Symmetry of the benzene molecule. The z axis, (a C_6 and S_6 symmetry axis) is perpendicular to the plane of the figure, which is the σ_h symmetry plane.

of which is coincident with the yz plane, and three "diagonal" planes, σ_d, one of which is coincident with the xz plane. The origin of the cartesian axis system is a center of inversion, i. Finally, the molecule is seen to possess two improper rotations of each of the types, S_3 and S_6. In summary, the symmetry of the molecule is that of the group $\mathfrak{D}_{6h}$ which consists of the following operations. E, $2C_6$, $2C_3$, C_2, $3C'_2$, $3C''_2$, i, $2S_3$, $2S_6$, σ_h, $3\sigma_d$, $3\sigma_v$.

10-2. Symmetry Species of Normal Coordinates and Internal Coordinates

Table 10-1 contains in addition to the character table for the group, $\mathfrak{D}_{6h}$, the analyses of the various coordinate representations into the irreducible representations or species. Thus the columns headed by the symbols $n_C^{(\gamma)}$ and $n_H^{(\gamma)}$ give the number in each species, γ, of (external) symmetry coordinates formed, respectively, from the cartesian displace-

TABLE 10-1. SYMMETRY SPECIES OF THE NORMAL COORDINATES OF BENZENE

$\mathfrak{D}_{6h}$	E	$2C_6$	$2C_3$	C_2	$3C_2'$	$3C_2''$	i	$2S_3$	$2S_6$	σ_h	$3\sigma_d$	$3\sigma_v$	$n_C^{(\gamma)}$	$n_H^{(\gamma)}$	$n_T^{(\gamma)}$	$n_R^{(\gamma)}$	$n^{(\gamma)}$
A_{1g}	1	1	1	1	1	1	1	1	1	1	1	1	1	1	0	0	2
A_{2g}	1	1	1	1	−1	−1	1	1	1	1	−1	−1	1	1	0	1	1
B_{1g}	1	−1	1	−1	1	−1	1	−1	1	−1	1	−1	0	0	0	0	0
B_{2g}	1	−1	1	−1	−1	1	1	−1	1	−1	−1	1	1	1	0	0	2
E_{1g}	2	1	−1	−2	0	0	2	1	−1	−2	0	0	1	1	0	1	1
E_{2g}	2	−1	−1	2	0	0	2	−1	−1	2	0	0	2	2	0	0	4
A_{1u}	1	1	1	1	1	1	−1	−1	−1	−1	−1	−1	0	0	0	0	0
A_{2u}	1	1	1	1	−1	−1	−1	−1	−1	−1	1	1	1	1	1	0	1
B_{1u}	1	−1	1	−1	1	−1	−1	1	−1	1	−1	1	1	1	0	0	2
B_{2u}	1	−1	1	−1	−1	1	−1	1	−1	1	1	−1	1	1	0	0	2
E_{1u}	2	1	−1	−2	0	0	−2	−1	1	2	0	0	2	2	1	0	3
E_{2u}	2	−1	−1	2	0	0	−2	1	1	−2	0	0	1	1	0	0	2

ment coordinates of the six equivalent carbon or hydrogen atoms. Likewise, the columns headed $n_T^{(\gamma)}$ and $n_R^{(\gamma)}$ give the corresponding quantities for translation and for rotation of the molecule as a whole. The column headed by $n^{(\gamma)}$ gives the number of genuine vibrations, which, in this case, is $n_C^{(\gamma)} + n_H^{(\gamma)} - n_T^{(\gamma)} - n_R^{(\gamma)}$. These numbers are all readily computed by employing the methods of Sec. 6-2.

In the further analysis of the vibrational motions of benzene, it will prove convenient to consider the in-plane and out-of-plane modes separately. The normal vibrations of species

$$\Gamma_i = 2A_{1g} + A_{2g} + 4E_{2g} + 2B_{1u} + 2B_{2u} + 3E_{1u} \tag{1}$$

constitute the in-plane modes, while those of species

$$\Gamma_o = 2B_{2g} + E_{1g} + A_{2u} + 2E_{2u} \tag{2}$$

are the out-of-plane modes. There should be $2N - 3 = 21$ in-plane modes and $N - 3 = 9$ out-of-plane modes, and this is indeed found to be the case upon multiplying each of the coefficients, $n^{(\gamma)}$, in (1) and (2), respectively, by $d^{(\gamma)}$, the dimension of the species, and then summing. The in-plane and out-of-plane modes are readily distinguished by reference to the character of σ_h, which should be positive in the first case and negative in the second.

Internal Coordinates for the In-plane Modes. The carbon-hydrogen (s) and the carbon-carbon (t) bond stretchings yield twelve of the twenty-one required coordinates. There are twelve more coordinates (ϕ) in the set of hydrogen-carbon-carbon bendings, so that three redundancies are expected. The first step is the determination of the number of symmetry coordinates of each species which can be formed from the above internal coordinates.

TABLE 10-2. CALCULATION OF SYMMETRY SPECIES FOR BENZENE IN TERMS OF INTERNAL COORDINATES

	E	$3C_2'$	$3C_2''$	σ_h	$3\sigma_d$	$3\sigma_v$	Species
s(CH)	6	2		6		2	$A_{1g} + E_{2g} + B_{1u} + E_{1u}$
t(CC)	6		2	6	2		$A_{1g} + E_{2g} + B_{2u} + E_{1u}$
ϕ(HCC)	12			12			$A_{1g} + A_{2g} + 2E_{2g} + B_{1u} + B_{2u} + 2E_{1u}$
$\alpha(+)$	6	2		6		2	$A_{1g} + E_{2g} + B_{1u} + E_{1u}$
$\beta(-)$	6	−2		6		−2	$A_{2g} + E_{2g} + B_{2u} + E_{1u}$
γ(CH)	6	−2		−6		2	$B_{2g} + E_{1g} + A_{2u} + E_{2u}$
δ(CCCC)	6		2	−6	−2		$B_{2g} + E_{1g} + A_{1u} + E_{2u}$

This can be done in several ways. The most elementary approach is that of Sec. 6-2. Table 10-2 summarizes the calculations. The figures in the table are the number of internal coordinates from a given equivalent set (such as the CH stretchings, s) which are left invariant (or reversed in sign) by a single operation of a given class. Only those classes of operations are included which give nonzero results for some set. A minus sign means that the operation reverses the sign of the coordinate. The rows are labeled by the symbol for the coordinate set. The α's and β's are plus and minus combinations of the ϕ's for each carbon atom, which will be introduced later in order to simplify the treatment. The γ and δ are out-of-plane coordinates discussed below.

The numbers in the table, combined with the characters of the species listed in Table 10-1, were substituted in Eq. (5), Sec. 6-2, to give the results for the species. These should be compared with the species given in Table 10-1, obtained by the cartesian method. It will be noted that s, t, and ϕ together give $3A_{1g}$ and $4E_{1u}$ instead of $2A_{1g}$ and $3E_{1u}$, the cartesian result. This indicates the location of the redundancies introduced among the HCC angles (ϕ) by the ring structure.

Internal Coordinates for the Out-of-plane Modes. Two sets of out-of-plane coordinates will be introduced. The first consists of the six equivalent bendings (γ) of a carbon-hydrogen bond out of the plane of the three nearest carbons. For the second, six torsions of the type CCCC will be employed (δ). Thus δ_1 is the dihedral angle between the planes of C_6-C_1-C_2 and C_1-C_2-C_3 and is therefore a measure of how much the partial double bond C_1-C_2 has been twisted.

The behavior of these sets of coordinates is given in Table 10-2 also. There are nine out-of-plane internal degrees of freedom but twelve γ's and δ's. There must be three redundancies. Comparison of the out-of-plane results of Tables 10-1 and 10-2 shows that there is a redundancy in A_{1u} for δ and a pair in E_{1g} either in γ or δ (or a combination of these). The ultimate decision and check on these will be given later, in connection with the **G** matrix.

10-3. Selection Rules

Species of the Electric Moment and the Polarizability. The first step is to determine the symmetry species of the electric moment and the polarizability in order to discuss the infrared and Raman spectra, respectively. For the electric moment, it is obvious that the z component, μ_z, is of species A_{2u}, since μ_z is sent into itself by all rotations around the sixfold axis and by the planes σ_d and σ_v, whereas μ_z goes into its negative under the twofold rotations C_2' and C_2'', the inversion i, the improper rotations S_3 and S_6, and reflection in the horizontal plane, σ_h. Inspection of Table 10-1 shows that the characters of A_{2u} describe precisely this behavior. The μ_x and μ_y components are degenerate, and they must have a character of -2 under i and $+2$ under σ_h, since these operations have the following effect on the components:

$$\mu_x \xrightarrow{i} -\mu_x$$
$$\mu_y \xrightarrow{i} -\mu_y$$
$$\mu_x \xrightarrow{\sigma_h} \mu_x \qquad (1)$$
$$\mu_y \xrightarrow{\sigma_h} \mu_y$$

These arguments show that μ_x and μ_y are of species E_{1u}. Therefore, the reduced representation of the electric moment is of the form

$$\Gamma_\mu = A_{2u} + E_{1u} \qquad (2)$$

which could also be found directly from Appendix X, using the fact that μ has the same species as the translations T_x, T_y, and T_z, which are given in the tables.

The symmetry properties of the polarizability components α_{xx}, α_{yy}, α_{zz}, α_{xy}, α_{xz}, and α_{yz} can similarly be obtained, using, if desired, the more powerful methods of Sec. 7-6. In that section it was shown that the character for the polarizability corresponded to the character of the second overtone of a triply degenerate frequency, which could be expressed in the form [Eq. (4), Sec. 7-6]:

$$\chi_R = 2 \cos \alpha_R(\pm 1 + 2 \cos \alpha_R) \qquad (3)$$

In (3), α_R is the angle of the (proper or improper) rotation by the group operation R, and the sign is positive for a proper, negative for an improper rotation. Table 10-3 summarizes the calculation for $\mathfrak{D}_{6h}$.

TABLE 10-3. CALCULATION OF THE CHARACTER OF THE POLARIZABILITY IN $\mathfrak{D}_{6h}$

	E	$2C_6$	$2C_3$	C_2	$3C_2'$	$3C_2''$	i	$2S_3$	$2S_6$	σ_h	$3\sigma_d$	$3\sigma_v$
α_R	0	$\frac{\pi}{3}$	$\frac{2\pi}{3}$	π	π	π	π	$\frac{2\pi}{3}$	$\frac{\pi}{3}$	0	0	0
$2 \cos \alpha_R(\pm 1 + 2 \cos \alpha_R)$	6	2	0	2	2	2	6	2	0	2	2	2

The reduction of the representation with characters as given in Table 10-3 leads to the following species:

$$\Gamma_\alpha = 2A_{1g} + E_{1g} + E_{2g} \qquad (4)$$

This result may be confirmed by inspection of the character table for $\mathfrak{D}_{6h}$ given in Appendix X.

Selection Rules for Fundamentals. It is now possible to determine which transitions are permitted by symmetry in the infrared or Raman spectra using (2) or (4), respectively. Recall first the general rule that the direct product of the species of the two vibrational states between which transitions are investigated, together with (2) or (4), must contain the totally symmetric species (A_{1g} in the present case). For the fundamental transitions, as pointed out in Sec. 7-7, this amounts to the requirement that the upper state must be of one of the species appearing in (2) or (4). By reference to Table 10-1, it may be concluded that one frequency of species A_{2u} and three of species E_{1u} may appear in the infrared spectrum; likewise in the Raman spectrum two of species A_{1g}, one of species E_{1g}, and four of species E_{2g} are allowed. These allowed fundamentals can, in principle, be further distinguished by band type in the infrared and polarization type (Chap. 3) in the Raman. Thus the single out-of-plane vibration, A_{2u}, should give rise to a parallel-type band, while the three fundamentals of species E_{1u} should give perpendicular-type bands. On the other hand, the two Raman fundamentals of species A_{1g} should yield polarized Raman lines, whereas the remaining five Raman lines are expected to be depolarized. Note that in this case the selection rules for fundamentals are mutually exclusive, that is, the infrared active fundamentals are Raman inactive, and vice versa. This is generally true of molecules possessing a center of inversion, since the dipole moment is antisymmetric to such an operation, whereas the polarizability is symmetric.

The frequencies of benzene are conveniently divided into CH stretching frequencies and others. The activity of the CH stretching frequencies is found by examining the species for the CH coordinates (s) in Table 10-2; that is, $A_{1g} + E_{2g} + B_{1u} + E_{1u}$. One therefore expects to find one polarized (A_{1g}) and one depolarized (E_{2g}) Raman line and one infrared band (E_{1u}) near 3,000 cm^{-1}, the region in which CH stretching frequencies occur. Similarly carbon-carbon double bonds generally give rise to reasonably characteristic frequencies, near 1,600 cm^{-1}. The bonds in benzene are not double bonds but rather hybrid double-single bonds so this type of frequency should be lower. Examination of the species of the CC coordinates in Table 10-2 shows that there should be one polarized (A_{1g}) and one depolarized (E_{2g}) Raman line and one infrared band (E_{1u}) in this general region. Because of the strong coupling between these CC bonds (and with the CH bending motions) this argument is not too useful, but these frequencies do fall between the single- and double-bond regions.

Selection Rules for Overtones and Combinations. The selection rules for overtone frequencies will next be considered. These are transitions between the ground vibrational state and an excited state in which one quantum number is greater than one and all other quantum numbers are zero. If the excited mode is degenerate, it is the sum of the quantum numbers for the degenerate components which must be greater than one. The properties of the overtones of the nondegenerate modes are easily described. Since all even overtones in such cases are totally symmetric (A_{1g}), the corresponding frequencies should be found (usually very weakly) in the Raman but not in the infrared spectrum. The odd overtones have the same species as the fundamentals, whose selection rules have been described above. The species for the first few overtones of the degenerate modes are not as obvious, but can be computed by the methods of Sec. 7-3 and are given in Table 10-4, which also indicates in which spectrum the transition may occur.

It might be noted that Table 10-4 together with the discussion given above predicts that no overtone frequencies with v even can occur in the infrared spectrum. This is a simple consequence of the presence of a center of symmetry.

TABLE 10-4. SPECIES OF THE OVERTONES OF DEGENERATE FUNDAMENTALS

v	E_{1g}	E_{2g}	E_{1u}	E_{2u}
2	$A_{1g} + E_{2g}(R)$	$A_{1g} + E_{2g}(R)$	$A_{1g} + E_{2g}(R)$	$A_{1g} + E_{2g}(R)$
3	$B_{1g} + B_{2g} + E_{1g}(R)$	$A_{1g} + A_{2g} + E_{2g}(R)$	$B_{1u} + B_{2u} + E_{1u}(IR)$	$A_{1u} + A_{2u} + E_{2u}(IR)$
4	$A_{1g} + 2E_{2g}(R)$	$A_{1g} + 2E_{2g}(R)$	$A_{1g} + 2E_{2g}(R)$	$A_{1g} + 2E_{2g}(R)$
5	$B_{1g} + B_{2g} + 2E_{1g}(R)$	$A_{1g} + A_{2g} + 2E_{2g}(R)$	$B_{1u} + B_{2u} + 2E_{1u}(IR)$	$A_{1u} + A_{2u} + 2E_{2u}(IR)$
6	$2A_{1g} + A_{2g} + 2E_{2g}(R)$	$2A_{1g} + A_{2g} + 2E_{2g}(R)$	$2A_{1g} + A_{2g} + 2E_{2g}(R)$	$2A_{1g} + A_{2g} + 2E_{2g}(R)$

Finally, the selection rules for simple combination frequencies will be considered briefly. Such rules may be obtained readily from Table 10-5, which is a multiplication table for the irreducible representations. Actually, Table 10-5 is given for the group $\mathfrak{D}_6$ in which there is no distinction between g and u species. In order to apply this table to $\mathfrak{D}_{6h}$, however, it is only necessary to add g or u subscripts to the species designations and observe the rule that products of the types $g \times g$ and $u \times u$ are g, while $g \times u$ and $u \times g$ are u.

TABLE 10-5. SYMMETRY SPECIES OF BINARY COMBINATIONS FOR $\mathfrak{D}_6$

	A_1	A_2	B_1	B_2	E_1	E_2
A_1	A_1	A_2	B_1	B_2	E_1	E_2
A_2		A_1	B_2	B_1	E_1	E_2
B_1			A_1	A_2	E_2	E_1
B_2				A_1	E_2	E_1
E_1					$A_1 + A_2 + E_2$	$B_1 + B_2 + E_1$
E_2						$A_1 + A_2 + E_2$

For example, the direct product $E_{1g} \times E_{2u} = B_{1u} + B_{2u} + E_{1u}$ which shows that a combination frequency corresponding to a transition between the ground state and that in which a Raman active fundamental, E_{1g}, and an inactive fundamental, E_{2u}, are each singly excited should be infrared active, since the resultant species, $B_{1u} + B_{2u} + E_{1u}$, contains E_{1u} which is one of the species occurring in the electric moment $(A_{2u} + E_{1u})$.

10-4. Isotopically Substituted Benzene and the Product Rule

Because of the large number of possible isotopic derivatives of benzene, the application of the product rule has been of great importance in making frequency assignments for this molecule. Moreover, the study of the vibrational spectra of such related molecules allows a large number of the force constants, in particular the interaction constants, to be evaluated.

Derivative of Same Symmetry. In case the isotopic substitution does not change the symmetry, the application of the product rule is, as pointed out in Sec. 8-5, relatively simple. Therefore the form of the product rule for such a case (C_6D_6 compared with C_6H_6) will be given first.

The general form of the product rule is

$$\frac{\omega_1'\omega_2' \cdots \omega_n'}{\omega_1\omega_2 \cdots \omega_n} = \left[\left(\frac{\mu_\alpha'}{\mu_\alpha}\right)^a \left(\frac{\mu_\beta'}{\mu_\beta}\right)^b \cdots \left(\frac{M'}{M}\right)^t \left(\frac{I_x'}{I_x}\right)^{r_x} \left(\frac{I_y'}{I_y}\right)^{r_y} \left(\frac{I_z'}{I_z}\right)^{r_z}\right]^{\frac{1}{2}} \quad (1)$$

When (1) is used for the vibrations belonging to a single irreducible representation, the exponents $a, b, \ldots, t, r_x, r_y$, and r_z are all immediately obtained from an analysis such as that given in Table 10-1. In the case of benzene, there are only two sets of equivalent atoms, the carbons and the hydrogens (or deuteriums). Therefore, if we put $\mu_\alpha' = \mu_\alpha = 1/m_C$, the first ratio on the right-hand side of (1) drops out and the second one becomes $\mu_\beta'/\mu_\beta = m_H/m_D$. The exponent b is then identified with n_H appearing in Table 10-1. Furthermore, since the molecule is a symmetric top, I_x and I_y always appear together in a degenerate species and are equal to each other and to the value of $I_\perp$, where $I_\perp$ is the moment of inertia computed with reference to any axis perpendicular to the principal symmetry axis, z. The exponent t is identified with n_T in Table 10-1 for any given species.

The appropriate values of r_x, r_y, and r_z are found as follows: rotation evidently accounts for three degrees of freedom, of species $A_{2g} + E_{1g}$ according to the column labeled $n_R^{(\gamma)}$ in Table 10-1. Since rotation about the z axis is clearly of species A_{2g}, $r_z = 1$ for this species and zero for all others. Furthermore, $r_x = 1$ for one of the degenerate E_{1g} components, and $r_y = 1$ for the other. Since the frequencies of the a and b components are equal only a single expression of the type (1) is written in which

$I_\perp = I_x = I_y$ appears to the first power, and $I_\perp$ will not appear in such an expression for any other species.

The product rule formulas can now be written out by inspection of Table 10-1. They have the following forms:[1]

$$A_{1g}: \qquad \frac{\omega_1'\omega_2'}{\omega_1\omega_2} = \left(\frac{m_{\mathrm{H}}}{m_{\mathrm{D}}}\right)^{\frac{1}{2}} \tag{2}$$

$$A_{2g}: \qquad \frac{\omega_3'}{\omega_3} = \left[\left(\frac{m_{\mathrm{H}}}{m_{\mathrm{D}}}\right)\frac{I_z'}{I_z}\right]^{\frac{1}{2}} \tag{3}$$

$$B_{2g}: \qquad \frac{\omega_4'\omega_5'}{\omega_4\omega_5} = \left(\frac{m_{\mathrm{H}}}{m_{\mathrm{D}}}\right)^{\frac{1}{2}} \tag{4}$$

$$E_{1g}: \qquad \frac{\omega_{10}'}{\omega_{10}} = \left[\left(\frac{m_{\mathrm{H}}}{m_{\mathrm{D}}}\right)\frac{I_\perp'}{I_\perp}\right]^{\frac{1}{2}} \tag{5}$$

$$E_{2g}: \qquad \frac{\omega_6'\omega_7'\omega_8'\omega_9'}{\omega_6\omega_7\omega_8\omega_9} = \frac{m_{\mathrm{H}}}{m_{\mathrm{D}}} \tag{6}$$

$$A_{2u}: \qquad \frac{\omega_{11}'}{\omega_{11}} = \left[\left(\frac{m_{\mathrm{H}}}{m_{\mathrm{D}}}\right)\frac{M'}{M}\right]^{\frac{1}{2}} \tag{7}$$

$$B_{1u}: \qquad \frac{\omega_{12}'\omega_{13}'}{\omega_{12}\omega_{13}} = \left(\frac{m_{\mathrm{H}}}{m_{\mathrm{D}}}\right)^{\frac{1}{2}} \tag{8}$$

$$B_{2u}: \qquad \frac{\omega_{14}'\omega_{15}'}{\omega_{14}\omega_{15}} = \left(\frac{m_{\mathrm{H}}}{m_{\mathrm{D}}}\right)^{\frac{1}{2}} \tag{9}$$

$$E_{1u}: \qquad \frac{\omega_{18}'\omega_{19}'\omega_{20}'}{\omega_{18}\omega_{19}\omega_{20}} = \frac{m_{\mathrm{H}}}{m_{\mathrm{D}}}\left(\frac{M'}{M}\right)^{\frac{1}{2}} \tag{10}$$

$$E_{2u}: \qquad \frac{\omega_{16}'\omega_{17}'}{\omega_{16}\omega_{17}} = \left(\frac{m_{\mathrm{H}}}{m_{\mathrm{D}}}\right)^{\frac{1}{2}} \tag{11}$$

Derivative of Lower Symmetry. In order to apply the product rule in the cases in which isotopic substitution lowers the symmetry, it is necessary to make use of the correlation tables between the irreducible representations of the group of the parent molecule and the subgroup which expresses the symmetry of the derivative. As an example, the case of p-dideuterobenzene, $\mathcal{K} = \mathcal{V}_h$, will be considered. If the z axis of benzene is correlated with the z axis of p-dideuterobenzene, the relation between the irreducible representations is as indicated in Table 10-6.

From this table the following facts are inferred. The degenerate vibrations of benzene are all split in such a way that, for example, ω_{18a}, ω_{19a}, and ω_{20a} of benzene (species E_{1u}) are of species B_{2u} in p-dideuterobenzene, whereas the b components of these degenerate pairs fall in the B_{3u} species of p-dideuterobenzene. Moreover, all the benzene vibrations which correlate with a given species of p-dideuterobenzene must appear in the same product: in the case of B_{2u}, the correlated benzene frequencies

[1] The numbering of the frequencies conforms with that first employed by Wilson.

will be ω_{12} and ω_{13} (B_{1u} in benzene) and the aforementioned ω_{18a}, ω_{19a}, and ω_{20a}.

TABLE 10-6. CORRELATION OF THE SPECIES FOR BENZENE ($\mathcal{G} = \mathfrak{D}_{6h}$) AND *p*-DIDEUTEROBENZENE ($\mathcal{H} = \mathcal{V}_h$)

$\mathfrak{D}_{6h}$	$\mathcal{V}_h$	$\mathfrak{D}_{6h}$	$\mathcal{V}_h$
A_{1g}	A_g	A_{1u}	A_u
A_{2g}	B_{1g}	A_{2u}	B_{1u}
B_{1g}	B_{2g}	B_{1u}	B_{2u}
B_{2g}	B_{3g}	B_{2u}	B_{3u}
E_{1g}	$B_{2g} + B_{3g}$	E_{1u}	$B_{2u} + B_{3u}$
E_{2g}	$A_g + B_{1g}$	E_{2u}	$A_u + B_{1u}$

The right-hand side of the product rule formula (1) must be modified inasmuch as *p*-dideuterobenzene contains both hydrogen and deuterium atoms and the molecule is no longer a symmetric rotor. All that is required, however, is the analogue of Table 10-1 for the case of *p*-dideuterobenzene and in particular, the columns n_D, n_T, $n_{R(x)}$, $n_{R(y)}$, and $n_{R(z)}$. Using the character table for $\mathcal{V}_h$ (Appendix X), one finds $n_D = 1$ for all species except A_u and B_{2g}, in which cases $n_D = 0$; $n_T = 1$ in B_{1u}, B_{2u}, and B_{3u}; $n_{R(x)} = 1$ in B_{3g}; $n_{R(y)} = 1$ in B_{2g}; $n_{R(z)} = 1$ in B_{1g}. Thus the following product rule equations are obtained:

$$A_g: \qquad \frac{\omega_1'\omega_2'\omega_{6a}'\omega_{7a}'\omega_{8a}'\omega_{9a}'}{\omega_1\omega_2\omega_6\omega_7\omega_8\omega_9} = \left(\frac{m_{\mathrm{H}}}{m_{\mathrm{D}}}\right)^{\frac{1}{2}} \tag{12}$$

$$B_{1g}: \qquad \frac{\omega_3'\omega_{6b}'\omega_{7b}'\omega_{8b}'\omega_{9b}'}{\omega_3\omega_6\omega_7\omega_8\omega_9} = \left[\frac{m_{\mathrm{H}}}{m_{\mathrm{D}}}\left(\frac{I_z'}{I_z}\right)\right]^{\frac{1}{2}} \tag{13}$$

$$B_{2g}: \qquad \frac{\omega_{10a}'}{\omega_{10}} = \left(\frac{I_y'}{I_y}\right)^{\frac{1}{2}} \tag{14}$$

$$B_{3g}: \qquad \frac{\omega_4'\omega_5'\omega_{10b}'}{\omega_4\omega_5\omega_{10}} = \left[\left(\frac{m_{\mathrm{H}}}{m_{\mathrm{D}}}\right)\frac{I_x'}{I_x}\right]^{\frac{1}{2}} \tag{15}$$

$$A_u: \qquad \frac{\omega_{16a}'\omega_{17a}'}{\omega_{16}\omega_{17}} = 1 \tag{16}$$

$$B_{1u}: \qquad \frac{\omega_{11}'\omega_{16b}'\omega_{17b}'}{\omega_{11}\omega_{16}\omega_{17}} = \left[\left(\frac{m_{\mathrm{H}}}{m_{\mathrm{D}}}\right)\frac{M'}{M}\right]^{\frac{1}{2}} \tag{17}$$

$$B_{2u}: \qquad \frac{\omega_{12}'\omega_{13}'\omega_{18a}'\omega_{19a}'\omega_{20a}'}{\omega_{12}\omega_{13}\omega_{18}\omega_{19}\omega_{20}} = \left[\left(\frac{m_{\mathrm{H}}}{m_{\mathrm{D}}}\right)\frac{M'}{M}\right]^{\frac{1}{2}} \tag{18}$$

$$B_{3u}: \qquad \frac{\omega_{14}'\omega_{15}'\omega_{18b}'\omega_{19b}'\omega_{20b}'}{\omega_{14}\omega_{15}\omega_{18}\omega_{19}\omega_{20}} = \left[\left(\frac{m_{\mathrm{H}}}{m_{\mathrm{D}}}\right)\frac{M'}{M}\right]^{\frac{1}{2}} \tag{19}$$

The case of Eq. (16) deserves some comment. Further examination reveals that not only the product of the frequencies, but their individual values should be identical in *p*-dideuterobenzene and ordinary benzene. Since the frequencies are inactive in both compounds, however, this result is not particularly useful.

10-5. Assignment of Observed Frequencies

Infrared Active Fundamentals. According to Sec. 10-3, the allowed active fundamentals consisted of one A_{2u} frequency and three E_{1u} frequencies in the infrared, and two A_{1g}, one E_{1g}, and four E_{2g} frequencies in the Raman spectrum. In the observed infrared spectrum, there are four bands which are appreciably stronger than any others, at 671, 1,037, 1,485, and in the region of 3,040 to 3,100 wave numbers (see Table 10-7). The last mentioned absorption falls in a range which is characteristic of CH stretching modes and can therefore be identified as a vibration of species E_{1u} (see Table 10-2).

In order to assign the remaining infrared frequencies, the frequencies of the C_6D_6 molecule together with the product rule will be employed. From Eqs. (7) and (10), Sec. 10-4,

$$\frac{\omega'_{11}}{\omega_{11}} = 0.74 \qquad A_{2u} \tag{1}$$

and

$$\frac{\omega'_{18}\omega'_{19}\omega'_{20}}{\omega_{18}\omega_{19}\omega_{20}} = 0.50 \qquad E_{1u} \tag{2}$$

TABLE 10-7. STRONGEST INFRARED ABSORPTION BANDS AND RAMAN LINES OF C_6H_6 AND C_6D_6
All values in cm^{-1}

Infrared bands (vapor)		Raman lines (liquid)	
C_6H_6	C_6D_6	C_6H_6	C_6D_6
671	503	605.6 (*d*)	576.7 (*d*)
1,037	813	848.9 (*d*)	661.2 (*d*)
1,485	1,333	991.6 (*p*)	944.7 (*p*)
3,045 }		1,178.0 (*d*)	867.2 (*d*)
3,073 }	2,293	1,584.8 } (*d*)	1,558.6 (*d*)
3,099 }		1,606.4 } (*d*)	
		3,046.8 (*d*)	2,263.9 (*d*)
		3,061.9 (*p*)	2,292.3 (*p*)

The C_6D_6 molecule has its strongest infrared frequencies at 503, 813, 1,333, and 2,293 cm^{-1}. The respective ratios (ω'/ω) are 0.75, 0.78, 0.89, and 0.74. The first and last of these agree most closely with the predicted value 0.74, but the last is clearly a CH stretching mode, whereas the single A_{2u} vibration is an out-of-plane CH bond bending. It is therefore concluded that 671 for C_6H_6 and 503 for C_6D_6 are the A_{2u} vibrations. This choice is supported by other arguments. The product of the three remaining ratios, 0.51, is in good agreement with (2), the discrepancy being in the direction to be expected on account of neglect of anharmonicity.

Raman Active Fundamentals. Turning to the Raman frequencies, it is seen in Table 10-7 that eight, rather than the theoretically predicted seven, strong (fundamental) lines are observed. Of these, 991.6 and 3,061.9 are polarized, and can clearly be assigned as ω_1 and ω_2 of the totally symmetric species, A_{1g}, which should include a symmetrical ring stretching mode and a symmetrical CH stretching mode. This assignment is confirmed by the product rule, using C_6D_6 frequencies of 944.7 and 2,292.3, which give a ratio of 0.71 compared with 0.71 from Eq. (2), Sec. 10-4.

The Raman fundamentals of species E_{1g}, namely, ω_{10} and ω'_{10}, should satisfy the simple ratio

$$\frac{\omega'_{10}}{\omega_{10}} = 0.78$$

from Eq. (5), Sec. 10-4, using the distances 1.39 A and 1.08 A, respectively, for the CC and CH bond lengths in computing the moment of inertia, $I_\perp$, with respect to an axis perpendicular to the sixfold symmetry axis. The fact that C_6D_6 exhibits only a single line, 1,558.6, which can be correlated with the doublet 1,584.8 and 1,606.4 in ordinary benzene, confirms the interpretation originally made by Wilson that the doublet is a case of Fermi resonance between a fundamental in this region and the combination $605.6 + 991.6 = 1{,}597.2$ which lies close to the mean value of the doublet, namely, 1,595.6. If one then takes 1,595 as an approximate unperturbed value of the fundamental, the five unassigned fundamentals give the observed ratios 0.95, 0.78, 0.74, 0.98, and 0.74. This clearly suggests, along with other information, that 848.9 in benzene and 661.2 in completely deuterated benzene (with ratio 0.78) are the vibrations of species E_{1g}. Therefore the remaining four lines are assigned to E_{2g} and give a product rule ratio of 0.51 in good agreement with the value predicted by Eq. (6), Sec. 10-4, namely, 0.50. Furthermore, the resonance interpretation of the benzene doublet is consistent with the selection rules, which allow only those motions of the same symmetry species to interact. The combination $605.6 + 991.6$ is of species

$$E_{2g} \times A_{1g} = E_{2g}$$

which is exactly the species to which the unperturbed fundamental at about 1,595 has just been assigned.

Reference to Tables 10-2 indicates, moreover, that in E_{2g}, vibrations which may roughly be described as CH stretching, two types of carbon ring deformation (CC stretch and α-type bending), and CH bond bending are anticipated. These can be identified, respectively, with the benzene frequencies 3,046.8, 1,595 and 605.6, and 1,178. Finally, the E_{1g} mode should be an out-of-plane bending of the CH bond (848.9). These assignments are summarized in Table 10-8.

TABLE 10-8. ASSIGNMENT OF ACTIVE FUNDAMENTALS IN C_6H_6 AND C_6D_6

Species	Active in:	Number	Type	C_6H_6	C_6D_6
A_{1g}	Raman	ω_2	s (CH)	3,061.9	2,292.3
		ω_1	t (CC)	991.6	944.7
A_{2u}	Infrared	ω_{11}	γ (HCC_2)	671	503
E_{1g}	Raman	ω_{10}	γ (HCC_2)	848.9	661.2
E_{1u}	Infrared	ω_{20}	s (CH)	3,080	2,293
		ω_{19}	t, α (CC, CCC)	1,485	1,333
		ω_{18}	β (HCC)	1,037	813
E_{2g}	Raman	ω_7	s (CH)	3,046.8	2,263.9
		ω_8	t (CC)	1,595	1,558.6
		ω_6	α (CCC)	605.6	576.7
		ω_9	β (HCC)	1,178	876.2

10-6. Potential and Kinetic Energy in Internal Coordinates

In this section, the most general expression for the potential energy will be described in terms of the internal coordinates whose symmetry properties were discussed in Sec. 10-2. At the same time, the corresponding **G** matrix elements (which give the kinetic energy as a quadratic function of the momenta conjugate to the internal coordinates) will be computed in terms of the atomic masses and the requisite geometrical parameters, using the methods of Chap. 4. It will be convenient to treat the in-plane and out-of-plane motions separately, since there is no interaction between these two sets of motions.

Modified In-plane Coordinates. The twelve CCH angles, ϕ, are not particularly convenient and it is better to use the linear combinations of them shown below.

$$\begin{aligned}
\alpha_1 &= -(\phi_1 + \phi_2) & \beta_1 &= \tfrac{1}{2}(\phi_1 - \phi_2)\\
\alpha_2 &= -(\phi_3 + \phi_4) & \beta_2 &= \tfrac{1}{2}(\phi_3 - \phi_4)\\
\alpha_3 &= -(\phi_5 + \phi_6) & \beta_3 &= \tfrac{1}{2}(\phi_5 - \phi_6)\\
\alpha_4 &= -(\phi_7 + \phi_8) & \beta_4 &= \tfrac{1}{2}(\phi_7 - \phi_8)\\
\alpha_5 &= -(\phi_9 + \phi_{10}) & \beta_5 &= \tfrac{1}{2}(\phi_9 - \phi_{10})\\
\alpha_6 &= -(\phi_{11} + \phi_{12}) & \beta_6 &= \tfrac{1}{2}(\phi_{11} - \phi_{12})
\end{aligned} \tag{1}$$

The advantage is that this breaks the equivalent set of ϕ's into two smaller equivalent sets α and β, each of which is such that no more than one degenerate set occurs in any one species, whereas, for example, there were two sets of ϕ's in $2E_{2g}$. As discussed in Sec. 6-7, it is desirable to avoid having more than one set in a species. The species for the α and β were listed in Table 10-2.

These coordinates α and β could have been introduced from the start on physical grounds. The α's are the changes in the CCC ring angles, while the β's are the angles between the CH bonds and the bisectors of the outer CCC angles.

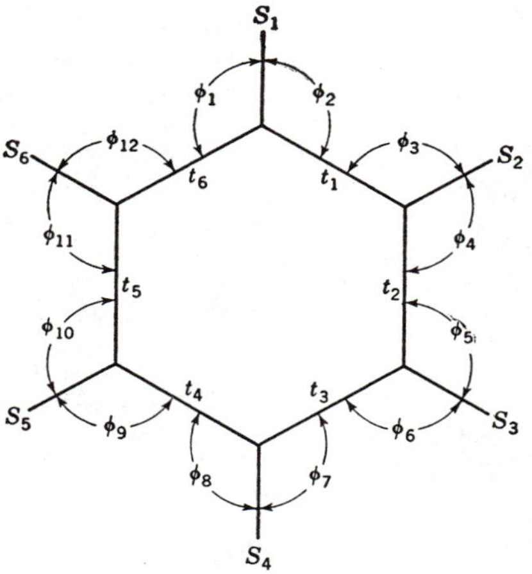

$\alpha_1 = -(\phi_1 + \phi_2), \alpha_2 = -(\phi_3 + \phi_4)$, etc.

$\beta_1 = \frac{1}{2}(\phi_1 - \phi_2), \beta_2 = \frac{1}{2}(\phi_3 - \phi_4)$, etc.

FIG. 10-2. Numbering of in-plane coordinates for the benzene molecule.

In-plane Potential Constants. Figure 10-2 shows the in-plane coordinates. With these, the quadratic force constants can be listed in a compact tabular form, Table 10-9, in which only the first row for each set of equivalent coordinates is given. The other rows consist of the same force constants in appropriately permuted orders determined by the symmetry. Furthermore, the constants in a given row are not all different. Thus the force constant associated with s_1 and α_2 is the same as that for s_1 and α_6 because the vertical plane through C_1-H_1 reflects α_2 and α_6 and leaves s_1 unaltered. Since the coordinates β can be positive or negative, minus signs occur with some of their force constants.

TABLE 10-9. FORCE CONSTANTS FOR IN-PLANE COORDINATES

	s_1	s_2	s_3	s_4	s_5	s_6	t_1	t_2	t_3	t_4	t_5	t_6
s_1	F_s^1	F_s^2	F_s^3	F_s^4	F_s^3	F_s^2	F_{st}^1	F_{st}^2	F_{st}^3	F_{st}^3	F_{st}^2	F_{st}^1
t_1							F_t^1	F_t^2	F_t^3	F_t^4	F_t^3	F_t^2

	α_1	α_2	α_3	α_4	α_5	α_6	β_1	β_2	β_3	β_4	β_5	β_6
s_1	$F_{s\alpha}^1$	$F_{s\alpha}^2$	$F_{s\alpha}^3$	$F_{s\alpha}^4$	$F_{s\alpha}^3$	$F_{s\alpha}^2$	0	$F_{s\beta}^1$	$F_{s\beta}^2$	0	$-F_{s\beta}^2$	$-F_{s\beta}^1$
t_1	$F_{t\alpha}^1$	$F_{t\alpha}^1$	$F_{t\alpha}^2$	$F_{t\alpha}^3$	$F_{t\alpha}^3$	$F_{t\alpha}^2$	$F_{t\beta}^1$	$-F_{t\beta}^1$	$F_{t\beta}^2$	$F_{t\beta}^3$	$-F_{t\beta}^3$	$-F_{t\beta}^2$
α_1	F_α^1	F_α^2	F_α^3	F_α^4	F_α^3	F_α^2	0	$F_{\alpha\beta}^1$	$F_{\alpha\beta}^2$	0	$-F_{\alpha\beta}^2$	$-F_{\alpha\beta}^1$
β_1							F_β^1	F_β^2	F_β^3	F_β^4	F_β^3	F_β^2

In-plane **G** *Matrix.* The **G** matrix elements which determine the kinetic energy will follow the same pattern as the force constants except that some of them will vanish because the two coordinates involved have no atoms in common. Thus, with a symbolism parallel to that employed for the force constants, it follows that G_s^2, G_s^3, G_s^4, G_{st}^2, G_{st}^3, $G_{s\alpha}^3$, $G_{s\alpha}^4$, $G_{s\beta}^2$, G_t^3, G_t^4, $G_{t\alpha}^3$, $G_{t\beta}^3$, G_α^4, and G_β^4 all vanish.

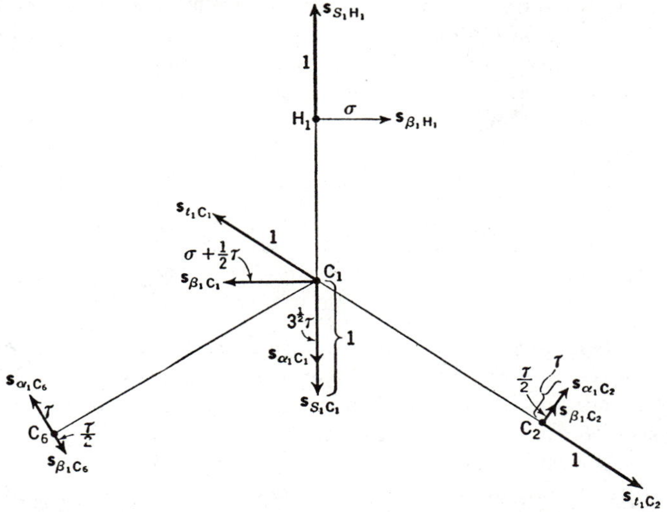

FIG. 10-3. Lengths and directions of **s** vectors for determination of in-plane $G_{tt'}$. Vectors are shown for s_1, t_1, α_1, and β_1 coordinates only; lengths are shown at side of vector with σ = reciprocal of equilibrium CH distance, τ = reciprocal of equilibrium CC distance.

The elements which do not vanish can be evaluated, either by the methods of Sec. 4-3 or from the tabulations of Appendix VI (see Fig. 10-3). The results are listed in Table 10-10. Here σ and τ represent the reciprocals of the CH and CC bond lengths, respectively, while μ_H and μ_C are the reciprocal masses of H and C. The bond angles are all 120°.

TABLE 10-10. IN-PLANE G MATRIX ELEMENTS FOR BENZENE

$G_s^1 = \mu_H + \mu_C$	$G_{t\beta}^1 = +\frac{1}{2}\,3^{\frac{1}{2}}(\sigma + \frac{1}{2}\tau)\mu_C$
$G_{st}^1 = -\frac{1}{2}\mu_C$	$G_{t\beta}^2 = -\frac{1}{4}\,3^{\frac{1}{2}}\tau\mu_C$
$G_{s\alpha}^1 = +3^{\frac{1}{2}}\tau\mu_C$	$G_\alpha^1 = 5\tau^2\mu_C$
$G_{s\alpha}^2 = -\frac{1}{2}\,3^{\frac{1}{2}}\tau\mu_C$	$G_\alpha^2 = -3\tau^2\mu_C$
$G_{s\beta}^1 = +\frac{1}{4}\,3^{\frac{1}{2}}\tau\mu_C$	$G_\alpha^3 = \frac{1}{2}\tau^2\mu_C$
	$G_\beta^1 = \sigma^2\mu_H + (\sigma^2 + \sigma\tau + \frac{3}{4}\tau^2)\mu_C$
$G_t^1 = 2\mu_C$	$G_\beta^2 = \frac{1}{2}\tau(\sigma + \frac{1}{2}\tau)\mu_C$
$G_t^2 = -\frac{1}{2}\mu_C$	$G_\beta^3 = -\frac{1}{8}\tau^2\mu_C$
$G_{t\alpha}^1 = -\frac{1}{2}\,3^{\frac{1}{2}}\tau\mu_C$	$G_{\alpha\beta}^1 = \frac{1}{2}\tau(2\tau + \sigma)\mu_C$
$G_{t\alpha}^2 = +\frac{1}{2}\,3^{\frac{1}{2}}\tau\mu_C$	$G_{\alpha\beta}^2 = -\frac{1}{4}\tau^2\mu_C$

Out-of-plane Vibrations. A figure showing the numbering of the γ and δ coordinates is omitted, since the numbering can be based upon that of the in-plane coordinates (Fig. 10-2) in a simple manner. The six γ's (bending of CH bond out of the plane of the adjacent CCC linkage) are numbered to correspond with the CH bond stretches (set s), while the six δ's (torsion of a CCCC linkage) are numbered to correspond to the CC stretches (set t) in such a way that the corresponding t coordinate forms the central bond, about which torsion occurs, in the linkage, CCCC. Using this scheme, the abridged **F** matrix assumes the form indicated in Table 10-11.

TABLE 10-11

	γ_1	γ_2	γ_3	γ_4	γ_5	γ_6	δ_1	δ_2	δ_3	δ_4	δ_5	δ_6
γ_1	F_γ^1	F_γ^2	F_γ^3	F_γ^4	F_γ^3	F_γ^2	$F_{\gamma\delta}^1$	$F_{\gamma\delta}^2$	$F_{\gamma\delta}^3$	$-F_{\gamma\delta}^3$	$-F_{\gamma\delta}^2$	$-F_{\gamma\delta}^1$
δ_1							F_δ^1	F_δ^2	F_δ^3	F_δ^4	F_δ^3	F_δ^2

The only **G** matrix element which vanishes because the coordinate pair involved possesses no atoms in common is G_γ^4. Since Appendix VI does not give general formulas for all **G** matrix elements of the present type, it is necessary to set up the **s** vectors as described in Chap. 4. All these vectors are perpendicular to the plane of the molecule, so that it is only necessary to determine their lengths and signs, which are found from Eq. (17), Sec. 4-1,

$$\begin{aligned} s_{\gamma\mathrm{H}} &= \sigma \\ s_{\gamma\mathrm{C}_1} &= -(\sigma + 2\tau) \\ s_{\gamma\mathrm{C}_2} &= \tau \\ s_{\gamma\mathrm{C}_6} &= \tau \end{aligned} \tag{2}$$

in which C_1 is the central carbon atom defining γ. Similarly, for the torsional coordinate, δ, using Eqs. (21) through (24), Sec. 4-1,

$$\begin{aligned} s_{\delta\mathrm{C}_1} &= -\tfrac{2}{3}\, 3^{\frac{1}{2}}\tau \\ s_{\delta\mathrm{C}_2} &= \tfrac{4}{3}\, 3^{\frac{1}{2}}\tau \\ s_{\delta\mathrm{C}_3} &= -\tfrac{4}{3}\, 3^{\frac{1}{2}}\tau \\ s_{\delta\mathrm{C}_4} &= \tfrac{2}{3}\, 3^{\frac{1}{2}}\tau \end{aligned} \tag{3}$$

in which the carbon atoms are numbered in order along the benzene ring. Recalling that the **G** matrix elements are given by the general formula,

$$G_{tt'} = \sum_{\alpha=1}^{N} \mu_\alpha \mathbf{s}_{t\alpha} \cdot \mathbf{s}_{t'\alpha} \tag{4}$$

and noting that the scalar products in this case become simple algebraic products taken from (2) and (3), the **G** matrix elements are readily computed. For example, G^1_γ is given by

$$\begin{aligned} G^1_\gamma &= \sigma^2\mu_H + [(\sigma + 2\tau)^2 + \tau^2 + \tau^2]\mu_C \\ &= \sigma^2\mu_H + (\sigma^2 + 4\sigma\tau + 6\tau^2)\mu_C \end{aligned} \tag{5}$$

Another example is G^4_δ, which involves the entire carbon ring, the two coordinates having in common two atoms situated diagonally opposite in the ring. In this case the matrix element has the value:

$$\begin{aligned} G^4_\delta &= (s_{\delta C_1} s_{\delta' C_4'} + s_{\delta C_4} s_{\delta' C_1'})\mu_C \\ &= [(-\tfrac{2}{3}\, 3^{\frac{1}{2}}\tau)(\tfrac{2}{3}\, 3^{\frac{1}{2}}\tau) + (\tfrac{2}{3}\, 3^{\frac{1}{2}}\tau)(-\tfrac{2}{3}\, 3^{\frac{1}{2}}\tau)]\mu_C \\ &= -\tfrac{8}{3}\tau^2\mu_C \end{aligned} \tag{6}$$

The complete set of **G** matrix elements is given in Table 10-12.

TABLE 10-12. G MATRIX ELEMENTS FOR OUT-OF-PLANE VIBRATIONS IN BENZENE

$$\begin{aligned}
G^1_\gamma &= \sigma^2\mu_H + (\sigma^2 + 4\sigma\tau + 6\tau^2)\mu_C \\
G^2_\gamma &= -2\tau(\sigma + 2\tau)\mu_C \\
G^3_\gamma &= \tau^2\mu_C \\
G^1_\delta &= \tfrac{40}{3}\tau^2\mu_C \\
G^2_\delta &= -\tfrac{32}{3}\tau^2\mu_C \\
G^3_\delta &= \tfrac{16}{3}\tau^2\mu_C \\
G^4_\delta &= -\tfrac{8}{3}\tau^2\mu_C \\
G^1_{\gamma\delta} &= -\frac{3^{\frac{1}{2}}}{3}\, 2\tau(2\sigma + 7\tau)\mu_C \\
G^2_{\gamma\delta} &= \frac{3^{\frac{1}{2}}}{3}\, 2\tau(\sigma + 4\tau)\mu_C \\
G^3_{\gamma\delta} &= -\frac{3^{\frac{1}{2}}}{3}\, 2\tau^2\mu_C
\end{aligned}$$

10-7. Symmetry Coordinates

In-plane Vibrations. The sets of six CH stretches (s) and six bending coordinates of type α have similar symmetry properties and will be considered first.

Reference to Table 10-2 shows that symmetry coordinates of the nondegenerate species A_{1g} and B_{1u} are required from the set s. The totally symmetric coordinate as usual is simply the sum of the internal coordinates of the given set

$$s_1 + s_2 + s_3 + s_4 + s_5 + s_6 \tag{1}$$

that of species B_{1u} differs by changing its sign under the group operations C^k_6 and S^k_6, where k is odd or under a rotation (C''_2) or reflection (σ_d) which bisects a pair of CC bonds, so that the signs of the CH stretches or CCC bendings must alternate going around the ring,

$$s_1 - s_2 + s_3 - s_4 + s_5 - s_6 \tag{2}$$

The degenerate symmetry coordinates for the same two sets are of species E_{1u} and E_{2g} (Table 10-2). Only one member of each degenerate pair will be constructed, using the simple method described in Sec. 6-4. Applying the general formula [Eq. (7), Sec. 6-4], namely,

$$\mathsf{S}^{(\gamma)} = \mathfrak{N} \sum_R \chi_R^{(\gamma)} R S_1 \tag{3}$$

to a representative CH stretch, s_1, one finds

$$\sum_R \chi_R^{(E_{1u})} R s_1 = 4s_1 + 2s_2 - 2s_3 - 4s_4 - 2s_5 + 2s_6 \tag{4}$$

and

$$\sum_R \chi_R^{(E_{2g})} R s_1 = 4s_1 - 2s_2 - 2s_3 + 4s_4 - 2s_5 - 2s_6 \tag{5}$$

Thus all s and α coordinates are accounted for, since the latter will be of the same form as those in s.

The CC stretching coordinates (t) clearly have very similar properties to those of the sets just discussed. The symmetry coordinate of species A_{1g} is just the sum of the six members of the set (suitably normalized), while that of species B_{2u} is like the B_{1u} s and α coordinates, in that the signs must alternate going around the ring. The latter statement can be seen from the fact that it is now the C_2' axis and σ_v plane (rather than C_2'' and σ_d) under which the symmetry coordinate is required to change sign; but this substitution exactly parallels the different orientations of the two sets and hence leads to formally identical expressions.

However, since the subgroup which leaves t_1 invariant is not the same as that which leaves s_1 invariant, the general formula (3) will not give correctly oriented degenerate coordinates if t_1 itself is used as S_1 in (3). To get correct orientation, one needs to use the linear combination $t_6 + t_1$ as the generating element in (3). This is not necessary for the nondegenerate combinations; for the degenerate ones, it leads to

$$\begin{aligned} \Sigma\chi_R^{(E_{1u})} R(t_6 + t_1) &= 4(t_6 + t_1) + 2(t_1 + t_2) - 2(t_2 + t_3) - 4(t_3 + t_4) \\ &\quad - 2(t_4 + t_5) + 2(t_5 + t_6) = 6t_1 - 6t_3 - 6t_4 + 6t_6 \qquad (6) \\ \Sigma\chi_R^{(E_{2g})} R(t_6 + t_1) &= t_1 - 2t_2 + t_3 + t_4 - 2t_5 + t_6 \qquad (7) \end{aligned}$$

The type β coordinates alone remain to be considered. The species of the symmetry coordinates required from this set were listed in Table 10-2. It is thus seen that

$$\Gamma_\beta = A_{2g} + E_{2g} + B_{2u} + E_{1u} \tag{8}$$

The symmetry coordinates of the nondegenerate species A_{2g} and B_{2u} may be constructed by using (3) again; namely,

$$A_{2g}: \quad (6)^{-\frac{1}{2}}(\beta_1 + \beta_2 + \beta_3 + \beta_4 + \beta_5 + \beta_6) \tag{9}$$

and

$$B_{2u}: \quad (6)^{-\frac{1}{2}}(\beta_1 - \beta_2 + \beta_3 - \beta_4 + \beta_5 - \beta_6) \tag{10}$$

In order to construct the degenerate β combinations, an appropriate generating combination to use in (3) is

$$\beta_2 - \beta_6 \tag{11}$$

because this combination will transform under $\mathfrak{C}_{2v}$ (E, C_2', σ_h, σ_v) in the same way as does s_1, or α_1, or $t_6 + t_1$. Consequently, use of the same coefficients as in (4) and (5) gives

$$E_{1u}: \quad 4(\beta_2 - \beta_6) + 2(\beta_3 - \beta_1) - 2(\beta_4 - \beta_2) - 4(\beta_5 - \beta_3) - 2(\beta_6 - \beta_4) + 2(\beta_1 - \beta_5) = 6\beta_2 + 6\beta_3 - 6\beta_5 - 6\beta_6 \tag{12}$$

$$E_{2g}: \quad \beta_2 - \beta_3 + \beta_5 - \beta_6 \tag{13}$$

In summary, the normalized combinations are given in Table 10-13.

TABLE 10-13

A_{1g}:	$6^{-\frac{1}{2}}(s_1 + s_2 + s_3 + s_4 + s_5 + s_6)$
	$6^{-\frac{1}{2}}(t_1 + t_2 + t_3 + t_4 + t_5 + t_6)$
	$6^{-\frac{1}{2}}(\alpha_1 + \alpha_2 + \alpha_3 + \alpha_4 + \alpha_5 + \alpha_6)$
A_{2g}:	$6^{-\frac{1}{2}}(\beta_1 + \beta_2 + \beta_3 + \beta_4 + \beta_5 + \beta_6)$
E_{2g}:	$12^{-\frac{1}{2}}(2s_1 - s_2 - s_3 + 2s_4 - s_5 - s_6)$
	$12^{-\frac{1}{2}}(t_1 - 2t_2 + t_3 + t_4 - 2t_5 + t_6)$
	$12^{-\frac{1}{2}}(2\alpha_1 - \alpha_2 - \alpha_3 + 2\alpha_4 - \alpha_5 - \alpha_6)$
	$\frac{1}{2}(\beta_2 - \beta_3 + \beta_5 - \beta_6)$
B_{1u}:	$6^{-\frac{1}{2}}(s_1 - s_2 + s_3 - s_4 + s_5 - s_6)$
	$6^{-\frac{1}{2}}(\alpha_1 - \alpha_2 + \alpha_3 - \alpha_4 + \alpha_5 - \alpha_6)$
B_{2u}:	$6^{-\frac{1}{2}}(t_1 - t_2 + t_3 - t_4 + t_5 - t_6)$
	$6^{-\frac{1}{2}}(\beta_1 - \beta_2 + \beta_3 - \beta_4 + \beta_5 - \beta_6)$
E_{1u}:	$12^{-\frac{1}{2}}(2s_1 + s_2 - s_3 - 2s_4 - s_5 + s_6)$
	$\frac{1}{2}(t_1 - t_3 - t_4 + t_6)$
	$12^{-\frac{1}{2}}(2\alpha_1 + \alpha_2 - \alpha_3 - 2\alpha_4 - \alpha_5 + \alpha_6)$
	$\frac{1}{2}(\beta_2 + \beta_3 - \beta_5 - \beta_6)$

Out-of-plane Vibrations. The methods employed are identical with those applied in the case of the in-plane vibrations. Reference to Table 10-2 shows the number of coordinates required of each species. The coordinate γ_1 is sent into itself by E and σ_v and into minus itself by C_2' and σ_h. The coordinate δ_1 does not share these properties but it is readily seen that $\delta_1 \xrightarrow{\sigma_v} -\delta_6$, $\delta_1 \xrightarrow{C_2'} +\delta_6$, $\delta_1 \xrightarrow{\sigma_h} -\delta_1$, and $\delta_6 \xrightarrow{\sigma_h} -\delta_6$ so that the linear combination $\delta_1 - \delta_6$ transforms exactly the same as γ_1 and is therefore the correct generating combination to use in place of S_1 in (3).

Table 10-14 gives the normalized results for the representative symmetry coordinates.

Table 10-14

B_{2g}:	$6^{-\frac{1}{2}}(\gamma_1 - \gamma_2 + \gamma_3 - \gamma_4 + \gamma_5 - \gamma_6)$
	$6^{-\frac{1}{2}}(\delta_1 - \delta_2 + \delta_3 - \delta_4 + \delta_5 - \delta_6)$
E_{1g}:	$12^{-\frac{1}{2}}(2\gamma_1 + \gamma_2 - \gamma_3 - 2\gamma_4 - \gamma_5 + \gamma_6)$
	$12^{-\frac{1}{2}}(\delta_1 + 2\delta_2 + \delta_3 - \delta_4 - 2\delta_5 - \delta_6)$
A_{1u}:	$6^{-\frac{1}{2}}(\delta_1 + \delta_2 + \delta_3 + \delta_4 + \delta_5 + \delta_6)$
A_{2u}:	$6^{-\frac{1}{2}}(\gamma_1 + \gamma_2 + \gamma_3 + \gamma_4 + \gamma_5 + \gamma_6)$
E_{2u}:	$12^{-\frac{1}{2}}(2\gamma_1 - \gamma_2 - \gamma_3 + 2\gamma_4 - \gamma_5 - \gamma_6)$
	$\frac{1}{2}(\delta_1 - \delta_3 + \delta_4 - \delta_6)$

10-8. Factored Potential and Kinetic Energy Matrices

Once the properly oriented and normalized symmetry coordinates have been obtained, the coefficients which give the potential energy in terms of symmetry coordinates can be written down immediately with the aid of the expressions given in Sec. 6-6, namely,

$$\mathsf{F}_{kk} = \frac{1}{U_{kt''}} \sum_{t'} U_{kt'} F_{t''t'} \tag{1}$$

for the diagonal terms, and

$$\mathsf{F}_{k'k} = \frac{1}{U_{k't''}} \sum_{t'} U_{kt'} F_{t''t'} \tag{2}$$

for the interaction terms. These two summations are extended normally over the first rows of suitable blocks of the **F** matrix, such as are exhibited in Tables 10-9 and 10-11. Using the U_{kt} given in Tables 10-13 and 10-14, the F elements will now be worked out taking one symmetry species at a time.

The Totally Symmetric Species, A_{1g}. Since the coefficients of each internal coordinate appearing in the three symmetry coordinates of this species are all equal, application of (1) and (2) yields[1]

$$\begin{aligned}
\mathsf{F}_{ss} &= F^1_s + 2F^2_s + 2F^3_s + F^4_s \\
\mathsf{F}_{tt} &= F^1_t + 2F^2_t + 2F^3_t + F^4_t \\
\mathsf{F}_{\alpha\alpha} &= F^1_\alpha + 2F^2_\alpha + 2F^3_\alpha + F^4_\alpha \\
\mathsf{F}_{st} &= 2(F^1_{st} + F^2_{st} + F^3_{st}) \\
\mathsf{F}_{s\alpha} &= F^1_{s\alpha} + 2F^2_{s\alpha} + 2F^3_{s\alpha} + F^4_{s\alpha} \\
\mathsf{F}_{t\alpha} &= 2(F^1_{t\alpha} + F^2_{t\alpha} + F^3_{t\alpha})
\end{aligned}$$

[1] The superscripts such as A_{1g} on $\mathsf{F}^{A_{1g}}_{ss}$ have been omitted here since each symmetry species is taken up separately.

The G elements are combined into factors in the same way, using exactly the same coefficients. Furthermore, expressions were given in Table 10-10 for the individual G elements. With these, the A_{1g} kinetic energy matrix becomes (only the elements on or above the diagonal are given)

A_{1g}	s	t	α
s	$\mu_H + \mu_C$	$-\mu_C$	0
t		μ_C	0
α			0

The fact that all elements involving α vanish shows that the redundancy predicted for this species in Sec. 10-2 is just the symmetry coordinate formed by taking the sum of the six α's.

The Carbon Ring versus Hydrogen Ring Libration, A_{2g}. The single symmetry coordinate of this in-plane species is simply the normalized sum of the β's. This mode is essentially a rigid rotation of the carbon ring opposed by a rigid rotation of the hydrogen "ring" in the opposite direction about the z axis. From Table 10-9 the single potential energy constant is

$$\mathsf{F}_{\beta\beta} = F_\beta^1 + 2F_\beta^2 + 2F_\beta^3 + F_\beta^4$$

while the corresponding kinetic energy coefficient is

$$\mathsf{G}_{\beta\beta} = \sigma^2\mu_H + (\sigma + \tau)^2\mu_C$$

The Degenerate Modes of Species E_{2g}. Four symmetry coordinates constructed, respectively, from the s, t, α, and β sets are shown in Table 10-13. Since β_1 does not appear in the β combination, it is necessary to visualize the form of the second row in the $\beta\beta$ block of the $\mathbf{F}$ matrix: by cyclic permutation it obviously consists of the elements F_β^2, F_β^1, F_β^2, F_β^3, F_β^4, and F_β^3. Thus the resultant potential constants are

$$\begin{aligned}
\mathsf{F}_{ss} &= F_s^1 - F_s^2 - F_s^3 + F_s^4 \\
\mathsf{F}_{tt} &= F_t^1 - F_t^2 - F_t^3 + F_t^4 \\
\mathsf{F}_{\alpha\alpha} &= F_\alpha^1 - F_\alpha^2 - F_\alpha^3 + F_\alpha^4 \\
\mathsf{F}_{\beta\beta} &= F_\beta^1 - F_\beta^2 - F_\beta^3 + F_\beta^4 \\
\mathsf{F}_{st} &= F_{st}^1 - 2F_{st}^2 + F_{st}^3 \\
\mathsf{F}_{s\alpha} &= F_{s\alpha}^1 - F_{s\alpha}^2 - F_{s\alpha}^3 + F_{s\alpha}^4 \\
\mathsf{F}_{s\beta} &= 3^{\frac{1}{2}}(F_{s\beta}^1 - F_{s\beta}^2) \\
\mathsf{F}_{t\alpha} &= F_{t\alpha}^1 - 2F_{t\alpha}^2 + F_{t\alpha}^3 \\
\mathsf{F}_{t\beta} &= -3^{\frac{1}{2}}(F_{t\beta}^1 + F_{t\beta}^3) \\
\mathsf{F}_{\alpha\beta} &= 3^{\frac{1}{2}}(F_{\alpha\beta}^1 - F_{\alpha\beta}^2)
\end{aligned}$$

By making the corresponding combinations of the appropriate kinetic energy coefficients, one finds

E_{2g}	s	t	α	β
s	$\mu_H + \mu_C$	$-\frac{1}{2}\mu_C$	$\frac{3}{2}3^{\frac{1}{2}}\tau\mu_C$	$\frac{3}{4}\tau\mu_C$
t		$\frac{5}{2}\mu_C$	$-\frac{3}{2}3^{\frac{1}{2}}\tau\mu_C$	$-\frac{3}{2}(\sigma + \frac{1}{2}\tau)\mu_C$
α			$\frac{15}{2}\tau^2\mu_C$	$3^{\frac{1}{2}}\tau(\frac{5}{4}\tau + \frac{1}{2}\sigma)\mu_C$
β				$\sigma^2\mu_H + (\sigma^2 + \frac{1}{2}\sigma\tau + \frac{5}{8}\tau^2)\mu_C$

The B_{1u} Modes. The potential constants here are

$$\begin{aligned}
\mathsf{F}_{ss} &= F_s^1 - 2F_s^2 + 2F_s^3 - F_s^4 \\
\mathsf{F}_{\alpha\alpha} &= F_\alpha^1 - 2F_\alpha^2 + 2F_\alpha^3 - F_\alpha^4 \\
\mathsf{F}_{s\alpha} &= F_{s\alpha}^1 - 2F_{s\alpha}^2 + 2F_{s\alpha}^3 - F_{s\alpha}^4
\end{aligned}$$

and the corresponding kinetic energy matrix becomes

B_{1u}	s	α
s	$\mu_H + \mu_C$	$2 \cdot 3^{\frac{1}{2}}\tau\mu_C$
α		$12\tau^2\mu_C$

The B_{2u} Modes

$$\begin{aligned}
\mathsf{F}_{tt} &= F_t^1 - 2F_t^2 + 2F_t^3 - F_t^4 \\
\mathsf{F}_{\beta\beta} &= F_\beta^1 - 2F_\beta^2 + 2F_\beta^3 - F_\beta^4 \\
\mathsf{F}_{t\beta} &= 2(F_{t\beta}^1 + F_{t\beta}^2 - F_{t\beta}^3)
\end{aligned}$$

B_{2u}	t	β
t	$3\mu_C$	$3^{\frac{1}{2}}\sigma\mu_C$
β		$\sigma^2(\mu_H + \mu_C)$

The E_{1u} Modes. The final species of the in-plane type contains contributions from the internal sets s, t, α, and β, although the cartesian treatment shows that only three vibrations of this species can occur. Thus the second (degenerate pair) redundancy should occur here. The potential constants are

$$\begin{aligned}
\mathsf{F}_{ss} &= F_s^1 + F_s^2 - F_s^3 - F_s^4 \\
\mathsf{F}_{tt} &= F_t^1 + F_t^2 - F_t^3 - F_t^4 \\
\mathsf{F}_{\alpha\alpha} &= F_\alpha^1 + F_\alpha^2 - F_\alpha^3 - F_\alpha^4 \\
\mathsf{F}_{\beta\beta} &= F_\beta^1 + F_\beta^2 - F_\beta^3 - F_\beta^4 \\
\mathsf{F}_{st} &= 3^{\frac{1}{2}}(F_{st}^1 - F_{st}^3) \\
\mathsf{F}_{s\alpha} &= F_{s\alpha}^1 + F_{s\alpha}^2 - F_{s\alpha}^3 - F_{s\alpha}^4 \\
\mathsf{F}_{s\beta} &= 3^{\frac{1}{2}}(F_{s\beta}^1 + F_{s\beta}^2) \\
\mathsf{F}_{t\alpha} &= 3^{\frac{1}{2}}(F_{t\alpha}^1 - F_{t\alpha}^3) \\
\mathsf{F}_{t\beta} &= -F_{t\beta}^1 + 2F_{t\beta}^2 + F_{t\beta}^3 \\
\mathsf{F}_{\alpha\beta} &= 3^{\frac{1}{2}}(F_{\alpha\beta}^1 + F_{\alpha\beta}^2)
\end{aligned}$$

The **G** matrix of this species has the form:

E_{1u}	s	t	α	β
s	$\mu_H + \mu_C$	$-\frac{3^{\frac{1}{2}}}{2}\mu_C$	$\frac{3^{\frac{1}{2}}}{2}\tau\mu_C$	$\frac{3}{4}\tau\mu_C$
t		$\frac{3}{2}\mu_C$	$-\frac{3}{2}\tau\mu_C$	$-\frac{3^{\frac{1}{2}}}{4}(2\sigma + 3\tau)\mu_C$
α			$\frac{3}{2}\tau^2\mu_C$	$\frac{3^{\frac{1}{2}}}{4}(2\sigma + 3\tau)\tau\mu_C$
β				$\sigma^2\mu_H + (\sigma^2 + \frac{3}{2}\sigma\tau + \frac{9}{8}\tau^2)\mu_C$

Since the column labeled α is exactly $-\tau$ times the column labeled t, the redundancy has the form $S_\alpha^{E_{1u}} + \tau S_t^{E_{1u}} = 0$. It can most simply be eliminated by omitting either the t coordinate or the α coordinate. The final secular equation will differ in form, depending upon which coordinate is retained, but the existence of the redundancy introduces an essential ambiguity into the force constants so that no physical difference can result (Sec. 8-1).

Out-of-plane Coordinates. As shown in Sec. 10-2, redundancies occur in A_{1u} and in E_{1g}. The former redundancy is the sum of the δ's shown in Table 10-14, and the latter is the δ combination appearing in E_{1g}, as can be seen by inspection of the appropriate **G** matrix which follows the potential energy matrices given below:

B_{2g}	γ	δ
γ	$F_\gamma^1 - 2F_\gamma^2 + 2F_\gamma^3 - F_\gamma^4$	$2(F_{\gamma\delta}^1 - F_{\gamma\delta}^2 + F_{\gamma\delta}^3)$
δ		$F_\delta^1 - 2F_\delta^2 + 2F_\delta^3 - F_\delta^4$

E_{1g}	γ	δ
γ	$F_\gamma^1 + F_\gamma^2 - F_\gamma^3 - F_\gamma^4$	$F_{\gamma\delta}^1 + 2F_{\gamma\delta}^2 + F_{\gamma\delta}^3$
δ		$F_\delta^1 + F_\delta^2 - F_\delta^3 - F_\delta^4$

A_{2u}	γ
γ	$F_\gamma^1 + 2F_\gamma^2 + 2F_\gamma^3 + F_\gamma^4$

E_{2u}	γ	δ
γ	$F_\gamma^1 - F_\gamma^2 - F_\gamma^3 + F_\gamma^4$	$\sqrt{3}\,(F_{\gamma\delta}^1 - F_{\gamma\delta}^3)$
δ		$F_\delta^1 - F_\delta^2 - F_\delta^3 + F_\delta^4$

The corresponding factors for the **G** matrix are

B_{2g}	γ	δ
γ	$\sigma^2\mu_H + (\sigma + 4\tau)^2\mu_C$	$-4\sqrt{3}\,\tau(\sigma + 4\tau)\mu_C$
δ		$48\tau^2\mu_C$

E_{1g}	γ	δ
γ	$\sigma^2\mu_H + (\sigma + \tau)^2\mu_C$	0
δ		0

A_{2u}	γ
γ	$\sigma^2\mu_H + \sigma^2\mu_C$

E_{2u}	γ	δ
γ	$\sigma^2\mu_H + (\sigma + 3\tau)^2\mu_C$	$-4\tau(\sigma + 3\tau)\mu_C$
δ		$16\tau^2\mu_C$

Checking. It is difficult to avoid errors in calculations by formal methods. Careful checking is therefore essential. Fortunately there are a number of special checks which can be used here. They are mostly based on certain invariant properties of a matrix under an orthogonal transformation (see Appendix V). Thus the diagonal elements must appear with unit coefficients in the factored matrices. The sum of the diagonal elements is unchanged, for each block. For example, the $\alpha\alpha$ block originally has a diagonal sum $6F_\alpha^1$. In factored form (**F**) the corresponding sum is

$$(F_\alpha^1 + 2F_\alpha^2 + 2F_\alpha^3 + F_\alpha^4) + 2(F_\alpha^1 - F_\alpha^2 - F_\alpha^3 + F_\alpha^4) \\ + (F_\alpha^1 - 2F_\alpha^2 + 2F_\alpha^3 - F_\alpha^4) + 2(F_\alpha^1 + F_\alpha^2 - F_\alpha^3 - F_\alpha^4) = 6F_\alpha^1$$

Note that redundant combinations must be included and degeneracy taken into account. The characteristic values of the matrices are unaltered by the transformation. If all force constants are put equal to unity, this property shows that all the elements should vanish except those involving coordinates which can be negative (β, δ, γ) and except those in A_{1g}. Thus for E_{2g}, $t\alpha$ one has

$$F_{t\alpha}^1 - 2F_{t\alpha}^2 + F_{t\alpha}^3$$

whose coefficients do add to zero.

The sum of all possible diagonal two-by-two determinantal minors for any double block (for example, the tt, $\alpha\alpha$, and $t\alpha$ blocks) is invariant.

By putting all but one off-diagonal force constant equal to zero and that one equal to unity, this test is not difficult to apply.

The above checks test only the transformation to symmetry coordinates and can be used on the **G** matrix as well as on **F**. They do not reveal errors in the original **G** elements. When factors of the secular equation are expanded, further tests are available.

10-9. Expansion of Secular Equation

Benzene has a number of two-by-two factors which are most conveniently handled by expansion into algebraic equations (quadratic).

Thus the A_{1g} factor immediately yields the equation

$$\lambda^2 - \lambda[(\mu_H + \mu_C)(F_s^1 + 2F_s^2 + 2F_s^3 + F_s^4) + \mu_C(F_t^1 + 2F_t^2 + 2F_t^3 + F_t^4) - 4\mu_C(F_{st}^1 + F_{st}^2 + F_{st}^3)] + \mu_H\mu_C[(F_s^1 + 2F_s^2 + 2F_s^3 + F_s^4) \times (F_t^1 + 2F_t^2 + 2F_t^3 + F_t^4) - 4(F_{st}^1 + F_{st}^2 + F_{st}^3)^2] = 0$$

The other quadratic factors can also be written down at once and will not be given here. It should be noted, however, that a check on the G elements arises from the fact that the last term will always involve $\mu_H\mu_C$ and never μ_H^2 or μ_C^2. This can be seen if the factored secular equation in terms of external symmetry coordinates (combinations of cartesians) is visualized. In terms of these the kinetic energy is diagonal, with the appropriate atomic masses as diagonal coefficients. The A_{1g} factor, according to Table 10-1, would involve one H mass and one C mass ($n_H^{(\gamma)} = 1$, $n_C^{(\gamma)} = 1$, $n_T^{(\gamma)} = n_R^{(\gamma)} = 0$) so that the product of the roots would involve $\mu_H\mu_C$.

It may be worthwhile to expand cubic factors, but the advantages of expansion diminish rather rapidly for factors of higher degree. In the next section some numerical solutions will be presented for illustration.

10-10. Calculation of Some Force Constants and Frequencies[1]

In principle, some of the inactive fundamental frequencies might be determined by computing force constants from the frequencies assigned in Sec. 10-5 and then using such force constants to compute the inactive frequencies. For example, if all off-diagonal force constants were neglected, there would be only the six constants, F_s^1, F_t^1, F_α^1, F_β^1, F_γ^1, and F_δ^1, to be computed from twenty-two frequencies as assigned. Perhaps the most obvious difficulty with this procedure is that none of the active frequencies involves the δ coordinate; therefore F_δ^1 cannot be determined. Furthermore, it will be seen that the assumption of a completely diagonal potential energy function is not justified on the basis of the observed frequencies. A common procedure is to introduce several interaction

[1] B. L. Crawford, Jr., and F. A. Miller, *J. Chem. Phys.*, **17:** 249 (1949); **14:** 282 (1946).

constants judiciously so as to obtain the best fit to all the observed frequencies. One could then proceed to an estimate of the inactive fundamentals somewhat more confidently.

Naturally, if the observed frequencies of other deuterobenzenes can be assigned with confidence, more equations become available for the determination of force constants. When the secular determinants involved are large, however, the problem is complicated by the occurrence of several sets of force constants which fit a given set of frequencies.

Obviously the best starting point in computing force constants is with the species A_{2u} and E_{1g} in each of which only a single frequency is involved. The secular determinants reduce, respectively, to the equations (see Sec. 10-8)

$$(F_\gamma^1 + 2F_\gamma^2 + 2F_\gamma^3 + F_\gamma^4)\sigma^2(\mu_H + \mu_C) - \lambda_{11} = 0 \tag{1}$$

and

$$(F_\gamma^1 + F_\gamma^2 - F_\gamma^3 - F_\gamma^4)[\sigma^2\mu_H + (\sigma + \tau)^2\mu_C] - \lambda_{10} = 0 \tag{2}$$

One may now substitute numerical values of $\sigma = 1/1.08$ A, $\tau = 1/1.39$ A, and values of the frequencies through the relation

$$\lambda = 4\pi^2\nu^2 = 4\pi^2c^2\omega^2 = \left(\frac{\omega}{1{,}302.9}\right)^2 \tag{3}$$

The last form of (3) is consistent with the following system of units: masses in atomic weight units, lengths in angstroms, stretching force constants in 10^5 dynes cm^{-1}, bending force constants in 10^{-11} dyne-cm, and stretch-bend interaction constants in 10^{-3} dyne. The results are

$$F_\gamma^1 + 2F_\gamma^2 + 2F_\gamma^3 + F_\gamma^4 = 0.288 \tag{4}$$
$$F_\gamma^1 + F_\gamma^2 - F_\gamma^3 - F_\gamma^4 = 0.394 \tag{5}$$

which immediately shows the inadequacy of the potential function in which the interaction constants F_γ^2, F_γ^3, and F_γ^4 are all neglected.

The A_{1g} factor can next be considered. Because of the wide separation of frequencies, the technique described in Sec. 4-8 may be applied, separating the problem into two uncoupled frequencies. The kinetic energy matrix has the form (Sec. 10-8)

$$\mathbf{G}^{(A_{1g})} = \begin{Vmatrix} \mu_H + \mu_C & -\mu_C \\ -\mu_C & \mu_C \end{Vmatrix} \tag{6}$$

In the present instance, after applying Eq. (2), Sec. 4-8, this separates into $\mu_H + \mu_C$ for the high-frequency part and

$$\mu_C - \frac{\mu_C{}^2}{\mu_H + \mu_C} = \frac{\mu_H\mu_C}{\mu_H + \mu_C}$$

for the low-frequency part. In this manner one obtains the equations

$$(F_s^1 + 2F_s^2 + 2F_s^3 + F_s^4)(\mu_H + \mu_C) \approx \lambda_2 \tag{7}$$

$$(F_t^1 + 2F_t^2 + 2F_t^3 + F_t^4)\,\frac{\mu_H\mu_C}{\mu_H + \mu_C} \approx \lambda_1 \tag{8}$$

After numerical substitution, the following values are obtained for the force constants:

$$F_s^1 + 2F_s^2 + 2F_s^3 + F_s^4 \approx 5.14 \tag{9}$$

$$F_t^1 + 2F_t^2 + 2F_t^3 + F_t^4 \approx 7.55 \tag{10}$$

Actually it is unnecessary to make this last approximation, since the C_6H_6 and C_6D_6 frequencies include four observed values from which three force constants can be obtained. As pointed out by Crawford *et al.*, this is just a sufficient number, since one of the relations derivable from the frequencies is the product rule, which gives no information on the individual values of the force constants.

A convenient procedure for the determination of the force constants utilizes the fact that in the present instance the coefficients c_1 and c_2 in the expanded secular equation

$$\lambda^2 + c_1\lambda + c_2 = 0 \tag{11}$$

are, respectively, minus the sum of the diagonal terms and the determinant of the matrix **FG**:

$$c_1 = -\sum_k (\mathbf{FG})_{kk} \tag{12}$$

$$c_2 = |\mathbf{FG}| = |\mathbf{F}|\,|\mathbf{G}| \tag{13}$$

On the other hand, the same coefficients are related to the frequencies through the equations

$$c_1 = -(\lambda_1 + \lambda_2) \tag{14}$$

$$c_2 = \lambda_1\lambda_2 \tag{15}$$

Expanding (12) and (13) and combining with (14) and (15), one obtains

$$\mathsf{F}_s(\mu_H + \mu_C) - 2\mathsf{F}_{st}\mu_C + \mathsf{F}_t\mu_C = \lambda_1 + \lambda_2 \tag{16}$$

$$[\mathsf{F}_s\mathsf{F}_t - (\mathsf{F}_{st})^2]\mu_H\mu_C = \lambda_1\lambda_2 \tag{17}$$

When the equations are put in this form, the remark made above becomes clear, since one can obtain two equations from (16) by using μ_H with the benzene frequencies and then substituting μ_D on the left and using the deuterobenzene frequencies, whereas the second equation yields only a value for $\mathsf{F}_s\mathsf{F}_t - (\mathsf{F}_{st})^2$ together with the product rule check.

Insertion of proper numerical values yields two sets of roots, of which only the following set is physically reasonable:

$$\begin{aligned} \mathsf{F}_s &= F_s^1 + 2F_s^2 + 2F_s^3 + F_s^4 = 5.003 \\ \mathsf{F}_t &= F_t^1 + 2F_t^2 + 2F_t^3 + F_t^4 = 7.832 \\ \mathsf{F}_{st} &= 2(F_{st}^1 + F_{st}^2 + F_{st}^3) = -0.4198 \end{aligned} \tag{18}$$

These values may be compared with the approximate results obtained in (9) and (10).

No attempt will be made here to compute further force constants; instead the reader is referred to the papers of Crawford and Miller cited above. However, as an example of the calculation of frequencies from a set of known constants, the E_{1u} factor (infrared active) will be worked out, using Crawford and Miller's constants. Although a problem of this size (three frequencies) can be carried through by direct expansion to the polynomial form, we shall not use such a method, but shall instead illustrate some of the matrix numerical methods which were described in Sec. 9-7. This method will also yield the transformation coefficients between symmetry and normal coordinates.

Use will be made of the fact that the force constants to be employed are chosen to fit the observed frequencies; namely, $\omega_{18} = 1{,}037$, $\omega_{19} = 1{,}485$, and $\omega_{20} = 3{,}080$. The corresponding values of λ are $\lambda_{18} = 0.63$, $\lambda_{19} = 1.30$, and $\lambda_{20} = 5.59$. As shown below, these approximate values can be used to make the iteration process converge upon any desired root.

The F and G matrices for the E_{1u} factor have the forms

$$\mathsf{F} = \left\Vert \begin{matrix} \mathsf{F}_s & \mathsf{F}_{st} & \mathsf{F}_{s\beta} \\ & \mathsf{F}_t & \mathsf{F}_{t\beta} \\ & & \mathsf{F}_\beta \end{matrix} \right\Vert \tag{19}$$

and

$$\mathsf{G} = \left\Vert \begin{matrix} \mu_H + \mu_C & -\frac{3^{\frac{1}{2}}}{2}\mu_C & \frac{3}{4}\tau\mu_C \\ & \frac{3}{2}\mu_C & -\frac{3^{\frac{1}{2}}}{4}(2\sigma + 3\tau)\mu_C \\ & & \sigma^2\mu_H + \left(\sigma^2 + \frac{3}{2}\sigma\tau + \frac{9}{8}\tau^2\right)\mu_C \end{matrix} \right\Vert \tag{20}$$

after elimination of the redundancy as described in Sec. 10-8. Note that the force constants appearing in (19) are in terms of symmetry coordinates, *e.g.*,

$$\mathsf{F}_s = F_s^1 + F_s^2 - F_s^3 - F_s^4$$

In order to find the numerical values of the force constants in terms of those of Crawford and Miller, we use the equations

$$\begin{aligned}
\mathsf{F}_s &= \Omega_4 = 5.147 \\
\mathsf{F}_t &= 2\Lambda_4 = 12.286 \\
\mathsf{F}_\beta &= \frac{1}{\sigma^2}\Gamma_4 = 1.01452 \\
\mathsf{F}_{st} &= 2^{\frac{1}{2}}\xi_4 = 0 \\
\mathsf{F}_{s\beta} &= \frac{1}{\sigma}\tau_4 = 0 \\
\mathsf{F}_{t\beta} &= \frac{2^{\frac{1}{2}}}{\sigma}\mu_4 = 2.13676
\end{aligned} \tag{21}$$

in which Ω_4, Λ_4, etc., are the constants given by Crawford and Miller. There are two reasons why the elements of the F matrix are not simply equal to Crawford and Miller's constants. First, the internal bending coordinates α and β are slightly different from those employed by Crawford and Miller, who used coordinates with the dimensions of length throughout. Second, Crawford and Miller used a slightly different method of eliminating the redundancy. These differences are of such a nature, however, that although the numerical values of the elements of the F matrix on the one hand, and of the elements of the G matrix on the other, will differ from Crawford and Miller's values, the elements of the product matrix FG will be identical.

We now proceed with the numerical evaluation of FG. From (21),

$$\mathsf{F} = \begin{Vmatrix} 5.14700 & 0 & 0 \\ 0 & 12.28600 & 2.13676 \\ 0 & 2.13676 & 1.01452 \end{Vmatrix} \tag{22}$$

while insertion of numerical values of μ_H, μ_C, σ, and τ yields:

$$\mathsf{G} = \begin{Vmatrix} 1.07532 & -0.07210 & 0.04492 \\ -0.07210 & 0.12489 & -0.14458 \\ 0.04492 & -0.14458 & 1.05357 \end{Vmatrix} \tag{23}$$

Matrix multiplication of (22) and (23) then gives:

$$\mathsf{H} = \mathsf{FG} = \begin{Vmatrix} 5.53467 & -0.37110 & 0.32699 \\ -0.78983 & 1.22548 & 0.67174 \\ -0.07671 & 0.08498 & 0.75995 \end{Vmatrix} \tag{24}$$

Note that H is not symmetric.

Since the approximate value of the largest $\lambda(\lambda_{20} \approx 5.59)$, as estimated from the observed frequencies above, is four times greater than the next largest estimated λ, one expects that the iteration process described in

Sec. 9-6 should converge rapidly upon λ_{20}, using $\mathbf{H}$ as it stands. It is reasonable to start the iteration upon a row matrix, $\mathbf{A}^{(0)}$, with components 1, 0, 0 and upon a column matrix, $\mathbf{B}^{(0)}$, with components 1, 0, 0, since in (24) the desired λ clearly is most closely approximated in the first row and column. The numerical details of the iteration are given in Table 10-15.

TABLE 10-15. ITERATIVE CALCULATION OF THE CHARACTERISTIC VALUE OF λ_{20} OF THE E_{1u} FACTOR OF BENZENE AND OF THE CORRESPONDING VECTORS

				$\mathbf{B}^{(0)}$	$\mathbf{B}^{(1)}$	$\mathbf{B}^{(2)}$	$\mathbf{B}^{(3)}$	$\mathbf{B}^{(4)}$
$\mathbf{A}^{(0)}$	1	0	0	1	5.535	30.9024	172.8555	967.2833
$\mathbf{A}^{(1)}$	5.535	−0.371	0.327	0	−0.790	−5.3916	−31.3845	−177.1685
$\mathbf{A}^{(2)}$	30.9023	−2.4809	1.8092	0	−0.077	−0.5502	−3.2468	−18.3942
$\mathbf{A}^{(3)}$	172.8547	−14.3544	9.8131					

$\lambda^{(0)} = 5.535$, $\lambda^{(1)} = 5.5936$, $\lambda^{(2)} = 5.5964$, $\lambda^{(3)} = 5.59656$

The successive values of λ are computed from

$$\lambda^{(p)} = \frac{\sum_k A_k^{(p)} B_k^{(p+1)}}{\sum_k A_k^{(p)} B_k^{(p)}} \tag{25}$$

Since, as shown in Sec. 9-8, the $\lambda^{(p)}$ are always less than the true value, it seems reasonable to take $\lambda_{20} = 5.5966$. Furthermore, the numbers listed in the final row for $\mathbf{A}$, after normalization according to Eq. (7), Sec. 4-7, give the transformation from normal to symmetry coordinates, while the final $\mathbf{B}$ column, normalized as described in Appendix VIII, yields the inverse transformation from symmetry to normal coordinates.

At this stage, λ_{18} and λ_{19} remain to be evaluated. The method adopted is that described in Sec. 9-8 which consists in using a modified $\mathbf{H}$ in such a way that a root corresponding to λ_{18} or to λ_{19} will become dominant. Suppose that iteration is used with

$$(\mathbf{H} - 5.5966\mathbf{E})(\mathbf{H} - 1.3\mathbf{E}) \tag{26}$$

The roots of the first factor of (26) are $\lambda_{18} - 5.5966$, $\lambda_{19} - 5.5966$, and $\lambda_{20} - 5.5966$. The last of these differences is zero; the first two are approximately $0.63 - 5.60 = -4.97$ and $1.30 - 5.60 = -4.30$, respectively, using estimates of λ_{18} and λ_{19} based on experimental frequencies as described above. Similarly, the second factor should have roots at approximately -0.67, 0, and -4.3. Therefore the matrix product in (26) should have roots at $-4.97 \times -0.67 = 3.33$, 0, and 0. Therefore the root corresponding to λ_{18} should be strongly dominant and the convergence rapid.

The numerical value of (26) is obtained by multiplying the matrices formed by subtracting 5.5966 and 1.3, respectively, from the diagonal elements of (24):

$$(\mathbf{H} - 5.5966\mathbf{E})(\mathbf{H} - 1.3\mathbf{E}) = \left\| \begin{matrix} 0.00577 & 0.07842 & -0.44612 \\ 0.05624 & 0.67593 & -3.55730 \\ -0.02094 & -0.38888 & 2.64403 \end{matrix} \right\| \tag{27}$$

In selecting the arbitrary components of $\mathbf{A}^{(0)}$ and $\mathbf{B}^{(0)}$ for iteration with (27), it must be borne in mind that the characteristic vectors of (27) are exactly the characteristic vectors of the original matrix (24). Since the root we are now seeking appears (approximately) in the last row and column of (24), it is reasonable to start the iteration with (27) on row or column vectors with components 0, 0, 1. A few preliminary trials, however, show that better initial choices are -0.01, -0.15, and 1.0, for $\mathbf{A}^{(0)}$ and -0.05, -0.4, and 0.3 for $\mathbf{B}^{(0)}$. The remaining numerical computations are summarized in Table 10-16.

TABLE 10-16. ITERATIVE CALCULATION OF THE CHARACTERISTIC VALUE $(\lambda_{18} - 5.5966)(\lambda_{18} - 1.3)$ OF THE MODIFIED E_{1u} FACTOR OF BENZENE

				$\mathbf{B}^{(0)}$	$\mathbf{B}^{(1)}$	$\mathbf{B}^{(2)}$	$\mathbf{B}^{(3)}$
$\mathbf{A}^{(0)}$	−0.01	−0.15	1.0	−0.05	−0.16549	−0.52980	−1.69423
$\mathbf{A}^{(1)}$	−0.02943	−0.49105	3.18209	−0.4	−1.34037	−4.29406	−13.73235
$\mathbf{A}^{(2)}$	−0.09442	−1.57167	10.17348	0.3	0.94981	3.03603	9.70832

$$(\lambda_{18}^{(0)} - 5.5966)(\lambda_{18}^{(0)} - 1.3) = 3.19700$$
$$(\lambda_{18}^{(1)} - 5.5966)(\lambda_{18}^{(1)} - 1.3) = 3.19774$$
$$(\lambda_{18}^{(2)} - 5.5966)(\lambda_{18}^{(2)} - 1.3) = 3.19775$$

Evidently the convergence is rapid, and the result in the second approximation (3.19775) leads to a value $\lambda_{18} = 0.65314$ (note that the other root of the quadratic equation for $\lambda_{18}^{(2)}$ is discarded).

One can now use the fact that

$$\lambda_{18} + \lambda_{19} + \lambda_{20} = \sum_k (\mathbf{FG})_{kk} = 7.2015 \tag{28}$$

to obtain $\lambda_{19} = 1.27036$. The iterative process which would lead to the characteristic vector associated with this root is omitted. Note that when λ_{19} is obtained by this method there is still an easy check available, namely, the sum of the diagonal terms of $(\mathbf{H} - 5.5966\mathbf{E})(\mathbf{H} - 1.3\mathbf{E})$ which should equal

$$(\lambda_{18} - 5.5966)(\lambda_{18} - 1.3) + (\lambda_{19} - 5.5966)(\lambda_{19} - 1.3) + (\lambda_{20} - 5.5966)(\lambda_{20} - 1.3) \tag{29}$$

One finds the sum of the diagonal terms to be 3.32596, while the expression (29) equals 3.32573, which is satisfactory.

The frequencies corresponding to the λ's just found are $\omega_{18} = 1{,}053$, $\omega_{19} = 1{,}468$, and $\omega_{20} = 3{,}082$, which are in good agreement with the experimental values.

The reader who plans to carry out similar numerical calculations is urged to check all these computations using a standard desk-type computing machine so as to gain a facility in the typical numerical operation involved, namely, matrix multiplication.

CHAPTER 11

THE SEPARATION OF ROTATION AND VIBRATION

In Sec. 2-1 there was given a qualitative discussion of the justification for treating the vibrational and rotational motions separately. In this chapter more detailed proofs will be worked out, and the nature of the approximations involved will be discussed.

11-1. Classical Kinetic Energy

The molecular model used in this book consists of point masses—the atoms—connected by forces which keep the atoms near their equilibrium positions.[1] A rigorous treatment of such a model must start with the determination of the kinetic energy in terms of a suitable set of coordinates. These coordinates consist of the cartesian coordinates X, Y, Z, which describe the position of the center of mass; the three Eulerian angles θ, ϕ, and χ (see Appendix I) which describe the orientation in space of a set of rotating coordinate axes x, y, z, whose origin coincides with the center of mass, and finally 3N − 6 normal coordinates which give the positions of the atoms relative to each other in the rotating axis system.

Vectorial methods are especially useful for this problem.[2] Let the position of the αth particle be given by the vector $\mathbf{r}_\alpha$ from a point O, which is the center of mass of the particles and the origin of a moving

[1] For a consideration of the effect of the electronic motion, see the following:
M. Born and J. R. Oppenheimer, *Ann. d. Physik*, **84:** 457 (1927).
R. Karplus, *J. Chem. Phys.*, **16:** 1170 (1948).
H. H. Nielsen, *Revs. Mod. Phys.*, **23:** 90 (1951).
Even when electronic effects are considered, Karplus shows that in most cases the molecule can still be treated as a set of atoms thought of as point masses, up to and including the correction for centrifugal distortion, but the moments of inertia I_x, etc., may have values slightly different from what would be calculated if the electronic structure were ignored.

[2] The treatment which follows is based on that in E. B. Wilson, Jr., and J. B. Howard, *J. Chem. Phys.*, **4:** 260 (1936), but has been corrected to conform to the criticism of B. T. Darling and D. M. Dennison, *Phys. Rev.*, **57:** 128 (1940). The vector technique used was adapted from the similar treatment of rigid bodies by L. Page, "Introduction to Theoretical Physics," 2d ed., Chap. 2, Van Nostrand, New York, 1934. See also H. Margenau and G. M. Murphy, "The Mathematics of Physics and Chemistry," Chap. 9, Van Nostrand, New York, 1943.

system of cartesian axes, to be specified later. Let x_α, y_α, and z_α be the components of $\mathbf{r}_\alpha$ in the moving system of axes. The position of O is given by another vector $\mathbf{R}$ whose origin is fixed in space. The equilibrium position of the αth particle of mass m_α, is described by $\mathbf{a}_\alpha$, a vector fixed to the moving axis system. The displacement vector $\boldsymbol{\rho}_\alpha$ is defined by the relation

$$\boldsymbol{\rho}_\alpha = \mathbf{r}_\alpha - \mathbf{a}_\alpha \tag{1}$$

If at any instant the rotating system of axes has the angular velocity $\boldsymbol{\omega}$ and if the vector $\mathbf{v}_\alpha$ is defined as the vector with the components $\dot{x}_\alpha$, $\dot{y}_\alpha$, and $\dot{z}_\alpha$ in the moving system, then the velocity of the αth particle in space is[1]

$$\dot{\mathbf{R}} + \boldsymbol{\omega} \times \mathbf{r}_\alpha + \mathbf{v}_\alpha \tag{2}$$

The kinetic energy T of the whole molecule is obtained from this velocity, and is

$$\begin{aligned} 2T = \dot{R}^2 \sum_\alpha m_\alpha + \sum_\alpha m_\alpha (\boldsymbol{\omega} \times \mathbf{r}_\alpha) \cdot (\boldsymbol{\omega} \times \mathbf{r}_\alpha) + \sum_\alpha m_\alpha v_\alpha^2 \\ + 2\dot{\mathbf{R}} \cdot \boldsymbol{\omega} \times \sum_\alpha m_\alpha \mathbf{r}_\alpha + 2\dot{\mathbf{R}} \cdot \sum_\alpha m_\alpha \mathbf{v}_\alpha + 2\boldsymbol{\omega} \cdot \sum_\alpha (m_\alpha \mathbf{r}_\alpha \times \mathbf{v}_\alpha) \end{aligned} \tag{3}$$

Since the point O is the center of gravity of the whole molecule, at every instant it must be true that

$$\sum_\alpha m_\alpha \mathbf{r}_\alpha = 0 \tag{4}$$

whence it follows that[2]

$$\sum_\alpha m_\alpha \mathbf{v}_\alpha = 0 \tag{5}$$

These conditions, however, do not suffice to define the rotating coordinate system completely. If the molecule were rigid, the rotating axes could be attached in some definite way to it, but the molecule is not rigid and all of its atoms can move about their equilibrium positions. The definition which is adopted was discussed in Sec. 2-1 and consists of the conditions[3]

[1] For further details concerning this step, see L. Page, "Introduction to Theoretical Physics," 2d ed., Chap. 2, Van Nostrand, New York, 1934.

[2] If $\sum_\alpha m_\alpha \mathbf{r}_\alpha = 0$, then $\sum_\alpha m_\alpha \dot{\mathbf{r}}_\alpha = \sum_\alpha m_\alpha[(\boldsymbol{\omega} \times \mathbf{r}_\alpha) + \mathbf{v}_\alpha] = 0$, or

$$\boldsymbol{\omega} \times \sum_\alpha m_\alpha \mathbf{r}_\alpha + \sum_\alpha m_\alpha \mathbf{v}_\alpha = \sum_\alpha m_\alpha \mathbf{v}_\alpha = 0.$$

[3] See C. Eckart, *Phys. Rev.*, **47**: 552 (1935).

$$\sum_\alpha m_\alpha \mathbf{a}_\alpha \times \mathbf{v}_\alpha = 0 \tag{6}$$

As explained in Sec. 2-1, this is almost but not quite equivalent to stating that there must be no angular momentum with respect to the rotating system of axes.[1]

By replacing $\mathbf{r}_\alpha$ by $\mathbf{a}_\alpha + \boldsymbol{\varrho}_\alpha$ in the last term of (3) and introducing the conditions in (4), (5), and (6), the kinetic energy becomes

$$2T = \dot{R}^2 \sum_\alpha m_\alpha + \sum m_\alpha(\boldsymbol{\omega} \times \mathbf{r}_\alpha) \cdot (\boldsymbol{\omega} \times \mathbf{r}_\alpha) + \sum m_\alpha v_\alpha^2 + 2\boldsymbol{\omega} \cdot \sum m_\alpha(\boldsymbol{\varrho}_\alpha \times \mathbf{v}_\alpha) \tag{7}$$

The first term is the translational energy of the molecule and will not be included hereafter because it can be separated in field-free problems. The second term is the rotational energy, the third the vibrational energy, and the last the coupling between rotation and vibration, the so-called *Coriolis* energy. When the terms of (7) are expanded by the standard methods of vector analysis,[2] they become

$$\begin{aligned} 2T = {} & I_{xx}\omega_x^2 + I_{yy}\omega_y^2 + I_{zz}\omega_z^2 - 2I_{xy}\omega_x\omega_y - 2I_{yz}\omega_y\omega_z - 2I_{zx}\omega_z\omega_x \\ & + \sum_\alpha m_\alpha v_\alpha^2 + 2\omega_x \sum_\alpha m_\alpha(\boldsymbol{\varrho}_\alpha \times \mathbf{v}_\alpha)_x + 2\omega_y \sum_\alpha m_\alpha(\boldsymbol{\varrho}_\alpha \times \mathbf{v}_\alpha)_y \\ & + 2\omega_z \sum_\alpha m_\alpha(\boldsymbol{\varrho}_\alpha \times \mathbf{v}_\alpha)_z \end{aligned} \tag{8}$$

Here I_{xx}, I_{yy}, I_{zz} are the instantaneous moments of inertia with respect to the moving x, y, z axes; I_{xy}, I_{yz}, and I_{zx} are the products of inertia. These quantities are not constants but functions of the positions of the

[1] It is perhaps better to use the condition

$$\sum_\alpha m_\alpha \mathbf{a}_\alpha \times \mathbf{r}_\alpha = 0 \tag{6a}$$

This implies Eq. (6) since, on differentiating,

$$\begin{aligned} 0 &= \sum m_\alpha \dot{\mathbf{a}}_\alpha \times \mathbf{r}_\alpha + \sum m_\alpha \mathbf{a}_\alpha \times \dot{\mathbf{r}}_\alpha \\ &= \sum m_\alpha(\boldsymbol{\omega} \times \mathbf{a}_\alpha) \times \mathbf{r}_\alpha + \sum m_\alpha \mathbf{a}_\alpha \times (\boldsymbol{\omega} \times \mathbf{r}_\alpha) + \sum m_\alpha \mathbf{a}_\alpha \times \mathbf{v}_\alpha \\ &= \sum m_\alpha \mathbf{a}_\alpha \times \mathbf{v}_\alpha \end{aligned}$$

from the properties of the triple vector product.

[2] See, for example, L. Page, "Introduction to Theoretical Physics," 2d ed., Introduction, Van Nostrand, New York, 1934, or H. Margenau and G. M. Murphy, "The Mathematics of Physics and Chemistry," Chap. 4, Van Nostrand, New York, 1943.

particles. ω_x, ω_y, ω_z are the components of the angular velocity $\boldsymbol{\omega}$ of the rotating system of axes.

Introduction of Normal Coordinates. The normal coordinates, Q_k, are defined in terms of the components Δx_α, etc., of $\boldsymbol{\varrho}_\alpha$ by the relations (see Secs. 2-4 and 2-5)

$$\begin{aligned}\Delta x_\alpha &= \sum_{k=1}^{3N-6} l'_{\alpha k} Q_k \\ \Delta y_\alpha &= \sum_{k=1}^{3N-6} m'_{\alpha k} Q_k \\ \Delta z_\alpha &= \sum_{k=1}^{3N-6} n'_{\alpha k} Q_k\end{aligned} \tag{9}$$

in which the constants $l'_{\alpha k}$, $m'_{\alpha k}$, and $n'_{\alpha k}$ are determined so that

$$\sum_{\alpha=1}^{N} m_\alpha v_\alpha^2 = \sum_{k=1}^{3N-6} \dot{Q}_k^2 \tag{10}$$

while

$$2V = \Sigma \lambda_k Q_k^2 + \text{higher terms} \tag{11}$$

The primes on l', m', and n' are used to distinguish them from the similar unprimed letters employed in Sec. 2-5, where mass-adjusted coordinates were used.

In terms of the normal coordinates, parts of the coupling terms become

$$\begin{aligned}\sum_\alpha m_\alpha(\boldsymbol{\varrho}_\alpha \times \mathbf{v}_\alpha)_x &= \sum_\alpha m_\alpha(\Delta y_\alpha\, \Delta \dot{z}_\alpha - \Delta z_\alpha\, \Delta \dot{y}_\alpha) \\ &= \sum_{k=1}^{3N-6} \mathfrak{X}_k \dot{Q}_k \\ \sum_\alpha m_\alpha(\boldsymbol{\varrho}_\alpha \times \mathbf{v}_\alpha)_y &= \sum_{k=1}^{3N-6} \mathfrak{Y}_k \dot{Q}_k \\ \sum_\alpha m_\alpha(\boldsymbol{\varrho}_\alpha \times \mathbf{v}_\alpha)_z &= \sum_{k=1}^{3N-6} \mathfrak{Z}_k \dot{Q}_k\end{aligned} \tag{12}$$

in which

$$\begin{aligned}\mathfrak{X}_k &= \sum_{\alpha,l} m_\alpha(m'_{\alpha l} n'_{\alpha k} - n'_{\alpha l} m'_{\alpha k}) Q_l \\ \mathfrak{Y}_k &= \sum_{\alpha,l} m_\alpha(n'_{\alpha l} l'_{\alpha k} - l'_{\alpha l} n'_{\alpha k}) Q_l \\ \mathfrak{Z}_k &= \sum_{\alpha,l} m_\alpha(l'_{\alpha l} m'_{\alpha k} - m'_{\alpha l} l'_{\alpha k}) Q_l\end{aligned} \tag{13}$$

The kinetic energy is therefore

$$2T = I_{xx}\omega_x^2 + I_{yy}\omega_y^2 + I_{zz}\omega_z^2 - 2I_{xy}\omega_x\omega_y - 2I_{yz}\omega_y\omega_z - 2I_{zx}\omega_z\omega_x + 2\omega_x \sum_k \mathfrak{X}_k \dot{Q}_k + 2\omega_y \sum_k \mathfrak{Y}_k \dot{Q}_k + 2\omega_z \sum_k \mathfrak{Z}_k \dot{Q}_k + \sum_k \dot{Q}_k^2 \tag{14}$$

11-2. Hamiltonian Form of the Kinetic Energy

In order to obtain the wave equation, it is necessary to have the kinetic energy expressed in terms of the angular momenta instead of the angular velocities. The angular momentum is a vector defined by the equation

$$\mathfrak{M} = \sum_\alpha m_\alpha \mathbf{r}_\alpha \times \dot{\mathbf{r}}_\alpha \tag{1}$$

This leads to the relation

$$\mathfrak{M} = \sum_\alpha m_\alpha[\mathbf{r}_\alpha \times (\boldsymbol{\omega} \times \mathbf{r}_\alpha)] + \sum_\alpha m_\alpha \mathbf{r}_\alpha \times \mathbf{v}_\alpha \tag{2}$$

Expansion of the vector products[1] yields the following expressions for the components of the angular momentum:

$$\begin{aligned}
\mathfrak{M}_x &= I_{xx}\omega_x - I_{xy}\omega_y - I_{zx}\omega_z + \sum_k \mathfrak{X}_k \dot{Q}_k = \frac{\partial T}{\partial \omega_x} \\
\mathfrak{M}_y &= -I_{xy}\omega_x + I_{yy}\omega_y - I_{yz}\omega_z + \sum_k \mathfrak{Y}_k \dot{Q}_k = \frac{\partial T}{\partial \omega_y} \\
\mathfrak{M}_z &= -I_{zx}\omega_x - I_{yz}\omega_y + I_{zz}\omega_z + \sum_k \mathfrak{Z}_k \dot{Q}_k = \frac{\partial T}{\partial \omega_z}
\end{aligned} \tag{3}$$

The momentum P_k conjugate to the normal coordinate Q_k is

$$P_k = \frac{\partial T}{\partial \dot{Q}_k} = \dot{Q}_k + \mathfrak{X}_k\omega_x + \mathfrak{Y}_k\omega_y + \mathfrak{Z}_k\omega_z \tag{4}$$

Equations (3) and (4) may be solved for ω_x, ω_y, ω_z and the $\dot{Q}_k$'s in terms of $\mathfrak{M}_x$, $\mathfrak{M}_y$, $\mathfrak{M}_z$ and the P_k's, and the results used in Eq. (14), Sec. 11-1, to obtain the energy in terms of the momenta; that is, the Hamiltonian form of the energy. The method of carrying out this substitution will be outlined below, but the complete Hamiltonian form for the kinetic energy

[1] See L. Page, "Introduction to Theoretical Physics," 2d ed., Introduction, Van Nostrand, New York, 1934.

thus found is

$$2T = \mu_{xx}(\mathfrak{M}_x - \mathfrak{m}_x)^2 + \mu_{yy}(\mathfrak{M}_y - \mathfrak{m}_y)^2 + \mu_{zz}(\mathfrak{M}_z - \mathfrak{m}_z)^2 + 2\mu_{xy}(\mathfrak{M}_x - \mathfrak{m}_x)(\mathfrak{M}_y - \mathfrak{m}_y) + 2\mu_{yz}(\mathfrak{M}_y - \mathfrak{m}_y)(\mathfrak{M}_z - \mathfrak{m}_z) + 2\mu_{zx}(\mathfrak{M}_z - \mathfrak{m}_z)(\mathfrak{M}_x - \mathfrak{m}_x) + \sum_k P_k^2 \quad (5)$$

Here $\mathfrak{m}_x$, $\mathfrak{m}_y$, and $\mathfrak{m}_z$ are the components of the vibrational angular momentum, given by the expressions

$$\mathfrak{m}_x = \sum_k \mathfrak{X}_k P_k \qquad \mathfrak{m}_y = \sum_k \mathfrak{Y}_k P_k \qquad \mathfrak{m}_z = \sum_k \mathfrak{Z}_k P_k \quad (6)$$

The coefficients μ_{xx}, etc., are functions of the normal coordinates only and arise as shown below.

The following steps were used to obtain (5). From Eq. (14), Sec. 11-1, and from (3) and (4) it is seen that

$$2T = \mathfrak{M}_x\omega_x + \mathfrak{M}_y\omega_y + \mathfrak{M}_z\omega_z + \sum_k P_k\dot{Q}_k \quad (7)$$

Substitution in this equation of the expression for $\dot{Q}_k$ obtained by solving (4) yields

$$2T = (\mathfrak{M}_x - \mathfrak{m}_x)\omega_x + (\mathfrak{M}_y - \mathfrak{m}_y)\omega_y + (\mathfrak{M}_z - \mathfrak{m}_z)\omega_z + \sum_k P_k^2 \quad (8)$$

when the definitions of $\mathfrak{m}_x$, $\mathfrak{m}_y$, and $\mathfrak{m}_z$ are introduced. The same expression for $\dot{Q}_k$ may be substituted in (3); the rearranged result is

$$\begin{aligned} \mathfrak{M}_x - \mathfrak{m}_x &= I'_{xx}\omega_x - I'_{xy}\omega_y - I'_{zx}\omega_z \\ \mathfrak{M}_y - \mathfrak{m}_y &= -I'_{xy}\omega_x + I'_{yy}\omega_y - I'_{yz}\omega_z \\ \mathfrak{M}_z - \mathfrak{m}_z &= -I'_{zx}\omega_x - I'_{yz}\omega_y + I'_{zz}\omega_z \end{aligned} \quad (9)$$

in which

$$\begin{aligned} I'_{xx} &= I_{xx} - \sum_k \mathfrak{X}_k^2 \qquad I'_{yy} = I_{yy} - \sum_k \mathfrak{Y}_k^2 \qquad I'_{zz} = I_{zz} - \sum_k \mathfrak{Z}_k^2 \\ I'_{xy} &= I_{xy} + \sum_k \mathfrak{X}_k\mathfrak{Y}_k \qquad I'_{yz} = I_{yz} + \sum_k \mathfrak{Y}_k\mathfrak{Z}_k \\ I'_{zx} &= I_{zx} + \sum_k \mathfrak{Z}_k\mathfrak{X}_k \end{aligned} \quad (10)$$

The inverse of (9) may be written

$$\begin{aligned} \omega_x &= \mu_{xx}(\mathfrak{M}_x - \mathfrak{m}_x) + \mu_{xy}(\mathfrak{M}_y - \mathfrak{m}_y) + \mu_{xz}(\mathfrak{M}_z - \mathfrak{m}_z) \\ \omega_y &= \mu_{yx}(\mathfrak{M}_x - \mathfrak{m}_x) + \mu_{yy}(\mathfrak{M}_y - \mathfrak{m}_y) + \mu_{yz}(\mathfrak{M}_z - \mathfrak{m}_z) \\ \omega_z &= \mu_{zx}(\mathfrak{M}_x - \mathfrak{m}_x) + \mu_{zy}(\mathfrak{M}_y - \mathfrak{m}_y) + \mu_{zz}(\mathfrak{M}_z - \mathfrak{m}_z) \end{aligned} \quad (11)$$

This set of equations defines the coefficients μ_{xx}, etc., and if these expressions for ω_x, ω_y, and ω_z are substituted in (8), the desired Hamiltonian form of (5) is obtained.

11-3. A General Theorem Concerning Quantum-mechanical Hamiltonians

It is next necessary to obtain the Schrödinger wave equation (in operator form) corresponding to the classical Hamiltonian given in Eq. (5), Sec. 11-2. The customary method of obtaining the wave equation from the classical Hamiltonian is not directly applicable to this case without slight extension, because momenta $\mathfrak{M}_x$, $\mathfrak{M}_y$, $\mathfrak{M}_z$ have been used which are not conjugate to any coordinates. Let the classical kinetic energy of a system be given by an expression

$$2T = \sum_{i,j} g_{ij}\dot{q}_i\dot{q}_j \tag{1}$$

in which the q's are generalized coordinates[1] and the g_{ij}'s are coefficients which may be functions of the coordinates. When expressed in terms of the momenta p_i which are conjugate to the coordinates q_i, this becomes

$$2T = \sum_{i,j} g^{ij}p_ip_j \tag{2}$$

in which the coefficients[2] g^{ij} are different from the g_{ij}. Then the wave-mechanical Hamiltonian operator is[3]

$$H' = \tfrac{1}{2}g^{\frac{1}{4}} \sum_{i,j} p_i g^{-\frac{1}{2}} g^{ij} p_j g^{\frac{1}{4}} + V \tag{3}$$

No distinction in notation will be made between the classical Hamiltonian H, the quantum-mechanical Hamiltonian operator H, or, later, the quantum-mechanical matrix H. Likewise, the symbol p_i will be used for the classical momentum, for the quantum operator $(h/2\pi i)(\partial/\partial q_i)$, and for the quantum-mechanical matrix for p_i. Any equation which is correct in terms of the latter two meanings of these symbols is also valid in the classical sense. The converse is not always true, however, because of the ambiguities introduced by the noncommuting nature of operators and matrices. In (3), g is the determinant of the quantities g^{ij}, and V is the potential energy.

[1] The symbol q_i is used here to represent a general coordinate for any system, and is not to be confused with the special use of q_i made elsewhere, especially in Chap. 2.

[2] The g^{ij} are the elements of the matrix which is the inverse of the matrix formed by the g_{ij}.

[3] See for example, E. C. Kemble, "The Fundamental Principles of Quantum Mechanics," Sec. 35*b*, McGraw-Hill, New York, 1937. Note, however, that his g is our $1/g$.

The above Hamiltonian is correct for the case in which the volume element used in normalizing the wave function is $dq_1\, dq_2 \ldots$, that is, no weight factor. In the present application Eulerian angles and normal coordinates will be used, and it is customary to normalize the wave functions with

$$\int \psi^* \psi \sin\theta \, d\theta \, d\phi \, d\chi \, dQ_1 \, dQ_2 \cdots \tag{4}$$

i.e., with respect to ordinary space, so that the volume element weight factor $\sin\theta$ enters. Let the weight factor (here $\sin\theta$) be generally represented by s' so that the wave function ψ' appropriate to the Hamiltonian (3) (in the sense that $H'\psi' = W\psi'$) is related to the function ψ in (4) by

$$\psi' = s'^{\frac{1}{2}}\psi \tag{5}$$

so that the Hamiltonian appropriate to ψ is given by

$$H = \frac{1}{2}\left(\frac{1}{s'}\right)^{\frac{1}{2}} g^{\frac{1}{2}} \sum_{i,j} p_i g^{-\frac{1}{2}} g^{ij} p_j g^{\frac{1}{2}} s'^{\frac{1}{2}} + V \tag{6}$$

Consider now a set of momenta $\mathfrak{M}'_m$, all the members of which are not conjugate to any coordinates but are defined in terms of the original momenta p_i by the linear equations

$$p_i = \sum_m s_{im} \mathfrak{M}'_m \tag{7}$$

In terms of the new momenta the kinetic energy becomes

$$2T = \sum_{m,n} G^{mn} \mathfrak{M}'_m \mathfrak{M}'_n \tag{8}$$

with

$$G^{mn} = \sum_{k,l} s_{km} g^{kl} s_{ln} \tag{9}$$

The question arises as to what conditions must be satisfied in order that the wave-mechanical Hamiltonian operator may be written in the form analogous to (3), that is,

$$H = \tfrac{1}{2} G^{\frac{1}{2}} \sum_{m,n} \mathfrak{M}'_m G^{mn} G^{-\frac{1}{2}} \mathfrak{M}'_n G^{\frac{1}{2}} + V \tag{10}$$

in which G is the determinant $|G^{mn}|$. To obtain the conditions which must be satisfied, substitute in (10) the expression

$$\mathfrak{M}'_m = \sum_i s^{mi} p_i \tag{11}$$

which is the inverse of the transformation (7). The result is that

$$H = \tfrac{1}{2}g^{\frac{1}{4}}s^{\frac{1}{2}} \sum_{ijkm} s^{mi}p_i s_{km} g^{kj} g^{-\frac{1}{2}} s^{-1} p_j g^{\frac{1}{4}} s^{\frac{1}{2}} + V \tag{12}$$

in which s is the determinant $|s_{im}|$ and use has been made of (9) and of the relation $G = s^2 g$, as well as of the equation $\sum_n s_{ln}s^{nj} = \delta_{lj}$. Equation (12) will reduce to (6): this is the condition which would justify the use of (10), provided the following relations hold:

$$s^{\frac{1}{2}} \sum_{im} s^{mi}p_i s_{km} s^{-1} = s^{-\frac{1}{2}} p_k \tag{13}$$

and $s = s'$.

11-4. The Quantum-mechanical Hamiltonian for a Molecule

The general theorem of the previous section will now be applied to the problem under discussion. If the kinetic energy were expressed in terms of the Eulerian angles θ, ϕ, χ and the normal coordinates Q_k, together with the conjugate momenta p_θ, p_ϕ, p_χ and P_k, the Hamiltonian operator would assume the form of Eq. (6), Sec. 11-3. The kinetic energy is, however, not so expressed, but is written in terms of the quantities $\mathfrak{M}_x - \mathfrak{m}_x$, $\mathfrak{M}_y - \mathfrak{m}_y$, $\mathfrak{M}_z - \mathfrak{m}_z$ and P_k. In order to apply Eq. (13), Sec. 11-3, so as to find out whether the Hamiltonian operator has the form, Eq. (10), Sec. 11-3, in terms of these momenta, it is first necessary to obtain the relations between $\mathfrak{M}_x$, $\mathfrak{M}_y$, $\mathfrak{M}_z$, P_k and p_θ, p_ϕ, p_χ and P_k.

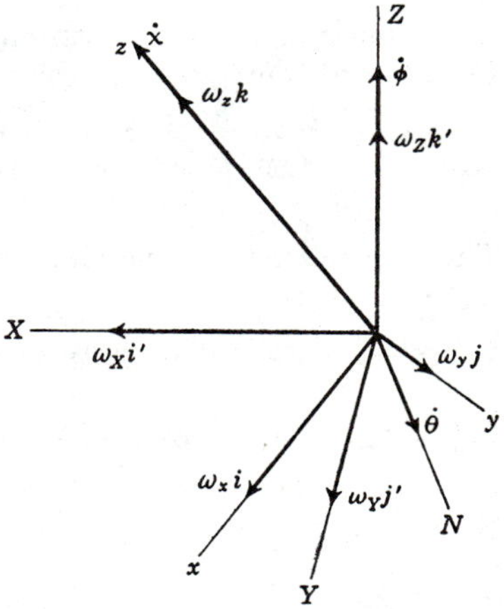

FIG. 11-1. The components of the angular velocity vector.

Figure 11-1 shows the various components of the angular velocity vector in terms of which these relations will be developed. The total angular velocity vector satisfies the following relations:

$$\boldsymbol{\omega} = \dot{\boldsymbol{\theta}} + \dot{\boldsymbol{\phi}} + \dot{\boldsymbol{\chi}} = \omega_x\mathbf{i} + \omega_y\mathbf{j} + \omega_z\mathbf{k} = \omega_X\mathbf{i}' + \omega_Y\mathbf{j}' + \omega_Z\mathbf{k}' \tag{1}$$

where the components are given by relations such as

$$\omega_x = \mathbf{i}\cdot\boldsymbol{\omega} = \mathbf{i}\cdot\dot{\boldsymbol{\theta}} + \mathbf{i}\cdot\dot{\boldsymbol{\phi}} + \mathbf{i}\cdot\dot{\boldsymbol{\chi}}, \text{ etc.} \tag{2}$$

Thus in terms of the amplitudes, $\dot{\theta}$, $\dot{\phi}$, $\dot{\chi}$, and the cosines of the angles between the x, y, z axes and the directions of $\dot{\boldsymbol{\theta}}$, $\dot{\boldsymbol{\phi}}$, $\dot{\boldsymbol{\chi}}$, the components in the rotating axes are readily shown to be

$$\begin{aligned} \omega_x &= \sin\chi \cdot \dot{\theta} - \sin\theta\cos\chi \cdot \dot{\phi} \\ \omega_y &= \cos\chi \cdot \dot{\theta} + \sin\theta\sin\chi \cdot \dot{\phi} \\ \omega_z &= \cos\theta \cdot \dot{\phi} + \dot{\chi} \end{aligned} \tag{3}$$

The transformation which is the inverse of (3) is

$$\begin{aligned} \dot{\theta} &= \sin\chi \cdot \omega_x + \cos\chi \cdot \omega_y \\ \dot{\phi} &= -\csc\theta\cos\chi \cdot \omega_x + \csc\theta\sin\chi \cdot \omega_y \\ \dot{\chi} &= \cot\theta\cos\chi \cdot \omega_x - \cot\theta\sin\chi \cdot \omega_y + \omega_z \end{aligned} \tag{4}$$

The components $\mathfrak{M}_x$, $\mathfrak{M}_y$, $\mathfrak{M}_z$ of the angular momentum were defined by Eq. (3), Sec. 11-2. Comparison with Eq. (14), Sec. 11-1, shows the following equation to be valid:

$$\begin{aligned} \mathfrak{M}_x &= \frac{\partial T}{\partial \omega_x} = \frac{\partial \dot{\theta}}{\partial \omega_x}\frac{\partial T}{\partial \dot{\theta}} + \frac{\partial \dot{\phi}}{\partial \omega_x}\frac{\partial T}{\partial \dot{\phi}} + \frac{\partial \dot{\chi}}{\partial \omega_x}\frac{\partial T}{\partial \dot{\chi}} \\ &= \frac{\partial \dot{\theta}}{\partial \omega_x} p_\theta + \frac{\partial \dot{\phi}}{\partial \omega_x} p_\phi + \frac{\partial \dot{\chi}}{\partial \omega_x} p_\chi \end{aligned} \tag{5}$$

with similar equations for $\mathfrak{M}_y$ and $\mathfrak{M}_z$. The values of $(\partial\dot{\theta}/\partial\omega_x)$, etc., can be obtained from (4), with the result that

$$\begin{aligned} \mathfrak{M}_x &= \sin\chi p_\theta - \csc\theta\cos\chi p_\phi + \cot\theta\cos\chi p_\chi \\ \mathfrak{M}_y &= \cos\chi p_\theta + \csc\theta\sin\chi p_\phi - \cot\theta\sin\chi\, p_\chi \\ \mathfrak{M}_z &= p_\chi \end{aligned} \tag{6}$$

These results may be combined with the definitions [Eq. (6), Sec. 11-2] of $\mathfrak{m}_x$, etc., to give

$$\begin{aligned} \mathfrak{M}'_x &= \mathfrak{M}_x - \mathfrak{m}_x = \sin\chi p_\theta - \csc\theta\cos\chi p_\phi + \cot\theta\cos\chi p_\chi - \sum_k \mathfrak{X}_k P_k \\ \mathfrak{M}'_y &= \mathfrak{M}_y - \mathfrak{m}_y = \cos\chi p_\theta + \csc\theta\sin\chi p_\phi - \cot\theta\sin\chi p_\chi - \sum_k \mathfrak{Y}_k P_k \\ \mathfrak{M}'_z &= \mathfrak{M}_z - \mathfrak{m}_z = p_\chi - \sum_k \mathfrak{Z}_k P_k \\ P_k &= P_k \end{aligned} \tag{7}$$

This is the equivalent of the transformation Eq. (11), Sec. 11-3, of the general theory and therefore provides the coefficients s^{mi}. The inverse equations are readily found to be

$$\begin{aligned} p_\theta &= \sin\chi\mathfrak{M}'_x + \cos\chi\mathfrak{M}'_y + \sum_k (\sin\chi\mathfrak{X}_k + \cos\chi\mathfrak{Y}_k)P_k \\ p_\phi &= -\sin\theta\cos\chi\mathfrak{M}'_x + \sin\theta\sin\chi\mathfrak{M}'_y + \cos\theta\mathfrak{M}'_z \\ &\quad + \sum_k (-\sin\theta\cos\chi\mathfrak{X}_k + \sin\theta\sin\chi\mathfrak{Y}_k + \cos\theta\mathfrak{Z}_k)P_k \\ p_\chi &= \mathfrak{M}'_z + \Sigma\mathfrak{Z}_k P_k \\ P_k &= P_k \end{aligned} \tag{8}$$

thus providing the inverse coefficients s_{im}, whose determinant has the value

$$s = \sin\theta \tag{9}$$

so that the condition $s' = s$ is satisfied. The other part of the condition, Eq. (13), Sec. 11-3, can be rewritten as

$$\sum_{ikm} s^{mi}(p_i s_{km} s^{-1}) = 0$$

where the parentheses are used here to indicate that the differential operator p_i operates only on the functions in the parentheses and not on any which follow. This reduces to

$$\sum_{ikm} s^{mi} s_{km}(p_i s^{-1}) + s^{-1} \sum_{ikm} s^{mi}\,(p_i s_{km}) = \sum_i (p_i s^{-1}) + s^{-1} \sum_{im} s^{mi}\left(p_i \sum_k s_{km}\right) = 0$$

This may now be applied to the case in hand and, with a little manipulation, it will be found that the above condition is in fact satisfied.

Consequently, the quantum-mechanical form of the Hamiltonian is that of Eq. (10), Sec. 11-3, namely,

$$H = \tfrac{1}{2}\mu^{\frac{1}{4}} \sum_{\gamma,\delta} (\mathfrak{M}_\gamma - \mathfrak{m}_\gamma)\mu_{\gamma\delta}\mu^{-\frac{1}{2}}(\mathfrak{M}_\delta - \mathfrak{m}_\delta)\mu^{\frac{1}{4}} + \tfrac{1}{2}\mu^{\frac{1}{4}} \sum_k P_k \mu^{-\frac{1}{2}} P_k \mu^{\frac{1}{4}} + V \tag{10}$$

in which γ, δ denote x, y, or z and μ is the determinant of the coefficients $\mu_{\gamma\delta}$, Eq. (11), Sec. 11-2. It is important to note that (10) is exact; no approximations whatsoever have been introduced in its derivation.[1] The quantities μ and $\mu_{\gamma\delta}$ are functions of the normal coordinates, so that no limit has been placed on the amplitudes of the vibrations. In practical applications, however, it is necessary to introduce approximations.

11-5. Practical Approximations

In many molecules, including most of the simple ones, the potential energy has a deep minimum at the equilibrium position so that classically the atoms will ordinarily vibrate with only small amplitudes about their equilibrium positions. In this case it is a good approximation to neglect the dependence of μ and $\mu_{\gamma\delta}$ on the normal coordinates. If these quantities are constants, they are not operated upon by $\mathfrak{m}_\gamma$ or P_k (they are

[1] Except the molecular model which ignores the electronic structure.

never affected by $\mathfrak{M}_\gamma$), so that the Hamiltonian reduces to the form

$$H = \tfrac{1}{2} \sum_{\gamma,\delta} \mu_{\gamma\delta}(\mathfrak{M}_\gamma - \mathfrak{m}_\gamma)(\mathfrak{M}_\delta - \mathfrak{m}_\delta) + \sum_k P_k^2 + V \tag{1}$$

Furthermore, the x, y, and z axes moving with the molecule can be chosen so as to coincide with the principal axes of inertia of the molecule in the equilibrium configuration, so that I_{xy}, I_{yz}, and I_{zx} vanish.[1] If $\sum_k \mathfrak{X}_k\mathfrak{Y}_k$, etc., are neglected because they depend on the squares of the vibrational displacements, then the coefficients I'_{xy}, I'_{yz}, and I'_{zx} in Eq. (9), Sec. 11-2, vanish and $I'_x \to I_x^0$, $I'_y \to I_y^0$, $I'_z \to I_z^0$, so that these equations are quickly inverted with the result that the inverse coefficients $\mu_{\gamma\delta}$ are $\mu_{xx} = 1/I_x^0$, $\mu_{yy} = 1/I_y^0$, $\mu_{zz} = 1/I_z^0$, and others are zero. The superscript zero indicates the equilibrium values.

The Hamiltonian operator is therefore approximately[2]

$$H = \frac{1}{2}\left[\frac{(\mathfrak{M}_x - \mathfrak{m}_x)^2}{I_x^0} + \frac{(\mathfrak{M}_y - \mathfrak{m}_y)^2}{I_y^0} + \frac{(\mathfrak{M}_z - \mathfrak{m}_z)^2}{I_z^0}\right] + \frac{1}{2}\sum_k P_k^2 + V \tag{2}$$

Here the terms in the brackets represent the rotational energy together with the interaction of the angular momenta of rotation and vibration. The other terms refer to vibration.

In many cases, some or all of the vibrational momenta $\mathfrak{m}_\gamma$ can be ignored. In this event, the energy operator takes on the form

$$H = \frac{1}{2}\left(\frac{\mathfrak{M}_x^2}{I_x^0} + \frac{\mathfrak{M}_y^2}{I_y^0} + \frac{\mathfrak{M}_z^2}{I_z^0}\right) + \frac{1}{2}\sum P_k^2 + V \tag{3}$$

which consists of the energy of a rigid, rotating body plus the vibrational energy of a nonrotating molecule.

[1] The principal values of the moments of inertia I_x, I_y, I_z are the values of λ which satisfy the secular equation

$$\begin{vmatrix} I_{xx} - \lambda & -I_{xy} & -I_{xz} \\ -I_{yx} & I_{yy} - \lambda & -I_{yz} \\ -I_{zx} & -I_{zy} & I_{zz} - \lambda \end{vmatrix} = 0$$

The coefficients of the transformation from arbitrary to principal axes may then be determined by a procedure analogous to that employed in the transformation to normal coordinates as described in Chap. 2.

[2] This expression has been justified in another way by J. H. Van Vleck, *Phys. Rev.*, **47**: 487 (1935).

APPENDIX I

COORDINATE SYSTEMS AND EULERIAN ANGLES

Throughout this book it has been necessary to refer to several cartesian coordinate systems described as follows:

1. Space-fixed, XYZ; the "observer's" axis system.
2. Nonrotating, XYZ; parallel to XYZ but with its origin translating with the molecular center of mass.
3. Rotating, xyz; the molecule-fixed axis system. The conditions specifying how this system is attached to a nonrigid molecule are discussed in Chaps. 2 and 11. For a rigid body, these axes may be taken coincident with the principal axes of inertia.

The translation of the molecule is readily separable as the motion of the molecular center of mass, O, in the XYZ coordinates (see Chap. 2).

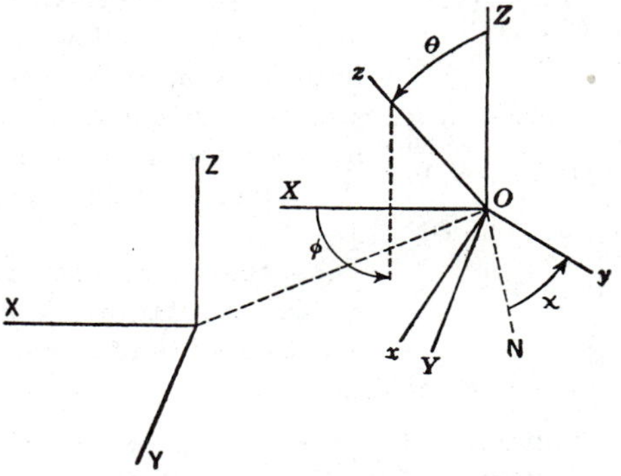

FIG. I-1. Coordinate axes and Eulerian angles.

For further analysis of rotation and vibration, the observer may imagine his coordinates to be the XYZ system, that is, assume the molecular center of mass to be fixed.

Unfortunately, there are numerous ways of specifying the so-called Eulerian angles describing the orientation of the x, y, z axes in the XYZ

TABLE I-1. DIRECTION COSINES RELATING ROTATING TO NONROTATING AXES AS FUNCTIONS OF EULERIAN ANGLES

	X	Y	Z
x	$\cos\theta\cos\phi\cos\chi - \sin\phi\sin\chi$	$\cos\theta\sin\phi\cos\chi + \cos\phi\sin\chi$	$-\sin\theta\cos\chi$
y	$-\cos\theta\cos\phi\sin\chi - \sin\phi\cos\chi$	$-\cos\theta\sin\phi\sin\chi + \cos\phi\cos\chi$	$\sin\theta\sin\chi$
z	$\sin\theta\cos\phi$	$\sin\theta\sin\phi$	$\cos\theta$

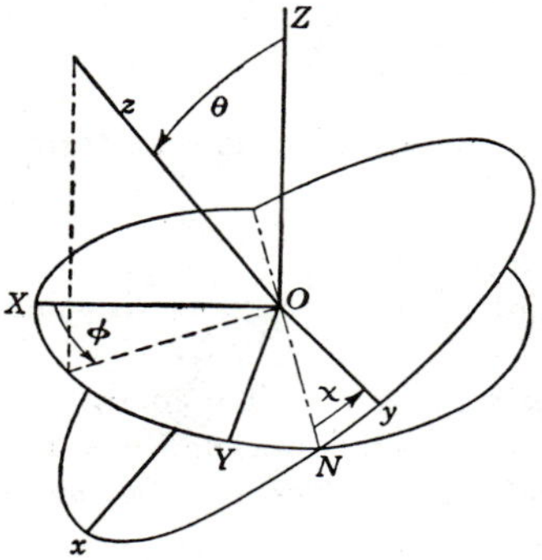

FIG. I-2. Definition of Eulerian angles.

ON = positive direction of line of nodes, the intersection of the XY and xy planes. Also positive sense of a rotation of OZ to Oz.

θ = angle from OZ to Oz $(0 \leq \theta \leq \pi)$.

ϕ = angle in XY plane from OX to the projection of Oz on the XY plane. Also angle from OY to ON $(0 \leq \phi \leq 2\pi)$.

χ = angle in xy plane from ON to Oy. Also angle from projection of $-Z$ on xy plane to Ox $(0 \leq \chi \leq 2\pi)$.

coordinate system. We have chosen the angles θ, ϕ, and χ so that θ and ϕ are the ordinary polar coordinates of the z axis in the XYZ system and χ is an angle in the xy plane measuring the rotation clockwise about the z axis. Figure I-1 shows schematically the above coordinate axes and Eulerian angles, while Fig. I-2 depicts the Eulerian angles here selected in greater detail.

The relations between the coordinates x, y, z and X, Y, Z of a point in space are given by Table I-1. Since the transformation is orthogonal, the table may be read either across or down. Read across it gives the coordinates x, y, z in terms of X, Y, Z; read down it gives X, Y, Z in terms of x, y, z. The entries in the table are, of course, the cosines of the angles between the various pairs of axes, *i.e.*, the direction cosines Φ_{gF} or Φ_{Fg} (Chap. 3).

For the determination of the symmetry species of rotational states, it is convenient to note the following effects of rotation by π about one of the moving axes.

Rotation of π about x:	$\theta \to \pi - \theta$	$\phi \to \pi + \phi$	$\chi \to -\chi$
Rotation of π about y:	$\theta \to \pi - \theta$	$\phi \to \pi + \phi$	$\chi \to \pi - \chi$
Rotation of π about z:	$\theta \to \theta$	$\phi \to \phi$	$\chi \to \pi + \chi$

In general, if the system is rotated by π about an axis OA in the xy plane and making an angle α with x, the effect is

$$\theta \to \pi - \theta \qquad \phi \to \pi + \phi \qquad \chi \to -2\alpha - \chi$$

Note that α is measured clockwise around z.

APPENDIX II

JUSTIFICATION OF THE PROCEDURE USED IN SECTION 2-2 FOR OBTAINING NORMAL VIBRATIONS

The arguments given in Chap. 11 show that the kinetic energy of a molecule which is moving as a whole, rotating, and vibrating can be written approximately in the form

$$T = f(\mathrm{X},\mathrm{Y},\mathrm{Z},\theta,\phi,\chi) + \tfrac{1}{2} \sum_{i,j=1}^{3N-6} t_{ij}\dot{S}_i\dot{S}_j \tag{1}$$

in which X, Y, Z are the coordinates in a space fixed system of axes of the center of mass; θ, ϕ, χ are the Eulerian angles of the rotating system of axes; and the coordinates S_i are some set of independent internal coordinates, $3N - 6$ in number. The function $f(\mathrm{X},\mathrm{Y},\mathrm{Z},\theta,\phi,\chi)$ is, to this approximation, independent of the vibrational coordinate S_i.

Consequently, $3N - 6$ of the Lagrange equations of motion will be of the form

$$\frac{d}{dt}\frac{\partial L_V}{\partial \dot{S}_i} - \frac{\partial L_V}{\partial S_i} = 0 \tag{2}$$

where $L_V = \tfrac{1}{2} \sum_{i,j=1}^{3N-6} t_{ij}\dot{S}_i\dot{S}_j - V$, with V the potential energy (a function of the S's only). The coefficients t_{ij} in the vibrational kinetic energy may be functions of the coordinates S_i.

If the six conditions [Eqs. (2) and (5), Sec. 2-1] on the cartesian coordinates are considered as dummy coordinates which are not involved in L_V, six additional Lagrange equations such as (2) will be satisfied for these dummy coordinates. With these six coordinates included, the transformation from internal coordinates to cartesian displacement coordinates ξ_j (or q_j) is unique and does not differ from ordinary transformations.

By the standard method[1] for showing that the Lagrange form of the equations of motion are the same in all coordinate systems, it follows

[1] See, for example, L. Pauling and E. B. Wilson, Jr., "Introduction to Quantum Mechanics," Secs. 1*b* and 1*c*, McGraw-Hill, New York, 1935.

directly that

$$\frac{d}{dt}\frac{\partial L}{\partial \dot{\xi}_j} - \frac{\partial L}{\partial \xi_j} = 0 \qquad j = 1, 2, \ldots, 3N \tag{3}$$

in which $\xi_1 = \Delta x_1$, $\xi_2 = \Delta y_1$, etc. Consequently, it was legitimate in Sec. 2-2 to use Eq. (3), App. I, even though the coordinates ξ_1, ξ_2, etc., are not independent.

To show that even though the coordinates q_i used in Sec. 2-2 are not all independent, the relations

$$\left(\frac{\partial V}{\partial q_i}\right)_0 = 0 \tag{4}$$

are nevertheless valid, use is made of the fact that

$$\frac{\partial V}{\partial q_i} = \sum_{j=1}^{3N-6} \frac{\partial V}{\partial S_j}\frac{\partial S_j}{\partial q_i} \qquad i = 1, 2, 3, \ldots, 3N \tag{5}$$

At the equilibrium position

$$\frac{\partial V}{\partial S_j} = 0 \tag{6}$$

and there is no uncertainty in this case because all the S's are independent. Insertion of these values in (5) shows that $(\partial V/\partial q_i)$ also is zero when evaluated at the equilibrium position, as previously stated.

APPENDIX III

HERMITE POLYNOMIALS AND SOME INTEGRALS INVOLVING THE HARMONIC OSCILLATOR WAVE FUNCTIONS

The following equations give several of the Hermite polynomials, $H_v(z)$, $z = \gamma^{\frac{1}{2}}Q$, in terms of which the harmonic oscillator wave functions are expressed in Eq. (1), Sec. 3-3. Additional explicit forms may be readily evaluated by use of the recursion formula, Eq. (4), Sec. 3-3.

$$\begin{aligned}
H_0(z) &= 1\\
H_1(z) &= 2z\\
H_2(z) &= 4z^2 - 2\\
H_3(z) &= 8z^3 - 12z\\
H_4(z) &= 16z^4 - 48z^2 + 12\\
H_5(z) &= 32z^5 - 160z^3 + 120z\\
H_6(z) &= 64z^6 - 480z^4 + 720z^2 - 120\\
H_7(z) &= 128z^7 - 1{,}344z^5 + 3{,}360z^3 - 1{,}680z\\
H_8(z) &= 256z^8 - 3{,}584z^6 + 13{,}440z^4 - 13{,}440z^2 + 1{,}680\\
H_9(z) &= 512z^9 - 9{,}216z^7 + 48{,}384z^5 - 80{,}640z^3 + 30{,}240z\\
H_{10}(z) &= 1{,}024z^{10} - 23{,}040z^8 + 161{,}280z^6 - 403{,}200z^4 + 302{,}400z^2\\
&\qquad - 30{,}240
\end{aligned}$$

In the equations below, the symbol $[f(P,Q)]_{v,v'}$ has the significance

$$[f(P,Q)]_{v,v'} = \int_{-\infty}^{+\infty} \psi_v{}^* f(\mathbf{P},\mathbf{Q})\psi_{v'}\, dQ \tag{1}$$

in which $f(P,Q)$ is some function of the coordinate Q and the conjugate momentum P. The boldface type represents the corresponding quantum-mechanical operator in which P has been replaced by

$$\mathbf{P} = \left(\frac{h}{2\pi i}\right)\frac{\partial}{\partial Q}$$

and Q by $\mathbf{Q} = Q$. ψ_v is the wave function for the vth state of the harmonic oscillator. In other words, $[f(P,Q)]_{v,v'}$ is the v,v'th element of the matrix for $f(P,Q)$ for the harmonic oscillator.

$$\gamma = \frac{4\pi^2\nu}{h} \tag{2}$$

$$(Q)_{v,v+1} = \left(\frac{v+1}{2\gamma}\right)^{\frac{1}{2}}$$

$$(Q)_{v,v-1} = \left(\frac{v}{2\gamma}\right)^{\frac{1}{2}}$$

$$(Q)_{v,v'} = 0, \qquad \text{if } v' \neq v \pm 1$$

$$(Q^2)_{v,v+2} = \frac{1}{2\gamma}\,[(v+1)(v+2)]^{\frac{1}{2}}$$

$$(Q^2)_{v,v} = \frac{v+\frac{1}{2}}{\gamma}$$

$$(Q^2)_{v,v-2} = \frac{1}{2\gamma}\,[v(v-1)]^{\frac{1}{2}}$$

$$(Q^2)_{v,v'} = 0 \qquad \text{if } v' \neq v, \text{ or } v \pm 2$$

$$(Q^3)_{v,v+3} = \left[\frac{(v+1)(v+2)(v+3)}{8\gamma^3}\right]^{\frac{1}{2}}$$

$$(Q^3)_{v,v+1} = 3\left[\frac{(v+1)^3}{8\gamma^3}\right]^{\frac{1}{2}}$$

$$(Q^3)_{v,v-1} = 3\left(\frac{v^3}{8\gamma^3}\right)^{\frac{1}{2}}$$

$$(Q^3)_{v,v-3} = \left[\frac{v(v-1)(v-2)}{8\gamma^3}\right]^{\frac{1}{2}}$$

$$(Q^3)_{v,v'} = 0 \qquad \text{if } v' \neq v \pm 1, \text{ or } v \pm 3$$

$$(Q^4)_{v,v+4} = \frac{1}{4\gamma^2}\,[(v+1)(v+2)(v+3)(v+4)]^{\frac{1}{2}}$$

$$(Q^4)_{v,v+2} = \frac{1}{2\gamma^2}\,(2v+3)[(v+1)(v+2)]^{\frac{1}{2}}$$

$$(Q^4)_{v,v} = \frac{3}{4\gamma^2}\,(2v^2+2v+1)$$

$$(Q^4)_{v,v-2} = \frac{1}{2\gamma^2}\,(2v-1)[(v-1)v]^{\frac{1}{2}}$$

$$(Q^4)_{v,v-4} = \frac{1}{4\gamma^2}\,[v(v-1)(v-2)(v-3)]^{\frac{1}{2}}$$

$$(Q^4)_{v,v'} = 0 \qquad \text{if } v' \neq v,\ v \pm 2, \text{ or } v \pm 4$$

$$(P)_{v,v+1} = \frac{h}{2\pi i}\left[\frac{1}{2}\,(v+1)\gamma\right]^{\frac{1}{2}}$$

$$(P)_{v,v-1} = -\frac{h}{2\pi i}\left(\frac{1}{2}\,v\gamma\right)^{\frac{1}{2}}$$

$$(P)_{v,v'} = 0 \qquad \text{if } v' \neq v \pm 1$$

$$(P^2)_{v,v+2} = -\frac{h^2}{8\pi^2}\,[(v+1)(v+2)\gamma^2]^{\frac{1}{2}}$$

$$(P^2)_{v,v} = \frac{h^2}{4\pi^2}(v + \tfrac{1}{2})\gamma$$

$$(P^2)_{v,v-2} = -\frac{h^2}{8\pi^2}[v(v-1)\gamma^2]^{\frac{1}{2}}$$

$$(P^2)_{v,v'} = 0 \qquad \text{if } v' \neq v, \text{ or } v \pm 2$$

$$(QP)_{v,v+2} = \frac{h}{4\pi i}[(v+1)(v+2)]^{\frac{1}{2}}$$

$$(QP)_{v,v} = -\frac{h}{4\pi i}$$

$$(QP)_{v,v-2} = -\frac{h}{4\pi i}[v(v-1)]^{\frac{1}{2}}$$

$$(QP)_{v,v'} = 0 \qquad \text{if } v' \neq v, \text{ or } v \pm 2$$

$$(PQ)_{v,v+2} = \frac{h}{4\pi i}[(v+1)(v+2)]^{\frac{1}{2}}$$

$$(PQ)_{v,v} = \frac{h}{4\pi i}$$

$$(PQ)_{v,v-2} = -\frac{h}{4\pi i}[v(v-1)]^{\frac{1}{2}}$$

$$(PQ)_{v,v'} = 0 \qquad \text{if } v' \neq v, \text{ or } v \pm 2$$

APPENDIX IV

AVERAGES OF DIRECTION COSINES OVER ALL ORIENTATIONS

In Sec. 3-6 the averages $\overline{\Phi_{Fi}^2\Phi_{F'i}^2}$ and $\overline{\Phi_{Fi}\Phi_{F'i}\Phi_{Fj}\Phi_{F'j}}$ were required, where Φ_{Fi} is the cosine of the angle between a nonrotating axis, $F = X, Y, Z$, and a principal axis of polarizability in the rotating molecule, $i = 1, 2, 3$. First consider $F = F'$; using polar coordinates

$$\overline{\Phi_{Fi}^4} = \overline{\cos^4 \theta} = \frac{1}{4\pi}\int_0^\pi \int_0^{2\pi} \cos^4 \theta \sin \theta \, d\theta \, d\phi = \frac{1}{5} \tag{1}$$

The Φ_{Fi} are coefficients in an orthogonal transformation, hence

$$\Phi_{Xi}^2 + \Phi_{Yi}^2 + \Phi_{Zi}^2 = 1 \tag{2}$$

By squaring and averaging (2) and using symmetry,

$$3\overline{\Phi_{Fi}^4} + 6\overline{\Phi_{Fi}^2\Phi_{F'i}^2} = 1 \qquad F \neq F' \tag{3}$$

from which it follows, using (1), that

$$\overline{\Phi_{Fi}^2\Phi_{F'i}^2} = \tfrac{1}{15} \tag{4}$$

The result of (4) is clearly also the value of $\overline{\Phi_{Fi}\Phi_{F'i}\Phi_{Fj}\Phi_{F'j}}$ when $i = j$. To compute $\overline{\Phi_{Fi}\Phi_{F'i}\Phi_{Fj}\Phi_{F'j}}$, $i \neq j$, use again the orthogonality condition, this time in the form

$$\Phi_{Xi}\Phi_{Xj} + \Phi_{Yi}\Phi_{Yj} + \Phi_{Zi}\Phi_{Zj} = 0 \tag{5}$$

Squaring (5), averaging, and using symmetry,

$$3\overline{\Phi_{Fi}^2\Phi_{Fj}^2} + 6\overline{\Phi_{Fi}\Phi_{F'i}\Phi_{Fj}\Phi_{F'j}} = 0 \tag{6}$$

from which, using (4), there is obtained finally

$$\overline{\Phi_{Fi}\Phi_{F'i}\Phi_{Fj}\Phi_{F'j}} = -\tfrac{1}{30} \qquad i \neq j \tag{7}$$

APPENDIX V

A SUMMARY OF MATRIX NOTATION[1]

If the coefficients $G_{tt'}$, in the expression for the kinetic energy of a molecule [Eq. (4), Sec. 4-2], are written in a tabular array as below, the whole set of numbers together can be represented by a single symbol **G**

$$\mathbf{G} = \begin{Vmatrix} G_{11} & G_{12} & \cdots & G_{1n} \\ G_{21} & G_{22} & \cdots & G_{2n} \\ \cdots & \cdots & \cdots & \cdots \\ G_{n1} & G_{n2} & \cdots & G_{nn} \end{Vmatrix} \tag{1}$$

It is to be emphasized that **G** is not a number but a collection of n^2 numbers, if the table has n rows and n columns. **G** is called a matrix and is often written

$$\mathbf{G} = \|G_{tt'}\| = (G_{tt'})$$

The numbers $G_{tt'}$ in the table are called the elements of the matrix.

In this book several other sets of quantities occur which can conveniently be treated in matrix language. Thus the table of force constants f_{ij} for a given molecule, in terms of any coordinate system, can be written in tabular form and the whole table called **F**, the force constant matrix. Similarly, the coefficients of various linear transformations and the quantum-mechanical integrals in Chap. 3 can be treated as elements of matrices.

Two matrices whose corresponding elements are equal are said to be equal; that is,

$$\text{If } A_{ij} = B_{ij} \text{ for all values of } i \text{ and } j,\ \mathbf{A} = \mathbf{B}$$

The *sum* **C** of two matrices **A** and **B** is, by definition, the matrix whose typical element C_{ij} is the sum of the corresponding elements of **A** and **B**;

[1] For more complete discussions of the properties of matrices, see the following:

A. C. Aitken, "Determinants and Matrices," Oliver & Boyd, Edinburgh and London, 1939.

R. A. Frazer, W. J. Duncan, and A. R. Collar, "Elementary Matrices," Cambridge, New York and London, 1938.

H. Margenau and G. M. Murphy, "The Mathematics of Physics and Chemistry," Van Nostrand, New York, 1943.

S. Perlis, "Theory of Matrices," Addison-Wesley, Cambridge, Mass., 1952.

that is,

$$\mathbf{C} = \mathbf{A} + \mathbf{B} \qquad \text{if } C_{ij} = A_{ij} + B_{ij} \tag{2}$$

The matrix all of whose elements are zero is called the *zero* or *null* matrix and is given the symbol **0**.

The product of a number c and a matrix **A** is defined as the matrix **B** whose elements are the elements of **A** multiplied by c; that is,

$$\mathbf{B} = c\mathbf{A} \qquad \text{if } B_{ij} = cA_{ij} \tag{3}$$

The *product* **C** of two matrices **A** and **B** is defined by the equations

$$\mathbf{C} = \mathbf{AB} \qquad \text{if } C_{ij} = \sum_k A_{ik}B_{kj} \tag{4}$$

the sum being over the number of columns of **A** (which must equal the number of rows of **B**). It is not necessarily true that $\mathbf{AB} = \mathbf{BA}$. If **AB** does equal **BA**, **A** and **B** are said to *commute*.

The ordinary matrix multiplication as just defined clearly yields a product whose number of rows is equal to the number of rows of the first matrix and whose number of columns is equal to the number of columns in the second matrix. While this kind of multiplication occurs much more frequently, a second kind of product is occasionally encountered, called the *direct product* and designated by the symbol $\mathbf{A} \times \mathbf{B}$; if **A** is a p-rowed square matrix and **B** is a q-rowed square matrix, then $\mathbf{A} \times \mathbf{B}$ is a square matrix with pq rows and columns, and its elements consist of the products of all possible pairs of elements, one from **A** and one from **B**,

$$(\mathbf{A} \times \mathbf{B})_{ik,jl} = A_{ij}B_{kl}$$

The rows and columns are usually arranged in "dictionary" order, which means that the row labeled ik precedes $i'k'$ if $i < i'$ or if $i = i'$ and $k < k'$.

A *diagonal* matrix is one with zeros everywhere except on the principal diagonal (running from upper left to lower right corner); that is $A_{ij} = 0$ unless $i = j$. The *unit* matrix **E** is a special diagonal matrix whose diagonal elements are each unity. Therefore $\mathbf{E} = ||\delta_{ij}||$. The unit matrix has the property that $\mathbf{EA} = \mathbf{AE} = \mathbf{A}$. A constant matrix, **C**, is a multiple of the unit matrix by a constant; that is $\mathbf{C} = c\mathbf{E}$.

A constant matrix, **C** ($= c\mathbf{E}$ and includes the null and unit matrices as special cases $c = 0$ and $c = 1$, respectively), commutes with any matrix, **A**, since

$$(\mathbf{CA})_{ij} = \sum_k c\delta_{ik}A_{kj} = cA_{ij} \tag{5}$$

and

$$(\mathbf{AC})_{ij} = \sum_k A_{ik}\delta_{kj}c = cA_{ij} \tag{6}$$

Any two diagonal matrices commute, for if

$$A_{ij} = \delta_{ij}A_i = \delta_{ij}A_j \qquad \text{and} \qquad B_{ij} = \delta_{ij}B_i = \delta_{ij}B_j$$

then

$$\sum_k A_{ik}B_{kj} = \sum_k \delta_{ik}\delta_{kj}A_iB_j = \delta_{ij}A_iB_j \tag{7}$$

and

$$\sum_k B_{ik}A_{kj} = \sum_k \delta_{ik}\delta_{kj}B_iA_j = \delta_{ij}B_iA_j \tag{8}$$

which show that **AB** and **BA** are diagonal matrices whose elements are equal to the products of the corresponding diagonal elements of **A** and **B**, and are consequently such that **AB** = **BA**.

The *inverse* (or *reciprocal*) of a matrix **A** is denoted by $\mathbf{A}^{-1}$ and defined by the relation

$$\mathbf{A}^{-1}\mathbf{A} = \mathbf{A}\mathbf{A}^{-1} = \mathbf{E} \tag{9}$$

Not all matrices possess inverses. The *determinant*[1] of a square matrix is the determinant whose elements are equal to the corresponding elements of the matrix. A matrix whose determinant vanishes is said to be *singular* and one whose determinant does not vanish is called a *nonsingular* matrix. Only square nonsingular matrices possess inverses, as can be seen by writing out (9) and attempting to solve for the elements of $\mathbf{A}^{-1}$.

The *transpose* of a matrix **A** is a matrix **A′** obtained from **A** by interchanging rows and columns; *i.e.*,

$$A'_{ij} = A_{ji} \tag{10}$$

The *complex conjugate* of **A** is the matrix **A*** whose elements are the complex conjugates of the corresponding elements of **A**. The conjugate transpose is the matrix **A†** which is the complex conjugate of the transpose of **A**, that is,

$$\mathbf{A}^\dagger = \mathbf{A}'^* \qquad \text{or} \qquad A^\dagger_{ij} = A^*_{ji} \tag{11}$$

Certain restricted classes of matrices have special names, as indicated in Table V-1.

[1] A determinant is a number expressed by an n-dimensional square array of quantities, a_{ij}, the expansion rule for which is

$$\Sigma\,(-1)^P a_{1j_1}a_{2j_2}a_{3j_3} \cdots a_{nj_n}$$

where P is the number of interchanges of pairs necessary to convert the sequence $j_1, j_2, j_3, \ldots, j_n$ into $1, 2, 3, \ldots, n$, and the summation extends over all possible permutations. For example, the three-dimensional determinant expands to

$$a_{11}a_{22}a_{33} + a_{12}a_{23}a_{31} + a_{13}a_{21}a_{32} - a_{13}a_{22}a_{31} - a_{12}a_{21}a_{33} - a_{11}a_{23}a_{32}$$

TABLE V-1

Name	Condition	Condition on elements
Real....................	$\mathbf{A}^* = \mathbf{A}$	Elements are real, $A^*_{ij} = A_{ij}$
Symmetric..............	$\mathbf{A}' = \mathbf{A}$	$A_{ji} = A_{ij}$
Hermitian................	$\mathbf{A}^\dagger = \mathbf{A}$ or $\mathbf{A}^* = \mathbf{A}'$	$A^*_{ji} = A_{ij}$
Orthogonal..............	$\mathbf{A}^{-1} = \mathbf{A}'$ or $\mathbf{AA}' = \mathbf{E}$	$(A^{-1})_{ji} = A_{ij}$
Unitary.................	$\mathbf{A}^{-1} = \mathbf{A}^\dagger$ or $\mathbf{AA}^\dagger = \mathbf{E}$	$(A^{-1})^*_{ji} = A_{ij}$

If **A** is real and *Hermitian* $\mathbf{A} = \mathbf{A}'$ so that *real Hermitian* is equivalent to *symmetric*. Similarly, *real unitary* is equivalent to *orthogonal*. Most discussions in this book deal with real matrices for which the *conjugate transpose* (†) is equivalent to the transpose ($'$) and the notation (†) is generally used although the expansions may be in terms of real elements. Thus, in this text, the notation ($'$) is available for other purposes, and, except in this appendix, does not signify the transpose.

The transpose of a product of several matrices is obtained by inverting the order of the matrices and taking the transpose of each; *i.e.*,

$$(\mathbf{ABC} \cdots)' = \cdots \mathbf{C}'\mathbf{B}'\mathbf{A}' \tag{12}$$

as is readily proved by use of the definition of the matrix product.

An exactly similar rule applies to the inverse of a product of non-singular square matrices; therefore

$$(\mathbf{ABC})^{-1} = \mathbf{C}^{-1}\mathbf{B}^{-1}\mathbf{A}^{-1} \tag{13}$$

as is proved by the result

$$\begin{aligned}(\mathbf{ABC})^{-1}(\mathbf{ABC}) &= \mathbf{C}^{-1}\mathbf{B}^{-1}(\mathbf{A}^{-1}\mathbf{A})\mathbf{BC} \\ &= \mathbf{C}^{-1}\mathbf{B}^{-1}\mathbf{EBC} = \mathbf{C}^{-1}\mathbf{B}^{-1}\mathbf{BC} = \mathbf{C}^{-1}\mathbf{C} = \mathbf{E}\end{aligned}$$

All the above considerations apply particularly to square matrices, but it is often useful to extend the concepts to rectangular arrays of elements. The definitions of *elements, equality, sum, null matrix, product, commutation, transpose, complex conjugate, conjugate transpose*, and *real matrix*, all apply equally well to rectangular matrices. Note that **AB** exists only if the number of columns of **A** equals the number of rows of **B**, whether either one is square or rectangular. Therefore, for rectangular matrices, if **AB** exists, **BA** may or may not exist.

There can be no *diagonal, unit, constant, inverse, symmetric, Hermitian, orthogonal*, or *unitary* rectangular matrices. The theorem on the transpose of a product does apply to rectangular matrices, but the corresponding theorem on the reciprocal of a product is meaningless in the rectangular case.

A special case of a rectangular matrix is the *column* matrix which has only one column of elements. This is often called a *vector*, an extension of this term not limited to three-dimensional cases. Similarly a *row* matrix has only one row. A linear transformation such as

$$\begin{aligned} y_1 &= A_{11}x_1 + A_{12}x_2 + A_{13}x_3 + A_{14}x_4 \\ y_2 &= A_{21}x_1 + A_{22}x_2 + A_{23}x_3 + A_{24}x_4 \\ y_3 &= A_{31}x_1 + A_{32}x_2 + A_{33}x_3 + A_{34}x_4 \\ y_4 &= A_{41}x_1 + A_{42}x_2 + A_{43}x_3 + A_{44}x_4 \end{aligned} \tag{14}$$

can be written in the compact form

$$\mathbf{y} = \mathbf{A}\mathbf{x} \tag{15}$$

in which

$$\mathbf{y} = \begin{Vmatrix} y_1 \\ y_2 \\ y_3 \\ y_4 \end{Vmatrix} \qquad \mathbf{x} = \begin{Vmatrix} x_1 \\ x_2 \\ x_3 \\ x_4 \end{Vmatrix}$$

$$\mathbf{A} = \begin{Vmatrix} A_{11} & A_{12} & A_{13} & A_{14} \\ A_{21} & A_{22} & A_{23} & A_{24} \\ A_{31} & A_{32} & A_{33} & A_{34} \\ A_{41} & A_{42} & A_{43} & A_{44} \end{Vmatrix} \tag{16}$$

and the regular rule for matrix multiplication has been used. The elements of a row matrix are often called *components*.

The transpose of a column matrix is clearly a row matrix. The generalization of an ordinary *scalar product* of three-dimensional vectors is conveniently written as

$$\mathbf{y}'\mathbf{x} = \sum_i y_i x_i \tag{17}$$

if $\mathbf{y}$ and $\mathbf{x}$ are column matrices with the same number of rows. $\mathbf{y}'$ is then a row matrix and the product of $\mathbf{y}'$ and $\mathbf{x}$ is a one-by-one matrix; *i.e.*, a single number. If $\mathbf{A}$ is square and $\mathbf{x}$ a column, then $\mathbf{Ax}$ is also a column; therefore

$$\mathbf{x}'\mathbf{A}\mathbf{x} = \sum_{ij} A_{ij}x_i x_j \tag{18}$$

This is also a one-row and one-column matrix; *i.e.*, a number, and is called a *quadratic form*.

The coefficient of $x_i x_j$ $(i \neq j)$ is $A_{ij} + A_{ji}$ and the value of the form will be unchanged if A is replaced by the symmetric matrix in which $(A_{ij} + A_{ji})/2$ replaces A_{ij} and A_{ji}.

If such a quadratic form is greater than zero for all real $x \neq 0$, $\mathbf{A}$ is said to be *positive definite.* If the form is merely nonnegative for all real $x \neq 0$, $\mathbf{A}$ is said to be *semidefinite.*

If $\mathbf{A}$ is Hermitian, the form $\mathbf{x}^\dagger\mathbf{A}\mathbf{x}$ is called a *Hermitian form,* and is positive or semidefinite under the same conditions as those described above, except that the components of $\mathbf{x}$ may now assume complex values.

Since matrices, square or rectangular, obey all the laws of ordinary algebra except the commutation law of multiplication, the letters representing matrices can be manipulated by all ordinary algebraic techniques except as follows. There is no operation of division, this being replaced by multiplication by a reciprocal, if one exists. The order of products becomes important and must be preserved. Two matrices may not be multiplied unless they match, *i.e.,* the columns of the first must equal in number the rows of the second. Such matrices are said to be *conformable.*

It is often very useful to *partition* matrices, either square or rectangular, into *submatrices,* as indicated by the examples below.

$$\mathbf{A} = \left\| \begin{array}{ccc:cc} A_{11} & A_{12} & A_{13} & A_{14} & A_{15} \\ A_{21} & A_{22} & A_{23} & A_{24} & A_{25} \\ \hdashline A_{31} & A_{32} & A_{33} & A_{34} & A_{35} \end{array} \right\| = \left\| \begin{array}{c:c} \mathbf{A}_{11} & \mathbf{A}_{12} \\ \hdashline \mathbf{A}_{21} & \mathbf{A}_{22} \end{array} \right\| \tag{19}$$

where light-faced letters have been used here for the elements of $\mathbf{A}$ and bold-faced letters with subscripts for the submatrices into which $\mathbf{A}$ has been partitioned by the dashed lines. Thus

$$\begin{aligned} \mathbf{A}_{11} &= \left\| \begin{array}{ccc} A_{11} & A_{12} & A_{13} \\ A_{21} & A_{22} & A_{23} \end{array} \right\| & \mathbf{A}_{12} &= \left\| \begin{array}{cc} A_{14} & A_{15} \\ A_{24} & A_{25} \end{array} \right\| \\ \mathbf{A}_{21} &= \left\| \begin{array}{ccc} A_{31} & A_{32} & A_{33} \end{array} \right\| & \mathbf{A}_{22} &= \left\| \begin{array}{cc} A_{34} & A_{35} \end{array} \right\| \end{aligned} \tag{20}$$

In other words by one or more dashed lines either vertical or horizontal, or both, a matrix may be divided into parts or submatrices which are themselves smaller matrices whose positions in the larger matrix are indicated by subscripts.

If two matrices are partitioned in the same way, their sum can be constructed by adding corresponding submatrices, or

$$\begin{aligned} \mathbf{A} + \mathbf{B} &= \left\| \begin{array}{c:c} \mathbf{A}_{11} & \mathbf{A}_{12} \\ \hdashline \mathbf{A}_{21} & \mathbf{A}_{22} \end{array} \right\| + \left\| \begin{array}{c:c} \mathbf{B}_{11} & \mathbf{B}_{12} \\ \hdashline \mathbf{B}_{21} & \mathbf{B}_{22} \end{array} \right\| \\ &= \left\| \begin{array}{c:c} \mathbf{A}_{11} + \mathbf{B}_{11} & \mathbf{A}_{12} + \mathbf{B}_{12} \\ \hdashline \mathbf{A}_{21} + \mathbf{B}_{21} & \mathbf{A}_{22} + \mathbf{B}_{22} \end{array} \right\| \end{aligned} \tag{21}$$

This follows from the definition of the sum and clearly requires that corresponding submatrices be of the same size.

If the partitioning of the columns of $\mathbf{A}$ matches the partitioning of the rows of $\mathbf{B}$, the product $\mathbf{AB}$ can be expressed in terms of the submatrices:

$$\mathbf{AB} = \left\| \mathbf{A}_{11} \quad \mathbf{A}_{12} \quad \mathbf{A}_{13} \right\| \left\| \begin{matrix} \mathbf{B}_{11} \\ \mathbf{B}_{21} \\ \mathbf{B}_{31} \end{matrix} \right\| = \left\| \mathbf{A}_{11}\mathbf{B}_{11} + \mathbf{A}_{12}\mathbf{B}_{21} + \mathbf{A}_{13}\mathbf{B}_{31} \right\| \tag{22}$$

This result follows if the basic definition of a product is applied.

Partitioning is useful where there is something special about certain rows or columns of a matrix so that these form submatrices with special properties.

The reader is warned that symbols such as $\mathbf{A}_{11}^{-1}$ are ambiguous when submatrices are involved and should be written as $(\mathbf{A}^{-1})_{11}$ or $(\mathbf{A}_{11})^{-1}$, depending on the sense intended.

An important concept is that of the *characteristic values*, or *eigenvalues* of a *square* matrix. Let $\mathbf{A}$ be a square matrix with elements A_{ij}. Then the determinantal equation

$$\begin{vmatrix} A_{11} - \lambda & A_{12} & \cdots & A_{1N} \\ A_{21} & A_{22} - \lambda & \cdots & A_{2N} \\ \cdots & \cdots & \cdots & \cdots \\ A_{N1} & A_{N2} & \cdots & A_{NN} - \lambda \end{vmatrix} = 0 \tag{23}$$

is equivalent to an algebraic equation of degree N in the unknown λ and therefore has N roots, some of which may be repeated. These roots are called the characteristic values of $\mathbf{A}$.

When the algebraic equation is obtained by expansion of the determinantal equation and the coefficients of the various powers of λ are collected, it is found that in

$$(-\lambda)^N + c_1(-\lambda)^{N-1} + c_2(-\lambda)^{N-2} + \cdots + c_{N-1}(-\lambda) + c_N = 0 \tag{24}$$

$$\begin{aligned} c_1 &= \sum_i A_{ii} \\ c_2 &= \sum_{j,i<j} (A_{ii}A_{jj} - A_{ij}A_{ji}) \\ &\cdots\cdots\cdots\cdots \\ c_N &= |\mathbf{A}| \end{aligned} \tag{25}$$

Moreover, in a polynomial equation of the form (24) it is known that

$$\begin{aligned} c_1 &= \sum_i \lambda_i \\ c_2 &= \sum_{j,i<j} \lambda_i\lambda_j \\ &\cdots\cdots\cdots \\ c_N &= \lambda_1\lambda_2 \cdots \lambda_N \end{aligned} \tag{26}$$

Comparison of (25) and (26) shows, for example, that $|\mathbf{A}|$ is equal to the product of the characteristic values of **A**; thus if **A** is nonsingular, no characteristic value can vanish.

The special kinds of matrices listed in the table earlier have special types of characteristic values. Thus Hermitian matrices have only real characteristic values, as proved below. By definition, if **A** is Hermitian

$$A_{ij}^* = A_{ji}$$

The result of taking the complex conjugate of (23) above is therefore equivalent to interchanging rows and columns and replacing λ by λ^*. But a determinant is unaltered by the interchanging rows and columns so the roots λ^* must be the same as the roots λ; that is, they must be real.

Another special class of matrix, the unitary matrix, $\mathbf{A}^{-1} = \mathbf{A}^\dagger$, has characteristic values of absolute value unity, $\lambda^*\lambda = 1$. Consider its secular equation, written in matrix form

$$|\mathbf{A} - \lambda\mathbf{E}| = 0 \tag{27}$$

Since multiplication of determinants and of matrices follows the same rules, multiplication of (27) by $-|\mathbf{A}^{-1}|$ will yield

$$-|\mathbf{A} - \lambda\mathbf{E}|\,|\mathbf{A}^{-1}| = |-\mathbf{A}\mathbf{A}^{-1} + \lambda\mathbf{A}^{-1}| = |-\mathbf{E} + \lambda\mathbf{A}^{-1}| = 0$$

Dividing every element by λ will then give

$$\left|\mathbf{A}^{-1} - \left(\frac{1}{\lambda}\right)\mathbf{E}\right| = 0 \tag{28}$$

provided $\lambda \neq 0$, which must be true if $\mathbf{A}^{-1}$ exists. But replacement of $\mathbf{A}^{-1}$ by $\mathbf{A}^\dagger$ in (28), followed by taking the complex conjugate, will result in

$$\left|\mathbf{A}' - \left(\frac{1}{\lambda^*}\right)\mathbf{E}\right| = 0 \tag{29}$$

Since transposing the rows and columns does not alter the value of a determinant, (29) shows $1/\lambda^*$ is identical with the characteristic value of **A** itself, $1/\lambda^* = \lambda$, or $\lambda^*\lambda = 1$, which was to be proved.

As indicated in Sec. 9-1, the characteristic equation usually arises as the condition that the simultaneous equations involving the components of the vector $\mathbf{x}_i$ as unknowns

$$\mathbf{A}\mathbf{x}_i = \lambda_i\mathbf{x}_i \tag{30}$$

should possess a solution other than $\mathbf{x} = \mathbf{0}$. When N distinct characteristic vectors have been found, they may be compounded in a square matrix, $\mathbf{X} = ||\mathbf{x}_1\mathbf{x}_2 \cdots \mathbf{x}_N||$, and (30) becomes

$$\mathbf{AX} = \mathbf{X\Lambda}$$

or

$$\mathbf{X}^{-1}\mathbf{AX} = \mathbf{\Lambda} \tag{31}$$

where $\mathbf{\Lambda}$ is a diagonal matrix whose elements are the characteristic values $\lambda_1, \lambda_2, \cdots, \lambda_N$ of $\mathbf{A}$. The transformation of $\mathbf{A}$ effected by multiplying it on the right by a matrix and on the left by the inverse of a matrix is called a *similarity transformation*, and when the *similarity transformation* yields the characteristic values, as in (31), $\mathbf{A}$ is said to be diagonalized by $\mathbf{X}$.

If $\mathbf{A}$ is symmetric, the matrix $\mathbf{X}$ which diagonalizes $\mathbf{A} = \mathbf{A}'$ can be orthogonal. To see this, merely take the transpose of (31):

$$(\mathbf{X}^{-1}\mathbf{A}\mathbf{X})' = \mathbf{X}'\mathbf{A}'(\mathbf{X}^{-1})' = \mathbf{X}'\mathbf{A}(\mathbf{X}')^{-1} = \mathbf{\Lambda}' = \mathbf{\Lambda}$$

Thus $(\mathbf{X}')^{-1}$ satisfies the same equation as $\mathbf{X}$ so that $\mathbf{X}' = \mathbf{X}^{-1}$. In a similar fashion, it can be shown that the matrix which diagonalizes either a Hermitian or unitary matrix can be unitary. In the Hermitian case, the proof is exactly parallel, except that conjugate transpose is used in place of transpose. In the unitary case, take the reciprocal conjugate transpose of (31), obtaining

$$\mathbf{X}^{\dagger}\mathbf{A}(\mathbf{X}^{\dagger})^{-1} = (\mathbf{\Lambda}^{\dagger})^{-1} = \mathbf{\Lambda}$$

in which the special property of the characteristic values of unitary matrices, $\lambda^* = \lambda^{-1}$, has been used. Evidently the same equations determine $\mathbf{X}$ and $(\mathbf{X}^{\dagger})^{-1}$ so that $\mathbf{X}$ can therefore be unitary.

If $\mathbf{A}$ is symmetric and positive definite, its characteristic values are real positive. This follows from the fact that

$$\mathbf{X}^{-1}\mathbf{A}\mathbf{X} = \mathbf{X}'\mathbf{A}\mathbf{X} = \mathbf{\Lambda}$$

($\mathbf{A}$ can be diagonalized by an orthogonal matrix) or

$$\lambda_i = \mathbf{x}_i'\mathbf{A}\mathbf{x}_i > 0$$

where $\mathbf{x}_i$ is a column of $\mathbf{X}$.

In exactly similar fashion, if $\mathbf{A}$ is Hermitian and positive definite, its characteristic values are real positive. If $\mathbf{A}$ is only semidefinite, $\lambda \geq 0$.

It may readily be shown that the matrices $\mathbf{A} = \mathbf{B}'\mathbf{B}$ and $\mathbf{C} = \mathbf{D}^{\dagger}\mathbf{D}$ are symmetric positive definite and Hermitian positive definite, respectively, for arbitrary nonsingular $\mathbf{B}$ and $\mathbf{D}$.

It is sometimes important to know the conditions under which several matrices can be simultaneously diagonalized by the same similarity transformation, that is, $\mathbf{X}^{-1}\mathbf{A}_1\mathbf{X} = \mathbf{\Lambda}_1$, $\mathbf{X}^{-1}\mathbf{A}_2\mathbf{X} = \mathbf{\Lambda}_2$, etc. If $\mathbf{A}_1$, $\mathbf{A}_2$, etc., are all symmetric, Hermitian, or unitary, respectively, a necessary and sufficient condition is that they should all commute with one another. The proof will be given for two matrices only and the reader will subsequently see how it may be extended to any number of matrices. The necessity of the condition is proved by noting that $\mathbf{\Lambda}_1$ and $\mathbf{\Lambda}_2$ must certainly commute, since they are both diagonal. But then

$$\begin{aligned} \mathbf{X}^{-1}\mathbf{A}_1\mathbf{X}\mathbf{X}^{-1}\mathbf{A}_2\mathbf{X} &= \mathbf{X}^{-1}\mathbf{A}_1\mathbf{A}_2\mathbf{X} \\ &= \mathbf{X}^{-1}\mathbf{A}_2\mathbf{X}\mathbf{X}^{-1}\mathbf{A}_1\mathbf{X} \\ &= \mathbf{X}^{-1}\mathbf{A}_2\mathbf{A}_1\mathbf{X} \end{aligned}$$

hence it must follow that

$$\mathbf{A}_1\mathbf{A}_2 = \mathbf{A}_2\mathbf{A}_1$$

To prove the sufficiency of the condition, suppose that $\mathbf{X}$ is a matrix which diagonalizes $\mathbf{A}_1$. Let $\mathbf{X}^{-1}\mathbf{A}_2\mathbf{X} = \mathbf{B}_2$; it is easy to show that if $\mathbf{A}_1\mathbf{A}_2 = \mathbf{A}_2\mathbf{A}_1$, then

$$\mathbf{\Lambda}_1\mathbf{B}_2 = \mathbf{B}_2\mathbf{\Lambda}_1$$

Expanding the ij element of this result,

$$(\lambda_1)_{ii}(B_2)_{ij} = (B_2)_{ij}(\lambda_1)_{jj}$$

or

$$[(\lambda_1)_{ii} - (\lambda_1)_{jj}](B_2)_{ij} = 0 \tag{32}$$

When $i \neq j$, either $(\lambda_1)_{ii} = (\lambda_1)_{jj}$ or $(B_2)_{ij} = 0$. Let the order of the rows and columns be so arranged that equal characteristic values of $\mathbf{A}_1$ occur adjacent to one another along the diagonal:

$$\mathbf{\Lambda}_1 = \left\| \begin{matrix} (\lambda_1)_1\mathbf{E}_1 & \mathbf{0} & \cdots & \cdots \\ \mathbf{0} & (\lambda_1)_2\mathbf{E}_2 & \cdots & \cdots \\ \cdots & \cdots & \cdots & \cdots \\ \cdots & \cdots & \cdots & (\lambda_1)_P\mathbf{E}_P \end{matrix} \right\|$$

where $\mathbf{E}_k$ is a unit matrix having a number of rows equal to the number of repetitions of the kth distinct λ. Then if $\mathbf{B}_2$ is partitioned to correspond to $\mathbf{\Lambda}_1$, it will have the form

$$\mathbf{B}_2 = \left\| \begin{matrix} \mathbf{B}_{11} & \mathbf{B}_{12} & \cdots \\ \mathbf{B}_{21} & \mathbf{B}_{22} & \cdots \\ \cdots & \cdots & \mathbf{B}_{PP} \end{matrix} \right\|$$

But then $\mathbf{B}_{ij} = \mathbf{0}$, when $i \neq j$ from (32). The diagonalization of $\mathbf{B}_2$ can then be completed by applying a similarity transformation with

$$\mathbf{Y} = \left\| \begin{matrix} \mathbf{Y}_1 & 0 & \cdots \\ 0 & \mathbf{Y}_2 & \cdots \\ \cdots & \cdots & \mathbf{Y}_P \end{matrix} \right\|$$

such that $\mathbf{Y}_i^{-1}\mathbf{B}_{ii}\mathbf{Y}_i$ is diagonal: such a transformation will not disturb the diagonal form of $\mathbf{\Lambda}_1$ since a similarity transformation on a constant matrix, such as $\mathbf{Y}_i^{-1}(\lambda_1)_i\mathbf{E}\mathbf{Y}_i$, yields the same constant matrix. Thus $\mathbf{Y}^{-1}\mathbf{X}^{-1}\mathbf{A}_1\mathbf{X}\mathbf{Y}$ and $\mathbf{Y}^{-1}\mathbf{X}^{-1}\mathbf{A}_2\mathbf{X}\mathbf{Y}$ will both be diagonal, proving the sufficiency of the commutation condition for two matrices.

The same arguments can be extended in a stepwise fashion to any number of commuting matrices.

APPENDIX VI

A TABULATION OF G MATRIX ELEMENTS

In this appendix, formulas for some of the more frequently used kinetic energy matrix elements will be tabulated. It is evident from Sec. 4-3 that these elements will, in general, depend upon the atomic masses and upon the equilibrium bond lengths and angles of the molecules. Since the masses and bond lengths frequently appear in the denominators, it will be convenient to introduce the symbols μ_α, for the reciprocal of the mass of the αth atom, and $\rho_{\alpha\beta}$ for the reciprocal of the α-β interatomic distance.

In order to systematize the tabulation of formulas, a system of classification and of notation for the different matrix elements must be devised. First, the matrix elements are classified in terms of the general types of coordinates involved. In what follows, only two types of coordinates will appear: bond stretching and valence angle bending.[1] These types have been described in Sec. 4-2.

A kinetic energy matrix element will be given a double subscript to indicate the general types of the two coordinates involved, as G_{rr}, $G_{r\phi}$, $G_{\phi\phi}$. Within each of these three categories, a subclassification is immediately possible on the basis of the number of atoms common to the two coordinates. For instance, $G_{r\phi}^2$ designates a matrix element involving a stretching and bending coordinate in such a way that two atoms are common; since no cyclic[2] configurations are included in this appendix, the symbol $G_{r\phi}^2$ is unique, the common atoms forming one side of the valence angle bending. The symbol $G_{r\phi}^1$, on the other hand, is not unique, since the common atom may be either an end atom or a central atom of the bending coordinate.

Figure VI-1 depicts all the **G** matrix elements for which formulas will subsequently be given. In these diagrams, the atoms common to both

[1] For more extensive tabulations of the **G** matrix elements including bond torsion and out-of-plane displacement, see J. C. Decius, *J. Chem. Phys.*, **16**: 1025 (1948).

For linear cases, see S. M. Ferigle and A. G. Meister, *J. Chem. Phys.*, **19**: 982 (1951).

[2] It can be shown that coordinate sets which are kinematically complete can be selected without involving any cyclic configurations. On the other hand, such sets will not be symmetrically complete in the case of cyclic symmetric molecules.

See J. C. Decius, *J. Chem. Phys.*, **17**: 1315 (1949).

coordinates are indicated by double circles to distinguish them from the noncommon atoms which are indicated by single circles. Furthermore, the common atoms are always put on a horizontal line.

The noncommon atoms are then indicated along 45° diagonals (regardless of the actual geometry of the molecule); those above the common set belong to the first coordinate, those below to the second.

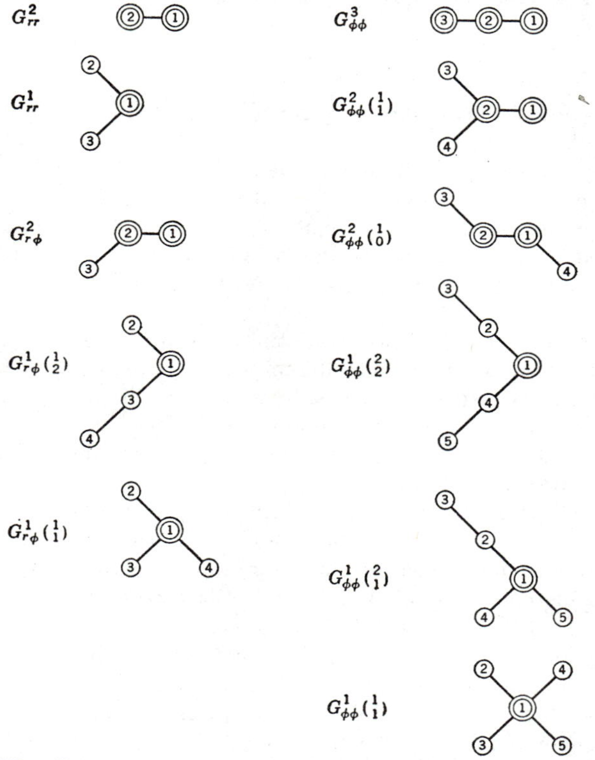

FIG. VI-1. Schematic representation of kinetic energy matrix elements.

This device makes it possible to provide a unique notation for most of the matrix elements. It is merely necessary to add to a symbol such as $G^1_{r\phi}$ a pair of numbers corresponding to the numbers of noncommon atoms along the upper left and lower left diagonals in the diagrams. Thus, when the single common atom is a terminal atom of the bending, the complete symbol is $G^1_{r\phi}\binom{1}{2}$, whereas in the other possible case (common atom a central atom of the bending) the symbol is $G^1_{r\phi}\binom{1}{1}$.

As a further standardization, a uniform atom numbering convention is employed for the coordinates. As indicated in the diagrams, the common atoms are numbered first, running from right to left. The non-common atoms are then numbered, running outwards along the upper left, lower left, upper right, and lower right diagonals in that order.

Table VI-1 gives completely general expressions for elements of the types G_{rr}, $G_{r\phi}$, and $G_{\phi\phi}$. The formulas involve certain angles given the symbol ψ in addition to the angles ϕ and τ which have been previously defined. The ψ's are dihedral angles and are related to the ϕ's according to the formula of spherical trigonometry:

$$\cos \psi_{\alpha\beta\gamma} = \frac{\cos \phi_{\alpha\delta\gamma} - \cos \phi_{\alpha\delta\beta} \cos \phi_{\beta\delta\gamma}}{\sin \phi_{\alpha\delta\beta} \sin \phi_{\beta\delta\gamma}}$$

(See Fig. VI-2.)

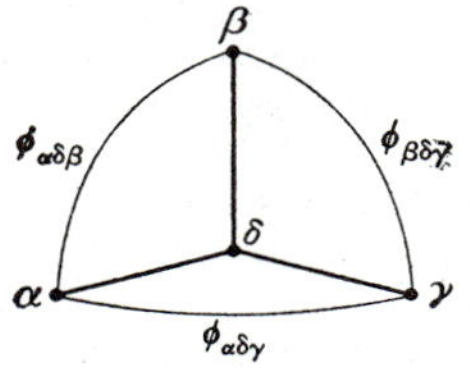

FIG. VI-2. Angles relating to the atomic positions α, β, γ, δ.

In Table VI-2, the values of the matrix elements are given for the special case that the valence bond angles are tetrahedral, 109°28′. In evaluating these elements, certain conventions of sign concerning the τ's must be observed (see Sec. 4-1).

Table VI-3 is similar to Table VI-2, but the valence bond angles have been given the special value of 120°.

TABLE VI-1. GENERAL FORMULAS FOR G_{rr}, $G_{r\phi}$, AND $G_{\phi\phi}$ ELEMENTS
($c\phi = \cos\phi$, $s\phi = \sin\phi$)

G^2_{rr}	$\mu_1 + \mu_2$
G^1_{rr}	$\mu_1 c\phi$
$G^2_{r\phi}$	$-\rho_{23}\mu_2 s\phi$
$G^1_{r\phi}\binom{1}{2}$	$\rho_{13}\mu_1 s\phi_1 c\tau$
$\binom{1}{1}$	$-(\rho_{13}s\phi_{213}c\psi_{234} + \rho_{14}s\phi_{214}c\psi_{243})\mu_1$
$G^3_{\phi\phi}$	$\rho^2_{12}\mu_1 + \rho^2_{23}\mu_3 + (\rho^2_{12} + \rho^2_{23} - 2\rho_{12}\rho_{23}c\phi)\mu_2$
$G^2_{\phi\phi}\binom{1}{1}$	$(\rho^2_{12}c\psi_{314})\mu_1 + [(\rho_{12} - \rho_{23}c\phi_{123} - \rho_{24}c\phi_{124})\rho_{12}c\psi_{314} +$
	$(s\phi_{123}s\phi_{124}s^2\psi_{314} + c\phi_{324}c\psi_{314})\rho_{23}\rho_{24}]\mu_2$
$\binom{1}{0}$	$-\rho_{12}c\tau[(\rho_{12} - \rho_{14}c\phi_1)\mu_1 + (\rho_{12} - \rho_{23}c\phi_2)\mu_2]$
$G^1_{\phi\phi}\binom{2}{2}$	$-(s\tau_{25}s\tau_{34} + c\tau_{25}c\tau_{34}c\phi_1)\rho_{12}\rho_{14}\mu_1$
$\binom{2}{1}$	$[(s\phi_{214}c\phi_{415}c\tau_{34} - s\phi_{215}c\tau_{35})\rho_{14} + (s\phi_{215}c\phi_{415}c\tau_{35} - s\phi_{214}c\tau_{34})\rho_{15}]\dfrac{\rho_{12}\mu_1}{s\phi_{415}}$
$\binom{1}{1}$	$[(c\phi_{415} - c\phi_{314}c\phi_{315} - c\phi_{214}c\phi_{215} + c\phi_{213}c\phi_{214}c\phi_{315})\rho_{12}\rho_{13}$
	$+ (c\phi_{413} - c\phi_{514}c\phi_{513} - c\phi_{214}c\phi_{213} + c\phi_{215}c\phi_{214}c\phi_{513})\rho_{12}\rho_{15}$
	$+ (c\phi_{215} - c\phi_{312}c\phi_{315} - c\phi_{412}c\phi_{415} + c\phi_{413}c\phi_{412}c\phi_{315})\rho_{14}\rho_{13}$
	$+ (c\phi_{213} - c\phi_{512}c\phi_{513} - c\phi_{412}c\phi_{413} + c\phi_{415}c\phi_{412}c\phi_{513})\rho_{14}\rho_{15}]\dfrac{\mu_1}{s\phi_{214}s\phi_{315}}$

TABLE VI-2. SPECIAL FORMULAS FOR G_{rr}, $G_{r\phi}$ AND $G_{\phi\phi}$, ELEMENTS
(All $\phi = 109°28'$)

G^2_{rr}	$\mu_1 + \mu_2$
G^1_{rr}	$-\dfrac{\mu_1}{3}$
$G^2_{r\phi}$	$-\dfrac{2^{\frac{3}{2}}}{3}\rho_{23}\mu_2$
$G^1_{r\phi}\binom{1}{2}$	$\dfrac{2^{\frac{3}{2}}}{3}\rho_{13}\mu_1 c\tau$
$\binom{1}{1}$	$\dfrac{2^{\frac{1}{2}}}{3}(\rho_{13} + \rho_{14})\mu_1$
$G^3_{\phi\phi}$	$\rho^2_{12}\mu_1 + \rho^2_{23}\mu_3 + \frac{1}{3}(3\rho^2_{12} + 3\rho^2_{23} + 2\rho_{12}\rho_{23})\mu_2$
$G^2_{\phi\phi}\binom{1}{1}$	$-\frac{1}{6}\{3\rho^2_{21}\mu_1 + [3\rho^2_{21} + (\rho_{23} + \rho_{24})\rho_{21} - 5\rho_{23}\rho_{24}]\mu_2\}$
$\binom{1}{0}$	$-\frac{1}{3}\rho_{12}c\tau[(3\rho_{12} + \rho_{14})\mu_1 + (3\rho_{12} + \rho_{23})\mu_2]$
$G^1_{\phi\phi}\binom{2}{2}$	$-\frac{1}{3}(3s\tau_{25}s\tau_{34} - c\tau_{25}c\tau_{34})\rho_{12}\rho_{14}\mu_1$
$\binom{2}{1}$	$-\frac{1}{3}[(3c\tau_{35} + c\tau_{34})\rho_{14} + (3c\tau_{34} + c\tau_{35})\rho_{15}]\rho_{12}\mu_1$
$\binom{1}{1}$	$-\frac{2}{3}(\rho_{12} + \rho_{14})(\rho_{13} + \rho_{15})\mu_1$

TABLE VI-3. SPECIAL FORMULAS FOR G_{rr}, $G_{r\phi}$, AND $G_{\phi\phi}$ ELEMENTS
(All $\phi = 120°$)

G^2_{rr}	$\mu_1 + \mu_2$
G^1_{rr}	$-\dfrac{\mu_1}{2}$
$G^2_{r\phi}$	$-\dfrac{3^{\frac{1}{2}}}{2}\rho_{23}\mu_2$
$G^1_{r\phi}\binom{1}{2}$	$\dfrac{3^{\frac{1}{2}}}{2}\rho_{13}\mu_1 c\tau$
$\binom{1}{1}$	$\dfrac{3^{\frac{1}{2}}}{2}(\rho_{13} + \rho_{14})\mu_1$
$G^3_{\phi\phi}$	$\rho^2_{12}\mu_1 + \rho^2_{23}\mu_3 + (\rho^2_{12} + \rho^2_{23} + \rho_{12}\rho_{23})\mu_2$
$G^2_{\phi\phi}\binom{1}{1}$	$-\frac{1}{2}\{2\rho^2_{12}\mu_1 + [\rho_{12}(2\rho_{12} + \rho_{23} + \rho_{24}) - \rho_{23}\rho_{24}]\mu_2\}$
$\binom{1}{0}$	$-\frac{1}{2}\rho_{12}c\tau[(2\rho_{12} + \rho_{14})\mu_1 + (2\rho_{12} + \rho_{23})\mu_2]$
$G^1_{\phi\phi}\binom{2}{2}$	$\frac{1}{2}(c\tau_{25}c\tau_{34} - 2s\tau_{25}s\tau_{34})\rho_{12}\rho_{14}\mu_1$
$\binom{2}{1}$	$\frac{1}{2}c\tau_{34}(\rho_{14} - \rho_{15})\rho_{12}\mu_1$
$\binom{1}{1}$	Element of this type impossible with all $\phi = 120°$

APPENDIX VII

CONNECTION BETWEEN G MATRIX AND KINETIC ENERGY

It is desired to prove that the matrix **G** whose elements are given by

$$G_{tt'} = \sum_j m_j^{-1} B_{tj} B_{t'j}^* \tag{1}$$

gives the kinetic energy in the form

$$2T = \mathbf{P}^\dagger \mathbf{G} \mathbf{P} \tag{2}$$

where **P** is the (column) matrix of the momenta, P_t, conjugate to the coordinates S_t.

In terms of mass-weighted cartesian coordinates, the kinetic energy is given by

$$2T = \dot{\mathbf{q}}^\dagger \dot{\mathbf{q}} \tag{3}$$

But if p_j is the momentum conjugate to q_j,

$$p_j = \frac{\partial T}{\partial \dot{q}_j} = \dot{q}_j \tag{4}$$

so that (3) in terms of the momenta is simply

$$2T = \mathbf{p}^\dagger \mathbf{p} \tag{5}$$

Let **S** be the column matrix of the internal coordinates (**S** may include redundant coordinates) and let the transformation from the mass-weighted cartesian displacement to internal coordinates be given by

$$\mathbf{S} = \mathbf{D}\mathbf{q} \tag{6}$$

Suppose that T is now considered as a function of the velocities in the internal coordinates: using the rules for partial differentiation,

$$p_j = \frac{\partial T}{\partial \dot{q}_j} = \sum_t \frac{\partial T}{\partial \dot{S}_t} \frac{\partial \dot{S}_t}{\partial \dot{q}_j} \tag{7}$$

But $\partial T/\partial \dot{S}_t = P_t$ and $\partial \dot{S}_t/\partial \dot{q}_j = \partial S_t/\partial q_j = D_{tj}$, and (7) becomes in matrix form,

$$\mathbf{p}^\dagger = \mathbf{P}^\dagger\mathbf{D} \tag{8}$$

From (8) it follows also that

$$\mathbf{p} = \mathbf{D}^\dagger\mathbf{P} \tag{9}$$

Substituting (8) and (9) in (5),

$$2T = \mathbf{P}^\dagger(\mathbf{D}\mathbf{D}^\dagger)\mathbf{P} \tag{10}$$

Now $D_{tj} = B_{tj}m_j^{-\frac{1}{2}}$, since $\mathbf{B}$ gives the transformation from the unweighted cartesian coordinates, ξ_i, to the internal coordinates; therefore,

$$\begin{aligned}(\mathbf{D}\mathbf{D}^\dagger)_{tt'} &= \sum_j D_{tj}D^\dagger_{jt'} = \sum_j D_{tj}D^*_{t'j} \\ &= \sum_j m_j^{-1}B_{tj}B^*_{t'j} = G_{tt'}\end{aligned} \tag{11}$$

or $\mathbf{D}\mathbf{D}^\dagger = \mathbf{G}$, which completes the proof.

If $|\mathbf{G}| \neq 0$, $\mathbf{G}^{-1}$ exists, and, by Hamilton's equations, $\dot{S}_t = \partial T/\partial P_t$, use of (2) gives

$$\dot{\mathbf{S}} = \mathbf{G}\mathbf{P} \tag{12}$$

which can be solved for $\mathbf{P}$:

$$\mathbf{P} = \mathbf{G}^{-1}\dot{\mathbf{S}} \tag{13}$$

Appropriate substitution in (2) yields the kinetic energy in terms of the velocities:

$$2T = \dot{\mathbf{S}}^\dagger\mathbf{G}^{-1}\dot{\mathbf{S}} \tag{14}$$

In case $|\mathbf{G}| = 0$, the reader is referred to Sec. 6-8. In any event $\mathbf{G}$ can always be defined as the matrix which gives the kinetic energy directly in terms of the momenta.

APPENDIX VIII

MATRIX TREATMENT OF NORMAL COORDINATE PROBLEM

In terms of the column matrices **S** and $\dot{\mathbf{S}}$ (made up of the internal coordinates and their time derivatives, respectively), the kinetic and potential energies of vibration are (see Sec. 2-2 and Appendixes VII and II)

$$2T = \dot{\mathbf{S}}^\dagger \mathbf{G}^{-1}\dot{\mathbf{S}} \qquad 2V = \mathbf{S}^\dagger \mathbf{F}\mathbf{S} \tag{1}$$

where **G** has been defined by Eq. (1), Sec. 4-3, and **F** is the force constant matrix (for internal coordinates). The normal coordinates (column matrix **Q**) are linearly related to the internal coordinates **S** by a transformation

$$\mathbf{S} = \mathbf{L}\mathbf{Q} \tag{2}$$

in which the (constant) transformation coefficients making up **L** are to be chosen so that the energies in terms of the normal coordinates have the diagonal forms

$$2V = \mathbf{Q}^\dagger \mathbf{L}^\dagger \mathbf{F}\mathbf{L}\mathbf{Q} = \mathbf{Q}^\dagger \mathbf{\Lambda}\mathbf{Q} \tag{3}$$

$$2T = \dot{\mathbf{Q}}^\dagger \mathbf{L}^\dagger \mathbf{G}^{-1}\mathbf{L}\dot{\mathbf{Q}} = \dot{\mathbf{Q}}^\dagger \mathbf{E}\dot{\mathbf{Q}} \tag{4}$$

where $\mathbf{\Lambda}$ is a diagonal matrix (whose diagonal elements will be the quantities $\lambda_k = 4\pi^2\nu_k^2$) while **E** is the unit matrix. Therefore

$$\mathbf{L}^\dagger \mathbf{F}\mathbf{L} = \mathbf{\Lambda} \qquad \text{and} \qquad \mathbf{L}^\dagger \mathbf{G}^{-1}\mathbf{L} = \mathbf{E} \tag{5}$$

or solving the second equation for $\mathbf{L}^\dagger = \mathbf{L}^{-1}\mathbf{G}$ and substituting in the first; then multiplying by **L** on the left,

$$\mathbf{G}\mathbf{F}\mathbf{L} = \mathbf{L}\mathbf{\Lambda} \tag{6}$$

This is a set of simultaneous equations which determine the transformation **L** (except for normalization). Multiplication on both left and right by $\mathbf{L}^{-1}$ gives similarly

$$\mathbf{L}^{-1}\mathbf{G}\mathbf{F} = \mathbf{\Lambda}\mathbf{L}^{-1} \tag{7}$$

whose transpose is

$$\mathbf{F}\mathbf{G}(\mathbf{L}^{-1})^\dagger = (\mathbf{L}^{-1})^\dagger \mathbf{\Lambda} \tag{8}$$

Written out, (6) and (8) become, respectively,

$$\sum_{t'} [(GF)_{tt'} - \delta_{tt'}\lambda_k]L_{t'k} = 0 \qquad k = 1, 2, \ldots, n \tag{9}$$

and

$$\sum_{t'} [(FG)_{tt'} - \delta_{tt'}\lambda_k](L^{-1})_{kt'} = 0 \qquad k = 1, 2, \ldots, n \tag{10}$$

in agreement with Eqs. (10) and (13), Sec. 4-7.

The condition of compatibility of (9) is

$$|\mathbf{GF} - \mathbf{E}\lambda_k| = 0 \tag{11}$$

which is one form of the secular equation [see Eq. (8), Sec. 4-4].

From (5) one obtains a relation between the transformation **L** and its inverse; *i.e.*,

$$\mathbf{L}^{-1} = \mathbf{L}^{\dagger}\mathbf{G}^{-1} \qquad \text{or} \qquad (\mathbf{L}^{-1})^{\dagger} = \mathbf{G}^{-1}\mathbf{L} \tag{12}$$

But Eq. (13), Sec. 4-4, can be written as

$$\mathbf{GA}' = -\mathbf{A}\boldsymbol{\sigma} \tag{13}$$

if **A** is the matrix of the amplitudes A_{tk}, A' is the matrix of the amplitudes A'_{tk}, and $\boldsymbol{\sigma}$ is a diagonal matrix with elements σ_k. Consequently,

$$\mathbf{A}' = -\mathbf{G}^{-1}\mathbf{A}\boldsymbol{\sigma} \tag{14}$$

But $\mathbf{L} = \mathbf{AN}$ where **N** is a diagonal matrix whose elements are the numbers required to normalize the amplitudes A_{tk}.

$$(\mathbf{L}^{-1})^{\dagger} = \mathbf{G}^{-1}\mathbf{L} = \mathbf{G}^{-1}\mathbf{AN} = -\mathbf{A}'\boldsymbol{\sigma}^{-1}\mathbf{N} \tag{15}$$

Consequently, the amplitudes A'_{tk} of Eq. (11), Sec. 4-4, are, except for multiplication by the appropriate constant $(-\sigma_k^{-1}N_k)$, the elements of the (transposed) inverse transformation $(\mathbf{L}^{-1})^{\dagger}$ as stated in Sec. 4-7.

APPENDIX IX

THE SEPARATION OF HIGH AND LOW FREQUENCIES

What is desired in order to prove the correctness of Eq. (2), Sec. 4-8, is the reciprocal of the matrix which results when certain rows and columns of $\mathbf{G}^{-1}$ are omitted. Let $\mathbf{G}^{-1}$ be written in terms of submatrices (Appendix V) $\mathbf{Y}_{11}$, $\mathbf{Y}_{12}$, etc., as follows:

$$\mathbf{G}^{-1} = \begin{Vmatrix} \mathbf{Y}_{11} & \mathbf{Y}_{12} \\ \mathbf{Y}_{21} & \mathbf{Y}_{22} \end{Vmatrix} \tag{1}$$

in which the subscript 1 includes those rows and columns of $\mathbf{G}^{-1}$ which are to be dropped. Let $\mathbf{G}$ be written in terms of submatrices $\mathbf{G}_{11}$, $\mathbf{G}_{12}$, etc., as follows:

$$\mathbf{G} = \begin{Vmatrix} \mathbf{G}_{11} & \mathbf{G}_{12} \\ \mathbf{G}_{21} & \mathbf{G}_{22} \end{Vmatrix} \tag{2}$$

Then since $\mathbf{G}\mathbf{G}^{-1} = \mathbf{E}$,

$$\mathbf{G}_{11}\mathbf{Y}_{12} + \mathbf{G}_{12}\mathbf{Y}_{22} = 0 \tag{3}$$

and

$$\mathbf{G}_{21}\mathbf{Y}_{12} + \mathbf{G}_{22}\mathbf{Y}_{22} = \mathbf{E}_{22} \tag{4}$$

where $\mathbf{E}_{22}$ is the unit matrix of appropriate size.

The quantity sought for is $(\mathbf{Y}_{22})^{-1}$. But (3) and (4) are two equations containing $\mathbf{Y}_{22}$ (and also $\mathbf{Y}_{12}$). Solve (3) for $\mathbf{Y}_{12}$ by multiplying by $(\mathbf{G}_{11})^{-1}$, the result being

$$\mathbf{Y}_{12} = -(\mathbf{G}_{11})^{-1}\mathbf{G}_{12}\mathbf{Y}_{22} \tag{5}$$

Substitute this in (4) and solve for $(\mathbf{Y}_{22})^{-1}$:

$$-\mathbf{G}_{21}(\mathbf{G}_{11})^{-1}\mathbf{G}_{12}\mathbf{Y}_{22} + \mathbf{G}_{22}\mathbf{Y}_{22} = \mathbf{E}_{22} \tag{6}$$

thus $(\mathbf{Y}_{22})^{-1} = \mathbf{G}_{22} - \mathbf{G}_{21}(\mathbf{G}_{11})^{-1}\mathbf{G}_{12}$.
This is equivalent to Eq. (2), Sec. 4-8, if it is realized that $(\mathbf{G}_{11})^{-1}$ is the matrix reciprocal to the *submatrix* $\mathbf{G}_{11}$ of $\mathbf{G}$.

APPENDIX X

SOME PROPERTIES OF GROUP REPRESENTATIONS: CHARACTER TABLES AND CORRELATION TABLES

It is the purpose of this appendix to derive and present in convenient form tables of the characters for the molecular symmetry groups described in Chap. 5. Certain additional information, such as the species (irreducible representations) of the molecular translational and rotational coordinates, and of the dipole moment and the polarizability, will also be indicated. Following the character tables proper, some information on the resolution of direct products will be given. The latter is useful in determining selection rules for overtone and combination transitions (see Chap. 7). Finally, the correlation tables mentioned in Chaps. 6 and 8 will be given.

A complete derivation of all the character tables will not be attempted, but it will be found instructive to derive the tables for the cyclic groups, $\mathcal{C}_n$, and the dihedral groups, $\mathcal{D}_n$, inasmuch as the results can then be extended to obtain the characters of all except the cubic groups, $\mathcal{T}$, $\mathcal{T}_d$, etc.

X-1. Derivation of Characters for the Cyclic Groups, $\mathcal{C}_n$[1]

The character tables of the groups, $\mathcal{C}_n$, which consist solely of the rotations about an n-fold axis, are particularly simple to derive. In such groups, every pair of operations commute:

$$C_n^k C_n^l = C_n^l C_n^k = C_n^{k+l} \tag{1}$$

Now consider the regular representation of such a group; such a representation was defined in Sec. 6-2 as that afforded by a set of coordinates consisting of exactly as many members as there are group operations ($g = d = n$). The matrices of this representation, by definition, must also satisfy (1),

$$\mathbf{C}_n^k \mathbf{C}_n^l = \mathbf{C}_n^l \mathbf{C}_n^k \tag{2}$$

that is, they all commute. But in Appendix V it is proved that any set

[1] The derivations given below are not necessary for use of the tables given later.

of (unitary) matrices which all commute with one another can be simultaneously diagonalized.

Therefore, when the regular representation is completely reduced, each matrix will be diagonal. But in Eq. (6), Sec. 6-2, it was proved that $n^{(\gamma)} = d_\gamma$ for the regular representation, which means that each species appears in the completely reduced form d_γ times. From these facts it then follows that $d_\gamma = 1$ for all species of such a group.

For one-dimensional species, the character of a product of group operations is the arithmetic product of the characters of the group operations

$$\chi_R \chi_{R'} = \chi_{RR'} \tag{3}$$

But since the character of the identity, E, must always be 1 for one-dimensional species, and since $(C_n)^n = E$, it follows that

$$(\chi_{C_n})^n = \chi_E = 1 \tag{4}$$

This last equation means that χ_{C_n} must be an nth root of unity; such numbers, which are in general complex, lie on the unit circle in the complex plane and have the form $e^{i\phi}$. Then, by (4),

$$(e^{i\phi})^n = 1 = e^{2\pi m i} \tag{5}$$

where m is an integer. But (5) implies that $n\phi = m2\pi$ or $\phi = 2\pi m/n$. The characters of the totally symmetric representation are obtained by putting $m = 0$; other possible values of $m = 1, 2, \ldots, n - 1$ yield $n - 1$ further sets of characters. If $m = n$, the totally symmetric representation would be obtained again, so that all possible characters (for the generating element, C_n) are given by $e^{2\pi i m/n}$, where $m = 0, 1, \ldots, n - 1$. Identifying γ with m, one could then express the character of any operation as

$$\chi^\gamma_{C_n{}^k} = (\chi^\gamma_{C_n})^k = e^{2\pi i \gamma k/n} \tag{6}$$

X-2. Character Orthogonality Relations

In order to derive character tables for more complicated groups, it is convenient to develop some general relations which must be satisfied by the characters of any group. From the orthogonality theorem [Eq. (1), App. XI] one such set of relations may be obtained directly. In the orthogonality theorem

$$\sum_R R^{\gamma *}_{ab} R^{\gamma'}_{a'b'} = \frac{\delta_{\gamma\gamma'}\delta_{aa'}\delta_{bb'}g}{d_\gamma} \tag{7}$$

put $a = b$ and $a' = b'$ and sum over a and a', obtaining

$$\sum_R \sum_a R_{aa}^{\gamma *} \sum_{a'} R_{a'a'}^{\gamma'} = \sum_R \chi_R^{\gamma *} \chi_R^{\gamma'}$$

$$= \delta_{\gamma\gamma'} \frac{g}{d_\gamma} \sum_{aa'} (\delta_{aa'})^2$$

$$= \delta_{\gamma\gamma'} \frac{g}{d_\gamma} d_\gamma = \delta_{\gamma\gamma'} g \tag{8}$$

This is an orthogonality relation involving the characters, and may also be written

$$\sum_{j=1}^{c} g_j \chi_j^{\gamma *} \chi_j^{\gamma'} = \delta_{\gamma\gamma'} g \tag{9}$$

where j is an index for a class, g_j is the number of symmetry operations in the jth class, and c is the number of classes.

It is also possible to derive orthogonality relations of the form

$$\sum_\gamma \chi_j^{\gamma *} \chi_{j'}^{\gamma} = \frac{\delta_{jj'} g}{g_j} \tag{10}$$

In order to prove (10) it is necessary to employ the concept of "class multiplication." The class product, $\mathcal{C}_j \mathcal{C}_{j'}$, is defined as the collection of products of all possible pairs of group operations, where one member of the pair belongs to the class $\mathcal{C}_j$, and the other to the class $\mathcal{C}_{j'}$. Such a product is itself a collection of complete classes and may be described symbolically by the equation

$$\mathcal{C}_j \mathcal{C}_{j'} = \sum_{j''} c_{jj'j''} \mathcal{C}_{j''} \tag{11}$$

The proof of this statement follows from the fact that if the left side of (11) is transformed by any group operation, R, the collection of group operations is unchanged (except in order of listing)

$$R \mathcal{C}_j R^{-1} R \mathcal{C}_{j'} R^{-1} = \mathcal{C}_j \mathcal{C}_{j'} \tag{12}$$

by the definition of a class. Since the right side is therefore unchanged for all R, it can only be a sum of complete classes, the coefficient $c_{jj'j''}$ denoting the number of appearances of the j''th class.

Now let $\mathbf{C}_j^\gamma$ be the sum of the matrices of the γth species over the jth class. Such matrix sums must also satisfy (11),

$$\mathbf{C}_j^\gamma \mathbf{C}_{j'}^\gamma = \sum_{j''} c_{jj'j''} \mathbf{C}_{j''}^\gamma \tag{13}$$

These matrix sums are necessarily constant matrices, $\mathbf{C}^\gamma = a_j^\gamma \mathbf{E}$, since they commute with any $\mathbf{R}^\gamma$ of the irreducible representation (Appendix XI). This is seen from

$$\mathbf{C}_j^\gamma = \mathbf{R}^\gamma \mathbf{C}_j^\gamma (\mathbf{R}^\gamma)^{-1}$$

or

$$\mathbf{C}_j^\gamma \mathbf{R}^\gamma = \mathbf{R}^\gamma \mathbf{C}_j^\gamma \tag{14}$$

Consider now the sum of the diagonal terms on each side of the matrix equation, (13). Since

$$\mathbf{C}_j^\gamma \mathbf{C}_{j'}^\gamma = a_j^\gamma a_{j'}^\gamma \mathbf{E} \tag{15}$$

it follows that

$$d_\gamma a_j^\gamma a_{j'}^\gamma = d_\gamma \sum_{j''} c_{jj'j''} a_{j''}^\gamma \tag{16}$$

But

$$d_\gamma a_j^\gamma = g_j \chi_j^\gamma \tag{17}$$

since each of the g_j matrices appearing in the sum $\mathbf{C}_j^\gamma$ has χ_j^γ as its sum of diagonal terms, and substitution of (17) in (16) therefore yields a relation involving the characters which may be written in the form:

$$g_j g_{j'} \chi_j^\gamma \chi_{j'}^\gamma = d_\gamma \sum_{j''} c_{jj'j''} g_{j''} \chi_{j''}^\gamma \tag{18}$$

To obtain the desired orthogonality relation (10), the next step is to sum (18) over γ:

$$\begin{aligned} g_j g_{j'} \sum_\gamma \chi_j^\gamma \chi_{j'}^\gamma &= \sum_{j''} c_{jj'j''} g_{j''} \sum_\gamma d_\gamma \chi_{j''}^\gamma \\ &= \sum_{j''} c_{jj'j''} g_{j''} g \delta_{1j''} \\ &= c_{jj'1} g \end{aligned} \tag{19}$$

in which use is made of the properties of the regular representation whose characters are:

$$\begin{aligned} \chi_E &= g = \sum_\gamma d_\gamma \chi_E^\gamma \\ \chi_j &= 0 = \sum_\gamma d_\gamma \chi_j^\gamma \qquad j \neq 1 \end{aligned} \tag{20}$$

The symbol $c_{jj'1}$ appearing in (19) has the following significance: it is the number of times that $\mathcal{C}_1$ (consisting solely of the identity) appears in the class product, $\mathcal{C}_j\mathcal{C}_{j'}$. Clearly, if $c_{jj'1} \neq 0$, an operation of $\mathcal{C}_j$ must be the inverse of some operation of $\mathcal{C}_{j'}$. It is a consequence of the group properties and the definition of classes that the inverses of the operations of a given class themselves constitute a class having the same number of members; a class and its inverse class may or may not be identical.

From these considerations it follows that $c_{jj'1} = 0$ unless j and j' refer to mutually inverse classes, in which case $c_{jj'1} = g_j = g_{j'}$. Thus it follows from (19) that the expression $\sum_\gamma \chi_j^\gamma \chi_{j'}^\gamma$ vanishes unless j and j' refer to inverse classes, in which case the sum is equal to $g/g_j = g/g_{j'}$. But since the representation matrices are unitary ($\mathbf{R}^{-1} = \mathbf{R}^\dagger$), it follows that the character of the inverse of an operation is equal to the complex conjugate of the character of the operation, so that the results just obtained may be expressed in the form

$$\sum_{\gamma=1}^{s} \chi_j^{\gamma*} \chi_{j'}^\gamma = \frac{\delta_{jj'} g}{g_j}$$

which is the desired result.

The two types of character orthogonality relations expressed by (9) and (10) taken together prove that $c = s$, that is, the number of classes equals the number of distinct species. For if a symbol, x_j^γ, is defined as a sort of "normalized" character by

$$x_j^\gamma = \left(\frac{g_j}{g}\right)^{\frac{1}{2}} \chi_j^\gamma \tag{21}$$

then the orthogonality equations (9) and (10) become

$$\sum_{j=1}^{c} x_j^{\gamma*} x_j^{\gamma'} = \delta_{\gamma\gamma'} \tag{22}$$

and

$$\sum_{\gamma=1}^{s} x_j^{\gamma*} x_{j'}^\gamma = \delta_{jj'} \tag{23}$$

In the first of these equation, $x_1^\gamma, x_2^\gamma, \ldots, x_c^\gamma$ can be regarded as the components of a c-dimensional vector, and since the number of mutually orthogonal vectors cannot exceed the number of components, $s \leq c$ Vice versa, in the second equation, $x_j^1, x_j^2, \ldots, x_j^s$ can be regarded as components, and analogously, $c \leq s$. Combining these facts, $c = s$, as was to be proved.

X-3. Characters of the Dihedral Groups, $\mathfrak{D}_n$

It is now quite feasible to derive the characters for the dihedral groups, which, in addition to a principal n-fold axis, possess n 2-fold axes in the plane perpendicular to the principal axis. In contrast to the cyclic groups, $\mathfrak{C}_n$, in which each operation is in a class by itself, the classes of the dihedral groups may involve more than one operation. The identity constitutes a class by itself (this is true in all groups). The operations C_n^k and its inverse, C_n^{n-k}, constitute a class, as a consequence of the relation

$$C_2 C_n^k = C_n^{n-k} C_2 \tag{24}$$

where C_2 is any one of the rotations about an axis perpendicular to the principal symmetry axis. Furthermore, the twofold axes themselves are either all in the same class, if n is odd, or fall into two classes, each containing $n/2$ members, if n is even.

Thus when n is odd, the classes are E, C_n and C_n^{n-1}, C_n^2 and C_n^{n-2}, . . . , $C_n^{(n-1)/2}$ and $C_n^{(n+1)/2}$; and nC_2. This gives a total number of classes

$$c = \frac{n+3}{2} \tag{25}$$

Furthermore, the order of the group is

$$g = 2n \tag{26}$$

On the other hand, when n is even, the classes are E, C_n and C_n^{n-1}, C_n^2 and C_n^{n-2}, . . . , $C_n^{n/2}C_2$; $(n/2)C_2'$, and $(n/2)C_2''$. The total number of classes is

$$c = \frac{n+6}{2} \tag{27}$$

The characters of the one-dimensional species must satisfy the same multiplication rules as the group operations. In particular, this observation immediately limits the possible characters of C_2 to ± 1. In the case of the cyclic groups, $\mathfrak{C}_n$, it was found that χ_{C_n} was an nth root of unity. This must likewise be true in the dihedral groups, but since C_n and $C_n^{n-1} = C_n^{-1}$ belong to the same class,

$$\chi_{C_n} = \chi_{C_n^{-1}} = \chi_{C_n}^* \tag{28}$$

which means that χ_{C_n} must be a real nth root of unity. Thus if n is even, χ_{C_n} can be ± 1, but if n is odd, χ_{C_n} can only be $+1$.

From these results, it readily follows that there are only two one-dimensional species of $\mathfrak{D}_n$ when n is odd (corresponding to the two choices for the character of C_2), and only four when n is even (by combining the two choices for the character of C_n with the two choices for the character of C_2'). The characters are given in Table X-1 below.

The next step in the derivation of the complete character tables of $\mathfrak{D}_n$ is the proof that the only other species are two-dimensional. Suppose

Table X-1. Characters for the General Dihedral Group

n odd:

	$\mathfrak{D}_n$	E	$2C_n$	$2C_n^2$	. . .	$2C_n^q$	nC_2'
$\alpha = 2\pi/n$	Γ^1	1	1	1	. . .	1	1
$q = (n-1)/2$	Γ^2	1	1	1	. . .	1	-1
	Γ^3	2	$2\cos\alpha$	$2\cos 2\alpha$	. . .	$2\cos q\alpha$	0
	Γ^4	2	$2\cos 2\alpha$	$2\cos 4\alpha$	. . .	$2\cos 2q\alpha$	0
	. . .	. . .	. . .	. . .	. . .	. . .	. . .
	Γ^s	2	$2\cos q\alpha$	$2\cos 2q\alpha$	. . .	$2\cos q^2\alpha$	0

n even:

	$\mathfrak{D}_n$	E	$2C_n$	. . .	$C_n^q = C_2$	$\frac{n}{2}C_2'$	$\frac{n}{2}C_2''$
$\alpha = 2\pi/n$	Γ^1	1	1	. . .	1	1	1
$q = n/2$	Γ^2	1	1	. . .	1	-1	-1
	Γ^3	1	-1	. . .	$(-1)^q$	1	-1
	Γ^4	1	-1	. . .	$(-1)^q$	-1	1
	Γ^5	2	$2\cos\alpha$	. . .	$2\cos q\alpha$	0	0
	Γ^6	2	$2\cos 2\alpha$	. . .	$2\cos 2q\alpha$	0	0
	. . .	. . .	. . .	. . .	. . .	. . .	. . .
	Γ^s	2	$2\cos(q-1)\alpha$	. . .	$2\cos q(q-1)\alpha$	0	0

s_1 is the number of distinct one-dimensional species, s_2 the number of distinct two-dimensional species, etc. Then since the total number of species equals the number of classes, it follows from (25) and (27) that

$$\begin{aligned} s_1 + s_2 + s_3 + s_4 + \cdots &= s = c = \frac{n+3}{2} \qquad n \text{ odd} \\ s_1 + s_2 + s_3 + s_4 + \cdots &= \frac{n+6}{2} \qquad n \text{ even} \end{aligned} \tag{29}$$

Moreover, since $\sum_\gamma d_\gamma^2 = g = 2n$, from (10), an additional equation involving s_1, etc., is obtained, namely,

$$s_1 + 4s_2 + 9s_3 + 16s_4 + \cdots = 2n \tag{30}$$

Substitution of the value $s_1 = 2$ for the case of n odd, followed by elimination of s_2 between (29) and (30), gives

$$5s_3 + 12s_4 + \cdots = 0$$

Since the s_i are nonnegative integers, this proves that there are no species of dimensions greater than two. Moreover,

$$s_2 = \frac{n-1}{2} \tag{31}$$

When n is even, similar arguments show that $s_i = 0$ for $i > 2$, and

$$s_2 = \frac{n-2}{2} \tag{32}$$

It will now be assumed that the two-dimensional representations, in addition to being unitary, are real. Since real unitary matrices are also real orthogonal, and since all real, two-dimensional, orthogonal matrices can be written in the form

$$\begin{Vmatrix} \cos\phi & -\sin\phi \\ \pm\sin\phi & \pm\cos\phi \end{Vmatrix}$$

the characters of the two-dimensional representations will be $2\cos\phi$ or zero.

The characters of the twofold rotations about axes perpendicular to the principal axis are necessarily zero. This assertion follows from the orthogonality relation (10), which is,

$$\begin{aligned} \sum_\gamma \chi_j^{\gamma *}\chi_j^\gamma &= \frac{g}{g_j} = \frac{2n}{n/2} = 4 \qquad n \text{ even} \\ &= \frac{2n}{n} \quad\;\; = 2 \qquad n \text{ odd} \end{aligned} \tag{33}$$

when $j = j'$ is the index for one of these classes. It has already been seen that the characters for these classes are ± 1 in the one-dimensional species, of which there are 4 when n is even and 2 when n is odd. One may therefore conclude that the summation in (33) has a value of zero when extended only over the two-dimensional species, and since each term in such a sum is a nonnegative number, it is necessarily zero.

Finally, the character of C_n remains to be determined. Since the nth power of the matrix representing C_n must be a unit matrix, and since

$$\begin{Vmatrix} \cos\phi & -\sin\phi \\ \sin\phi & \cos\phi \end{Vmatrix}^n = \begin{Vmatrix} \cos n\phi & -\sin n\phi \\ \sin n\phi & \cos n\phi \end{Vmatrix}$$

it follows that

$$n\phi = 2\pi l$$

or

$$\phi = \frac{2\pi l}{n} \tag{34}$$

where l is an integer.

The values of l which give rise to distinct representations are, when n is even, $l = 1, 2, \ldots, (n/2) - 1$, since when $l > n/2$, duplications of characters will occur, and when $l = n/2$, the representation is reducible (diagonal). When n is odd, the values of l are restricted to $l = 1, 2, \ldots, (n-1)/2$.

These results are summarized in the character tables for the general dihedral group (Table X-1).

X-4. Isomorphic and Direct Product Groups

In addition to the cyclic and dihedral groups, there are to be considered groups of the types $\mathfrak{D}_{nh}$, $\mathfrak{C}_{nh}$, $\mathfrak{C}_{nv}$, $\mathfrak{S}_n$, and $\mathfrak{D}_{nd}$. Two additional concepts will make it very easy to obtain the character tables of these groups from those of the cyclic or dihedral groups. First, it may happen that two groups consisting of different geometrical symmetry operations may nevertheless possess identical multiplication tables. As an example, $\mathfrak{D}_3$ and $\mathfrak{C}_{3v}$ will be found to possess identical multiplication tables, the operations σ_v, etc., in $\mathfrak{C}_{3v}$ playing the same role as C_2, etc., in $\mathfrak{D}_3$. Such groups are said to be *isomorphic*. Since the characters are completely determined by the group multiplication tables, ismorphic groups will have identical character tables.

The isomorphism between $\mathfrak{D}_3$ and $\mathfrak{C}_{3v}$ is merely a special case of a general isomorphism between $\mathfrak{D}_n$ and $\mathfrak{C}_{nv}$. Further isomorphisms which are useful in obtaining character tables are the following:

(i) $\mathfrak{D}_{2n}$ and $\mathfrak{D}_{nd}$ when n is even
(ii) $\mathfrak{D}_{2n}$ and $\mathfrak{D}_{nh}$ when n is odd

The second principle which may be used to obtain further character tables is concerned with groups which are expressible as *direct products* of two groups. If one group consists of the operations $E = R_1, R_2, \ldots, R_g$ and the other of the operations $E = R'_1, R'_2, \ldots, R'_{g'}$, then the direct product group is the set of all products of the type $R_iR'_j = R'_jR_i$. Note that it is explicitly assumed that all elements of the first group commute with all elements of the second group.

Suppose that $\mathbf{R}_i$ and $\mathbf{R}'_j$ are representation matrices for the respective group operations. Then the direct product matrix, $\mathbf{R}_i \times \mathbf{R}'_j$ (not to be confused with an ordinary matrix product; see Appendix V), is a representation matrix for the direct product group. The proof that this is true may be given by demonstrating that if the group operations satisfy a relation of the type

$$(R_iR'_j)(R_kR'_l) = R_mR'_n \tag{35}$$

then the direct product matrices satisfy the corresponding matrix equation

$$(\mathbf{R}_i \times \mathbf{R}'_j)(\mathbf{R}_k \times \mathbf{R}'_l) = \mathbf{R}_m \times \mathbf{R}'_n \tag{36}$$

where the symbols enclosed in parentheses are combined by ordinary matrix multiplication.

For the construction of character tables, the following theorem is important:

$$\chi_{R_iR_j'} = \chi_{R_i}\chi_{R_j'} \tag{37}$$

By definition,

$$\chi_{R_i} = \sum_a (R_i)_{aa} \tag{38}$$

The character of the direct product matrix is

$$\begin{aligned}\chi_{R_iR_j'} &= \sum_a \sum_{a'} (R_i)_{aa}(R_j')_{a'a'} = \sum_a (R_i)_{aa} \sum_{a'} (R_j')_{a'a'} \\ &= \chi_{R_i}\chi_{R_j'}\end{aligned} \tag{39}$$

The direct product representation may be either reducible or irreducible but it is always irreducible if either or both of the two representations entering into the direct product are one-dimensional. Thus the construction of a direct product character table is particularly simple when one of the two groups being combined possesses only one-dimensional species. This will indeed be true for all cases which will be considered here, since the direct products involve either $\mathfrak{C}_i = (E,i)$ or $\mathfrak{C}_s = (E,\sigma_h)$ which are isomorphic and have characters 1, 1 and 1, −1.

The groups of the type $\mathfrak{D}_{nh}$ (n even) are all expressible as a direct product of $\mathfrak{D}_n$ with $\mathfrak{C}_i$ which is usually indicated by $\mathfrak{D}_n \times i$, emphasizing the element i which generates the new group. Similarly, $\mathfrak{C}_{nh} = \mathfrak{C}_n \times i$, when n is even.

TABLE X-2. ISOMORPHISMS AND DIRECT PRODUCTS RELATING CERTAIN POINT GROUPS TO $\mathfrak{C}_n$ AND $\mathfrak{D}_n$

	n odd	n even
$\mathfrak{D}_{nh}$	$\mathfrak{D}_{2n}$	$\mathfrak{D}_n \times i$
$\mathfrak{C}_{nh}$	$\mathfrak{C}_n \times \sigma_h$	$\mathfrak{C}_n \times i$
$\mathfrak{C}_{nv}$	$\mathfrak{D}_n$	$\mathfrak{D}_n$
$\mathfrak{S}_{2n}$	$\mathfrak{C}_n \times i$	. . .
$\mathfrak{D}_{nd}$	$\mathfrak{D}_n \times i$	$\mathfrak{D}_{2n}$

On the other hand, when n is odd, $\mathfrak{C}_{nh} = \mathfrak{C}_n \times \sigma_h$, which means that $\mathfrak{C}_{nh}$ is the direct product of $\mathfrak{C}_n$ and $\mathfrak{C}_s$. Finally, the groups $\mathfrak{S}_{2n}$ and $\mathfrak{D}_{nd}$ (n odd) may be expressed as direct products $\mathfrak{S}_{2n} = \mathfrak{C}_n \times i$ and $\mathfrak{D}_{nd} = \mathfrak{D}_n \times i$. These isomorphisms and direct product relationships are summarized in Table X-2.

The derivation of the character tables of the remaining groups, $\mathfrak{T}$, $\mathfrak{T}_d$, $\mathfrak{T}_h$, $\mathfrak{O}$, $\mathfrak{O}_h$, $\mathfrak{I}$, and $\mathfrak{I}_h$, is more complicated and will not be attempted here. The interested reader may refer to the treatise by Littlewood.[1]

Although molecular symmetries are not limited to the cases $n = 1, 2, 3, 4$, and 6 as are crystal symmetries, molecules possessing 7-fold and higher symmetries are presumably rare. For this reason, as a matter of convenience, explicit character tables will be given below only for $n = 2$ through 6 together with the cubic groups and the infinite groups, $\mathfrak{C}_{\infty v}$ and $\mathfrak{D}_{\infty h}$, which express the symmetry of linear molecules.

X-5. Species Notation

It is customary to designate the one-dimensional species by the letters A or B, the two-dimensional species by E, three-dimensional species by F. A and B distinguish the species which are respectively symmetric or antisymmetric with respect to the generating operation C_n in the groups $\mathfrak{D}_n$, or to the corresponding operations in groups isomorphic with $\mathfrak{D}_n$. Subscripts 1 or 2 are used with A or B to designate the species which are symmetric or antisymmetric under one of the twofold rotations about an axis perpendicular to the principal symmetry axis in $\mathfrak{D}_n$, or its analogue, such as a vertical plane, σ_v in a group like $\mathfrak{C}_{nv}$ which is isomorphic with $\mathfrak{D}_n$. Finally, when a group such as $\mathfrak{D}_{nh}$ is constructed by forming a direct product, all the species are doubled. The species which are symmetric under i are given a subscript g, while the antisymmetric species are denoted with a subscript u. If the direct product employs the horizontal plane, σ_h, as the generating element, the species symmetric under σ_h are primed and the antisymmetric species are double primed.

Numerical subscripts are also employed in the case of degenerate species, E, F, etc. The subscripts on E may be identified with the index l of (34); loosely speaking, one might say that E_l represents the transformation properties of coordinates which rotate l times as fast as the cartesian coordinates, x and y, which are always taken in the plane perpendicular to the principal symmetry axis.

Although the cyclic groups have one-dimensional species, the characters are frequently complex numbers. It is sometimes convenient to group the species by pairs where the characters of one member of the pair are the complex conjugates of the corresponding characters of the other member of the pair. Then if these pairs are regarded as doubly degenerate species, the resultant characters (sum of a number and its complex conjugate) will all be real. In the following tables the complex characters of each representation of the pair are given, but the species notation E is used for the pair to emphasize the frequency degeneracy.

[1] D. E. Littlewood, "The Theory of Group Characters," Oxford, New York and London, 1940.

TABLE X-3. CHARACTER TABLES OF $\mathcal{C}_s$, $\mathcal{C}_i$, AND THE CYCLIC GROUPS $\mathcal{C}_n$ ($n = 2, 3, 4, 5, 6$)

$\mathcal{C}_s$	E	σ_h		
A'	1	1	$T_x, T_y; R_z$	$\alpha_{xx}, \alpha_{yy}, \alpha_{zz}, \alpha_{xy}$
A''	1	-1	$T_z; R_x, R_y$	α_{yz}, α_{zx}

$\mathcal{C}_i$	E	i		
A_g	1	1	$\mathbf{R}$	$\boldsymbol{\alpha}$
A_u	1	-1	$\mathbf{T}$	

$\mathcal{C}_2$	E	C_2		
A	1	1	$T_z; R_z$	$\alpha_{xx}, \alpha_{yy}, \alpha_{zz}, \alpha_{xy}$
B	1	-1	$T_x, T_y; R_x, R_y$	α_{yz}, α_{zx}

$\mathcal{C}_3$	E	C_3	C_3^2		$\epsilon = e^{\frac{2\pi i}{3}}$
A	1	1	1	$T_z; R_z$	$\alpha_{xx} + \alpha_{yy}, \alpha_{zz}$
E	1	ϵ	ϵ^*	$(T_x,T_y); (R_x,R_y)$	$(\alpha_{xx} - \alpha_{yy}, \alpha_{xy}), (\alpha_{yz},\alpha_{zx})$
	1	ϵ^*	ϵ		

$\mathcal{C}_4$	E	C_4	C_2	C_4^3		
A	1	1	1	1	$T_z; R_z$	$\alpha_{xx} + \alpha_{yy}, \alpha_{zz}$
B	1	-1	1	-1		$\alpha_{xx} - \alpha_{yy}, \alpha_{xy}$
E	1	i	-1	$-i$	$(T_x,T_y); (R_x,R_y)$	$(\alpha_{yz},\alpha_{zx})$
	1	$-i$	-1	i		

$\mathcal{C}_5$	E	C_5	C_5^2	C_5^3	C_5^4		$\epsilon = e^{\frac{2\pi i}{5}}$
A	1	1	1	1	1	T_z, R_z	$\alpha_{xx} + \alpha_{yy}, \alpha_{zz}$
E_1	1	ϵ	ϵ^2	ϵ^{2*}	ϵ^*	$(T_x,T_y); (R_x,R_y)$	$(\alpha_{yz}, \alpha_{zx})$
	1	ϵ^*	ϵ^{2*}	ϵ^2	ϵ		
E_2	1	ϵ^2	ϵ^*	ϵ	ϵ^{2*}		$(\alpha_{xx} - \alpha_{yy}, \alpha_{xy})$
	1	ϵ^{2*}	ϵ	ϵ^*	ϵ^2		

$\mathcal{C}_6$	E	C_6	C_3	C_2	C_3^2	C_6^5		$\epsilon = e^{\frac{2\pi i}{6}}$
A	1	1	1	1	1	1	$T_z; R_z$	$\alpha_{xx} + \alpha_{yy}, \alpha_{zz}$
B	1	-1	1	-1	1	-1		
E_1	1	ϵ	$-\epsilon^*$	-1	$-\epsilon$	ϵ^*	$(T_x,T_y); (R_x,R_y)$	$(\alpha_{yz}, \alpha_{zx})$
	1	ϵ^*	$-\epsilon$	-1	$-\epsilon^*$	ϵ		
E_2	1	$-\epsilon^*$	$-\epsilon$	1	$-\epsilon^*$	$-\epsilon$		$(\alpha_{xx} - \alpha_{yy}, \alpha_{xy})$
	1	$-\epsilon$	$-\epsilon^*$	1	$-\epsilon$	$-\epsilon^*$		

TABLE X-4. CHARACTER TABLES OF THE DIHEDRAL GROUPS $\mathfrak{D}_n$ (n = 2, 3, 4, 5, 6)

$\mathfrak{D}_2$ ‡ = $\mathfrak{V}$	E	$C_2(z)$	$C_2(y)$	$C_2(x)$		
A	1	1	1	1		$\alpha_{xx}, \alpha_{yy}, \alpha_{zz}$
B_1	1	1	−1	−1	T_z, R_z	α_{xy}
B_2	1	−1	1	−1	T_y, R_y	α_{xz}
B_3	1	−1	−1	1	T_x, R_x	α_{yz}

$\mathfrak{D}_3$	E	$2C_3$	$3C_2$		
A_1	1	1	1		$\alpha_{xx} + \alpha_{yy}, \alpha_{zz}$
A_2	1	1	−1	T_z; R_z	
E	2	−1	0	(T_x,T_y); (R_x,R_y)	$(\alpha_{xx} - \alpha_{yy}, \alpha_{xy})$; $(\alpha_{yz},\alpha_{zx})$

$\mathfrak{D}_4$	E	$2C_4$	$C_4^2 = C_2$	$2C_2'$	$2C_2''$		
A_1	1	1	1	1	1		$\alpha_{xx} + \alpha_{yy}, \alpha_{zz}$
A_2	1	1	1	−1	−1	T_z, R_z	
B_1	1	−1	1	1	−1		$\alpha_{xx} - \alpha_{yy}$
B_2	1	−1	1	−1	1		α_{xy}
E	2	0	−2	0	0	(T_x,T_y); (R_x,R_y)	$(\alpha_{yz},\alpha_{zx})$

$\mathfrak{D}_5$	E	$2C_5$	$2C_5^2$	$5C_2$		
A_1	1	1	1	1		$\alpha_{xx} + \alpha_{yy}, \alpha_{zz}$
A_2	1	1	1	−1	T_z; R_z	
E_1	2	2 cos 72°	2 cos 144°	0	(T_x,T_y); (R_x,R_y)	$(\alpha_{yz},\alpha_{zx})$
E_2	2	2 cos 144°	2 cos 72°	0		$(\alpha_{xx} - \alpha_{yy}, \alpha_{xy})$

$\mathfrak{D}_6$	E	$2C_6$	$2C_3$	C_2	$3C_2'$	$3C_2''$		
A_1	1	1	1	1	1	1		$\alpha_{xx} + \alpha_{yy}, \alpha_{zz}$
A_2	1	1	1	1	−1	−1	T_z; R_z	
B_1	1	−1	1	−1	1	−1		
B_2	1	−1	1	−1	−1	1		
E_1	2	1	−1	−2	0	0	(T_x,T_y); (R_x,R_y)	$(\alpha_{yz},\alpha_{zx})$
E_2	2	−1	−1	2	0	0		$(\alpha_{xx} - \alpha_{yy}, \alpha_{xy})$

‡ Because of the complete equivalence of the three twofold axes, the species notation is modified in this group as compared with the other $\mathfrak{D}_n$

TABLE X-5. CHARACTER TABLES OF THE GROUPS $\mathcal{C}_{nv}$ $(n = 2, 3, 4, 5, 6)$

$\mathcal{C}_{2v}$	E	C_2	$\sigma_v(zx)$	$\sigma_v(yz)$		
A_1	1	1	1	1	T_z	$\alpha_{xx}, \alpha_{yy}, \alpha_{zz}$
A_2	1	1	−1	−1	R_z	α_{xy}
B_1	1	−1	1	−1	$T_x; R_y$	α_{zx}
B_2	1	−1	−1	1	$T_y; R_x$	α_{yz}

$\mathcal{C}_{3v}$	E	$2C_3$	$3\sigma_v$		
A_1	1	1	1	T_z	$\alpha_{xx} + \alpha_{yy}, \alpha_{zz}$
A_2	1	1	−1	R_z	
E	2	−1	0	$(T_x,T_y); (R_x,R_y)$	$(\alpha_{xx} - \alpha_{yy}, \alpha_{xy}); (\alpha_{yz},\alpha_{zx})$

$\mathcal{C}_{4v}$	E	$2C_4$	C_2	$2\sigma_v$	$2\sigma_d$		
A_1	1	1	1	1	1	T_z	$\alpha_{xx} + \alpha_{yy}, \alpha_{zz}$
A_2	1	1	1	−1	−1	R_z	
B_1	1	−1	1	1	−1		$\alpha_{xx} - \alpha_{yy}$
B_2	1	−1	1	−1	1		α_{xy}
E	2	0	−2	0	0	$(T_x,T_y); (R_x,R_y)$	$(\alpha_{yz},\alpha_{zx})$

$\mathcal{C}_{5v}$	E	$2C_5$	$2C_5^2$	$5\sigma_v$		
A_1	1	1	1	1	T_z	$\alpha_{xx} + \alpha_{yy}, \alpha_{zz}$
A_2	1	1	1	−1	R_z	
E_1	2	2 cos 72°	2 cos 144°	0	$(T_x,T_y); (R_x,R_y)$	$(\alpha_{yz},\alpha_{zx})$
E_2	2	2 cos 144°	2 cos 72°	0		$(\alpha_{xx} - \alpha_{yy}, \alpha_{xy})$

$\mathcal{C}_{6v}$	E	$2C_6$	$2C_3$	C_2	$3\sigma_v$	$3\sigma_d$		
A_1	1	1	1	1	1	1	T_z	$\alpha_{xx} + \alpha_{yy}, \alpha_{zz}$
A_2	1	1	1	1	−1	−1	R_z	
B_1	1	−1	1	−1	1	−1		
B_2	1	−1	1	−1	−1	1		
E_1	2	1	−1	−2	0	0	$(T_x,T_y); (R_x,R_y)$	$(\alpha_{yz},\alpha_{zx})$
E_2	2	−1	−1	2	0	0		$(\alpha_{xx} - \alpha_{yy}, \alpha_{xy})$

TABLE X-6. CHARACTER TABLES OF THE GROUPS $\mathcal{C}_{nh}$ ($n = 2, 3, 4, 5, 6$)

$\mathcal{C}_{2h}$	E	C_2	i	σ_h		
A_g	1	1	1	1	R_z	$\alpha_{xx}, \alpha_{yy}, \alpha_{zz}, \alpha_{xy}$
B_g	1	−1	1	−1	R_x, R_y	α_{yz}, α_{zx}
A_u	1	1	−1	−1	T_z	
B_u	1	−1	−1	1	T_x, T_y	

$\mathcal{C}_{3h}$	E	C_3	C_3^2	σ_h	S_3	S_3^5		$\epsilon = e^{\frac{2\pi i}{3}}$
A'	1	1	1	1	1	1	R_z	$\alpha_{xx} + \alpha_{yy}, \alpha_{zz}$
E'	1	ϵ	ϵ^*	1	ϵ	ϵ^*	(T_x, T_y)	$(\alpha_{xx} - \alpha_{yy}, \alpha_{xy})$
	1	ϵ^*	ϵ	1	ϵ^*	ϵ		
A''	1	1	1	−1	−1	−1	T_z	
E''	1	ϵ	ϵ^*	−1	$-\epsilon$	$-\epsilon^*$	(R_x, R_y)	$(\alpha_{yz}, \alpha_{zx})$
	1	ϵ^*	ϵ	−1	$-\epsilon^*$	$-\epsilon$		

$\mathcal{C}_{4h}$	E	C_4	C_2	C_4^3	i	S_4^3	σ_h	S_4		
A_g	1	1	1	1	1	1	1	1	R_z	$\alpha_{xx} + \alpha_{yy}, \alpha_{zz}$
B_g	1	−1	1	−1	1	−1	1	−1		$\alpha_{xx} - \alpha_{yy}, \alpha_{xy}$
E_g	1	i	−1	$-i$	1	i	−1	$-i$	(R_x, R_y)	$(\alpha_{yz}, \alpha_{zx})$
	1	$-i$	−1	i	1	$-i$	−1	i		
A_u	1	1	1	1	−1	−1	−1	−1	T_z	
B_u	1	−1	1	−1	−1	1	−1	1		
E_u	1	i	−1	$-i$	−1	$-i$	1	i	(T_x, T_y)	
	1	$-i$	−1	i	−1	i	1	$-i$		

$\mathcal{C}_{5h}$	E	C_5	C_5^2	C_5^3	C_5^4	σ_h	S_5	S_5^7	S_5^3	S_5^9		$\epsilon = e^{\frac{2\pi i}{5}}$
A'	1	1	1	1	1	1	1	1	1	1	R_z	$\alpha_{xx} + \alpha_{yy}, \alpha_{zz}$
E_1'	1	ϵ	ϵ^2	ϵ^{2*}	ϵ^*	1	ϵ	ϵ^2	ϵ^{2*}	ϵ^*	(T_x, T_y)	
	1	ϵ^*	ϵ^{2*}	ϵ^2	ϵ	1	ϵ^*	ϵ^{2*}	ϵ^2	ϵ		
E_2'	1	ϵ^2	ϵ^*	ϵ	ϵ^{2*}	1	ϵ^2	ϵ^*	ϵ	ϵ^{2*}		$(\alpha_{xx} - \alpha_{yy}, \alpha_{xy})$
	1	ϵ^{2*}	ϵ	ϵ^*	ϵ^2	1	ϵ^{2*}	ϵ	ϵ^*	ϵ^2		
A''	1	1	1	1	1	−1	−1	−1	−1	−1	T_z	
E_1''	1	ϵ	ϵ^2	ϵ^{2*}	ϵ^*	−1	$-\epsilon$	$-\epsilon^2$	$-\epsilon^{2*}$	$-\epsilon^*$	(R_x, R_y)	α_{yz}, α_{zx}
	1	ϵ^*	ϵ^{2*}	ϵ^2	ϵ	−1	$-\epsilon^*$	$-\epsilon^{2*}$	$-\epsilon^2$	$-\epsilon$		
E_2''	1	ϵ^2	ϵ^*	ϵ	ϵ^{2*}	−1	$-\epsilon^2$	$-\epsilon^*$	$-\epsilon$	$-\epsilon^{2*}$		
	1	ϵ^{2*}	ϵ	ϵ^*	ϵ^2	−1	$-\epsilon^{2*}$	$-\epsilon$	$-\epsilon^*$	$-\epsilon^2$		

$\mathcal{C}_{6h}$	E	C_6	C_3	C_2	C_3^2	C_6^5	i	S_3^5	S_6^5	σ_h	S_6	S_3		$\epsilon = e^{\frac{2\pi i}{6}}$
A_g	1	1	1	1	1	1	1	1	1	1	1	1	R_z	$\alpha_{xx} + \alpha_{yy}, \alpha_{zz}$
B_g	1	−1	1	−1	1	−1	1	−1	1	−1	1	−1		
E_{1g}	1	ϵ	$-\epsilon^*$	−1	$-\epsilon$	ϵ^*	1	ϵ	$-\epsilon^*$	−1	$-\epsilon$	ϵ^*	(R_x, R_y)	$(\alpha_{yz}, \alpha_{zx})$
	1	ϵ^*	$-\epsilon$	−1	$-\epsilon^*$	ϵ	1	ϵ^*	$-\epsilon$	−1	$-\epsilon^*$	ϵ		
E_{2g}	1	$-\epsilon^*$	$-\epsilon$	1	$-\epsilon^*$	$-\epsilon$	1	$-\epsilon^*$	$-\epsilon$	1	$-\epsilon^*$	$-\epsilon$		$(\alpha_{xx} - \alpha_{yy}, \alpha_{xy})$
	1	$-\epsilon$	$-\epsilon^*$	1	$-\epsilon$	$-\epsilon^*$	1	$-\epsilon$	$-\epsilon^*$	1	$-\epsilon$	$-\epsilon^*$		
A_u	1	1	1	1	1	1	−1	−1	−1	−1	−1	−1	T_z	
B_u	1	−1	1	−1	1	−1	−1	1	−1	1	−1	1		
E_{1u}	1	ϵ	$-\epsilon^*$	−1	$-\epsilon$	ϵ^*	−1	$-\epsilon$	ϵ^*	1	ϵ	$-\epsilon^*$	(T_x, T_y)	
	1	ϵ^*	$-\epsilon$	−1	$-\epsilon^*$	ϵ	−1	$-\epsilon^*$	ϵ	1	ϵ^*	$-\epsilon$		
E_{2u}	1	$-\epsilon^*$	$-\epsilon$	1	$-\epsilon^*$	$-\epsilon$	−1	ϵ^*	ϵ	−1	ϵ^*	ϵ		
	1	$-\epsilon$	$-\epsilon^*$	1	$-\epsilon$	$-\epsilon^*$	−1	ϵ	ϵ^*	−1	ϵ	ϵ^*		

TABLE X-7. TABLES OF CHARACTERS OF THE GROUPS $\mathfrak{D}_{nh}$ (n = 2, 3, 4, 5, 6)

$\mathfrak{D}_{2h} = \mathfrak{V}_h$	E	$C_2(z)$	$C_2(y)$	$C_2(x)$	i	$\sigma(xy)$	$\sigma(zx)$	$\sigma(yz)$		
A_g	1	1	1	1	1	1	1	1		α_{xx}, α_{yy}, α_{zz}
B_{1g}	1	1	−1	−1	1	1	−1	−1	R_z	α_{xy}
B_{2g}	1	−1	1	−1	1	−1	1	−1	R_y	α_{zx}
B_{3g}	1	−1	−1	1	1	−1	−1	1	R_x	α_{yz}
A_u	1	1	1	1	−1	−1	−1	−1		
B_{1u}	1	1	−1	−1	−1	−1	1	1	T_z	
B_{2u}	1	−1	1	−1	−1	1	−1	1	T_y	
B_{3u}	1	−1	−1	1	−1	1	1	−1	T_x	

$\mathfrak{D}_{3h}$	E	$2C_3$	$3C_2$	σ_h	$2S_3$	$3\sigma_v$		
A_1'	1	1	1	1	1	1		$\alpha_{xx} + \alpha_{yy}$, α_{zz}
A_2'	1	1	−1	1	1	−1	R_z	
E'	2	−1	0	2	−1	0	(T_x, T_y)	$(\alpha_{xx} - \alpha_{yy}, \alpha_{xy})$
A_1''	1	1	1	−1	−1	−1		
A_2''	1	1	−1	−1	−1	1	T_z	
E''	2	−1	0	−2	1	0	(R_x, R_y)	$(\alpha_{yz}, \alpha_{zx})$

$\mathfrak{D}_{4h}$	E	$2C_4$	C_2	$2C_2'$	$2C_2''$	i	$2S_4$	σ_h	$2\sigma_v$	$2\sigma_d$		
A_{1g}	1	1	1	1	1	1	1	1	1	1		$\alpha_{xx} + \alpha_{yy}$, α_{zz}
A_{2g}	1	1	1	−1	−1	1	1	1	−1	−1	R_z	
B_{1g}	1	−1	1	1	−1	1	−1	1	1	−1		$\alpha_{xx} - \alpha_{yy}$
B_{2g}	1	−1	1	−1	1	1	−1	1	−1	1		α_{xy}
E_g	2	0	−2	0	0	2	0	−2	0	0	(R_x, R_y)	$(\alpha_{yz}, \alpha_{zx})$
A_{1u}	1	1	1	1	1	−1	−1	−1	−1	−1		
A_{2u}	1	1	1	−1	−1	−1	−1	−1	1	1	T_z	
B_{1u}	1	−1	1	1	−1	−1	1	−1	−1	1		
B_{2u}	1	−1	1	−1	1	−1	1	−1	1	−1		
E_u	2	0	−2	0	0	−2	0	2	0	0	(T_x, T_y)	

$\mathfrak{D}_{5h}$	E	$2C_5$	$2C_5^2$	$5C_2$	σ_h	$2S_5$	$2S_5^3$	$5\sigma_v$		
A_1'	1	1	1	1	1	1	1	1		$\alpha_{xx} + \alpha_{yy}$, α_{zz}
A_2'	1	1	1	−1	1	1	1	−1	R_z	
E_1'	2	2 cos 72°	2 cos 144°	0	2	2 cos 72°	2 cos 144°	0	(T_x, T_y)	
E_2'	2	2 cos 144°	2 cos 72°	0	2	2 cos 144°	2 cos 72°	0		$(\alpha_{xx} - \alpha_{yy}, \alpha_{xy})$
A_1''	1	1	1	1	−1	−1	−1	−1		
A_2''	1	1	1	−1	−1	−1	−1	1	T_z	
E_1''	2	2 cos 72°	2 cos 144°	0	−2	−2 cos 72°	−2 cos 144°	0	(R_x, R_y)	$(\alpha_{yz}, \alpha_{zx})$
E_2''	2	2 cos 144°	2 cos 72°	0	−2	−2 cos 144°	−2 cos 72°	0		

TABLE X-7. TABLES OF CHARACTERS OF THE GROUPS $\mathfrak{D}_{nh}$ ($n = 2, 3, 4, 5, 6$)
(*Continued*)

$\mathfrak{D}_{6h}$	E	$2C_6$	$2C_3$	C_2	$3C_2'$	$3C_2''$	i	$2S_3$	$2S_6$	σ_h	$3\sigma_d$	$3\sigma_v$		
A_{1g}	1	1	1	1	1	1	1	1	1	1	1	1		$\alpha_{xx} + \alpha_{yy}, \alpha_{zz}$
A_{2g}	1	1	1	1	−1	−1	1	1	1	1	−1	−1	R_z	
B_{1g}	1	−1	1	−1	1	−1	1	−1	1	−1	1	−1		
B_{2g}	1	−1	1	−1	−1	1	1	−1	1	−1	−1	1		
E_{1g}	2	1	−1	−2	0	0	2	1	−1	−2	0	0	(R_x,R_y)	$(\alpha_{yz},\alpha_{zx})$
E_{2g}	2	−1	−1	2	0	0	2	−1	−1	2	0	0		$(\alpha_{xx} - \alpha_{yy}, \alpha_{xy})$
A_{1u}	1	1	1	1	1	1	−1	−1	−1	−1	−1	−1		
A_{2u}	1	1	1	1	−1	−1	−1	−1	−1	−1	1	1	T_z	
B_{1u}	1	−1	1	−1	1	−1	−1	1	−1	1	−1	1		
B_{2u}	1	−1	1	−1	−1	1	−1	1	−1	1	1	−1		
E_{1u}	2	1	−1	−2	0	0	−2	−1	1	2	0	0	(T_x,T_y)	
E_{2u}	2	−1	−1	2	0	0	−2	1	1	−2	0	0		

TABLE X-8. CHARACTER TABLES OF THE GROUPS $\mathfrak{D}_{nd}$ ($n = 2, 3, 4, 5, 6$)

$\mathfrak{D}_{2d} = \mathfrak{V}_d$	E	$2S_4$	C_2	$2C_2'$	$2\sigma_d$		
A_1	1	1	1	1	1		$\alpha_{xx} + \alpha_{yy}, \alpha_{zz}$
A_2	1	1	1	−1	−1	R_z	
B_1	1	−1	1	1	−1		$\alpha_{xx} - \alpha_{yy}$
B_2	1	−1	1	−1	1	T_z	α_{xy}
E	2	0	−2	0	0	(T_x,T_y); (R_x,R_y)	$(\alpha_{yz},\alpha_{zx})$

$\mathfrak{D}_{3d}$	E	$2C_3$	$3C_2$	i	$2S_6$	$3\sigma_d$		
A_{1g}	1	1	1	1	1	1		$\alpha_{xx} + \alpha_{yy}, \alpha_{zz}$
A_{2g}	1	1	−1	1	1	−1	R_z	
E_g	2	−1	0	2	−1	0	(R_x, R_y)	$(\alpha_{xx} - \alpha_{yy}, \alpha_{xy})$, $(\alpha_{yz},\alpha_{zx})$
A_{1u}	1	1	1	−1	−1	−1		
A_{2u}	1	1	−1	−1	−1	1	T_z	
E_u	2	−1	0	−2	1	0	(T_x, T_y)	

$\mathfrak{D}_{4d}$	E	$2S_8$	$2C_4$	$2S_8^3$	C_2	$4C_2'$	$4\sigma_d$		
A_1	1	1	1	1	1	1	1		$\alpha_{xx} + \alpha_{yy}, \alpha_{zz}$
A_2	1	1	1	1	1	−1	−1	R_z	
B_1	1	−1	1	−1	1	1	−1		
B_2	1	−1	1	−1	1	−1	1	T_z	
E_1	2	$\sqrt{2}$	0	$-\sqrt{2}$	−2	0	0	(T_x,T_y)	
E_2	2	0	−2	0	2	0	0		$(\alpha_{xx} - \alpha_{yy}, \alpha_{xy})$
E_3	2	$-\sqrt{2}$	0	$\sqrt{2}$	−2	0	0	(R_x,R_y)	$(\alpha_{yz},\alpha_{zx})$

TABLE X-8. CHARACTER TABLES OF THE GROUPS $\mathfrak{D}_{nd}$ (n = 2, 3, 4, 5, 6)
(*Continued*)

$\mathfrak{D}_{5d}$	E	$2C_5$	$2C_5^2$	$5C_2$	i	$2S_{10}^3$	$2S_{10}$	$5\sigma_d$		
A_{1g}	1	1	1	1	1	1	1	1		$\alpha_{xx} + \alpha_{yy}, \alpha_{zz}$
A_{2g}	1	1	1	−1	1	1	1	−1	R_z	
E_{1g}	2	2 cos 72°	2 cos 144°	0	2	2 cos 72°	2 cos 144°	0	(R_x, R_y)	$(\alpha_{yz}, \alpha_{zx})$
E_{2g}	2	2 cos 144°	2 cos 72°	0	2	2 cos 144°	2 cos 72°	0		$(\alpha_{xx} - \alpha_{yy}, \alpha_{xy})$
A_{1u}	1	1	1	1	−1	−1	−1	−1		
A_{2u}	1	1	1	−1	−1	−1	−1	1	T_z	
E_{1u}	2	2 cos 72°	2 cos 144°	0	−2	−2 cos 72°	−2 cos 144°	0	(T_x, T_y)	
E_{2u}	2	2 cos 144°	2 cos 72°	0	−2	−2 cos 144°	−2 cos 72°	0		

$\mathfrak{D}_{6d}$	E	$2S_{12}$	$2C_6$	$2S_4$	$2C_3$	$2S_{12}^5$	C_2	$6C_2'$	$6\sigma_d$		
A_1	1	1	1	1	1	1	1	1	1		$\alpha_{xx} + \alpha_{yy}, \alpha_{zz}$
A_2	1	1	1	1	1	1	1	−1	−1	R_z	
B_1	1	−1	1	−1	1	−1	1	1	−1		
B_2	1	−1	1	−1	1	−1	1	−1	1	T_z	
E_1	2	$\sqrt{3}$	1	0	−1	$-\sqrt{3}$	−2	0	0	(T_x, T_y)	
E_2	2	1	−1	−2	−1	1	2	0	0		$(\alpha_{xx} - \alpha_{yy}, \alpha_{xy})$
E_3	2	0	−2	0	2	0	−2	0	0		
E_4	2	−1	−1	2	−1	−1	2	0	0		
E_5	2	$-\sqrt{3}$	1	0	−1	$\sqrt{3}$	−2	0	0	(R_x, R_y)	$(\alpha_{yz}, \alpha_{zx})$

TABLE X-9. CHARACTER TABLES OF THE GROUPS S_n (n = 4, 6, 8)

S_4	E	S_4	C_2	S_4^3		
A	1	1	1	1	R_z	$\alpha_{xx} + \alpha_{yy}, \alpha_{zz}$
B	1	−1	1	−1	T_z	$\alpha_{xx} - \alpha_{yy}, \alpha_{xy}$
E	{1	i	−1	$-i$}	(T_x, T_y); (R_x, R_y)	$(\alpha_{yz}, \alpha_{zx})$
	{1	$-i$	−1	i}		

S_6	E	C_3	C_3^2	i	S_6^5	S_6		$\mathrm{S}_6 = \mathrm{C}_3 \times i \qquad \epsilon = e^{\frac{2\pi i}{3}}$
A_g	1	1	1	1	1	1	R_z	$\alpha_{xx} + \alpha_{yy}, \alpha_{zz}$
E_g	{1	ϵ	ϵ^*	1	ϵ	ϵ^*}	(R_x, R_y)	$(\alpha_{xx} - \alpha_{yy}, \alpha_{xy})$; $(\alpha_{yz}, \alpha_{zx})$
	{1	ϵ^*	ϵ	1	ϵ^*	ϵ}		
A_u	1	1	1	−1	−1	−1	T_z	
E_u	{1	ϵ	ϵ^*	−1	$-\epsilon$	$-\epsilon^*$}	(T_x, T_y)	
	{1	ϵ^*	ϵ	−1	$-\epsilon^*$	$-\epsilon$}		

S_8	E	S_8	C_4	S_8^3	C_2	S_8^5	C_4^3	S_8^7		$\epsilon = e^{\frac{2\pi i}{8}}$
A	1	1	1	1	1	1	1	1	R_z	$\alpha_{xx} + \alpha_{yy}, \alpha_{zz}$
B	1	−1	1	−1	1	−1	1	−1	T_z	
E_1	{1	ϵ	i	$-\epsilon^*$	−1	$-\epsilon$	$-i$	ϵ^*}	(T_x, T_y); (R_x, R_y)	
	{1	ϵ^*	$-i$	$-\epsilon$	−1	$-\epsilon^*$	i	ϵ}		
E_2	{1	i	−1	$-i$	1	i	−1	$-i$}		$(\alpha_{xx} - \alpha_{yy}, \alpha_{xy})$
	{1	$-i$	−1	i	1	$-i$	−1	i}		
E_3	{1	$-\epsilon^*$	$-i$	ϵ	−1	ϵ^*	i	$-\epsilon$}		$(\alpha_{yz}, \alpha_{zx})$
	{1	$-\epsilon$	i	ϵ^*	−1	ϵ	$-i$	$-\epsilon^*$}		

Table X-10. Character Tables of the Cubic Groups, $\mathfrak{T}$, $\mathfrak{T}_h$, $\mathfrak{T}_d$, $\mathfrak{O}$, $\mathfrak{O}_h$, $\mathfrak{I}$, $\mathfrak{I}_h$

$\mathfrak{T}$	E	$4C_3$	$4C_3^2$	$3C_2$		$\epsilon = e^{\frac{2\pi i}{3}}$
A	1	1	1	1		$\alpha_{xx} + \alpha_{yy} + \alpha_{zz}$
E	$\begin{cases} 1 \\ 1 \end{cases}$	ϵ ϵ^*	ϵ^* ϵ	$\begin{matrix} 1 \\ 1 \end{matrix}\Big\}$		$(\alpha_{xx} + \alpha_{yy} - 2\alpha_{zz},\ \alpha_{xx} - \alpha_{yy})$
F	3	0	0	-1	**T, R**	$(\alpha_{xy}, \alpha_{yz}, \alpha_{zx})$

$\mathfrak{T}_h = \mathfrak{T} \times i$; **T** in F_u; **R** in F_g; α in A_g, E_g, F_g

$\mathfrak{T}_d$	E	$8C_3$	$3C_2$	$6S_4$	$6\sigma_d$		
$\mathfrak{O}$	E	$8C_3$	$3C_2$	$6C_4$	$6C_2'$		
A_1	1	1	1	1	1		$\alpha_{xx} + \alpha_{yy} + \alpha_{zz}$
A_2	1	1	1	-1	-1		
E	2	-1	2	0	0		$(\alpha_{xx} + \alpha_{yy} - 2\alpha_{zz},\ \alpha_{xx} - \alpha_{yy})$
F_1	3	0	-1	1	-1	**R**; **T** in $\mathfrak{O}$	
F_2	3	0	-1	-1	1	**T** in $\mathfrak{T}_d$	$(\alpha_{xy}, \alpha_{yz}, \alpha_{zx})$

$\mathfrak{O}_h = \mathfrak{O} \times i$; **T** in F_{1u}; **R** in F_{1g}; α in A_{1g}, E_g, F_{2g}

$\mathfrak{I}$	E	$12C_5$	$12C_5^2$	$20C_3$	$15C_2$		
A	1	1	1	1	1		$\alpha_{xx} + \alpha_{yy} + \alpha_{zz}$
F_1	3	$\frac{1+\sqrt{5}}{2}$	$\frac{1-\sqrt{5}}{2}$	0	-1	**T, R**	
F_2	3	$\frac{1-\sqrt{5}}{2}$	$\frac{1+\sqrt{5}}{2}$	0	-1		
G	4	-1	-1	1	0		
H	5	0	0	-1	1		$(\alpha_{xx} + \alpha_{yy} - 2\alpha_{zz},\ \alpha_{xx} - \alpha_{yy},$ $\alpha_{xy},\ \alpha_{yz},\ \alpha_{zx})$

$\mathfrak{I}_h = \mathfrak{I} \times i$; **T** in F_{1u}; **R** in F_{1g}; α in A_g, H_g

Table X-11. Character Tables of the Groups, $\mathfrak{C}_{\infty v}$ and $\mathfrak{D}_{\infty h}$, for Linear Molecules

$\mathfrak{C}_{\infty v}$	E	$2C_\infty^\phi$	. . .	$\infty\sigma_v$		
$A_1 \equiv \Sigma^+$	1	1	. . .	1	T_z	$\alpha_{xx} + \alpha_{yy},\ \alpha_{zz}$
$A_2 \equiv \Sigma^-$	1	1	. . .	-1	R_z	
$E_1 \equiv \Pi$	2	$2\cos\phi$	. . .	0	(T_x, T_y); (R_x, R_y)	$(\alpha_{yz}, \alpha_{zx})$
$E_2 \equiv \Delta$	2	$2\cos 2\phi$	. . .	0		$(\alpha_{xx} - \alpha_{yy},\ \alpha_{xy})$
$E_3 \equiv \Phi$	2	$2\cos 3\phi$	. . .	0		
. . .	. . .	. . .	. . .	. . .		

$\mathfrak{D}_{\infty h}$	E	$2C_\infty^\phi$	. . .	$\infty\sigma_v$	i	$2S_\infty^\phi$	. . .	∞C_2		
Σ_g^+	1	1	. . .	1	1	1	. . .	1		$\alpha_{xx} + \alpha_{yy},\ \alpha_{zz}$
Σ_g^-	1	1	. . .	-1	1	1	. . .	-1	R_z	
Π_g	2	$2\cos\phi$	. . .	0	2	$-2\cos\phi$	. . .	0	(R_x, R_y)	$(\alpha_{yz}, \alpha_{zx})$
Δ_g	2	$2\cos 2\phi$	. . .	0	2	$2\cos 2\phi$	. . .	0		$(\alpha_{xx} - \alpha_{yy},\ \alpha_{xy})$
. . .	. . .	. . .	. . .	. . .	. . .	. . .	. . .	. . .		
Σ_u^+	1	1	. . .	1	-1	-1	. . .	-1	T_z	
Σ_u^-	1	1	. . .	-1	-1	-1	. . .	1		
Π_u	2	$2\cos\phi$	. . .	0	-2	$2\cos\phi$	. . .	0	(T_x, T_y)	
Δ_u	2	$2\cos 2\phi$	. . .	0	-2	$-2\cos 2\phi$	. . .	0		
. . .	. . .	. . .	. . .	. . .	. . .	. . .	. . .	. . .		

X-6. Symmetry Species of Combinations: Direct Products of Irreducible Representations

In Sec. 7-3 it was shown that the symmetry species of combination levels could be found by taking the *direct product* of two irreducible representations. It is possible to summarize the results of such calculations in the form of some simple rules about products of the types $A \times A = A$, $A \times B = B$, etc. This is done in Table X-12.

TABLE X-12. MULTIPLICATION PROPERTIES OF IRREDUCIBLE REPRESENTATIONS

General rules:

$A \times A = A$, $B \times B = A$‡, $A \times B = B$, $A \times E = E$, $B \times E = E$, $A \times F = F$, $B \times F = F$; $g \times g = g$, $u \times u = g$, $u \times g = u$; $' \times ' = '$, $'' \times '' = '$, $' \times '' = ''$

$A \times E_1 = E_1$, $A \times E_2 = E_2$, $B \times E_1 = E_2$, $B \times E_2 = E_1$

Subscripts on A or B:

$1 \times 1 = 1$, $2 \times 2 = 1$, $1 \times 2 = 2$, except for $\mathfrak{D}_2 = \mathfrak{V}$ and $\mathfrak{D}_{2h} = \mathfrak{V}_h$, where $1 \times 2 = 3$, $2 \times 3 = 1$, $1 \times 3 = 2$

Doubly degenerate representations:

For $\mathfrak{C}_3$, $\mathfrak{C}_{3h}$, $\mathfrak{C}_{3v}$, $\mathfrak{D}_3$, $\mathfrak{D}_{3h}$ $\mathfrak{D}_{3d}$, $\mathfrak{C}_6$, $\mathfrak{C}_{6h}$, $\mathfrak{C}_{6v}$, $\mathfrak{D}_6$, $\mathfrak{D}_{6h}$, $\mathfrak{S}_6$, $\mathfrak{O}$, $\mathfrak{O}_h$, $\mathfrak{T}$, $\mathfrak{T}_d$, $\mathfrak{T}_h$:

$$E_1 \times E_1 = E_2 \times E_2 = A_1 + A_2 + E_2$$
$$E_1 \times E_2 = B_1 + B_2 + E_1$$

For $\mathfrak{C}_4$, $\mathfrak{C}_{4v}$, $\mathfrak{C}_{4h}$, $\mathfrak{D}_{2d}$, $\mathfrak{D}_4$, $\mathfrak{D}_{4h}$, $\mathfrak{S}_4$: $E \times E = A_1 + A_2 + B_1 + B_2$

For groups in above lists which have symbols A, B, or E without subscripts, read $A_1 = A_2 = A$, etc.

Triply degenerate representations:

For $\mathfrak{T}_d$, $\mathfrak{O}$, $\mathfrak{O}_h$: $E \times F_1 = E \times F_2 = F_1 + F_2$

$$F_1 \times F_1 = F_2 \times F_2 = A_1 + E + F_1 + F_2$$
$$F_1 \times F_2 = A_2 + E + F_1 + F_2$$

For $\mathfrak{T}$, $\mathfrak{T}_h$: Drop subscripts 1 and 2 from A and F

Linear molecules ($\mathfrak{C}_{\infty v}$ and $\mathfrak{D}_{\infty h}$):

$$\Sigma^+ \times \Sigma^+ = \Sigma^- \times \Sigma^- = \Sigma^+; \quad \Sigma^+ \times \Sigma^- = \Sigma^-$$
$$\Sigma^+ \times \Pi = \Sigma^- \times \Pi = \Pi; \quad \Sigma^+ \times \Delta = \Sigma^- \times \Delta = \Delta; \quad \text{etc.}$$
$$\Pi \times \Pi = \Sigma^+ + \Sigma^- + \Delta$$
$$\Delta \times \Delta = \Sigma^+ + \Sigma^- + \Gamma$$
$$\Pi \times \Delta = \Pi + \Phi$$

As an example of the use of Table X-12, consider the resolution of the direct product $A_{2g} \times E_{1g} \times E_{2u}$ in $\mathfrak{D}_{6h}$. Since $E_1 \times E_2 = B_1 + B_2 + E_1$ and since $g \times u = u$, $E_{1g} \times E_{2u} = B_{1u} + B_{2u} + E_{1u}$. Employment of the rules, $A_2 \times B_1 = B_2$, $A_2 \times B_2 = B_1$, and $A_2 \times E_1 = E_1$, leads to

‡ Exception: in the groups $\mathfrak{D}_2$, $\mathfrak{D}_{2h}$, $B_1 \times B_2 = B_3$, $B_2 \times B_3 = B_1$, $B_3 \times B_1 = B_2$.

the final result

$$A_{2g} \times E_{1g} \times E_{2u} = A_{2g} \times (B_{1u} + B_{2u} + E_{1u})$$
$$= B_{2u} + B_{1u} + E_{1u}$$

X-7. Overtones of Degenerate Fundamentals

The method of determining the symmetry species of such levels was described in Sec. 7-3. Table X-13 summarizes the results of such calculations for the more important groups.

TABLE X-13. SYMMETRY SPECIES OF OVERTONES OF DEGENERATE FUNDAMENTALS

General rules:

$(g)^v = g$; $(u)^v = g$ if v is even, $= u$ if v is odd; $(')^v = (')$; $('')^v = (')$ if v is even, $= ('')$ if v is odd

Doubly degenerate fundamentals:

For $\mathcal{C}_3, \mathcal{C}_{3v}, \mathcal{C}_{3h}, \mathcal{D}_3, \mathcal{D}_{3h}, \mathcal{T}, \mathcal{T}_d, \mathcal{T}_h, \mathcal{O}, \mathcal{O}_h$:

v even

Let $\dfrac{v-2}{2} = 3p + q$

where $p = 0, 1, 2, 3, \ldots$

$q = 0, 1, 2$

$(E)^v = A_1 + E + p(A_1 + A_2 + 2E)$

$+E$ if $q = 1$

or

$+A_1 + A_2 + E$ if $q = 2$

v odd

Let $\dfrac{v+1}{2} = 3p + q$

where $p = 0, 1, 2, 3, \ldots$

$q = 0, 1, 2$

$(E)^v = p(A_1 + A_2 + 2E)$

$+E$ if $q = 1$

or

$+A_1 + A_2 + E$ if $q = 2$

For $\mathcal{C}_6, \mathcal{C}_{6v}, \mathcal{C}_{6h}, \mathcal{D}_6, \mathcal{D}_{3d}, \mathcal{D}_{6h}, \mathcal{S}_6$: Use same rules as for C_3, etc., with following modifications:

v even

$(E_1)^v$: Put subscript 2 on E

$(E_2)^v$: Put subscript 2 on E

v odd

$(E_1)^v$: Change A to B; put subscript 1 on E

$(E_2)^v$: Put subscript 2 on E

For $\mathcal{C}_4, \mathcal{C}_{4v}, \mathcal{C}_{4h}, \mathcal{D}_4, \mathcal{D}_{2d}, \mathcal{D}_{4h}, \mathcal{S}_4$:

v even

Let $\dfrac{v}{2} = 2p + q$

where $p = 0, 1, 2, 3, \ldots$

$q = 0, 1$

$(E)^v = A_1 + p(A_1 + A_2 + B_1 + B_2)$

$+ q(B_1 + B_2)$

v odd

$(E)^v = \dfrac{v+1}{2} E$

For Linear molecules ($\mathcal{C}_{\infty v}$ or $\mathcal{D}_{\infty h}$):

v even

$(E_1)^v = A_1 + E_2 + E_4 + E_6 + \cdots + E_v$

v odd

$(E_1)^v = E_1 + E_3 + E_5 + \cdots + E_v$

where $A_1 = \Sigma^+$, $E_1 = \Pi$, $E_2 = \Delta$, $E_3 = \Phi$, $E_4 = \Gamma$, etc.

TABLE X-13. SYMMETRY SPECIES OF OVERTONES OF DEGENERATE FUNDAMENTALS (*Continued*)

Triply degenerate fundamentals:

For $\mathfrak{T}_d$, $\mathfrak{O}$, $\mathfrak{O}_h$:

v even	*v odd*
Let $\frac{v}{2} = 6p + q$	Let $\frac{v+1}{2} = 6p + q$
where $p = 0, 1, 2, 3, 4, \ldots$	where $p = 0, 1, 2, 3, \ldots$
$q = 0, 1, 2, 3, 4, 5$	$q = 0, 1, 2, 3, 4, 5$
$(F_1)^v = p\Gamma + p(3p + q - 3)\Gamma' + \Gamma_q$ if $q \neq 0$	$(F_1)^v = p\Gamma + p(3p + q - 3)\Gamma' + \Gamma_q$ if $q \neq 0$
$\Gamma = 7A_1 + 3A_2 + 9E + 9F_1 + 12F_2$	$\Gamma = A_1 + 4A_2 + 5E + 12F_1 + 9F_2$
$\Gamma' = A_1 + A_2 + 2E + 3F_1 + 3F_2$	$\Gamma' = A_1 + A_2 + 2E + 3F_1 + 3F_2$
$\Gamma_1 = A_1 + E + F_2$	$\Gamma_1 = F_1$
$\Gamma_2 = 2A_1 + 2E + F_1 + 2F_2$	$\Gamma_2 = A_2 + 2F_1 + F_2$
$\Gamma_3 = 3A_1 + A_2 + 3E + 2F_1 + 4F_2$	$\Gamma_3 = A_2 + E + 4F_1 + 2F_2$
$\Gamma_4 = 4A_1 + A_2 + 5E + 4F_1 + 6F_2$	$\Gamma_4 = 2A_2 + 2E + 6F_1 + 4F_2$
$\Gamma_5 = 5A_1 + 2A_2 + 7E + 6F_1 + 9F_2$	$\Gamma_5 = A_1 + 3A_2 + 3E + 9F_1 + 6F_2$
$(F_2)^v = (F_1)^v$	$(F_2)^v$: Same as $(F_1)^v$ but with subscripts 1 and 2 permuted

For $\mathfrak{T}$, $\mathfrak{T}_h$: Use the same rules as for $\mathfrak{T}_d$, etc., but drop the subscripts 1 and 2.

X-8. Correlation Tables

The methods of constructing symmetry coordinates described in Chap. 6 and the analysis of the symmetry species of the vibrations of a molecule whose symmetry is altered by isotopic substitution as discussed in Chap. 8 both made use of the relation between the species of a given group and those of its subgroups. For convenience, many such relations are given below in Table X-14.

In some instances, more than one correlation exists between a given pair of groups. Thus the correlation between $\mathfrak{D}_{6h}$ and $\mathfrak{C}_{2v}$ can be made in at least two ways, depending upon the choice of the twofold axis, which may be either C_2' or C_2'' (*e.g.*, a symmetry axis for a CH bond or a CC bond in benzene).

When more than one correlation exists, the situation is defined by specifying the choice of symmetry operation from the larger group. Examples are $\mathfrak{C}_{4v} \rightarrow \mathfrak{C}_s$ where σ_v or σ_d written over the column listing the species of the subgroup means that the respective planes become the (sole) plane of symmetry of $\mathfrak{C}_s$; for the correlation $\mathfrak{D}_{2h} \rightarrow \mathfrak{C}_{2v}$, the symbols $C_2(z)$, $C_2(y)$, $C_2(x)$ indicate which of the three twofold axes of $\mathfrak{D}_{2h}$ becomes the twofold axis of $\mathfrak{C}_{2v}$.

Ambiguities also occur with groups such as $\mathfrak{C}_{2v}$ in which the species B_1 and B_2 are determined only by some conventional specification of the two vertical planes, and with the "four" groups, $\mathfrak{D}_2 = \mathfrak{V}$, $\mathfrak{D}_{2d} = \mathfrak{V}_d$, and $\mathfrak{D}_{2h} = \mathfrak{V}_h$, for which the species B_1, B_2, and B_3 are determined only by establishing a convention as to the directions of the (twofold) x, y, and z axes.

TABLE X-14. CORRELATION TABLES FOR THE SPECIES OF A GROUP AND ITS SUBGROUPS

$\mathfrak{C}_4$	$\mathfrak{C}_2$
A	A
B	A
E	$2B$

$\mathfrak{C}_6$	$\mathfrak{C}_3$	$\mathfrak{C}_2$
A	A	A
B	A	B
E_1	E	$2B$
E_2	E	$2A$

$\mathfrak{D}_2$	$\mathfrak{C}_2$	$\mathfrak{C}_2$	$\mathfrak{C}_2$
A	A	A	A
B_1	A	B	B
B_2	B	A	B
B_3	B	B	A

$\mathfrak{D}_3$	$\mathfrak{C}_3$	$\mathfrak{C}_2$
A_1	A	A
A_2	A	B
E	E	$A + B$

$\mathfrak{D}_4$	$\mathfrak{C}_4$	$\mathfrak{C}_2$	C_2' $\mathfrak{C}_2$	C_2'' $\mathfrak{C}_2$
A_1	A	A	A	A
A_2	A	A	B	B
B_1	B	A	A	B
B_2	B	A	B	A
E	E	$2B$	$A + B$	$A + B$

$\mathfrak{D}_5$	$\mathfrak{C}_5$	$\mathfrak{C}_2$
A_1	A	A
A_2	A	B
E_1	E_1	$A + B$
E_2	E_2	$A + B$

$\mathfrak{D}_6$	$\mathfrak{C}_6$	C_2' $\mathfrak{D}_3$	C_2'' $\mathfrak{D}_3$	$\mathfrak{D}_2$	$\mathfrak{C}_3$	$\mathfrak{C}_2$	C_2' $\mathfrak{C}_2$	C_2'' $\mathfrak{C}_2$
A_1	A	A_1	A_1	A	A	A	A	A
A_2	A	A_2	A_2	B_1	A	A	B	B
B_1	B	A_1	A_2	B_2	A	B	A	B
B_2	B	A_2	A_1	B_3	A	B	B	A
E_1	E_1	E	E	$B_2 + B_3$	E	$2B$	$A + B$	$A + B$
E_2	E_2	E	E	$A + B_1$	E	$2A$	$A + B$	$A + B$

$\mathfrak{C}_{2v}$	$\mathfrak{C}_2$	$\sigma(zx)$ $\mathfrak{C}_s$	$\sigma(yz)$ $\mathfrak{C}_s$
A_1	A	A'	A'
A_2	A	A''	A''
B_1	B	A'	A''
B_2	B	A''	A'

$\mathfrak{C}_{3v}$	$\mathfrak{C}_3$	$\mathfrak{C}_s$
A_1	A	A'
A_2	A	A''
E	E	$A' + A''$

$\mathfrak{C}_{4v}$	$\mathfrak{C}_4$	σ_v $\mathfrak{C}_{2v}$	σ_d $\mathfrak{C}_{2v}$	$\mathfrak{C}_2$	σ_v $\mathfrak{C}_s$	σ_d $\mathfrak{C}_s$
A_1	A	A_1	A_1	A	A'	A'
A_2	A	A_2	A_2	A	A''	A''
B_1	B	A_1	A_2	A	A'	A''
B_2	B	A_2	A_1	A	A''	A'
E	E	$B_1 + B_2$	$B_1 + B_2$	$2B$	$A' + A''$	$A' + A''$

TABLE X-14. CORRELATION TABLES FOR THE SPECIES OF A GROUP AND ITS SUBGROUPS (*Continued*)

$\mathcal{C}_{5v}$	$\mathcal{C}_5$	$\mathcal{C}_s$
A_1	A	A'
A_2	A	A''
E_1	E_1	$A + A''$
E_2	E_2	$A' + A''$

$\mathcal{C}_{6v}$	$\mathcal{C}_6$	σ_v $\mathcal{C}_{3v}$	σ_d $\mathcal{C}_{3v}$	$\sigma_v \rightarrow \sigma(zx)$ $\mathcal{C}_{2v}$	$\mathcal{C}_3$	$\mathcal{C}_2$	σ_v $\mathcal{C}_s$	σ_d $\mathcal{C}_s$
A_1	A	A_1	A_1	A_1	A	A	A'	A'
A_2	A	A_2	A_2	A_2	A	A	A''	A''
B_1	B	A_1	A_2	B_1	A	B	A'	A''
B_2	B	A_2	A_1	B_2	A	B	A''	A'
E_1	E_1	E	E	$B_1 + B_2$	E	$2B$	$A' + A''$	$A' + A''$
E_2	E_2	E	E	$A_1 + A_2$	E	$2A$	$A' + A''$	$A' + A''$

$\mathcal{C}_{2h}$	$\mathcal{C}_2$	$\mathcal{C}_s$	$\mathcal{C}_i$
A_g	A	A'	A_g
B_g	B	A''	A_g
A_u	A	A''	A_u
B_u	B	A'	A_u

$\mathcal{C}_{3h}$	$\mathcal{C}_3$	$\mathcal{C}_s$
A'	A	A'
E'	E	$2A'$
A''	A	A''
E''	E	$2A''$

$\mathcal{C}_{4h}$	$\mathcal{C}_4$	$\mathcal{S}_4$	$\mathcal{C}_{2h}$	$\mathcal{C}_2$	$\mathcal{C}_s$	$\mathcal{C}_i$
A_g	A	A	A_g	A	A'	A_g
B_g	B	B	A_g	A	A'	A_g
E_g	E	E	$2B_g$	$2B$	$2A''$	$2A_g$
A_u	A	B	A_u	A	A''	A_u
B_u	B	A	A_u	A	A''	A_u
E_u	E	E	$2B_u$	$2B$	$2A'$	$2A_u$

$\mathcal{C}_{5h}$	$\mathcal{C}_5$	$\mathcal{C}_s$
A'	A	A'
E_1'	E_1	$2A'$
E_2'	E_2	$2A'$
A''	A	A''
E_1''	E_1	$2A''$
E_2''	E_2	$2A''$

$\mathcal{C}_{6h}$	$\mathcal{C}_6$	$\mathcal{C}_{3h}$	$\mathcal{S}_6$	$\mathcal{C}_{2h}$	$\mathcal{C}_3$	$\mathcal{C}_2$	$\mathcal{C}_s$	$\mathcal{C}_i$
A_g	A	A'	A_g	A_g	A	A	A'	A_g
B_g	B	A''	A_g	B_g	A	B	A''	A_g
E_{1g}	E_1	E''	E_g	$2B_g$	E	$2B$	$2A''$	$2A_g$
E_{2g}	E_2	E'	E_g	$2A_g$	E	$2A$	$2A'$	$2A_g$
A_u	A	A''	A_u	A_u	A	A	A''	A_u
B_u	B	A'	A_u	B_u	A	B	A'	A_u
E_{1u}	E_1	E'	E_u	$2B_u$	E	$2B$	$2A'$	$2A_u$
E_{2u}	E_2	E''	E_u	$2A_u$	E	$2A$	$2A''$	$2A_u$

$\mathcal{D}_{2h}$	$\mathcal{D}_2$	$C_2(z)$ $\mathcal{C}_{2v}$	$C_2(y)$ $\mathcal{C}_{2v}$	$C_2(x)$ $\mathcal{C}_{2v}$	$C_2(z)$ $\mathcal{C}_{2h}$	$C_2(y)$ $\mathcal{C}_{2h}$	$C_2(x)$ $\mathcal{C}_{2h}$	$C_2(z)$ $\mathcal{C}_2$	$C_2(y)$ $\mathcal{C}_2$	$C_2(x)$ $\mathcal{C}_2$	$\sigma(xy)$ $\mathcal{C}_s$	$\sigma(zx)$ $\mathcal{C}_s$	$\sigma(yz)$ $\mathcal{C}_s$
A_g	A	A_1	A_1	A_1	A_g	A_g	A_g	A	A	A	A'	A'	A'
B_{1g}	B_1	A_2	B_2	B_1	A_g	B_g	B_g	A	B	B	A'	A''	A''
B_{2g}	B_2	B_1	A_2	B_2	B_g	A_g	B_g	B	A	B	A''	A'	A''
B_{3g}	B_3	B_2	B_1	A_2	B_g	B_g	A_g	B	B	A	A''	A''	A'
A_u	A	A_2	A_2	A_2	A_u	A_u	A_u	A	A	A	A''	A''	A''
B_{1u}	B_1	A_1	B_1	B_2	A_u	B_u	B_u	A	B	B	A''	A'	A'
B_{2u}	B_2	B_2	A_1	B_1	B_u	A_u	B_u	B	A	B	A'	A''	A'
B_{3u}	B_3	B_1	B_2	A_1	B_u	B_u	A_u	B	B	A	A'	A'	A''

TABLE X-14. CORRELATION TABLES FOR THE SPECIES OF A GROUP AND ITS SUBGROUPS (*Continued*)

$\mathfrak{D}_{3h}$	$\mathfrak{C}_{3h}$	$\mathfrak{D}_3$	$\mathfrak{C}_{3v}$	$\sigma_h \to \sigma_v(zy)$ $\mathfrak{C}_{2v}$	$\mathfrak{C}_3$	$\mathfrak{C}_2$	σ_h $\mathfrak{C}_s$	σ_v $\mathfrak{C}_s$
A_1'	A'	A_1	A_1	A_1	A	A	A'	A'
A_2'	A'	A_2	A_2	B_2	A	B	A'	A''
E'	E'	E	E	$A_1 + B_2$	E	$A + B$	$2A'$	$A' + A''$
A_1''	A''	A_1	A_2	A_2	A	A	A''	A''
A_2''	A''	A_2	A_1	B_1	A	B	A''	A'
E''	E''	E	E	$A_2 + B_1$	E	$A + B$	$2A''$	$A' + A''$

$\mathfrak{D}_{4h}$	$\mathfrak{D}_4$	$C_2' \to C_2'$ $\mathfrak{D}_{2d}$	$C_2'' \to C_2'$ $\mathfrak{D}_{2d}$	$\mathfrak{C}_{4v}$	$\mathfrak{C}_{4h}$	C_2' $\mathfrak{D}_{2h}$	C_2'' $\mathfrak{D}_{2h}$	$\mathfrak{C}_4$	$\mathfrak{S}_4$
A_{1g}	A_1	A_1	A_1	A_1	A_g	A_g	A_g	A	A
A_{2g}	A_2	A_2	A_2	A_2	A_g	B_{1g}	B_{1g}	A	A
B_{1g}	B_1	B_1	B_2	B_1	B_g	A_g	B_{1g}	B	B
B_{2g}	B_2	B_2	B_1	B_2	B_g	B_{1g}	A_g	B	B
E_g	E	E	E	E	E_g	$B_{2g} + B_{3g}$	$B_{2g} + B_{3g}$	E	E
A_{1u}	A_1	B_1	B_1	A_2	A_u	A_u	A_u	A	B
A_{2u}	A_2	B_2	B_2	A_1	A_u	B_{1u}	B_{1u}	A	B
B_{1u}	B_1	A_1	A_2	B_2	B_u	A_u	B_{1u}	B	A
B_{2u}	B_2	A_2	A_1	B_1	B_u	B_{1u}	A_u	B	A
E_u	E	E	E	E	E_u	$B_{2u} + B_{3u}$	$B_{2u} + B_{3u}$	E	E

$\mathfrak{D}_{4h}$ (*cont.*)	C_2' $\mathfrak{D}_2$	C_2'' $\mathfrak{D}_2$	C_2, σ_v $\mathfrak{C}_{2v}$	C_2, σ_d $\mathfrak{C}_{2v}$	C_2' $\mathfrak{C}_{2v}$	C_2'' $\mathfrak{C}_{2v}$
A_{1g}	A	A	A_1	A_1	A_1	A_1
A_{2g}	B_1	B_1	A_2	A_2	B_1	B_1
B_{1g}	A	B_1	A_1	A_2	A_1	B_1
B_{2g}	B_1	A	A_2	A_1	B_1	A_1
E_g	$B_2 + B_3$	$B_2 + B_3$	$B_1 + B_2$	$B_1 + B_2$	$A_2 + B_2$	$A_2 + B_2$
A_{1u}	A	A	A_2	A_2	A_2	A_2
A_{2u}	B_1	B_1	A_1	A_1	B_2	B_2
B_{1u}	A	B_1	A_2	A_1	A_2	B_2
B_{2u}	B_1	A	A_1	A_2	B_2	A_2
E_u	$B_2 + B_3$	$B_2 + B_3$	$B_1 + B_2$	$B_1 + B_2$	$A_1 + B_1$	$A_1 + B_1$

$\mathfrak{D}_{4h}$ (*cont.*)	C_2 $\mathfrak{C}_{2h}$	C_2' $\mathfrak{C}_{2h}$	C_2'' $\mathfrak{C}_{2h}$	C_2 $\mathfrak{C}_2$	C_2' $\mathfrak{C}_2$	C_2'' $\mathfrak{C}_2$	σ_h $\mathfrak{C}_s$	σ_v $\mathfrak{C}_s$	σ_d $\mathfrak{C}_s$	$\mathfrak{C}_i$
A_{1g}	A_g	A_g	A_g	A	A	A	A'	A'	A'	A_g
A_{2g}	A_g	B_g	B_g	A	B	B	A'	A''	A''	A_g
B_{1g}	A_g	A_g	B_g	A	A	B	A'	A'	A''	A_g
B_{2g}	A_g	B_g	A_g	A	B	A	A'	A''	A'	A_g
E_g	$2B_g$	$A_g + B_g$	$A_g + B_g$	$2B$	$A + B$	$A + B$	$2A''$	$A' + A''$	$A' + A''$	$2A_g$
A_{1u}	A_u	A_u	A_u	A	A	A	A''	A''	A''	A_u
A_{2u}	A_u	B_u	B_u	A	B	B	A''	A'	A'	A_u
B_{1u}	A_u	A_u	B_u	A	A	B	A''	A''	A'	A_u
B_{2u}	A_u	B_u	A_u	A	B	A	A''	A'	A''	A_u
E_u	$2B_u$	$A_u + B_u$	$A_u + B_u$	$2B$	$A + B$	$A + B$	$2A'$	$A' + A''$	$A' + A''$	$2A_u$

TABLE X-14. CORRELATION TABLES FOR THE SPECIES OF A GROUP AND ITS SUBGROUPS (*Continued*)

$\mathfrak{D}_{5h}$	$\mathfrak{D}_5$	$\mathfrak{C}_{5v}$	$\mathfrak{C}_{5h}$	$\mathfrak{C}_5$	$\sigma_h \to \sigma(zx)$ $\mathfrak{C}_{2v}$	$\mathfrak{C}_2$	σ_h $\mathfrak{C}_s$	σ_v $\mathfrak{C}_s$
A_1'	A_1	A_1	A'	A	A_1	A	A'	A'
A_2'	A_2	A_2	A'	A	B_1	B	A'	A''
E_1'	E_1	E_1	E_1'	E_1	$A_1 + B_1$	$A + B$	$2A'$	$A' + A''$
E_2'	E_2	E_2	E_2'	E_2	$A_1 + B_1$	$A + B$	$2A'$	$A' + A''$
A_1''	A_1	A_2	A''	A	A_2	A	A''	A''
A_2''	A_2	A_1	A''	A	B_2	B	A''	A'
E_1''	E_1	E_1	E_1''	E_1	$A_2 + B_2$	$A + B$	$2A''$	$A' + A''$
E_2''	E_2	E_2	E_2''	E_2	$A_2 + B_2$	$A + B$	$2A''$	$A' + A''$

$\mathfrak{D}_{6h}$	$\mathfrak{D}_6$	C_2' $\mathfrak{D}_{3h}$	C_2'' $\mathfrak{D}_{3h}$	$\mathfrak{C}_{6v}$	$\mathfrak{C}_{6h}$	C_2'' $\mathfrak{D}_{3d}$	C_2' $\mathfrak{D}_{3d}$	$\sigma_h \to \sigma(xy)$ $\sigma_v \to \sigma(yz)$ $\mathfrak{D}_{2h}$	$\mathfrak{C}_6$	$\mathfrak{C}_{3h}$	C_2' $\mathfrak{D}_3$	C_2'' $\mathfrak{D}_3$	σ_v $\mathfrak{C}_{3v}$	σ_d $\mathfrak{C}_{3v}$	$\mathfrak{S}_6$	$\mathfrak{D}_2$
A_{1g}	A_1	A_1'	A_1'	A_1	A_g	A_{1g}	A_{1g}	A_g	A	A'	A_1	A_1	A_1	A_1	A_g	A
A_{2g}	A_2	A_2'	A_2'	A_2	A_g	A_{2g}	A_{2g}	B_{1g}	A	A'	A_2	A_2	A_2	A_2	A_g	B_1
B_{1g}	B_1	A_1''	A_2''	B_2	B_g	A_{2g}	A_{1g}	B_{2g}	B	A''	A_1	A_2	A_2	A_1	A_g	B_2
B_{2g}	B_2	A_2''	A_1''	B_1	B_g	A_{1g}	A_{2g}	B_{3g}	B	A''	A_2	A_1	A_1	A_2	A_g	B_3
E_{1g}	E_1	E''	E''	E_1	E_{1g}	E_g	E_g	$B_{2g} + B_{3g}$	E_1	E''	E	E	E	E	E_g	$B_2 + B_3$
E_{2g}	E_2	E'	E'	E_2	E_{2g}	E_g	E_g	$A_g + B_{1g}$	E_2	E'	E	E	E	E	E_g	$A + B_1$
A_{1u}	A_1	A_1''	A_1''	A_2	A_u	A_{1u}	A_{1u}	A_u	A	A''	A_1	A_1	A_2	A_2	A_u	A
A_{2u}	A_2	A_2''	A_2''	A_1	A_u	A_{2u}	A_{2u}	B_{1u}	A	A''	A_2	A_2	A_1	A_1	A_u	B_1
B_{1u}	B_1	A_1'	A_2'	B_1	B_u	A_{2u}	A_{1u}	B_{2u}	B	A'	A_1	A_2	A_1	A_2	A_u	B_2
B_{2u}	B_2	A_2'	A_1'	B_2	B_u	A_{1u}	A_{2u}	B_{3u}	B	A'	A_2	A_1	A_2	A_1	A_u	B_3
E_{1u}	E_1	E'	E'	E_1	E_{1u}	E_u	E_u	$B_{2u} + B_{3u}$	E_1	E'	E	E	E	E	E_u	$B_2 + B_3$
E_{2u}	E_2	E''	E''	E_2	E_{2u}	E_u	E_u	$A_u + B_{1u}$	E_2	E''	E	E	E	E	E_u	$A + B_1$

$\mathfrak{D}_{6h}$ (*cont.*)	C_2' $\mathfrak{C}_{2v}$	C_2'' $\mathfrak{C}_{2v}$	C_2 $\mathfrak{C}_{2h}$	C_2' $\mathfrak{C}_{2h}$	C_2'' $\mathfrak{C}_{2h}$	$\mathfrak{C}_3$	C_2 $\mathfrak{C}_2$
A_{1g}	A_1	A_1	A_g	A_g	A_g	A	A
A_{2g}	B_1	B_1	A_g	B_g	B_g	A	A
B_{1g}	A_2	B_2	B_g	A_g	B_g	A	B
B_{2g}	B_2	A_2	B_g	B_g	A_g	A	B
E_{1g}	$A_2 + B_2$	$A_2 + B_2$	$2B_g$	$A_g + B_g$	$A_g + B_g$	E	$2B$
E_{2g}	$A_1 + B_1$	$A_1 + B_1$	$2A_g$	$A_g + B_g$	$A_g + B_g$	E	$2A$
A_{1u}	A_2	A_2	A_u	A_u	A_u	A	A
A_{2u}	B_2	B_2	A_u	B_u	B_u	A	A
B_{1u}	A_1	B_1	B_u	A_u	B_u	A	B
B_{2u}	B_1	A_1	B_u	B_u	A_u	A	B
E_{1u}	$A_1 + B_1$	$A_1 + B_1$	$2B_u$	$A_u + B_u$	$A_u + B_u$	E	$2B$
E_{2u}	$A_2 + B_2$	$A_2 + B_2$	$2A_u$	$A_u + B_u$	$A_u + B_u$	E	$2A$

TABLE X-14. CORRELATION TABLES FOR THE SPECIES OF A GROUP AND ITS SUBGROUPS (*Continued*)

	C_2'	C_2''	σ_h	σ_d	σ_v	
$\mathfrak{D}_{6h}$ (*cont.*)	$\mathfrak{C}_2$	$\mathfrak{C}_2$	$\mathfrak{C}_s$	$\mathfrak{C}_s$	$\mathfrak{C}_s$	$\mathfrak{C}_i$
A_{1g}	A	A	A'	A'	A'	A_g
A_{2g}	B	B	A'	A''	A''	A_g
B_{1g}	A	B	A''	A'	A''	A_g
B_{2g}	B	A	A''	A''	A'	A_g
E_{1g}	$A + B$	$A + B$	$2A''$	$A' + A''$	$A' + A''$	$2A_g$
E_{2g}	$A + B$	$A + B$	$2A'$	$A' + A''$	$A' + A''$	$2A_g$
A_{1u}	A	A	A''	A''	A''	A_u
A_{2u}	B	B	A''	A'	A'	A_u
B_{1u}	A	B	A'	A''	A'	A_u
B_{2u}	B	A	A'	A'	A''	A_u
E_{1u}	$A + B$	$A + B$	$2A'$	$A' + A''$	$A' + A''$	$2A_u$
E_{2u}	$A + B$	$A + B$	$2A''$	$A' + A''$	$A' + A''$	$2A_u$

		$C_2 \rightarrow C_2(z)$		C_2	C_2'	
$\mathfrak{D}_{2d}$	$\mathfrak{S}_4$	$\mathfrak{D}_2$	$\mathfrak{C}_{2v}$	$\mathfrak{C}_2$	$\mathfrak{C}_2$	$\mathfrak{C}_s$
A_1	A	A	A_1	A	A	A'
A_2	A	B_1	A_2	A	B	A''
B_1	B	A	A_2	A	A	A''
B_2	B	B_1	A_1	A	B	A'
E	E	$B_2 + B_3$	$B_1 + B_2$	$2B$	$A + B$	$A' + A''$

$\mathfrak{D}_{3d}$	$\mathfrak{D}_3$	$\mathfrak{C}_{3v}$	$\mathfrak{S}_6$	$\mathfrak{C}_3$	$\mathfrak{C}_{2h}$	$\mathfrak{C}_2$	$\mathfrak{C}_s$	$\mathfrak{C}_i$
A_{1g}	A_1	A_1	A_g	A	A_g	A	A'	A_g
A_{2g}	A_2	A_2	A_g	A	B_g	B	A''	A_g
E_g	E	E	E_g	E	$A_g + B_g$	$A + B$	$A' + A''$	$2A_g$
A_{1u}	A_1	A_2	A_u	A	A_u	A	A''	A_u
A_{2u}	A_2	A_1	A_u	A	B_u	B	A'	A_u
E_u	E	E	E_u	E	$A_u + B_u$	$A + B$	$A' + A''$	$2A_u$

						C_2	C_2'	
$\mathfrak{D}_{4d}$	$\mathfrak{D}_4$	$\mathfrak{C}_{4v}$	$\mathfrak{S}_8$	$\mathfrak{C}_4$	$\mathfrak{C}_{2v}$	$\mathfrak{C}_2$	$\mathfrak{C}_2$	$\mathfrak{C}_s$
A_1	A_1	A_1	A	A	A_1	A	A	A'
A_2	A_2	A_2	A	A	A_2	A	B	A''
B_1	A_1	A_2	B	A	A_2	A	A	A''
B_2	A_2	A_1	B	A	A_1	A	B	A'
E_1	E	E	E_1	E	$B_1 + B_2$	$2B$	$A + B$	$A' + A''$
E_2	$B_1 + B_2$	$B_1 + B_2$	E_2	$2B$	$A_1 + A_2$	$2A$	$A + B$	$A' + A''$
E_3	E	E	E_3	E	$B_1 + B_2$	$2B$	$A + B$	$A' + A''$

TABLE X-14. CORRELATION TABLES FOR THE SPECIES OF A GROUP AND ITS SUBGROUPS (*Continued*)

$\mathfrak{D}_{5d}$	$\mathfrak{D}_5$	$\mathfrak{C}_{5v}$	$\mathfrak{C}_5$	$\mathfrak{C}_2$	$\mathfrak{C}_s$	$\mathfrak{C}_i$
A_{1g}	A_1	A_1	A	A	A'	A_g
A_{2g}	A_2	A_2	A	B	A''	A_g
E_{1g}	E_1	E_1	E_1	$A + B$	$A' + A''$	$2A_g$
E_{2g}	E_2	E_2	E_2	$A + B$	$A' + A''$	$2A_g$
A_{1u}	A_1	A_2	A	A	A''	A_u
A_{2u}	A_2	A_1	A	B	A'	A_u
E_{1u}	E_1	E_1	E_1	$A + B$	$A' + A''$	$2A_u$
E_{2u}	E_2	E_2	E_2	$A + B$	$A' + A''$	$2A_u$

$\mathfrak{D}_{6d}$	$\mathfrak{D}_6$	$\mathfrak{C}_{6v}$	$\mathfrak{C}_6$	$\mathfrak{D}_{2d}$	$\mathfrak{D}_3$	$\mathfrak{C}_{3v}$
A_1	A_1	A_1	A	A_1	A_1	A_1
A_2	A_2	A_2	A	A_2	A_2	A_2
B_1	A_1	A_2	A	B_1	A_1	A_2
B_2	A_2	A_1	A	B_2	A_2	A_1
E_1	E_1	E_1	E_1	E	E	E
E_2	E_2	E_2	E_2	$B_1 + B_2$	E	E
E_3	$B_1 + B_2$	$B_1 + B_2$	$2B$	E	$A_1 + A_2$	$A_1 + A_2$
E_4	E_2	E_2	E_2	$A_1 + A_2$	E	E
E_5	E_1	E_1	E_1	E	E	E

$\mathfrak{D}_{6d}$ (*cont.*)	$\mathfrak{D}_2$	$\mathfrak{C}_{2v}$	$\mathfrak{S}_4$	$\mathfrak{C}_3$	C_2 $\mathfrak{C}_2$	C_2' $\mathfrak{C}_2$	$\mathfrak{C}_s$
A_1	A	A_1	A	A	A	A	A'
A_2	B_1	A_2	A	A	A	B	A''
B_1	A	A_2	B	A	A	A	A''
B_2	B_1	A_1	B	A	A	B	A'
E_1	$B_2 + B_3$	$B_1 + B_2$	E	E	$2B$	$A + B$	$A' + A''$
E_2	$A + B_1$	$A_1 + A_2$	$2B$	E	$2A$	$A + B$	$A' + A''$
E_3	$B_2 + B_3$	$B_1 + B_2$	E	$2A$	$2B$	$A + B$	$A' + A''$
E_4	$A + B_1$	$A_1 + A_2$	$2A$	E	$2A$	$A + B$	$A' + A''$
E_5	$B_2 + B_3$	$B_1 + B_2$	E	E	$2B$	$A + B$	$A' + A''$

$\mathfrak{S}_4$	$\mathfrak{C}_2$
A	A
B	A
E	$2B$

$\mathfrak{S}_6$	$\mathfrak{C}_3$	$\mathfrak{C}_i$
A_g	A	A_g
E_g	E	$2A_g$
A_u	A	A_u
E_u	E	$2A_u$

$\mathfrak{S}_8$	$\mathfrak{C}_4$	$\mathfrak{C}_2$
A	A	A
B	A	A
E_1	E	$2B$
E_2	$2B$	$2A$
E_3	E	$2B$

$\mathfrak{T}$	$\mathfrak{D}_2$	$\mathfrak{C}_3$	$\mathfrak{C}_2$
A	A	A	A
E	$2A$	E	$2A$
F	$B_1 + B_2 + B_3$	$A + E$	$A + 2B$

TABLE X-14. CORRELATION TABLES FOR THE SPECIES OF A GROUP AND ITS SUBGROUPS (*Continued*)

$\mathfrak{T}_h$	$\mathfrak{T}$	$\mathfrak{D}_{2h}$	$\mathfrak{S}_6$	$\mathfrak{D}_2$
A_g	A	A_g	A_g	A
E_g	E	$2A_g$	E_g	$2A$
F_g	F	$B_{1g} + B_{2g} + B_{3g}$	$A_g + E_g$	$B_1 + B_2 + B_3$
A_u	A	A_u	A_u	A
E_u	E	$2A_u$	E_u	$2A$
F_u	F	$B_{1u} + B_{2u} + B_{3u}$	$A_u + E_u$	$B_1 + B_2 + B_3$

$\mathfrak{T}_h$ (*cont.*)	$\mathfrak{C}_{2v}$	$\mathfrak{C}_{2h}$	$\mathfrak{C}_3$	$\mathfrak{C}_2$	$\mathfrak{C}_s$	$\mathfrak{C}_i$
A_g	A_1	A_g	A	A	A'	A_g
E_g	$2A_1$	$2A_g$	E	$2A$	$2A'$	$2A_g$
F_g	$A_2 + B_1 + B_2$	$A_g + 2B_g$	$A + E$	$A + 2B$	$A' + 2A''$	$3A_g$
A_u	A_2	A_u	A	A	A''	A_u
E_u	$2A_2$	$2A_u$	E	$2A$	$2A''$	$2A_u$
F_u	$A_1 + B_1 + B_2$	$A_u + 2B_u$	$A + E$	$A + 2B$	$2A' + A''$	$3A_u$

$\mathfrak{T}_d$	$\mathfrak{T}$	$\mathfrak{D}_{2d}$	$\mathfrak{C}_{3v}$	$\mathfrak{S}_4$	$\mathfrak{D}_2$	$\mathfrak{C}_{2v}$	$\mathfrak{C}_3$	$\mathfrak{C}_2$	$\mathfrak{C}_s$
A_1	A	A_1	A_1	A	A	A_1	A	A	A'
A_2	A	B_1	A_2	B	A	A_2	A	A	A''
E	E	$A_1 + B_1$	E	$A + B$	$2A$	$A_1 + A_2$	E	$2A$	$A' + A''$
F_1	F	$A_2 + E$	$A_2 + E$	$A + E$	$B_1 + B_2 + B_3$	$A_2 + B_1 + B_2$	$A + E$	$A + 2B$	$A' + 2A''$
F_2	F	$B_2 + E$	$A_1 + E$	$B + E$	$B_1 + B_2 + B_3$	$A_1 + B_1 + B_2$	$A + E$	$A + 2B$	$2A' + A''$

$\mathfrak{O}$	$\mathfrak{T}$	$\mathfrak{D}_4$	$\mathfrak{D}_3$	$\mathfrak{C}_4$	$3C_2$ $\mathfrak{D}_2$	$C_2, 2C_2'$ $\mathfrak{D}_2$	$\mathfrak{C}_3$	$\mathfrak{C}_2$	$\mathfrak{C}_2$
A_1	A	A_1	A_1	A	A	A	A	A	A
A_2	A	B_1	A_2	B	A	B_1	A	A	B
E	E	$A_1 + B_1$	E	$A + B$	$2A$	$A + B_1$	E	$2A$	$A + B$
F_1	F	$A_2 + E$	$A_2 + E$	$A + E$	$B_1 + B_2 + B_3$	$B_1 + B_2 + B_3$	$A + E$	$A + 2B$	$A + 2B$
F_2	F	$B_2 + E$	$A_1 + E$	$B + E$	$B_1 + B_2 + B_3$	$A + B_2 + B_3$	$A + E$	$A + 2B$	$2A + B$

$\mathfrak{O}_h$‡	$\mathfrak{O}$	$\mathfrak{T}_d$	$\mathfrak{T}_h$	$\mathfrak{D}_{4h}$	$\mathfrak{D}_{3d}$
A_{1g}	A_1	A_1	A_g	A_{1g}	A_{1g}
A_{2g}	A_2	A_2	A_g	B_{1g}	A_{2g}
E_g	E	E	E_g	$A_{1g} + B_{1g}$	E_g
F_{1g}	F_1	F_1	F_g	$A_{2g} + E_g$	$A_{2g} + E_g$
F_{2g}	F_2	F_2	F_g	$B_{2g} + E_g$	$A_{1g} + E_g$
A_{1u}	A_1	A_2	A_u	A_{1u}	A_{1u}
A_{2u}	A_2	A_1	A_u	B_{1u}	A_{2u}
E_u	E	E	E_u	$A_{1u} + B_{1u}$	E_u
F_{1u}	F_1	F_2	F_u	$A_{2u} + E_u$	$A_{2u} + E_u$
F_{2u}	F_2	F_1	F_u	$B_{2u} + E_u$	$A_{1u} + E_u$

‡ To find correlations with smaller subgroups, carry out the correlation in two steps, for example, if the correlation of $\mathfrak{O}_h$ with $\mathfrak{C}_{2v}$ is desired, use the above table to pass from $\mathfrak{O}_h$ to $\mathfrak{T}_d$ and then employ the table for $\mathfrak{T}_d$ to go on to $\mathfrak{C}_{2v}$.

APPENDIX XI

ORTHOGONALITY OF IRREDUCIBLE REPRESENTATION MATRICES

The principal purpose of this appendix is to prove the orthogonality theorem which was used repeatedly in Chap. 6, namely,

$$\sum_R R^{\gamma}_{aa'}{}^* R^{\gamma'}_{a''a'''} = \frac{\delta_{\gamma\gamma'}\delta_{aa''}\delta_{a'a'''}g}{d_\gamma} \tag{1}$$

in which $R^{\gamma}_{aa'}$ is an element of the γth irreducible representation matrix expressing the effect of the symmetry operation, R. The δ's vanish unless their subscripts are equal, in which case they have a value of unity; g is the order (number of symmetry operations) of the group and d_γ is the dimension (number of coordinates) of the γth irreducible representation.

The meaning of this theorem may be clarified if the reader will imagine a stack of cards, one card representing each of the g symmetry operations, R. On each card, the set of irreducible representation matrices is written out in some standard arrangement so that corresponding elements of corresponding representations are superimposed when the cards are stacked. Now let the cards be run through a machine commanded to multiply two numbers[1] from arbitrarily selected positions on the top card and then add up the products from identical positions on all the cards. The theorem simply asserts that such sums would vanish in all cases except those in which the two numbers were selected from one and the same position, in which cases the sum would be the order of the group divided by the dimension of the irreducible representation involved.

The proof of this theorem requires several steps, one of which is useful in proving that the introduction of symmetry coordinates factors the secular determinant (Sec. 6-3 and Appendix XII). These steps are as follows:

Step I. A unitary representation (orthogonal if real) is said to be *reducible* if there exists a transformation of coordinates which simul-

[1] If the numbers have imaginary parts, the machine must also perform the operation of taking the complex conjugate of the first number.

taneously brings all matrices of the group into the form

$$\mathbf{R} = \begin{Vmatrix} \mathbf{R}_1 & \mathbf{0} \\ \mathbf{0} & \mathbf{R}_2 \end{Vmatrix} \tag{2}$$

in which $\mathbf{R}_1$ and $\mathbf{R}_2$ are square matrices of dimensions d_1 and d_2, respectively ($d_1 + d_2 = d$). It will be shown that a representation is reducible if there exists any set of n independent vectors where $0 < n < d$, such that any vector of the set is sent into a linear combination of the members of the set by all matrices of the representation.

Step II. Second, it will be shown that if R^γ and $R^{\gamma'}$ are two *irreducible* representations, and if there is any matrix $\mathbf{M}$ which satisfies the equation

$$\mathbf{R}^\gamma\mathbf{M} = \mathbf{M}\mathbf{R}^{\gamma'} \tag{3}$$

for all $\mathbf{R}$ in the group, then $\mathbf{M}$ is necessarily either a zero matrix or a square matrix ($d_\gamma = d_{\gamma'}$) whose determinant does not vanish. This second step allows an immediate proof of the factoring of the secular equation described above.

Step III. The orthogonality theorem is finally proved making use of the above theorem.

Proof of Step I. Let $\mathbf{V}_1, \mathbf{V}_2, \ldots, \mathbf{V}_n$ be the set of independent vectors which by hypothesis are sent into linear combinations of themselves

$$\mathbf{R}\mathbf{V}_i = \sum_{j=1}^{n} a_{ij}\mathbf{V}_j \qquad i = 1, 2, \ldots, n \tag{4}$$

for all $\mathbf{R}$ in the group (the a_{ij} can vary with $\mathbf{R}$). It is possible to construct from the $\mathbf{V}_j$ an orthonormal set $\mathbf{U}_1, \mathbf{U}_2, \ldots, \mathbf{U}_n$ by the method described in Sec. 9-2, namely,

$$\begin{aligned} \mathbf{U}_1 &= C_{11}\mathbf{V}_1 \\ \mathbf{U}_2 &= C_{21}\mathbf{U}_1 + C_{22}\mathbf{V}_2 \\ &\text{etc.} \end{aligned}$$

where the C's are chosen so that

$$\mathbf{U}_i^\dagger\mathbf{U}_j = \delta_{ij} \qquad i, j = 1, 2, \ldots, n \tag{5}$$

which is the analogue of the ordinary orthogonality condition for vectors when there is a possibility of complex components. Such a set of vectors can then be augmented by a set of $d - n$ additional vectors, $\mathbf{U}_{n+1}$, $\mathbf{U}_{n+2}, \ldots, \mathbf{U}_d$, which are, by construction, orthonormal to one another and to the first n vectors. The first n of the vectors of the new set will have the same property as that exhibited by the $\mathbf{V}$'s, namely, that any one of them is transformed into a linear combination of the first n vectors only, when operated upon by **R**:

$$\mathbf{R}\mathbf{U}_i = \sum_{j=1}^{n} b_{ij}\mathbf{U}_j \qquad i = 1, 2, \ldots, n \tag{6}$$

Suppose now that the (column) vectors $\mathbf{U}_1, \mathbf{U}_2, \ldots, \mathbf{U}_d$ are organized as a square matrix,

$$\mathbf{U} = ||\mathbf{U}_1\mathbf{U}_2 \cdots \mathbf{U}_d|| \tag{7}$$

and that this matrix is used to transform the **R** according to the matrix transformation

$$\mathbf{R} \rightarrow \mathsf{R} = \mathbf{U}^\dagger\mathbf{R}\mathbf{U} \tag{8}$$

The new matrices have the following elements in their first n columns:

$$\mathsf{R}_{ki} = \mathbf{U}_k^\dagger\mathbf{R}\mathbf{U}_i = \sum_{j=1}^{n} b_{ij}\mathbf{U}_k^\dagger\mathbf{U}_j = b_{ik} \tag{9}$$

provided k is one of the first n rows; if $k > n$, then none of the $\mathbf{U}_j$ appearing in the summation in (9) will be the same as $\mathbf{U}_k$ so that $R_{ki} = 0$. In other words, this transformation will have simultaneously introduced a $d - n$ by n block of zeros into the lower left-hand corners of all the new **R**. Likewise it can be shown that an n by $d - n$ block of zeros is simultaneously introduced into the upper right-hand corner of the new **R**'s. By taking the transpose conjugate of (6), one obtains

$$(\mathbf{R}\mathbf{U}_i)^\dagger = \mathbf{U}_i^\dagger\mathbf{R}^\dagger = \mathbf{U}_i^\dagger\mathbf{R}^{-1} = \sum_{j=1}^{n} b_{ij}^*\mathbf{U}_j^\dagger \tag{10}$$

where the second equality follows from the assumption that the **R**'s are unitary. Therefore the new matrix element in the ith row and kth column becomes

$$(\mathsf{R}^{-1})_{ik} = \mathbf{U}_i^\dagger(\mathbf{R}^{-1})\mathbf{U}_k = \sum_{j=1}^{n} b_{ij}^*\mathbf{U}_j^\dagger\mathbf{U}_k = b_{ik}^* \tag{11}$$

provided $k \leq n$; when $k > n$, $(R^{-1})_{ik} = 0$. But due to the group properties, the set of inverses is identical with the set of **R**'s except for a possible change in order, so that proof of Step I is completed.

Proof of Step II. It is supposed that $\mathbf{R}^\gamma$ and $\mathbf{R}^{\gamma'}$ are irreducible representations and that **M** is a matrix which satisfies

$$\mathbf{R}^\gamma\mathbf{M} = \mathbf{M}\mathbf{R}^{\gamma'} \tag{3}$$

for all **R** in the group. Let $\mathbf{M}_i$ be the ith column of **M**. Then (3) may be written:

$$\mathbf{R}^\gamma \mathbf{M}_i = \left\| \begin{matrix} \sum_{j=1}^{d_{\gamma'}} M_{1j} R_{ji}^{\gamma'} \\ \sum_{j=1}^{d_{\gamma'}} M_{2j} R_{ji}^{\gamma'} \\ \cdots\cdots\cdot \\ \sum_{j=1}^{d_{\gamma'}} M_{dj} R_{ji}^{\gamma'} \end{matrix} \right\| = \sum_{j=1}^{d_{\gamma'}} R_{ji}^{\gamma'} \mathbf{M}_j \tag{12}$$

Since $\mathbf{R}^\gamma$ is irreducible, by Step I, $d_{\gamma'}$, the number of independent vectors $\mathbf{M}_j$, must be either zero or at least as large as d_γ. Therefore, aside from the possibility that $\mathbf{M}$ may be a null matrix [in which case (3) can of course be satisfied by any values of $\mathbf{R}^\gamma$ and $\mathbf{R}^{\gamma'}$ whatsoever], one may assume that $d_{\gamma'} \geq d_\gamma$. Next, by taking the transpose conjugate of (3), by similar reasoning, one may draw the conclusion that either $\mathbf{M}^\dagger = \mathbf{0}$ or that $d_\gamma \geq d_{\gamma'}$. Assembling these alternatives, it follows that either $\mathbf{M} = \mathbf{0}$ or that $d_\gamma = d_{\gamma'}$, that is, $\mathbf{M}$ is square. But this is not all that can be said about $\mathbf{M}$: if $\mathbf{M}$ is not $\mathbf{0}$, its determinant does not vanish. This follows from the fact that the $d_\gamma = d_{\gamma'}$ vectors $\mathbf{M}_i$ are linearly independent, which is precisely the condition that the determinant of the matrix $|\mathbf{M}|$ composed of the $\mathbf{M}_i$ as columns should be nonvanishing. But if $|\mathbf{M}| \neq 0$, $\mathbf{M}^{-1}$ exists so that (3) may be written

$$\mathbf{R}^\gamma = \mathbf{M}\mathbf{R}^{\gamma'}\mathbf{M}^{-1}$$

which shows that $\mathbf{R}^\gamma$ and $\mathbf{R}^{\gamma'}$ are equivalent representations, differing only in the "orientation" of their coordinate axes (see Sec. 6-4).

When $\mathbf{R}^\gamma = \mathbf{R}^{\gamma'}$, it is possible to go further and show that $\mathbf{M} = \lambda\mathbf{E}$, that is, $\mathbf{M}$ is a constant matrix. For if $\mathbf{M}$ commutes with $\mathbf{R}^\gamma$, so does the matrix $\mathbf{M} - \lambda\mathbf{E}$, where λ is any constant. But then $\mathbf{M} - \lambda\mathbf{E}$ is subject to the same alternatives as $\mathbf{M}$, namely, $\mathbf{M} - \lambda\mathbf{E} = \mathbf{0}$, or $|\mathbf{M} - \lambda\mathbf{E}| \neq 0$. Since the latter alternative cannot hold for every λ, some value of which must satisfy $|\mathbf{M} - \lambda\mathbf{E}| = 0$, the conclusion is reached that $\mathbf{M} - \lambda\mathbf{E} = \mathbf{0}$ or $\mathbf{M} = \lambda\mathbf{E}$, as was to be proved.

At this point the reader is referred to Appendix XII for a proof of the theorem that the introduction of symmetry coordinates factors the secular determinant.

Proof of Step III. The preceding results now allow a proof of the orthogonality theorem to be set up immediately. Consider the d_γ by $d_{\gamma'}$ matrix $\mathbf{A}$ defined by

$$\mathbf{A} = \sum_T \mathbf{T}^\gamma \mathbf{B} \mathbf{T}^{\gamma'\dagger} \tag{13}$$

in which **T** now stands for a typical group operation, the sum being extended over the whole group; **B** is a completely arbitrary d_γ by $d_{\gamma'}$ matrix which makes possible a matrix multiplication involving the two irreducible representations indicated by γ and γ'. It will now be shown that **A** satisfies all the requirements on the **M** of (3). For

$$\mathbf{R}^\gamma\mathbf{A} = \sum_T \mathbf{R}^\gamma\mathbf{T}^\gamma\mathbf{B}\mathbf{T}^{\gamma'\dagger} = \left[\sum_T (\mathbf{R}^\gamma\mathbf{T}^\gamma)\mathbf{B}\mathbf{T}^{\gamma'\dagger}\mathbf{R}^{\gamma'\dagger}\right]\mathbf{R}^{\gamma'}$$
$$= \left[\sum_T (\mathbf{RT})^\gamma\mathbf{B}(\mathbf{RT})^{\gamma'\dagger}\right]\mathbf{R}^{\gamma'} = \mathbf{A}\mathbf{R}^{\gamma'} \tag{14}$$

The steps of this demonstration depend upon the facts that $\mathbf{R}^{\gamma'}$ is unitary (so that $\mathbf{R}^{\gamma'\dagger}\mathbf{R}^{\gamma'} = \mathbf{E}$), that the transpose conjugate of a product is the product of the transpose conjugates of the individual factors, taken in reverse order, and that summation over T of some matrix function of **RT** is equivalent to summation over the same matrix function of **T** because of the group properties.

Suppose now that $\mathbf{R}^\gamma$ and $\mathbf{R}^{\gamma'}$ are distinct (nonequivalent) irreducible representations. Then, by Theorem II, $\mathbf{A} = \mathbf{0}$. Since **B** is completely arbitrary, it may be chosen in such a way that every element except, say, the $a'a'''$th, vanishes, the exceptional element being assigned the value of unity. Then, by (13), replacing the symbol T by R,

$$\sum_R (R^\gamma)_{aa'}(R^{\gamma'\dagger})_{a'''a''} = 0$$

or, upon taking the complex conjugate,

$$\sum_R R^{\gamma*}_{aa'}R^{\gamma'}_{a''a'''} = 0 \qquad (\gamma \neq \gamma') \tag{15}$$

Note that this result holds for every a and a'' (since **A** vanishes) as well as for every a' and a''' (because of the completely arbitrary character of **B**).

If in contrast to the assumption of the previous paragraph, $\mathbf{R}^\gamma$ and $\mathbf{R}^{\gamma'}$ are equal representations, the matrix **A** of (14) is necessarily a constant matrix. This follows from the fact that (14) implies that **A** *commutes* with all **R** when $\mathbf{R}^\gamma$ and $\mathbf{R}^{\gamma'}$ are identical.

This being the case, the same choice of **B** as was employed in arriving at (15) will lead to

$$\sum_R (R^\gamma)_{aa'}(R^{\gamma\dagger})_{a'''a''} = \delta_{aa''}\lambda \tag{16}$$

Putting $a = a''$ and summing (16) over a,

$$\sum_R \sum_{a=1}^{d_\gamma} (R^{\gamma\dagger})_{a'''a}(R^\gamma)_{aa'} = \sum_R (E)_{a'''a'} = g\delta_{a'''a'} = \sum_{a=1}^{d_\gamma} \delta_{aa}\lambda = d_\gamma\lambda \tag{17}$$

This determines the value of $\lambda = (g/d_\gamma)\delta_{a'''a'}$. Inserting this value of λ in (16), one obtains, after taking the complex conjugate,

$$\sum_R R^{\gamma*}_{aa'} R^{\gamma'}_{a''a'''} = \delta_{aa''}\delta_{a'a'''}\frac{g}{d_\gamma} \qquad (\gamma \neq \gamma')$$

or, upon combining this result with (15), the complete orthogonality theorem

$$\sum_R R^{\gamma*}_{aa'} R^{\gamma'}_{a''a'''} = \delta_{\gamma\gamma'}\delta_{aa''}\delta_{a'a'''}\frac{g}{d_\gamma} \tag{18}$$

APPENDIX XII

PROOF OF THE FACTORING OF THE SECULAR DETERMINANT

In Chap. 6 it was asserted that the introduction of symmetry coordinates would factor the secular equation. A general proof will now be given.

Consider first the potential energy, which, in terms of the symmetry coordinates, $S_{ka}^{(\gamma)}$, assumes the following (matrix) form:

$$2V = \mathbf{S}^{\dagger}\mathbf{F}\mathbf{S} \tag{1}$$

In (1), $\mathbf{S}$ is a column matrix whose elements are the $S_{ka}^{(\gamma)}$, $\mathbf{F}$ is the (square) matrix of the force constants, $\mathbf{S}^{\dagger}$ is the transpose (conjugate) of $\mathbf{S}$.

Now imagine the molecule subjected to a symmetry operation such that

$$\mathbf{S} \rightarrow \mathbf{S}' = \mathbf{R}\mathbf{S} \tag{2}$$

that is, $\mathbf{R}$, a square matrix, represents the transformation of the symmetry coordinates under the general group operation, R. Now, the definition of symmetry coordinates given in the text required that:

1. They constitute a basis for a completely reduced unitary representation of the point group, $\mathcal{G}$, of the molecule.

2. Sets of coordinates of the same degenerate species (irreducible representation) must have identical transformation coefficients.

These requirements mean that $\mathbf{R}$ is a matrix in diagonal block form such that $\mathbf{R}$ contains $\mathbf{R}^{(\gamma)}$ n_γ times along the diagonal if $\mathbf{R}^{(\gamma)}$ corresponds with the γth irreducible representation. For example, if $n_1 = 1, n_2 = 0, n_3 = 0, n_4 = 2$, then

$$\mathbf{R} = \begin{Vmatrix} \mathbf{R}^{(1)} & 0 & 0 \\ 0 & \mathbf{R}^{(4)} & 0 \\ 0 & 0 & \mathbf{R}^{(4)} \end{Vmatrix} \tag{3}$$

The effect of the symmetry operation R upon the potential energy may be expressed in the form

$$R(2V) = \mathbf{S}^{\dagger}\mathbf{R}^{\dagger}\mathbf{F}\mathbf{R}\mathbf{S} \tag{4}$$

but the fundamental property of invariance of the potential energy under all group operations means that

$$R(2V) = 2V \tag{5}$$

Equating the right-hand side of (4) to that of (1) then shows that

$$\mathbf{R}^\dagger \mathsf{F} \mathbf{R} = \mathsf{F} \tag{6}$$

or

$$\mathsf{F}\mathbf{R} = \mathbf{R}\mathsf{F} \tag{7}$$

since $\mathbf{RR}^\dagger$ is the identity when $\mathbf{R}$ is a unitary matrix. The result of (7) may be expressed as follows: the matrix of the potential energy commutes with all group representation matrices.

Now let F be partitioned in blocks such that a given block contains all degenerate components of a given set of symmetry coordinates. In other words, F will be partitioned to correspond with the diagonal blocks of $\mathbf{R}$ and, in the example of $\mathbf{R}$ illustrated in (3), would take the form

$$\mathsf{F} = \begin{Vmatrix} \mathsf{F}_{11} & \mathsf{F}_{12} & \mathsf{F}_{13} \\ \mathsf{F}_{21} & \mathsf{F}_{22} & \mathsf{F}_{23} \\ \mathsf{F}_{31} & \mathsf{F}_{32} & \mathsf{F}_{33} \end{Vmatrix} \tag{8}$$

From the commutation equation (7), it may now be seen first of all, that each diagonal block F_{ii} must commute with the corresponding $\mathbf{R}^{(\gamma)}$; furthermore, the off-diagonal blocks, $\mathsf{F}_{ii'}$, satisfy conditions of the type:

$$\mathbf{R}^{(\gamma)}\mathsf{F}_{ii'} = \mathsf{F}_{ii'}\mathbf{R}^{(\gamma')} \tag{9}$$

But as shown in Appendix XI, F_{ii} must be a *constant* matrix, as also must $\mathsf{F}_{ii'}$ *if* $\mathbf{R}^{(\gamma)}$ and $\mathbf{R}^{(\gamma')}$ are identical irreducible representations. If, on the other hand, $\mathbf{R}^{(\gamma)}$ and $\mathbf{R}^{(\gamma')}$ are nonequivalent irreducible representations, then necessarily $\mathsf{F}_{ii'} = \mathbf{0}$. Thus, in (8), F_{11}, F_{22}, F_{23}, and F_{33} would each be constant matrices, while F_{12}, F_{13} (also F_{21} and F_{31}) would vanish. In this same example, if $\mathbf{R}^{(1)}$ were nondegenerate and $\mathbf{R}^{(4)}$ triply degenerate ($d_1 = 1$, $d_4 = 3$), F would have the detailed form

$$\mathsf{F} = \left\Vert\begin{array}{c|ccc|ccc} a & 0 & 0 & 0 & 0 & 0 & 0 \\ \hline 0 & b & 0 & 0 & d & 0 & 0 \\ 0 & 0 & b & 0 & 0 & d & 0 \\ 0 & 0 & 0 & b & 0 & 0 & d \\ \hline 0 & d & 0 & 0 & c & 0 & 0 \\ 0 & 0 & d & 0 & 0 & c & 0 \\ 0 & 0 & 0 & d & 0 & 0 & c \end{array}\right\Vert \tag{10}$$

But any such **F** could be rearranged to the form

$$\mathbf{F} = \left\|\begin{array}{c|cc|cc|cc}
a & 0 & 0 & 0 & 0 & 0 & 0 \\
\hline
0 & b & d & 0 & 0 & 0 & 0 \\
0 & d & c & 0 & 0 & 0 & 0 \\
\hline
0 & 0 & 0 & b & d & 0 & 0 \\
0 & 0 & 0 & d & c & 0 & 0 \\
\hline
0 & 0 & 0 & 0 & 0 & b & d \\
0 & 0 & 0 & 0 & 0 & d & c
\end{array}\right\|$$

by merely permuting rows and columns, so that it is clear that in the final form the factor corresponding to $R^{(\gamma)}$ will appear identically d_γ times. Since entirely analogous arguments can be applied to the kinetic energy, the **G** matrix will factor in a similar fashion, which immediately shows that any of the various forms of the secular equation factor likewise.

APPENDIX XIII

REDUCTION OF AN IRREDUCIBLE REPRESENTATION IN A SUBGROUP

It is the purpose of this appendix to prove a theorem used in Eq. (20), Sec. 6-4, namely,

$$\sum_{\text{coset } i} R_{aa'}^{(\gamma)} = \delta_{aa'} \sum_{\text{coset } i} \chi_R^{(\gamma)} \tag{1}$$

In general, the elements $R_{aa'}^{(\gamma)}$ of degenerate representations are not uniquely determined. There is, however, a special choice of the $R_{aa'}^{(\gamma)}$, which is important for the construction of symmetry coordinates. This is the choice that corresponds to the reduction of $\Gamma^{(\gamma)}$ in the subgroup $\mathcal{H}$. Suppose that $n^{(\gamma)} = n_\gamma^{(\eta')} = 1$, and that $\Gamma^{(\eta')}$ is totally symmetric. If $\Gamma^{(\gamma)}$ has been reduced in $\mathcal{H}$, the elements $T_{aa'}^{(\gamma)}$ are the elements of certain irreducible representations $\Gamma^{(\eta)}$ of $\mathcal{H}$, provided T is an operation of $\mathcal{H}$. Therefore

$$\sum_T T_{aa'}^{(\gamma)} = \sum_T T_{a''a'''}^{(\eta)} \tag{2}$$

According to the orthogonality theorem for the subgroup

$$\sum_T T_{a''a'''}^{(\eta)} T_{11}^{(\eta')} = \sum_T T_{a''a'''}^{(\eta)} = \delta_{\eta\eta'} h \tag{3}$$

where $T_{11}^{(\eta')} = 1$ because $\Gamma^{(\eta')}$ is assumed to be totally symmetric. Thus it is seen that the sums of the $T_{aa'}^{(\gamma)}$ over the subgroup vanish unless $T_{aa'}^{(\gamma)} = T_{a''a'''}^{(\eta)}$ corresponds to the special species, $\Gamma^{(\eta')}$, which by hypothesis occurs just once. Assuming that the special species occurs in the aa position, it is then true that

$$\sum_T T_{aa}^{(\gamma)} = h \tag{4}$$

and all other sums vanish:

$$\sum_T T_{ab}^{(\gamma)} = 0 \qquad \sum_T T_{bb}^{(\gamma)} = 0, \text{ etc.}$$

In particular, since the sums in the bb and cc positions vanish

$$\sum_T T_{aa}^{(\gamma)} = \sum_T (T_{aa}^{(\gamma)} + T_{bb}^{(\gamma)} + T_{cc}^{(\gamma)}) = \sum_T \chi_T^{(\gamma)} \tag{5}$$

This proves (1) for the special case where $i = 1$. To prove it for other i, it is necessary to note that *all* operations which send S_1 into S_i can be expressed in the form X_iT, where T is any operation of the subgroup which sends S_1 into itself and X_i is *some fixed* operation which sends S_1 into S_i. Then

$$\sum_{\text{coset } i} R = \sum_T X_iT = X_i \sum_T T \tag{6}$$

Expanding (6) to find the aa' matrix element,

$$\begin{aligned}\sum_{\text{coset } i} R_{aa'}^{(\gamma)} &= (X_i^{(\gamma)})_{aa} \sum_T T_{aa'}^{(\gamma)} + (X_i^{(\gamma)})_{ab} \sum_T T_{ba'}^{(\gamma)} + (X_i^{(\gamma)})_{ac} \sum_T T_{ca'}^{(\gamma)} \\ &= (X_i^{(\gamma)})_{aa} \sum_T T_{aa'}^{(\gamma)} = \delta_{aa'} h (X_i^{(\gamma)})_{aa}\end{aligned} \tag{7}$$

Since $\sum_{\text{coset } i} R_{bb}^{(\gamma)}$ and $\sum_{\text{coset } i} R_{cc}^{(\gamma)}$ would not contain $\sum_T T_{aa}^{(\gamma)}$ upon similar expansion, it follows that they vanish and therefore

$$\sum_{\text{coset } i} R_{aa}^{(\gamma)} = \sum_{\text{coset } i} (R_{aa}^{(\gamma)} + R_{bb}^{(\gamma)} + R_{cc}^{(\gamma)}) = \sum_{\text{coset } i} \chi_R^{(\gamma)} \tag{8}$$

Combining (7) and (8), the result is

$$\sum_{\text{coset } i} R_{aa'}^{(\gamma)} = \delta_{aa'} \sum_{\text{coset } i} \chi_R^{(\gamma)}$$

which is the desired theorem.

In case $\Gamma^{(\eta')}$ is not totally symmetric, (1) should be replaced by

$$\sum_T (X_iT)_{aa'}^{(\gamma)} \chi_T^{(\eta')} = \delta_{aa'} \sum_T \chi_{X_iT}^{(\gamma)} \chi_T^{(\eta')}$$

APPENDIX XIV

SYMMETRY SPECIES FOR THE OVERTONES OF DEGENERATE FUNDAMENTALS

The basis of the alternate method mentioned in Sec. 7-3 is the use of polar rather than cartesian degenerate normal coordinates. It is shown in treatises on quantum mechanics[1] that the solution of the wave equation for the doubly degenerate harmonic oscillator expressed in terms of coordinates ρ and ϕ defined by

$$Q_a = \rho \cos \phi \tag{1}$$

$$Q_b = \rho \sin \phi \tag{2}$$

is of the form

$$\psi_{vl} = \mathfrak{N} e^{-\gamma\rho^2/2} F_{v|l|}(\gamma^{\frac{1}{2}}\rho) e^{il\phi} \tag{3}$$

and that the energy of the corresponding stationary state is

$$W_v = (v + 1)h\nu_0 \tag{4}$$

In (3), $F_{v|l|}$ is a polynomial containing only the following powers of ρ: $v, v - 2, v - 4, \ldots, |l|$. The quantum number v assumes the values $0, 1, 2, \ldots$, while l is restricted to the values $0, \pm 2, \pm 4, \ldots, \pm v$, if v is even, or to $\pm 1, \pm 3, \pm 5, \ldots, \pm v$, if v is odd. The total degeneracy of the vibrational level with quantum number v is therefore $v + 1$

The character of the degenerate set of wave functions

$$\psi_{v,\pm v},\ \psi_{v,\pm(v-2)},\ \psi_{v,\pm(v-4)},\ \text{etc.}$$

may be determined when the effect of the symmetry operations upon the coordinates ρ and ϕ is found. Since $\rho^2 = Q_a^2 + Q_b^2$ and all R's leave $Q_a^2 + Q_b^2$ invariant, ρ can only change to $\pm\rho$. Actually, ρ, as a radial polar coordinate, is restricted to positive values, and hence is completely invariant. Therefore, it suffices to determine the effect of the group operations upon the angular coordinate, ϕ.

For certain doubly degenerate species of the dihedral groups, namely, those which represent translation in the plane perpendicular to the principal symmetry axis (T_x and T_y), the transformation of ϕ is readily

[1] See L. Pauling and E. B. Wilson, Jr., "Introduction to Quantum Mechanics," pp. 105–111, McGraw-Hill, New York, 1935.

obtained by visualizing the effect of the group operation in question on Q_a and Q_b. Thus a rotation by an angle α_R about the principal axis sends $\phi \rightarrow \phi + \alpha_R$. If, on the other hand, the symmetry operation sends $Q_a \rightarrow +Q_a$ and $Q_b \rightarrow -Q_b$ (this might be the result of a vertical plane or a horizontal twofold axis), inspection of (1) and (2) shows that $\phi \rightarrow -\phi$. In general, for a species representing translation

$$\phi \xrightarrow{R} \pm(\phi + \alpha_R) \tag{5}$$

Now consider the character of the level $v = 1$ for such a species. The ϕ dependent parts of the wave functions $\psi_{1,+1}$ and $\psi_{1,-1}$ are $e^{i\phi}$ and $e^{-i\phi}$. From (5) it is clear that

$$\begin{aligned} e^{i\phi} &\xrightarrow{R} e^{i\alpha_R}e^{i\phi}, \text{ or } e^{-i\alpha_R}e^{-i\phi} \\ e^{-i\phi} &\xrightarrow{R} e^{-i\alpha_R}e^{-i\phi}, \text{ or } e^{i\alpha_R}e^{i\phi} \end{aligned} \tag{6}$$

where the alternative transforms depend upon whether the sign in (5) is positive or negative, respectively. In the former case,

$$\chi_1(R) = e^{i\alpha_R} + e^{-i\alpha_R} = 2\cos\alpha_R \tag{7}$$

while in the latter case,

$$\chi_1(R) = 0 \tag{8}$$

If, on the other hand, a level with $v > 1$ were under consideration, the character could be expressed for each allowed absolute value of l (constituting a doubly degenerate sublevel) as

$$\chi_{v|l|}(R) = e^{il\alpha_R} + e^{-il\alpha_R} = 2\cos l\alpha_R \tag{9}$$

or

$$\chi_{v|l|}(R) = 0 \tag{10}$$

since the ϕ dependent parts of the corresponding wave functions ψ_{vl} and ψ_{v-l} are $e^{il\phi}$ and $e^{-il\phi}$. Furthermore, the character for the complete level would be

$$\chi_v(R) = \sum_{|l|} \chi_{v|l|}(R) \tag{11}$$

where $|l|$ is summed over the values allowed, as described above. The above formulas are appropriate to all values of $|l| \neq 0$; when $l = 0$, it is clear that the sublevel is non-degenerate and, in fact, totally symmetric, since ϕ does not enter at all.

Exactly the same arguments apply to doubly degenerate species other than that which expresses the transformation of the cartesian coordinates x and y, except that the angle α_R is no longer visualizable in a direct fashion. As shown in Appendix X, the transformation coefficients (for the Q's) are in general of the form

$$\left\| \begin{matrix} \cos \alpha_R & -\sin \alpha_R \\ \pm \sin \alpha_R & \pm \cos \alpha_R \end{matrix} \right\| \tag{12}$$

so that the character of the species in question is $2 \cos \alpha_R$, or zero. It is not hard to see that these values are identical with $\chi_1(R)$, and that (9) and (10) give the characters of the overtones for these species as well as for the translation species. This follows because the transformation of the Q's by (12) has exactly the same effect on the angular coordinate ϕ as that expressed in (5) although α_R is no longer determinable by inspection. If the second choice of signs applies in (12) it is clear that $\chi_{v|l|}(R) = 0$ for all values of $|l|$, except zero, for which $\chi_{v|0|}(R) = 1$, so that it is unnecessary to determine α_R. Therefore, the problem of determining the characters $\chi_v(R)$ reduces to:

1. Determining which choice of signs applies in (12).
2. If the first choice applies, determining the value of α_R. This can be found immediately from the formula

$$\chi_1(R) = \chi_R^{(k)} = 2 \cos \alpha_R \tag{13}$$

To determine which choice of signs applies in (12), note that $\chi_R^{(k)} = 0$ is a necessary, but not a sufficient, condition that the second choice be correct, for if $\alpha_R = \pi/2$, $\chi_R^{(k)}$ would vanish for either choice. But if the first choice applied, the square of the transformation matrix (12) would be

$$\left\| \begin{matrix} \cos \frac{\pi}{2} & -\sin \frac{\pi}{2} \\ \sin \frac{\pi}{2} & \cos \frac{\pi}{2} \end{matrix} \right\|^2 = \left\| \begin{matrix} \cos \pi & -\sin \pi \\ \sin \pi & \cos \pi \end{matrix} \right\| \tag{14}$$

of character -2, whereas for the second choice

$$\left\| \begin{matrix} \cos \frac{\pi}{2} & -\sin \frac{\pi}{2} \\ -\sin \frac{\pi}{2} & -\cos \frac{\pi}{2} \end{matrix} \right\|^2 = \left\| \begin{matrix} 1 & 0 \\ 0 & 1 \end{matrix} \right\| \tag{15}$$

of character $+2$. Therefore, the criterion for the choice of signs is that the second choice applies if and only if $\chi_R^{(k)} = 0$ and $\chi_{R^2}^{(k)} = 2$.

As a further assistance in determining α_R in the dihedral groups, it may be noted that

$$\alpha_R = \alpha_R^{(T)} q \tag{16}$$

where $\alpha_R^{(T)}$ is the appropriate angle for the same operation in the species which represents translation and q is identical with the numerical subscript designating the species E_q.

For illustration, the example given in Table 7-1 will be repeated. In Table XIV-1, the first two rows give the classes and characters of E. In the next two rows the choice of sign and the values of α_R are given. The

TABLE XIV-1. SYMMETRY SPECIES OF THE OVERTONE $v = 4$ OF E IN $\mathfrak{T}_d$

R	E	C_3	C_2	σ_d	S_4	Species
$\chi_{1(R)} = \chi_R^{(E)}$	2	−1	2	0	0	E
Sign	+	+	+	−	−	
α_R	0	$2\frac{\pi}{3}$	0			
$\chi_{4\|0\|(R)}$	1	1	1	1	1	A_1
$\chi_{4\|2\|(R)}$	2	−1	2	0	0	E
$\chi_{4\|4\|(R)}$	2	−1	2	0	0	E
$\chi_{4(R)}$	5	−1	5	1	1	$A_1 + 2E$

sign for σ_d is negative since $\chi_{\sigma_d}^{(E)} = 0$ and $\chi_{\sigma_d^2}^{(E)} = \chi_E^{(E)} = +2$. The total character for the level is the sum for the sublevels $|l| = 0, 2,$ and 4, which are, for convenience, displayed separately. The total character can be given in closed form, however, since

$$\chi_v(R) = \sum_l e^{il\alpha_R} \tag{17}$$

where l runs over the allowed values. By regarding the sum as that of a geometric progression, the following results are obtained,

$$\chi_v(R^+) = \frac{\sin (v + 1)\alpha_{R^+}}{\sin \alpha_{R^+}} \tag{18}$$

$$\chi_v(R^-) = \begin{cases} 0 & v \text{ odd} \\ 1 & v \text{ even} \end{cases} \tag{19}$$

in which R^+ stands for an operation for which the sign in (12) is positive, R^- an operation with negative sign. The reader may apply (18) and (19) to the last row of Table XIV-1, noting that

$$\lim_{\alpha \to 0} \frac{\sin (v + 1)\alpha}{\sin \alpha} = v + 1$$

The triply degenerate species may be handled in an entirely analogous fashion. If

$$Q_a = r \sin \theta \cos \phi \tag{20}$$

$$Q_b = r \sin \theta \sin \phi \tag{21}$$

$$Q_c = r \cos \theta \tag{22}$$

the wave functions are

$$\psi_{vlm} = \mathfrak{N}e^{-\gamma r^2/2}G_{vl}(\gamma^{\frac{1}{2}}r)P_l^{|m|}(\cos\theta)e^{im\phi} \tag{23}$$

in which the quantum numbers have the ranges $v = 0, 1, 2, \ldots$; $l = 0, 2, 4, \ldots, v$, if v is even, or $l = 1, 3, 5, \ldots, v$, if v is odd; $m = 0, \pm 1, \pm 2, \ldots, \pm l$. The total number of wave functions for a level of energy

$$W_v = (v + \tfrac{3}{2})h\nu_0$$

is $(v + 1)(v + 2)/2$ which corresponds with the degeneracy found with cartesian coordinates.

To determine the characters, it is necessary to find the effect of the symmetry operations upon θ and ϕ, noting that the associated Legendre function, $P_l^{|m|}(\cos\theta)$, can be written in the form

$$P_l^{|m|}(\cos\theta) = \sin^{|m|}\theta \frac{d^{|m|}P_l(\cos\theta)}{(d\cos\theta)^{|m|}} \tag{24}$$

where P_l is a polynomial containing only those powers of $\cos\theta$ of the same parity as l. The effect of a group operation upon Q_a, Q_b, and Q_c may be readily visualized by orienting these coordinates in such a fashion that Q_c lies along the axis about which the rotation occurs. This, of course, cannot be done simultaneously for all group operations, but since we are interested only in characters, it is permissible. The transformation matrix for the Q's then has the form

$$\begin{Vmatrix} \cos\alpha_R & -\sin\alpha_R & 0 \\ \sin\alpha_R & \cos\alpha_R & 0 \\ 0 & 0 & \pm 1 \end{Vmatrix} \tag{25}$$

where α_R is the angle through which Q_a and Q_b are rotated, and the choice of signs depends upon whether $Q_c \to +Q_c$ or $Q_c \to -Q_c$. It is now clear that $\theta \xrightarrow{R} \theta$ if $+1$ appears in (25), or $\theta \xrightarrow{R} \pi - \theta$ if -1 appears. Furthermore, $\phi \xrightarrow{R} \phi + \alpha_R$. The effect of these transformations of angles upon the functions defined in (23) and (24) may be expressed in the form,

$$\psi_{vlm} \xrightarrow{R^+} e^{im\alpha_{R^+}}\psi_{vlm} \tag{26}$$

$$\psi_{vlm} \xrightarrow{R^-} (-1)^{l-|m|}e^{im\alpha_{R^-}}\psi_{vlm} \tag{27}$$

since, if $\theta \xrightarrow{R} \theta$, $P_l^{|m|}(\cos\theta) \xrightarrow{R} P_l^{|m|}(\cos\theta)$, if $\theta \xrightarrow{R} \pi - \theta$, $\cos\theta \xrightarrow{R} -\cos\theta$, $\sin\theta \xrightarrow{R} \sin\theta$, and $P_l^{|m|}(\cos\theta) \xrightarrow{R} (-1)^{l-|m|}P_l^{|m|}(\cos\theta)$, the last transformation following, since $P_l^{|m|}(-\cos\theta)$ changes sign if $l - |m|$ is odd by (24). Furthermore,

$$e^{im\phi} \longrightarrow e^{im(\phi+\alpha_R)} = e^{im\alpha_R} \cdot e^{im\phi}$$

From (26) and (27), the characters[1] of various sublevels of v can be deduced immediately. For example,

$$\chi_{vl}(R^+) = \sum_{m=-l}^{+l} e^{im\alpha_{R^+}} = \frac{\sin [(2l + 1)\alpha_{R^+}/2]}{\sin (\alpha_{R^+}/2)} \tag{28}$$

$$\chi_{vl}(R^-) = \sum_{m=-l}^{+l} (-1)^{l-|m|} e^{im\alpha_{R^-}} = \frac{\cos [(2l + 1)\alpha_{R^-}/2]}{\cos (\alpha_{R^-}/2)} \tag{29}$$

These summations can also be extended over all allowed values of l with the following results:

$$\chi_v(R^+) = \frac{\sin [(v + 1)\alpha_{R^+}/2] \sin [(v + 2)\alpha_{R^+}/2]}{\sin (\alpha_{R^+}/2) \sin \alpha_{R^+}} \tag{30}$$

$$\chi_v(R^-) = \frac{\sin [(v + 1)\alpha_{R^-}/2] \cos [(v + 2)\alpha_{R^-}/2]}{\cos (\alpha_{R^-}/2) \sin \alpha_{R^-}} \qquad v \text{ odd} \tag{31}$$

$$\chi_v(R^-) = \frac{\cos [(v + 1)\alpha_{R^-}/2] \sin [(v + 2)\alpha_{R^-}/2]}{\cos (\alpha_{R^-}/2) \sin \alpha_{R^-}} \qquad v \text{ even} \tag{32}$$

As in the doubly degenerate case, the determination of the sign in (25) and the evaluation of α_R depend upon the character

$$\chi_1(R) = \chi_R^{(F)} = \pm 1 + 2 \cos \alpha_R \tag{33}$$

To determine the sign, note that the character of R^2 is the sum of the diagonal elements of

$$\left\| \begin{matrix} \cos \alpha_R & -\sin \alpha_R & 0 \\ \sin \alpha_R & \cos \alpha_R & 0 \\ 0 & 0 & \pm 1 \end{matrix} \right\|^2 = \left\| \begin{matrix} \cos 2\alpha_R & -\sin 2\alpha_R & 0 \\ \sin 2\alpha_R & \cos 2\alpha_R & 0 \\ 0 & 0 & 1 \end{matrix} \right\|$$

that is, $1 + 2 \cos 2\alpha_R = 1 + (4 \cos^2 \alpha_R - 2)$, and that the square of the character of R is $1 \pm 4 \cos \alpha_R + 4 \cos^2 \alpha_R$. Therefore,

$$\chi_1(R^2) = -1 + 4 \cos^2 \alpha_R = [\chi_1(R)]^2 \mp 2\chi_1(R) \tag{34}$$

where the negative sign in (34) corresponds with the positive sign in (33), and vice versa. In case $\chi_1(R) = 0$, this criterion fails, but an extension of this sort of argument shows that the sign in (33) must be positive if $\chi_1(R) = 0$ and $\chi_1(R^3) = +3$.

[1] The reader should note that the consequence of choosing different orientations of Q_a, Q_b, and Q_c for convenience in determining the characters is that the ψ_{vlm} are apparently nonmixing according to (26) and (27). This would imply that each ψ_{vlm} transformed as the basis of a nondegenerate representation, which is clearly impossible. It is only when a suitable summation over m is carried out that proper characters of sublevels are obtained.

In Table XIV-2, the cases $v = 10$ of F_{1g} and F_{1u} in the group $\mathcal{O}_h$ are worked out as an example.

TABLE XIV-2. SYMMETRY SPECIES OF THE OVERTONE $v = 10$ OF F_{1g} AND F_{1u} IN $\mathcal{O}_h$

R	E	$8C_3$	$6C_2'$	$6C_4$	$3C_2$	i	$8S_6$	$6\sigma_d$	$6S_4$	$3\sigma_h$
$\chi_{1(R)} = \chi_R^{(F_{1g})}$	+3	0	−1	+1	−1	+3	0	−1	+1	−1
Sign	+	+	+	+	+	+	+	+	+	+
α_R	0	$2\frac{\pi}{3}$	π	$\frac{\pi}{2}$	π	0	$2\frac{\pi}{3}$	π	$\frac{\pi}{2}$	π
$\chi_{10(R)}$	66	0	6	0	6	66	0	6	0	6
$\chi_{1(R)} = \chi_R^{(F_{1u})}$	+3	0	−1	+1	−1	−3	0	+1	−1	+1
Sign	+	+	+	+	+	−	−	−	−	−
α_R	0	$2\frac{\pi}{3}$	π	$\frac{\pi}{2}$	π	π	$\frac{\pi}{3}$	0	$\frac{\pi}{2}$	0
$\chi_{10(R)}$	66	0	6	0	6	66	0	6	0	6

In this example, as is the case generally, even overtones of corresponding g and u species turn out to have the same structure, which in the present instance is

$$\Gamma = 5A_{1g} + 2A_{2g} + 7E_g + 6F_{1g} + 9F_{2g}$$

The signs for the group operation i are determined, using (34), by the fact that

$$\chi_{i^2}^{(F_{1g})} = \chi_E^{(F_{1g})} = (\chi_i^{(F_{1g})})^2 - 2\chi_i^{(F_{1g})}$$

and

$$\chi_{i^2}^{(F_{1u})} = \chi_E^{(F_{1u})} = (\chi_i^{(F_{1u})})^2 + 2\chi_i^{(F_{1u})}$$

Equation (33) is used to compute the values of α_R, (30) to compute $\chi_{10}(R)$ for the overtone of F_{1g}, while (30) and (32) are used to evaluate $\chi_{10}(R)$ for the overtone of F_{1u}.

APPENDIX XV

TRANSFORMATION OF THE POLARIZABILITY COMPONENTS, α_{xx}, α_{xy}, ETC.

In Sec. 7-6 it was asserted that α_{xx}, α_{xy}, etc., transform under symmetry operations like x^2, xy, etc. Write the equations which define the polarizability [Eq. (1), Sec. 7-6] in matrix-vector form

$$\boldsymbol{\mu} = \boldsymbol{\alpha}\boldsymbol{\mathcal{E}} \tag{1}$$

Let the transformation matrix be $\mathbf{R}$; the dipole moment and electric field are

$$\boldsymbol{\mu}' = \mathbf{R}\boldsymbol{\mu} \tag{2}$$

and

$$\boldsymbol{\mathcal{E}}' = \mathbf{R}\boldsymbol{\mathcal{E}} \tag{3}$$

respectively, after the transformation. If $\boldsymbol{\alpha}'$ is the polarizability matrix after the transformation

$$\boldsymbol{\mu}' = \boldsymbol{\alpha}'\boldsymbol{\mathcal{E}}' \tag{4}$$

But, from (2) and (3),

$$\mathbf{R}\boldsymbol{\mu} = \boldsymbol{\alpha}'\mathbf{R}\boldsymbol{\mathcal{E}}$$

or

$$\boldsymbol{\mu} = \mathbf{R}^{-1}\boldsymbol{\alpha}'\mathbf{R}\boldsymbol{\mathcal{E}}$$

hence

$$\boldsymbol{\alpha} = \mathbf{R}^{-1}\boldsymbol{\alpha}'\mathbf{R} \tag{5}$$

From (5) it follows that

$$\boldsymbol{\alpha}' = \mathbf{R}\boldsymbol{\alpha}\mathbf{R}^{-1} \tag{6}$$

Since $\mathbf{R}$ is an orthogonal matrix, $(R^{-1})_{ij} = R_{ji}$, and $\boldsymbol{\alpha}$ is symmetric, the component equations of (6) have the form

$$\alpha'_{ij} = \sum_{kl} R_{ik}\alpha_{kl}(R^{-1})_{lj} = \sum_{kl} R_{ik}R_{jl}\alpha_{kl} \tag{7}$$

namely,

$$\begin{aligned} \alpha'_{xx} &= R_{xx}^2\alpha_{xx} + R_{xy}^2\alpha_{yy} + R_{xz}^2\alpha_{zz} \\ &\qquad + 2R_{xx}R_{xy}\alpha_{xy} + 2R_{xx}R_{xz}\alpha_{xz} + 2R_{xy}R_{xz}\alpha_{yz} \qquad (8) \\ \alpha'_{xy} &= R_{xx}R_{yx}\alpha_{xx} + R_{xy}R_{yy}\alpha_{yy} + R_{xz}R_{yz}\alpha_{zz} + (R_{xx}R_{yy} + R_{xy}R_{yx})\alpha_{xy} \\ &\qquad + (R_{xx}R_{yz} + R_{xz}R_{yx})\alpha_{xz} + (R_{xy}R_{yz} + R_{xz}R_{yy})\alpha_{yz} \end{aligned}$$

etc.

But since

$$\begin{aligned} X' &= R_{xx}X + R_{xy}Y + R_{xz}Z \\ Y' &= R_{yx}X + R_{yy}Y + R_{yz}Z \\ Z' &= R_{zx}X + R_{zy}Y + R_{zz}Z \end{aligned} \tag{9}$$

$$\begin{aligned} (X')^2 &= R_{xx}^2X^2 + R_{xy}^2Y^2 + R_{xz}^2Z^2 \\ &\qquad + 2R_{xx}R_{xy}XY + 2R_{xx}R_{xz}XZ + 2R_{xy}R_{xz}YZ \quad (10) \\ X'Y' &= R_{xx}R_{yx}X^2 + R_{xy}R_{yy}Y^2 + R_{xz}R_{yz}Z^2 + (R_{xx}R_{yy} + R_{xy}R_{yx})XY \\ &\qquad + (R_{xx}R_{yz} + R_{xz}R_{yx})XZ + (R_{xy}R_{yz} + R_{xz}R_{yy})YZ \\ &\text{etc.} \end{aligned}$$

which proves the assertion.

The sum of the three diagonal components of $\boldsymbol{\alpha}$ is invariant since

$$\begin{aligned} \sum_i \alpha'_{ii} &= \sum_i \sum_{kl} R_{ik}R_{il}\alpha_{kl} = \sum_{kl} \alpha_{kl} \sum_i R_{ik}R_{il} \\ &= \sum_{kl} \alpha_{kl}\delta_{kl} = \sum_k \alpha_{kk} \end{aligned}$$

because $\mathbf{R}$ is orthogonal.

For a given group operation, R, involving a rotation about the z axis (with $+$ denoting a proper and $-$ an improper rotation), the character of the transformation of the polarizability components may be evaluated as

$$\begin{aligned} \chi(R) &= R_{xx}^2 + R_{yy}^2 + R_{zz}^2 + (R_{xx}R_{yy} + R_{xy}R_{yx}) \\ &\qquad + (R_{xx}R_{zz} + R_{xz}R_{zx}) + (R_{yy}R_{zz} + R_{yz}R_{zy}) \\ &= \cos^2 \alpha_R^{\pm} + \cos^2 \alpha_R^{\pm} + 1 + (\cos^2 \alpha_R^{\pm} - \sin^2 \alpha_R^{\pm}) \\ &\qquad \pm \cos \alpha_R^{\pm} \pm \cos \alpha_R^{\pm} \\ &= 2 \cos \alpha_R^{\pm}(\pm 1 + 2 \cos \alpha_R^{\pm}) \end{aligned} \tag{11}$$

This agrees with Eq. (4), Sec. 7-6, which was obtained through use of the analogy with the overtone $v = 2$ of a triply degenerate frequency.

APPENDIX XVI

ROTATIONAL ENERGIES AND SELECTION RULES

The interpretation of the vibrational spectrum, and in particular the assignment of gaseous-phase infrared or Raman frequencies to the different symmetry species, is sometimes facilitated by consideration of the shape of the envelope of the band, which is determined by the rotational energy levels and selection rules. In this appendix a brief résumé of the rotational theory will be developed.

According to Eq. (3), Sec. 11-5, the rotational energy may be separated from vibrational energy in the approximation in which change of moment of inertia with vibration and the interaction of angular momenta of rotation and vibration are neglected. In such a case, the part of the Hamiltonian corresponding to the rotational energy is given by

$$H = \frac{1}{2}\left(\frac{\mathfrak{M}_x^2}{I_x^0} + \frac{\mathfrak{M}_y^2}{I_y^0} + \frac{\mathfrak{M}_z^2}{I_z^0}\right)$$

in which $\mathfrak{M}_x$, $\mathfrak{M}_y$, $\mathfrak{M}_z$ are the components of angular momentum resolved along the principal axes of the molecule and I_x^0, I_y^0, I_z^0 are the principal moments of inertia. A system with the Hamiltonian given above is known as a rigid rotor. The rigid rotor forms the basis for the treatment of rotational spectra, but this approximation is invalid for vibration-rotation bands involving degenerate vibrations because of the neglect of the coupling of rotational and vibrational angular momentum. The effect of this coupling is discussed later.

Classification of Rotor Type. The solutions of the rotational energy problem are classified according to the rotor type, *i.e.*, according to relations between the principal moments of inertia. A *linear* molecule has $I_x^0 = I_y^0$ and $I_z^0 = 0$ since all atoms lie on the z axis. A *spherical* rotor has $I_x^0 = I_y^0 = I_z^0$. A *symmetric* rotor has $I_x^0 = I_y^0 \neq I_z^0 \neq 0$. Finally, an *asymmetric* rotor has three different principal moments of inertia.

Molecules may have two or three equal moments of inertia (and hence be symmetric or spherical tops) either by symmetry or accidentally. The following elements of symmetry are sufficient conditions to ensure that the rotors with three nonzero moments are symmetrical or spherical:

1. If there is one symmetry axis of threefold or higher symmetry, the molecule will be a symmetric rotor (groups $\mathcal{C}_n$, $\mathcal{C}_{nv}$, $\mathcal{C}_{nh}$, $\mathcal{S}_n$, $\mathcal{D}_n$, $\mathcal{D}_{nd}$, $\mathcal{D}_{nh}$ with $n > 2$, and $\mathcal{V}_d$).

2. If there are two or more axes of greater than twofold symmetry, the molecule will be a spherical rotor (groups $\mathfrak{T}$, $\mathfrak{T}_h$, $\mathfrak{T}_d$, $\mathcal{O}$, $\mathcal{O}_h$, $\mathcal{I}$).

Energy Levels and Degeneracies of the Rigid Rotor. The levels and degeneracies for the four rotor types are summarized in Table XVI-1.

TABLE XVI-1. ENERGY LEVELS AND DEGENERACIES FOR THE FOUR TYPES OF RIGID ROTORS
($J = 0, 1, 2, \ldots ; K = 0, \pm 1, \pm 2, \ldots , \pm J$)

Type	Condition on moments	Energy	Degeneracy
Linear	$I_x^0 = I_y^0 = \frac{h^2}{8\pi^2 b}$; $I_z^0 = 0$	$bJ(J+1)$	$2J+1$
Spherical ...	$I_x^0 = I_y^0 = I_z^0 = \frac{h^2}{8\pi^2 b}$	$bJ(J+1)$	$(2J+1)^2$
Symmetric ..	$I_x^0 = I_y^0 = \frac{h^2}{8\pi^2 b} \neq I_z^0 = \frac{h^2}{8\pi^2 a}$	$bJ(J+1) + (a-b)K^2$	$2J+1$ if $K = 0$; $2(2J+1)$ if $\|K\| > 0$
Asymmetric.	$I_a^0 = \frac{h^2}{8\pi^2 a} < I_b^0 = \frac{h^2}{8\pi^2 b} < I_c^0 = \frac{h^2}{8\pi^2 c}$	$\frac{a+c}{2} J(J+1) + \frac{a-c}{2} E_\tau^J(\kappa)$	$2J+1$

In the table, the quantum number J in each case specifies the *square of the total angular momentum*, equal to $(h^2/4\pi^2)J(J+1)$. Moreover, the quantum number K gives the *component of angular momentum* along the z axis of the molecule, equal to $(h/2\pi)K$. The energy of the asymmetric rotor cannot, in general, be expressed in closed form, but can be obtained from tabulated values of the function[1] $E_\tau^J(\kappa)$ in which the index $\tau = 0, \pm 1, \pm 2, \ldots , \pm J$; and κ, the *asymmetry parameter*, is given by

$$\kappa = \frac{2b - a - c}{a - c} \qquad (1)$$

where a, b, c are defined in Table XVI-1.

If the axes in the asymmetric rotor are chosen in accordance with the convention that $a \geq b \geq c$ it is apparent that $-1 < \kappa < +1$, and that

[1] G. W. King, R. M. Hainer, and P. C. Cross, *J. Chem. Phys.*, **11**: 27 (1943). See also T. E. Turner, B. L. Hicks, and G. Reitwiesner, *Ballistic Research Lab. Rept.* **878**, Aberdeen Proving Ground.

$\kappa = -1 (b = c)$ and $\kappa = +1$ $(b = a)$ correspond to a prolate and to an oblate symmetric rotor respectively.

The rotational states require a third quantum number (in addition to J and K or J and τ) for their complete specification. This number is M which gives the component of angular momentum, equal to $(h/2\pi)M$, along the Z axis of the nonrotating coordinate system. Since M, like K, can assume the values $M = 0, \pm 1, \pm 2, \ldots, \pm J$, there are $2J + 1$ states which differ only in the direction of the axis about which rotation occurs. These states are degenerate in the absence of external electromagnetic fields. In addition, the spherical rotor is $2J + 1$ = fold degenerate since its energy does not depend upon K. The symmetric rotor also has a twofold degeneracy in the sense that $+K$ and $-K$ lead to equal energies.

Selection Rules; Infrared Absorption. In Chap. 3 it was shown that the possibility of an infrared absorption accompanying a simultaneous change in vibrational and rotational energy depended upon the simultaneous nonvanishing of the integrals

$$\int \psi^*_{R''} \Phi_{Fg} \psi_{R'} \, d\tau_R$$

and

$$\int \psi^*_{V''} \mu_g \psi_{V'} \, d\tau_V$$

where $F = X, Y, Z$, and g, which must be the same in the two integrals, is x, y, or z. From the character tables it is easy to determine, for any particular vibrational transition, which component of dipole moment is involved.

For the linear rigid rotor, the fundamentals associated with vibration along the molecular axis give rise to integrals involving μ_z; the (degenerate) bending vibrations give rise to integrals involving μ_x and μ_y. If these bands are designated as parallel (||) and perpendicular ($\perp$) bands, respectively, the rotational transitions which accompany the change in vibrational energy will be determined by the following integrals:

$$\| \text{ bands } \int \psi^*_{R''} \Phi_{Fz} \psi_{R'} \, d\tau$$
$$\perp \text{ bands } \int \psi^*_{R''} \Phi_{Fx} \psi_{R'} \, d\tau$$
$$\int \psi^*_{R''} \Phi_{Fy} \psi_{R'} \, d\tau$$

It may be shown[1] that the integral involving Φ_{Fz} vanishes unless $\Delta J = \pm 1$ while that involving Φ_{Fx} or Φ_{Fy} vanishes unless $\Delta J = 0, \pm 1$. The vibration-rotation band of parallel (||) type should therefore consist of the following frequencies,

$$\nu = \nu_v - \frac{2b}{h} J \qquad J = 1, 2, 3, \ldots \tag{2}$$

[1] F. C. Cross, R. M. Hainer, and G. W. King, *J. Chem. Phys.*, **12**: 210 (1944).

or

$$\nu = \nu_v + \frac{2b}{h}(J + 1) \qquad J = 0, 1, 2, \ldots \tag{3}$$

while that of the perpendicular ($\perp$) band also contains the frequency $\nu = \nu_v$, which should have an intensity comparable with that of the sum of all the lines given by (2) and (3).

The lines corresponding to $\Delta J = -1$ constitute the so-called P branch, the collection of lines corresponding to $\Delta J = 0$ (missing in the parallel ($\|$) band) is the Q branch, and the lines corresponding to $\Delta J = +1$ form the R branch.

Selection rules for several types of rotors are summarized in Table XVI-2.

TABLE XVI-2. INFRARED ROTATIONAL SELECTION RULES

Rotor type	Band type	Selection rules
Linear	$\|(\mu_z)$	$\Delta J = \pm 1$
	$\perp(\mu_x, \mu_y)$	$\Delta J = 0, \pm 1$
Spherical	(μ_x, μ_y, μ_z)	$\Delta J = 0, \pm 1$
Symmetric	$\|(\mu_z)$	$\Delta J = 0, \pm 1; \Delta K = 0$ if $K \neq 0$
		$\Delta J = \pm 1; \Delta K = 0$ if $K = 0$
	$\perp(\mu_x, \mu_y)$	$\Delta J = 0, \pm 1; \Delta K = \pm 1$
Asymmetric[a]	$\|_x(\mu_x)$	
	$\|_y(\mu_y)$	$\Delta J = 0, \pm 1$
	$\|_z(\mu_z)$	

[a] G. Herzberg, "Infrared and Raman Spectra of Polyatomic Molecules," pp. 468–491, Van Nostrand, New York, 1945.

Evidently the rotational structure of the band of a spherical rigid rotor is similar to that of a perpendicular ($\perp$) band of a linear rigid rotor. The two types of bands of a symmetric rotor are more complex. The parallel ($\|$) type of band, for which $\Delta K = 0$, will, however, be similar in structure to that of a spherical rotor or a perpendicular ($\perp$) band in a linear rotor. For a perpendicular ($\perp$) band, however, the position of the Q branch is dependent both upon the initial value of K in the transition and upon the sign of ΔK, as can be seen from the following expressions:[1]

$$Q \text{ branch:} \qquad \Delta J = 0 \qquad \nu = \nu_v + \frac{a - b}{h}(\pm 2K + 1) = \nu_Q(\pm K) \tag{4}$$

[1] The reader is reminded that these equations are for the rigid rotor and seldom are valid approximations for real molecules because of the coupling of angular momenta.

P branch: $\Delta J = -1$

$$\nu = \nu_Q(\pm K) - \frac{2b}{h} J \qquad J = 1, 2, 3, \ldots \quad (5)$$

R branch: $\Delta J = 1$

$$\nu = \nu_Q(\pm K) + \frac{2b}{h}(J + 1) \qquad J = 0, 1, 2, \ldots \quad (6)$$

In the above expressions, $+K$ is chosen in case $\Delta K = +1$, and K can then assume the values $0, 1, 2, \ldots, J - 1$ whereas $-K$ is chosen in case $\Delta K = -1$, in which case $K = 1, 2, \ldots, J$. Here again the intensity of each *Q* branch should be comparable with the integrated intensity of the *P* and *R* branch; in the present case, therefore, the most prominent feature of the band will be the various *Q* branches separated by the frequency interval $2(a - b)/h$.

No attempt will be made here to discuss the very complex band types of the asymmetric rotor.[1]

Selection Rules; Raman Effect. Since the intensities of Raman scattering are determined by the integrals

$$\int \psi^*_{R''} \Phi_{Fg} \Phi_{F'g'} \psi_{R'} \, d\tau_R \int \psi^*_{V''} \alpha_{gg'} \psi_{V'} \, d\tau_V$$

according to Sec. 3-7, it is apparent that the rotational selection rules will depend upon the vanishing or nonvanishing of the integrals involving quadratic terms in the direction cosines, and in particular, those with indices g and g' corresponding to a nonvanishing vibrational integral of $\alpha_{gg'}$. Just as in the case of infrared absorption, the character tables can be used to find the species of the various components of $\alpha_{gg'}$. Using the linear molecule again as an example (groups $\mathcal{C}_{\infty v}$ and $\mathcal{D}_{\infty h}$) it is seen that $\alpha_{xx} + \alpha_{yy}$ and α_{zz} appear in the totally symmetric species (Σ^+ or Σ_g^+), while α_{yz} and α_{zx} constitute a degenerate pair in Π or Π_g species. Since there are no fundamentals of species Δ, the components $\alpha_{xx} - \alpha_{yy}$ and α_{xy} need not be further considered in a discussion of the fundamental Raman transitions.

The integrals involving the appropriate $\Phi_{Fg}\Phi_{F'g'}$ have been discussed,[2] and the results are summarized in Table XVI-3.

From Table XVI-3, several statements about the rotational fine structure in the Raman effect may be made immediately. The polarized, totally symmetric band of a linear rigid rotor will resemble a perpendicular ($\perp$) infrared band of the same type of rotor, except that the line spacing is twice as great ($|\Delta J| = 2$ instead of $|\Delta J| = 1$). The degenerate Raman band, on the other hand, will more nearly resemble a parallel ($\|$) infrared

[1] G. Herzberg, "Infrared and Raman Spectra of Polyatomic Molecules," pp. 468–491, Van Nostrand, New York, 1945.

[2] G. Placzek and E. Teller, *Z. Physik*, **81**: 209 (1933).

TABLE XVI-3. RAMAN ROTATIONAL SELECTION RULES

Rotor type	Band type			Selection rules
	Species	Polarizability component	Polarization	
Linear	Totally symmetric	$\alpha_{xx} + \alpha_{yy}; \alpha_{zz}$	Polarized	$\Delta J = 0, \pm 2$
	Degenerate (Π)	$(\alpha_{yz}, \alpha_{zx})$	Depolarized	$\Delta J = 0^{a}, \pm 1, \pm 2$
Spherical	Totally symmetric	$\alpha_{xx} + \alpha_{yy} + \alpha_{zz}$	Polarized	$\Delta J = 0^{b}$
	Doubly degenerate	$(\alpha_{xx} + \alpha_{yy} - 2\alpha_{zz}, \alpha_{xx} - \alpha_{yy})$	Depolarized	$\Delta J = 0, \pm 1, \pm 2;$ $J' + J'' \geq 2$
	Triply degenerate	$(\alpha_{xy}, \alpha_{yz}, \alpha_{zx})$	Depolarized	$\Delta J = 0, \pm 1, \pm 2;$ $J' + J'' \geq 2$
Symmetric	Totally symmetric	$\alpha_{xx} + \alpha_{yy}; \alpha_{zz}$	Polarized	$\Delta J = 0^{b}, \pm 1, \pm 2;$ $J' + J'' \geq 2; \Delta K = 0$
	Nontotally symmetric,[c] nondegenerate	$\alpha_{xx} - \alpha_{yy}; \alpha_{xy}$	Depolarized	$\Delta J = 0, \pm 1, \pm 2;$ $J' + J'' \geq 2; \Delta K = \pm 2$
	Degenerate	$(\alpha_{xx} - \alpha_{yy}; \alpha_{xy})^{d}$	Depolarized	$\Delta J = 0, \pm 1, \pm 2;$ $J' + J'' \geq 2; \Delta K = \pm 2$
		$(\alpha_{yz}, \alpha_{zx})^{d}$	Depolarized	$\Delta J = 0, \pm 1, \pm 2;$ $J' + J'' \geq 2; \Delta K = \pm 1$
Asymmetric[e]				$\Delta J = 0, \pm 1, \pm 2;$ $J' + J'' \geq 2$

[a] Weak.

[b] Strong.

[c] Only in molecules with fourfold axes.

[d] For some symmetry groups, both polarizability components occur in a given species, while for others, only one occurs.

[e] See G. Herzberg, "Infrared and Raman Spectra of Polyatomic Molecules," pp. 468–491, Van Nostrand, New York, 1945.

band, since the Q branch, although allowed, is very weak. The O and S branches, corresponding to $\Delta J = -2$ and $\Delta J = +2$, will also modify the intensity distribution somewhat.

For the spherical rotor, for example, CH_4, CCl_4, SF_6, the totally symmetric, polarized "band" should be extremely sharp since only the Q branch is allowed.

The symmetric rotor fine structure for a totally symmetric vibration (polarized) should be similar in the Raman effect to the parallel (||) infrared band of the same type of molecule, except that O and S branches occur and that the Q branch is very strong. The nontotally symmetric and degenerate modes in the Raman spectrum should have a fine structure resembling that of a perpendicular ($\perp$) infrared band, since changes in K occur.

Band Envelopes. Except where experimental apparatus and conditions permit high resolution (less than a few wave numbers) and except for molecules possessing quite small moments of inertia, such as CH_4, H_2O, NH_3, etc., where the only atoms at appreciable distances from the center of gravity are hydrogens, the fine structure will be unresolved. The actual form of the envelope will depend upon quantitative line intensities, the Boltzmann factor governing the population in excited rotational states, and indeed upon pressure and the experimental slit width. Quantitative calculations of the theoretical band envelopes, especially with reference to the relative intensities of the O, P, Q, R, and S branches, have been made.[1] Note, however, that for real molecules the coupling of rotational and vibrational angular momentum must be taken into account. The reader is warned, therefore, that critical judgment must be exercised in attempting to make vibrational assignments on the basis of the band contours.

Coupling of Angular Momenta. The preceding discussion, which neglects the vibration-rotation interaction (Coriolis effect), is especially inadequate for doubly degenerate vibrational states of symmetric rotors and degenerate states of spherical rotors, where the vibrational angular momentum causes a first order perturbation which splits the degeneracy. Still neglecting the change of moment of inertia with vibration, the Hamiltonian can be written as

$$H = \frac{1}{2}\left[\frac{(\mathfrak{M}_x - \mathfrak{m}_x)^2}{I_x^0} + \frac{(\mathfrak{M}_y - \mathfrak{m}_y)^2}{I_y^0} + \frac{(\mathfrak{M}_z - \mathfrak{m}_z)^2}{I_z^0}\right] + H_V$$

in which H_V is the purely vibrational Hamiltonian.

[1] R. M. Badger and L. R. Zumwalt, *J. Chem. Phys.*, **6**: 711 (1938).
S. L. Gerhard and D. M. Dennison, *Phys. Rev.*, **43**: 197 (1933).
E. Teller, *Hand- u. Jahrb. chem. Physik*, **9**(II): 43 (1934).

For a symmetric rotor, in the present approximation, only the z component of $\mathfrak{m}$, the vibrational angular momentum, needs to be considered. The problem may be treated as a perturbation employing zero-order wave functions which are products of rigid rotor and harmonic oscillator functions. When the molecule is in a state such that $v_{ka} + v_{kb} = 1$, where Q_{ka} and Q_{kb} are degenerate, it is necessary to solve the secular determinant

$$\begin{vmatrix} H'_{AA} - W' & H'_{AB} \\ H'_{BA} & H'_{BB} - W' \end{vmatrix} = 0$$

in order to find the energy W'. The zero-order wave functions can be written as

$$\psi_A = \psi_R\psi_1(Q_{ka})\psi_0(Q_{kb})\Pi'$$
$$\psi_B = \psi_R\psi_0(Q_{ka})\psi_1(Q_{kb})\Pi'$$

where Π' is the product of harmonic oscillator wave functions for all vibrational modes other than Q_{ka} and Q_{kb}.

The perturbation operator to be employed is

$$H' = -\frac{\mathfrak{M}_z\mathfrak{m}_z}{I_z^0}$$

because the term $\mathfrak{m}_z^2/2I_z^0$ does not involve the rotational functions. Since

$$\int\psi_R\mathfrak{M}_z\psi_R\, d\tau_R = \frac{h}{2\pi}K$$

for a symmetric rotor, a constant factor of $(h/2\pi)K$ can be extracted from each element of H'. The operator $\mathfrak{m}_z$ was defined in Eq. (13), Sec. 11-1, and Eq. (6), Sec. 11-2, giving

$$\begin{aligned} \mathfrak{m}_z &= \sum_k \mathfrak{Z}_k P_k \\ &= \sum_{kl}\sum_\alpha m_\alpha(l'_{\alpha l}m'_{\alpha k} - m'_{\alpha l}l'_{\alpha k})Q_l P_k \\ &= \sum_{k\neq l}\zeta_{kl}^{(z)}Q_l P_k \end{aligned}$$

in which $\zeta_{kl}^{(z)}(= 0$ if $k = l)$ depends upon the masses of the atoms and the force constants through the coefficients $m'_{\alpha l}$, etc. Vibrational integrals of the type

$$\int\psi_v(Q_k)Q_k\psi_v(Q_k)\, dQ_k \qquad \text{and} \qquad \int\psi_v(Q_k)P_k\psi_v(Q_k)\, dQ_k$$

vanish, with the result that $H'_{AA} = H'_{BB} = 0$ and

$$H'_{AB} = -H'_{BA} = \frac{h}{2\pi}K\frac{h}{2\pi i}\frac{\zeta_k}{I_z^0}$$

in which ζ_k is an abbreviation for $\zeta^{(z)}_{ka,kb}$. The energy is finally given by

$$W' = \mp \frac{h}{2\pi} K \frac{h}{2\pi} \frac{\zeta_k}{I_z^0} = \mp 2a\zeta_k K$$

Since a transition from the ground vibrational state to $W' = -2a\zeta_k K$ is allowed if $\Delta K = +1$, and to $W' = 2a\zeta_k K$ if $\Delta K = -1$, the formula for the frequencies of lines in a perpendicular band which replaces (4) is

$$Q \text{ branch:} \qquad \nu = \nu_v + \frac{[a(1 - 2\zeta_k) - b]}{h} \pm 2\frac{[a(1 - \zeta_k) - b]K}{h}$$
$$= \nu_Q(\pm K)$$

so that the (equidistant) line spacing is $2[a(1 - \zeta_k) - b]/h$ instead of $2(a - b)/h$, and is therefore in general different for each perpendicular-type band.

A similar treatment of the spherical rotor can be carried out with the result that the triply degenerate vibrational levels are split by the energies

$$W'_+ = 2b\zeta_k(J + 1)$$
$$W'_0 = 0$$
$$W'_- = -2b\zeta_k J$$

The selection rules show that for $\Delta J = +1$, the transition must occur to W'_-, for $\Delta J = 0$ to W'_0, and for $\Delta J = -1$ to W'_+ so that the expressions for P and R branches become

$$P \text{ branch:} \qquad \nu = \nu_v - \frac{2b(1 - \zeta_k)}{h} J$$

$$R \text{ branch:} \qquad \nu = \nu_v + \frac{2b(1 - \zeta_k)}{h} (J + 1)$$

so that in this case again the theory predicts a spacing of $2b(1 - \zeta_k)/h$ which may be different for different vibrational states.

Other Complications. When the fine structure is resolved, a number of refinements of the vibration-rotation theory are usually required. In the first place, since the molecule is not really a rigid rotor, the variation of *effective moments of inertia* with vibrational state must be considered. This introduces the possibility that the rotational constants a, b, and c are not the same in upper and lower vibrational states, and would change the simple expression for the fine structure of a parallel infrared band of a linear molecule to (R branch)

$$\nu = \nu_v + \frac{2b'}{h} + \frac{3b' - b''}{h} J + \frac{b' - b''}{h} J^2$$

where prime and double prime distinguish the two vibrational states.

The Coriolis forces,[1] which in the case of degenerate vibrations were shown above to give rise to a first-order perturbation, are also responsible for second-order perturbations between any two vibrations for which the direct product of the symmetry species contains the species of a rotation.

Finally, attention should be directed to the phenomenon of *centrifugal distortion*, in which the moment of inertia varies with rotational quantum number. For example, increased angular momentum about the symmetry axis of ammonia may be expected to spread out the hydrogen atoms slightly, leading to a greater I_c or decreased rotational constant, **c**. In fact, this centrifugal distortion should give rise to a term of the form $-dK^4$ in the energy, where d is a positive constant, very small compared with a.

These effects greatly increase the complexity of the rotational structure of the various types of bands, but a full discussion is beyond the scope of this book.

[1] C. Eckart, *Phys. Rev.*, **47**: 552 (1935).
H. A. Jahn, *Ann. Physik*, **23**: 529 (1935).
M. Johnston and D. M. Dennison, *Phys. Rev.*, **48**: 868 (1935).
E. Teller and L. Tisza, *Z. Physik*, **73**: 791 (1932).
E. Teller, *Hand- u. Jahrb. chem. Physik*, **9**(II): 125 (1934).
J. H. Van Vleck, *Phys. Rev.*, **47**: 487 (1935).
E. B. Wilson, Jr., *J. Chem. Phys.*, **4**: 313 (1936).

NAME INDEX

SUBJECT INDEX

A CATALOGUE OF SELECTED DOVER BOOKS IN ALL FIELDS OF INTEREST

THE PHILOSOPHY OF HISTORY, Georg W. Hegel. Great classic of Western thought develops concept that history is not chance but a rational process, the evolution of freedom. 457pp. 5⅜ x 8½. 20112-0 Pa. $4.50

LANGUAGE, TRUTH AND LOGIC, Alfred J. Ayer. Famous, clear introduction to Vienna, Cambridge schools of Logical Positivism. Role of philosophy, elimination of metaphysics, nature of analysis, etc. 160pp. 5⅜ x 8½. (Available in U.S. only) 20010-8 Pa. $2.00

A PREFACE TO LOGIC, Morris R. Cohen. Great City College teacher in renowned, easily followed exposition of formal logic, probability, values, logic and world order and similar topics; no previous background needed. 209pp. 5⅜ x 8½. 23517-3 Pa. $3.50

REASON AND NATURE, Morris R. Cohen. Brilliant analysis of reason and its multitudinous ramifications by charismatic teacher. Interdisciplinary, synthesizing work widely praised when it first appeared in 1931. Second (1953) edition. Indexes. 496pp. 5⅜ x 8½. 23633-1 Pa. $6.50

AN ESSAY CONCERNING HUMAN UNDERSTANDING, John Locke. The only complete edition of enormously important classic, with authoritative editorial material by A. C. Fraser. Total of 1176pp. 5⅜ x 8½. 20530-4, 20531-2 Pa., Two-vol. set $16.00

HANDBOOK OF MATHEMATICAL FUNCTIONS WITH FORMULAS, GRAPHS, AND MATHEMATICAL TABLES, edited by Milton Abramowitz and Irene A. Stegun. Vast compendium: 29 sets of tables, some to as high as 20 places. 1,046pp. 8 x 10½. 61272-4 Pa. $14.95

MATHEMATICS FOR THE PHYSICAL SCIENCES, Herbert S. Wilf. Highly acclaimed work offers clear presentations of vector spaces and matrices, orthogonal functions, roots of polynomial equations, conformal mapping, calculus of variations, etc. Knowledge of theory of functions of real and complex variables is assumed. Exercises and solutions. Index. 284pp. 5⅝ x 8¼. 63635-6 Pa. $5.00

THE PRINCIPLE OF RELATIVITY, Albert Einstein et al. Eleven most important original papers on special and general theories. Seven by Einstein, two by Lorentz, one each by Minkowski and Weyl. All translated, unabridged. 216pp. 5⅜ x 8½. 60081-5 Pa. $3.50

THERMODYNAMICS, Enrico Fermi. A classic of modern science. Clear, organized treatment of systems, first and second laws, entropy, thermodynamic potentials, gaseous reactions, dilute solutions, entropy constant. No math beyond calculus required. Problems. 160pp. 5⅜ x 8½. 60361-X Pa. $3.00

ELEMENTARY MECHANICS OF FLUIDS, Hunter Rouse. Classic undergraduate text widely considered to be far better than many later books. Ranges from fluid velocity and acceleration to role of compressibility in fluid motion. Numerous examples, questions, problems. 224 illustrations. 376pp. 5⅝ x 8¼. 63699-2 Pa. $5.00

THE COMPLETE WOODCUTS OF ALBRECHT DURER, edited by Dr. W. Kurth. 346 in all: "Old Testament," "St. Jerome," "Passion," "Life of Virgin," Apocalypse," many others. Introduction by Campbell Dodgson. 285pp. 8½ x 12¼. 21097-9 Pa. $7.50

DRAWINGS OF ALBRECHT DURER, edited by Heinrich Wolfflin. 81 plates show development from youth to full style. Many favorites; many new. Introduction by Alfred Werner. 96pp. 8⅛ x 11. 22352-3 Pa. $5.00

THE HUMAN FIGURE, Albrecht Dürer. Experiments in various techniques—stereometric, progressive proportional, and others. Also life studies that rank among finest ever done. Complete reprinting of *Dresden Sketchbook*. 170 plates. 355pp. 8⅜ x 11¼. 21042-1 Pa. $7.95

OF THE JUST SHAPING OF LETTERS, Albrecht Dürer. Renaissance artist explains design of Roman majuscules by geometry, also Gothic lower and capitals. Grolier Club edition. 43pp. 7⅞ x 10¾ 21306-4 Pa. $3.00

TEN BOOKS ON ARCHITECTURE, Vitruvius. The most important book ever written on architecture. Early Roman aesthetics, technology, classical orders, site selection, all other aspects. Stands behind everything since. Morgan translation. 331pp. 5⅜ x 8½. 20645-9 Pa. $4.50

THE FOUR BOOKS OF ARCHITECTURE, Andrea Palladio. 16th-century classic responsible for Palladian movement and style. Covers classical architectural remains, Renaissance revivals, classical orders, etc. 1738 Ware English edition. Introduction by A. Placzek. 216 plates. 110pp. of text. 9½ x 12¾. 21308-0 Pa. $10.00

HORIZONS, Norman Bel Geddes. Great industrialist stage designer, "father of streamlining," on application of aesthetics to transportation, amusement, architecture, etc. 1932 prophetic account; function, theory, specific projects. 222 illustrations. 312pp. 7⅞ x 10¾. 23514-9 Pa. $6.95

FRANK LLOYD WRIGHT'S FALLINGWATER, Donald Hoffmann. Full, illustrated story of conception and building of Wright's masterwork at Bear Run, Pa. 100 photographs of site, construction, and details of completed structure. 112pp. 9¼ x 10. 23671-4 Pa. $5.50

THE ELEMENTS OF DRAWING, John Ruskin. Timeless classic by great Viltorian; starts with basic ideas, works through more difficult. Many practical exercises. 48 illustrations. Introduction by Lawrence Campbell. 228pp. 5⅜ x 8½. 22730-8 Pa. $3.75

GIST OF ART, John Sloan. Greatest modern American teacher, Art Students League, offers innumerable hints, instructions, guided comments to help you in painting. Not a formal course. 46 illustrations. Introduction by Helen Sloan. 200pp. 5⅜ x 8½. 23435-5 Pa. $4.00

THE DEPRESSION YEARS AS PHOTOGRAPHED BY ARTHUR ROTHSTEIN, Arthur Rothstein. First collection devoted entirely to the work of outstanding 1930s photographer: famous dust storm photo, ragged children, unemployed, etc. 120 photographs. Captions. 119pp. 9¼ x 10¾.
23590-4 Pa. $5.00

CAMERA WORK: A PICTORIAL GUIDE, Alfred Stieglitz. All 559 illustrations and plates from the most important periodical in the history of art photography, *Camera Work* (1903-17). Presented four to a page, reduced in size but still clear, in strict chronological order, with complete captions. Three indexes. Glossary. Bibliography. 176pp. 8⅜ x 11¼.
23591-2 Pa. $6.95

ALVIN LANGDON COBURN, PHOTOGRAPHER, Alvin L. Coburn. Revealing autobiography by one of greatest photographers of 20th century gives insider's version of Photo-Secession, plus comments on his own work. 77 photographs by Coburn. Edited by Helmut and Alison Gernsheim. 160pp. 8⅛ x 11.
23685-4 Pa. $6.00

NEW YORK IN THE FORTIES, Andreas Feininger. 162 brilliant photographs by the well-known photographer, formerly with *Life* magazine, show commuters, shoppers, Times Square at night, Harlem nightclub, Lower East Side, etc. Introduction and full captions by John von Hartz. 181pp. 9¼ x 10¾.
23585-8 Pa. $6.95

GREAT NEWS PHOTOS AND THE STORIES BEHIND THEM, John Faber. Dramatic volume of 140 great news photos, 1855 through 1976, and revealing stories behind them, with both historical and technical information. Hindenburg disaster, shooting of Oswald, nomination of Jimmy Carter, etc. 160pp. 8¼ x 11.
23667-6 Pa. $5.00

THE ART OF THE CINEMATOGRAPHER, Leonard Maltin. Survey of American cinematography history and anecdotal interviews with 5 masters—Arthur Miller, Hal Mohr, Hal Rosson, Lucien Ballard, and Conrad Hall. Very large selection of behind-the-scenes production photos. 105 photographs. Filmographies. Index. Originally *Behind the Camera.* 144pp. 8¼ x 11.
23686-2 Pa. $5.00

DESIGNS FOR THE THREE-CORNERED HAT (LE TRICORNE), Pablo Picasso. 32 fabulously rare drawings—including 31 color illustrations of costumes and accessories—for 1919 production of famous ballet. Edited by Parmenia Migel, who has written new introduction. 48pp. 9⅜ x 12¼. (Available in U.S. only)
23709-5 Pa. $5.00

NOTES OF A FILM DIRECTOR, Sergei Eisenstein. Greatest Russian filmmaker explains montage, making of *Alexander Nevsky,* aesthetics; comments on self, associates, great rivals (Chaplin), similar material. 78 illustrations. 240pp. 5⅜ x 8½.
22392-2 Pa. $4.50

CATALOGUE OF DOVER BOOKS